# FIELD GUIDE TO THE BIRDS OF THE UNITED STATES AND CANADA EAST

SECOND EDITION

# FIELD GUIDE TO THE BIRDS OF THE UNITED STATES AND CANADA EAST

SECOND EDITION

TED FLOYD

NATIONAL GEOGRAPHIC
WASHINGTON, D.C.

# CONTENTS



Opposite: An American Robin feasts on dogwood berries in early winter. Page 2: Great Blue Herons

## INTRODUCTION

Welcome to the second edition of the *National Geographic Field Guide to the Birds of the United States and Canada—East*. It has been more than 15 years since the first edition of this book was published. In that time, both the science of ornithology and the experience of birding have changed tremendously. This new edition reflects and responds to those changes.

Advances in genetic science are informing a new understanding of bird evolution and taxonomy, redrawing the relationships among species. Crowdsourced data are expanding our knowledge of bird status and distribution and driving new maps like those integral to this edition, which were created in partnership with the Cornell Lab of Ornithology. And careful field research is greatly refining our knowledge of avian behavior and ecology.

Smartphones are now ubiquitous in the birder's toolkit, offering access to massive databases with visual and auditory aids for identification and online platforms where individuals can record their finds. Today digital cameras, powerful and affordable, can be as important in bird identification as binoculars.

In this new era, where does a book—a traditional field guide like this one—fit in? Here you can flip through, hold up, compare, bookmark, and annotate descriptions and illustrations of all 586 species likely to be encountered in the eastern regions of the United States and Canada. You can read informative entries on every single species, learn of their relationships to others, and see telling characteristics pointed out in art on every spread.

By browsing through this book before you go birding, you can get a notion of what to look for. Thumbing through it afterward will help you confirm identifications and enrich your understanding of what you have observed. And of course this book is intended for quick reference in the field as well. The guiding principle behind this volume, whether you use it at home or in the field, is efficiency of presentation. Species accounts, art, and maps have been created to succinctly convey the essentials for accurate identification of all the bird species occurring regularly in the East.

Birding takes people to spectacular scenes and places, like this dramatic rock outcropping overlooking a Northern Gannet rookery in Newfoundland.

Field guides can help readers become aware of birds they didn't know existed before, like this tiny Northern Parula, widespread in the East during the warmer months.

## ORGANIZATION OF THIS BOOK

This book includes all the birds of the eastern United States and Canada (see p. 12 for a map defining this region), focusing especially on those most likely to be observed. The accounts that form the bulk of the book, pages 20–409, represent the 586 species that regularly occur in the region. Longer accounts represent the 498 more widespread and common species. Shorter accounts add another 88 species; while these reliably occur within the region, they are either highly localized or regionally rare. Special sections in the back of the book list 241 species found less than annually in the East; nine species presumed to be extinct are described on pages 425–427.

This book's contents are organized taxonomically, driven by what we know today about the evolutionary relationships among bird species. Biologists classify birds in a hierarchical, or tiered, system that groups them according to those relationships. All birds belong to the class Aves, which is divided into about 40 orders. Those orders are in turn subdivided into about 250 families; the families are further organized in about 2,200 genera (plural of genus); and, within those genera, there are more than 11,000 species of birds worldwide. Many species are further divided into subspecies, or groups of closely related yet morphologically distinct individuals from different geographic regions.

The order of species in this book follows the checklist of the American Ornithological Society (AOS), the definitive authority for bird taxonomy in the Americas, as of July 2023. Birds in the same family appear together, sometimes on a single spread and sometimes on a suite of pages. An indication of the family or families on each spread, both common and scientific names, appears in the upper corner of each right-hand page. A visual index of bird families found in this book appears inside the front and back covers.

Every bird species has both a common name (English in our region) and a scientific name, the latter usually derived from Latin or Greek, and recognized universally by scientists working in all languages. The scientific name indicates the bird's genus (first word, capitalized and italicized) and species (second word, lowercase and italicized). The scientific name of the American Crow, for example, is *Corvus brachyrhynchos*. Species in the same genus are each other's closest relatives. Often, the scientific or common names carry interesting etymologies, which we share when useful. The tradition among ornithologists and birders is to capitalize the common name, as we do here.

In 2023, the AOS announced its intention to assign new common names to all bird species in the area under its jurisdiction that are currently named after people, beginning with 70 to 80 species in the United States and Canada. By removing these names, many with disturbing historical associations, the AOS hopes to create a more welcoming and inclusive space for all those who care about birds. Any changes made by press time are reflected in this edition.

The accounts that follow are organized as two-page spreads, each representing a group of related species. The left-hand pages describe the species, and the right-hand pages mirror those descriptions with annotated art.

## ABOUT THIS BOOK

At the top of each left-hand page, a short passage introduces the species featured on the spread and explains why they are grouped together. The species most likely to be observed are represented by longer accounts; the others, present but less common, are represented by shorter accounts. Whether full-length or shorter, each account and its accompanying art capture the key identifying characteristics of the bird described.

### ON THE LEFT

Every account begins with the bird's common and scientific names, the typical size of an adult, and its four-letter banding code—an abbreviation widely used as shorthand in the birding community. For species currently listed by the International Union for Conservation of Nature (IUCN) as critically endangered or endangered, we have included a CR or EN symbol, respectively; national, state, and provincial authorities were consulted for the status of subspecies. Each entry then portrays that species' natural history and describes its distinctive behavior, its role in the broader environment, or even its significance in human affairs. A range map accompanies every full-length species account.

The longer accounts are divided into sections that address the big three questions for field ID: What does the bird look like? What does it sound like? Where and when does it occur?

**Appearance** covers physical features, such as size, shape, color, and pattern, with emphasis on field marks—the visual details key to field ID.

**Vocalizations** describes what the bird sounds like, both shorter calls and more complex songs. Nonvocal sounds (drumming, bill snapping, etc.) are noted when relevant to identification.

**Populations** details where and when birds can be seen, on both a larger scale (their geographic

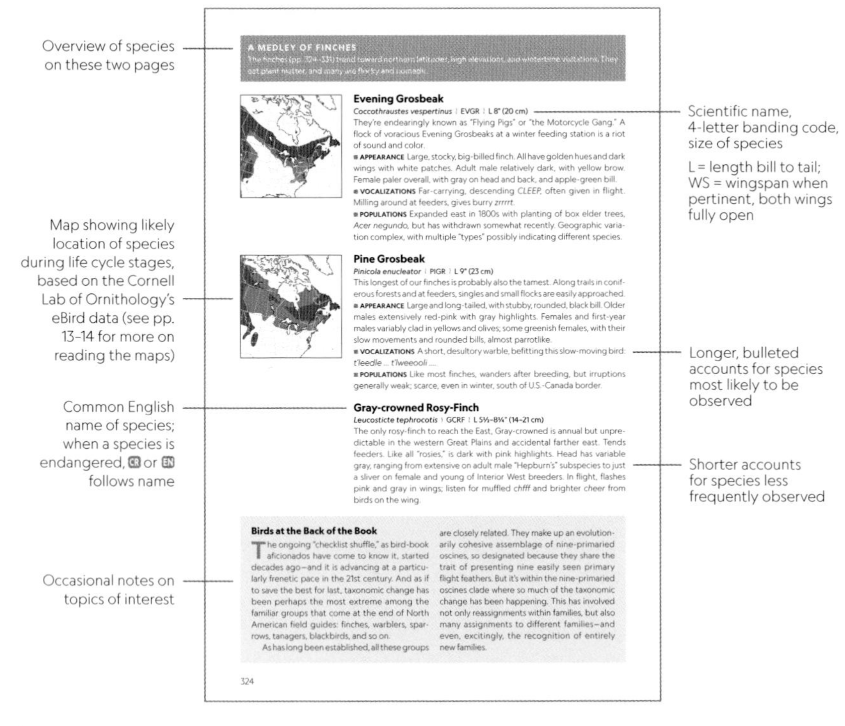


A MEDLEY OF FINCHES

**Evening Grosbeak**
*Coccothraustes vespertinus* | EVGR | L 8" (20 cm)
They're endearingly known as "Flying Pigs" or "the Motorcycle Gang." A flock of voracious Evening Grosbeaks at a winter feeding station is a riot of sound and color.
■ **APPEARANCE** Large, stocky, big-billed finch. All have golden hues and dark wings with white patches. Adult male relatively dark, with yellow brow. Female paler overall, with gray on head and back, and apple-green bill.
■ **VOCALIZATIONS** Far-carrying, descending *CLEEP*, often given in flight. Milling around at feeders, gives burry *zrrrrt*.
■ **POPULATIONS** Expanded east in 1800s with planting of box elder trees, *Acer negundo*, but has withdrawn somewhat recently. Geographic variation complex, with multiple "types" possibly indicating different species.

**Pine Grosbeak**
*Pinicola enucleator* | PIGR | L 9" (23 cm)
This longest of our finches is probably also the tamest. Along trails in coniferous forests and at feeders, singles and small flocks are easily approached.
■ **APPEARANCE** Large and long-tailed, with stubby, rounded, black bill. Older males extensively red-pink with gray highlights. Females and first-year males variably clad in yellows and olives; some greenish females, with their slow movements and rounded bills, almost parrotlike.
■ **VOCALIZATIONS** A short, desultory warble, befitting this slow-moving bird: *t'leedle ... t'lweeooli ....*
■ **POPULATIONS** Like most finches, wanders after breeding, but irruptions generally weak; scarce, even in winter, south of U.S.-Canada border.

**Gray-crowned Rosy-Finch**
*Leucosticte tephrocotis* | GCRF | L 5½–8¼" (14–21 cm)
The only rosy-finch to reach the East, Gray-crowned is annual but unpredictable in the western Great Plains and accidental farther east. Tends feeders. Like all "rosies," is dark with pink highlights. Head has variable gray, ranging from extensive on adult male "Hepburn's" subspecies to just a sliver on female and young of Interior West breeders. In flight, flashes pink and gray in wings; listen for muffled *chfff* and brighter *cheer* from birds on the wing.

**Birds at the Back of the Book**
The ongoing "checklist shuffle," as bird-book aficionados have come to know it, started decades ago—and it is advancing at a particularly frenetic pace in the 21st century. And as if to save the best for last, taxonomic change has been perhaps the most extreme among the familiar groups that come at the end of North American field guides: finches, warblers, sparrows, tanagers, blackbirds, and so on.
As has long been established, all these groups are closely related. They make up an evolutionarily cohesive assemblage of nine-primaried oscines, so designated because they share the trait of presenting nine easily seen primary flight feathers. But it's within the nine-primaried oscines clade where so much of the taxonomic change has been happening. This has involved not only reassignments within families, but also many assignments to different families—and even, excitingly, the recognition of entirely new families.

324

range) and a smaller scale (their habitat). This section may also mention subspecies and occurrence patterns not captured in the range maps. Season, geography, and habitat can be as important to identifying a bird as appearance and vocalizations.

The shorter accounts describe a bird's appearance, vocalizations, and where and when it can be found, but without accompanying range maps, since these species have either a very limited range or an irregular pattern of occurrence.

Sidebars discussing taxonomic science, identification nuances, and other ideas meant to enrich your birding experience often appear at the bottom of these pages.

### ON THE RIGHT

Right-hand pages match the facing full-length and shorter accounts with expert color illustrations of the species. The artists whose work is collected in this field guide are esteemed for their knowledge about birds and their behavior. Based on that knowledge, they have represented species in postures and plumages likely to be found in ordinary field encounters. Illustrations such as these are somewhat idealized, and individual birds in the field will not always look exactly like the art: Birds in real life look different from season to season and even minute to minute. Light and shadow hugely influence what we see (and don't see) in the field, and birds' feathers and bare parts vary greatly with ordinary wear and tear.

Annotations alongside the illustrations draw attention to certain important details, especially those that characterize a species or distinguish it from close look-alikes. Labels in bold signal differences in sex, life stage, or molt and plumage within a species. When multiple subspecies are pictured, they are indicated in bold (for common name or geographic region) and italic (for scientific name).

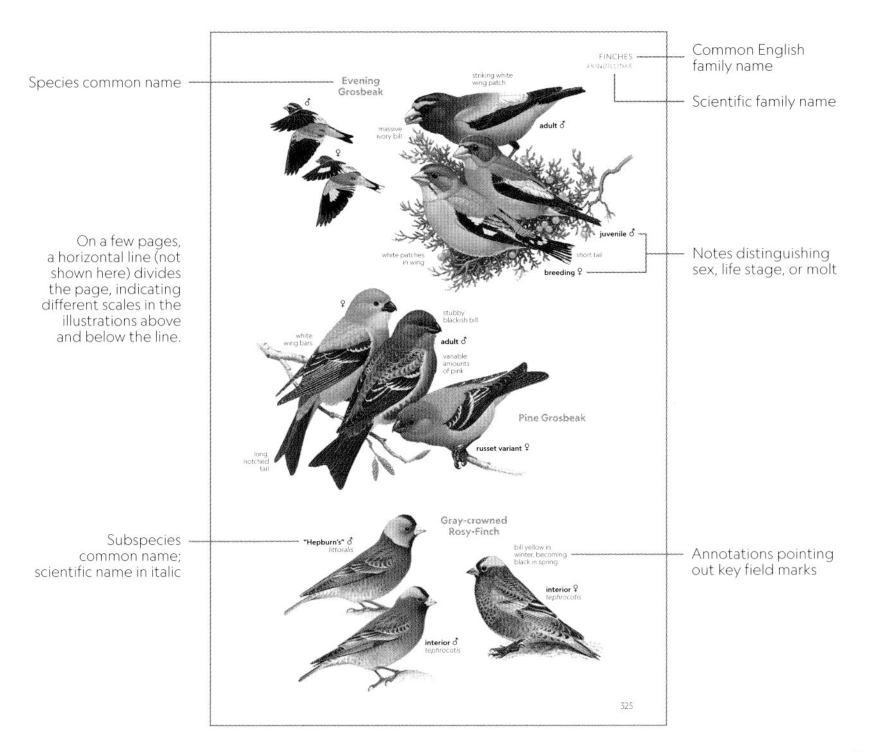

## PARTS OF A BIRD

Every detail of a bird's anatomy has a technical name, but even serious field ornithologists often use casual, descriptive language to describe the parts of a bird. The language in this book's species accounts tends toward the conversational, with some technical terms used for clarity. The diagrams here help define many of those terms, and they appear in the glossary, pp. 428–432, as well.

There is one case in which some amount of terminological precision is essential, however, and that is for the feathers involved in flight. A bird's wing comprises two main tracts of flight feathers: an outer tract, or primaries, and an inner tract, or secondaries. The innermost secondaries may be called tertials. Together, these three groups of feathers are called the remiges (singular, remex). All the flight feathers of a bird's tail are called rectrices (singular, rectrix). The feathers covering the remiges and the rectrices are called coverts, and they often differ in color and pattern from the flight feathers.

Field identification also requires noticing a bird's bare parts—the eyes, bill, and feet. The color of the bare parts is, in some instances, as important for identification as the color of the feathers. The size and shape of the bill and feet, as well as the size and position of the eyes, can also be important factors in identifying a bird.

Since birds molt—or grow new feathers—once, twice, or sometimes even more times each year, the appearance of a bird's feathers can, and often does, change drastically through the seasons. Bare parts can change in color through the year as well. All in all, a bird's total appearance—the bare parts plus all the feather tracts—is an integrated whole, reflecting the bird's age and sex and the overall effects of diet, season, and climate, not to mention intrinsic variation among individuals.

Here we offer for reference annotated diagrams of a sparrow—representing the passerines, or songbirds, the largest group of birds in the East and indeed all the world—in addition to representative diagrams of a hummingbird, a shorebird, and a gull.

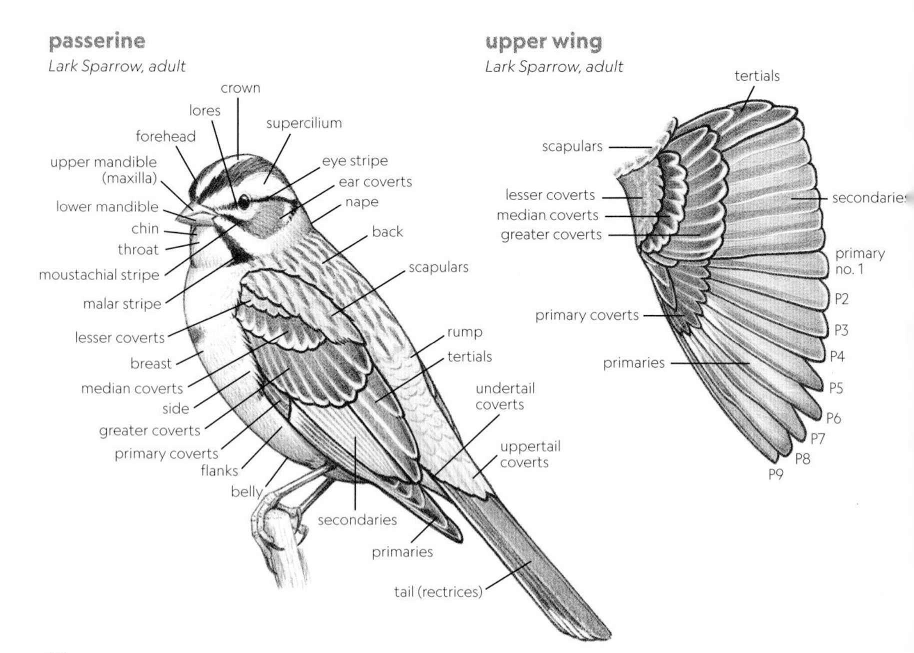

**hummingbird**
*Rufous Hummingbird, adult male*

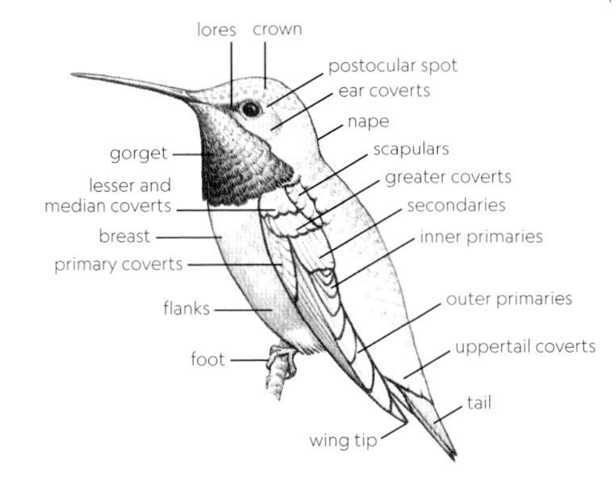


**shorebird**
*Semipalmated Sandpiper, juvenile*

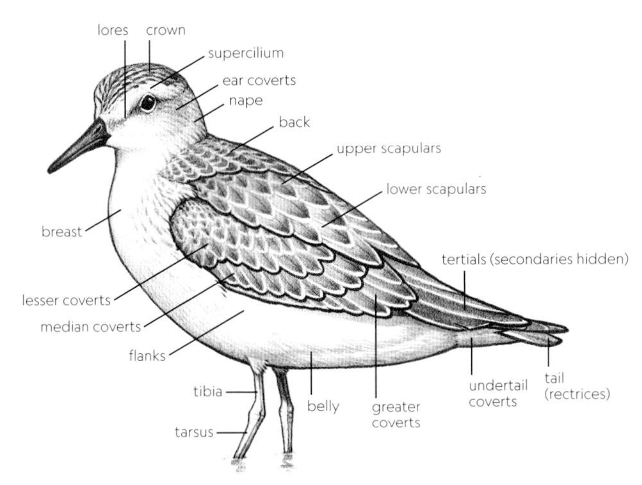


**gull**
*Herring Gull, breeding adult*

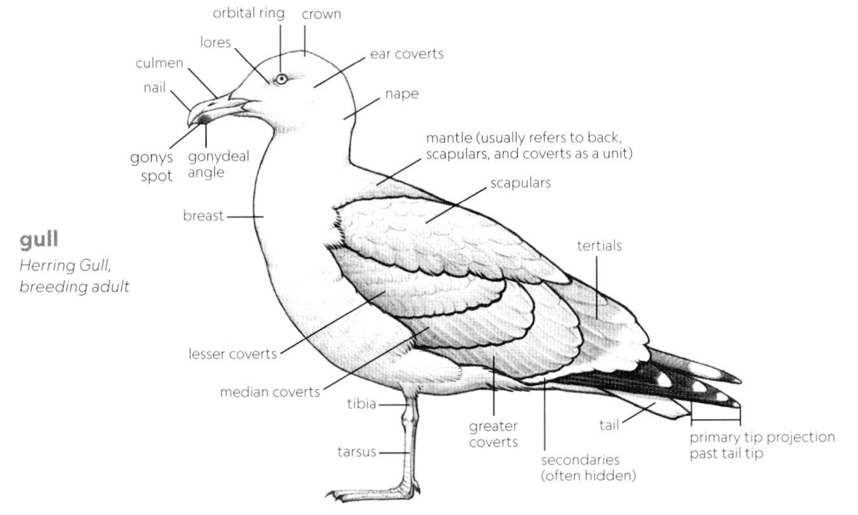

The stirring calls of Sandhill Cranes, especially during migration, thrill many birders, no matter their skill level.

## DEFINING THE EAST

In keeping with current biogeography, we define the region of this book to include the United States and Canada from the Atlantic Ocean in the east to near the base of the Rocky Mountains in the west. This region's western edge follows that range of mountains roughly diagonally, from northwest to southeast.

This biogeographical boundary line can delineate dramatically different regions, as in northern Colorado, where the steep foothills of the Rockies are sharply demarcated from the tall-grass prairie of the Great Plains. In many other regions, though, the division is far more gradual. And a few outposts of western montane habitat, notably in Nebraska, South Dakota, and Saskatchewan, are included in the East. Texas is a special case: The avifauna of Texas west of the Pecos River, in the more arid and mountainous Trans-Pecos region, is assigned to the West, and described in this book's companion volume, *National Geographic Field Guide to the Birds of the United States and Canada—West*. The birds found in the rest of Texas are included here.

The biological division between East and West is better thought of as a blurry band than a precise boundary.

Most of the species featured here nest every year somewhere in the East. Others breed outside the region but appear annually, usually in winter, and frequently enough to be included. Some breed in the West and wander in fall to the East; others breed in the Old World, yet manage to find their way to the eastern United States or Canada every year. And a few breed in the Southern Hemisphere, then spend the austral winter (northern summer) in our region.

The East is home to some of the most stirring ornithological spectacles on the continent and indeed anywhere on the planet. Winter along the Gulf Coast is brilliant, summer in boreal and Arctic Canada even more so. But it is the phenomenon of migration that is the ne plus ultra of eastern birding: hawks cruising Appalachian ridgetops in October, cranes staging on the Platte River in March, the mesmerizing nocturnal migration of songbirds in September, and, most of all, the staggering May migration of warblers through the entire region, considered by many birders to be the Greatest Show on Earth.

## NEW APPROACH TO RANGE MAPS

To capture the most current knowledge about bird distribution throughout the East, National Geographic has partnered with the Cornell Lab of Ornithology to present data-driven range maps with key innovations for understanding how bird populations occur in space and time. These maps are made possible because of data from eBird, a collaborative enterprise with dozens of partner organizations, thousands of regional experts, and nearly a million birders whose observations represent one of the most valuable resources for birding and conservation science in the digital age. These contributions from birders include more than a billion bird observations, photos, and sound recordings from around the world.

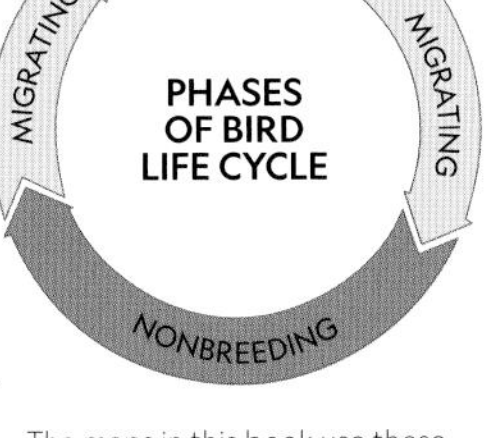

The maps in this book use these colors to represent key phases in each species' life cycle.

The Cornell Lab has used the past 15 years of observations from 400,000 eBird users to power the statistical analyses for its eBird Status and Trends data products and maps, including those created for this book. Experts have reviewed our maps against current knowledge, fine-tuning them to create the most accurate range maps in any field guide for this region.

Different from those found in other field guides, these maps use a new and conceptually powerful approach to mapping bird ranges. Rather than depicting the time of year when a bird can be found in a certain location, the maps in this book show the distributions of birds according to the phases of their life cycle: their nonbreeding, migration, and breeding seasons.

A given species breeds at roughly the same time every year, but the timing of breeding varies from one species to another. Most birds found in our region breed in the Northern Hemisphere during our spring and summer. After breeding, migratory birds typically travel to their nonbreeding grounds; they then migrate back to their breeding grounds the following year. Other bird species remain in the same region year-round.

For pelagic, or open ocean, species that breed in the Southern Hemisphere—several of which can be found in the East—seasonal cycles are reversed from our point of view here in the Northern Hemisphere. These birds breed far to our south during our winter (the austral summer), and they are found in our area during their nonbreeding season, our summer.

A species' annual cycle is depicted on the maps with four colors. *Purple* indicates where a species occurs year-round, with breeding usually occurring spring to summer and nonbreeding fall to winter—although some individuals may remain as nonbreeders throughout the year. *Red* indicates where a species is normally present only during its breeding season. *Blue* indicates that a species is normally present in that area only during its nonbreeding season. *Yellow* typically shows where a species occurs on migration between its breeding and nonbreeding seasons, but yellow can also indicate other movements, including dispersal of juveniles or molt migration.

Keep in mind that not all individual birds will breed, even when present in the area indicated during the breeding season. Additionally, migration timing and distances vary among species, and not all species migrate. In general, these maps depict breeding and nonbreeding seasons as periods during which bird populations are relatively stationary, and migrations as times when birds are generally on the

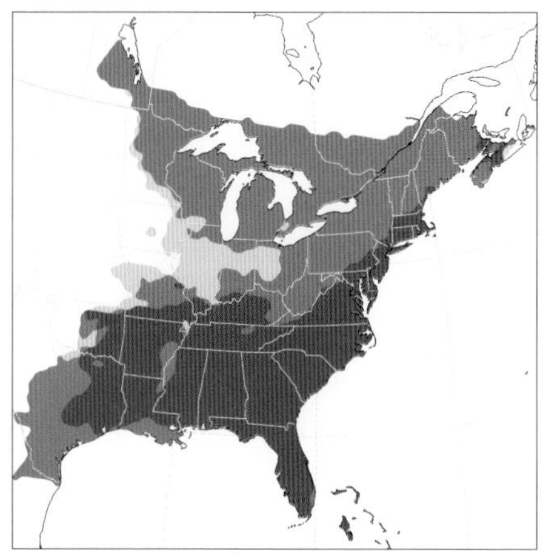

**Pine Warbler**

Birds' ranges are indicated by one to four colors corresponding to different components of the annual cycle. The area where the Pine Warbler can be found during its breeding season is indicated in red; blue indicates where it goes during the nonbreeding season; yellow indicates places where it is found only on migration; and purple indicates where it occurs year-round. For most species in this book, like the one shown here, the annual cycle corresponds closely to Northern Hemisphere seasonality, with breeding taking place during our spring and summer.

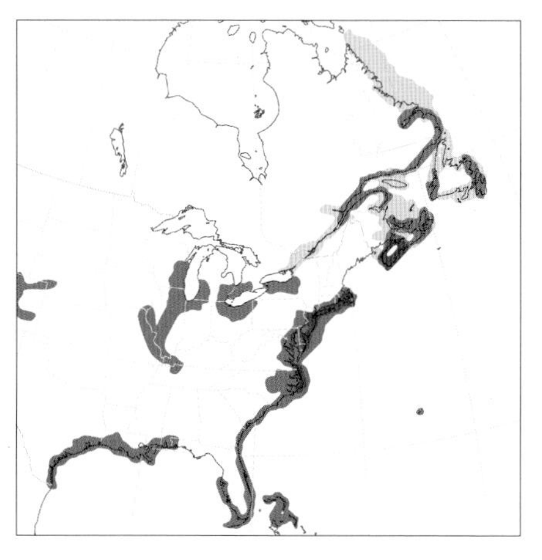

**Lesser Black-backed Gull**

The Lesser Black-backed Gull does not regularly breed in North America, yet nonbreeders occur widely in the East. As indicated by red, individuals can be found in eastern Canada during the species' breeding season—even though they do not breed there. Along the mid-Atlantic coast, individuals are found year-round, shown as purple—but, again, they do not breed there. Around the Great Lakes and the southern coasts, this gull is found only at those times of the year when it does not breed, shown as blue. And it is detected on active migration—shown as yellow—in much of coastal eastern Canada and parts of the Midwest.

move. But at times, migratory birds can be on the move within their breeding (red) and nonbreeding (blue) seasons; conversely, many bird species engage in prolonged sojourns, called stopovers, within their migratory (yellow) seasons.

While lines on a printed map stay permanent, within and across years, species' ranges may expand, contract, or change in other ways. For seabirds in particular, ranges often shift annually as ocean conditions change due to warming waters and food availability. In many years, "irruptive" land birds stray far from their normal ranges, often driven by regional failures in food supplies, such as conifer seeds. "Vagrants," or individual birds that wander far from the normal limits of their occurrence, may be found hundreds or even thousands of miles out of range. Be sure to consult the accompanying text for additional nuances of seasonal and geographic occurrence, using a mix of caution and open-mindedness when you see a bird apparently outside its expected range.

You can explore the latest sightings and contribute your own data to eBird to help shape the next generation of maps for field guides, research, and conservation by downloading the eBird app or visiting *eBird.org*.

## NOTICING THE DETAILS

Any bird, whether it is a tiny Ruby-throated Hummingbird or a titanic American White Pelican, has feathers and bare parts (eyes, feet, and bill). All extant species in the East can fly most of the year, almost all can walk, many swim, and all make sounds. Add all those things up—color and pattern, behavior, calls and songs—and you are well on your way to putting the right name on the bird.

Learning to identify birds requires precision in noticing the parts of a bird—for example, distinguishing between the coverts and flight feathers (remiges) of a bird's wings, or distinguishing between the "eyebrow" (technically, the supercilium) and the "eyeline" (or transocular) of a warbler or sparrow. The same goes for calls and songs: Listen carefully and learn to distinguish between high-pitched and low-pitched, rising and falling, whistled (pure-tone) and buzzy (modulated).

Also keep in mind that birds, like humans, differ from individual to individual. Birds often display reliable and recognizable differences between the sexes, among different age classes, and through the annual cycle. Most juvenile birds, especially among smaller species, are more likely to be seen in summer and early autumn than in winter or spring. Adults in fresh plumage are more likely to be seen in fall, right after the annual molt, than in early summer, when many are quite bedraggled.

A small, conical bill distinguishes this Indigo Bunting from a Blue Grosbeak, a similarly colored species with a larger, thicker bill.

Birds also display situational differences having to do with behavior, health, and, importantly, the bird's location relative to the observer. A bird puffed up in cold weather looks quite different from one of the same species singing from a

Neotropical migrants like the Canada Warbler, at left, and the Baltimore Oriole, at right, are present in the East on migration and during their breeding seasons.

Some birds are straightforward to identify, like this adult White Ibis, but knowing what to expect in a given habitat can help narrow down the possibilities with trickier species.

summer perch. A bird infected with feather mites might not resemble others of its species. Even the context within which a bird is spotted—out in the open, in bright sunlight, as opposed to in the shadows of the forest understory, for instance—may influence an individual's appearance in ways that a field guide simply cannot cover.

## THE BIG PICTURE

Adding up all these details in the service of identifying birds in the wild is a gratifying mental exercise. A "mostly blue bird with orange and white patches below" could be an Eastern Bluebird, a Lazuli Bunting, or even a Belted Kingfisher. A birder would rarely if ever confuse them, though, despite their superficial similarities. For one thing, their body structures are different, but even without that knowledge, these birds' behaviors alone distinguish them. The kingfisher perches on snags over waterways, it hovers in flight, and it plunges into a pond or stream for fish. Bluebirds and buntings don't do those things at all. Bluebirds forage in the open, often on the ground, in meadows with scattered perches; buntings prefer brushy habitats, glean food from foliage, and often visit feeders. All these distinctive behaviors say so much more about these species than the shared color of their feathers.

Getting good at bird identification requires embracing the idea that the whole is greater than the sum of its parts. Feathers, vocalizations, and body structure are all a part of it, but behavior and habitat are critically important as well, along with knowledge of when and where certain species occur. All these features are dynamic and interactive, coalescing in the living creature—and in our human recognition and appreciation of that creature in its environment.

## A BIRDER'S TOOLS

It is possible—indeed it can be wonderfully rewarding—to go out in the field and simply watch and wonder at birds without any equipment at all. But a few items of hardware can contribute valuably to the experience, especially if the emphasis is on identification.

Most birders use binoculars to watch birds. There is, unsurprisingly, a direct relationship between the price of a binocular and how good it is. The best are technological marvels, exquisitely bright and clear, and they are very expensive. Regardless of the price of the binocular you use, give thought to some basic considerations of performance and ergonomics. Weight (compact vs. full-size) can make a big difference if you are in the field for any amount of time. Lower magnifications, all things considered, deliver brighter images than higher magnifications—less is more. And the size and feel of the binoculars in your own hands, although personal and subjective, importantly affect the experience of birding. Research what you buy. Shop around, talk to friends, and, ideally, test a pair in the field (not just in a store) before purchasing.

Binoculars were until recently considered obligatory for birding, but that is changing. Today's digital cameras are small, powerful, and far more affordable than the best binoculars. They are superb for identifying birds in the field, and they preserve the experience as digital images—something binoculars, at least given present technology, cannot do. Just as with binoculars, shop around and ask questions before making a purchase; the variety of digital cameras out there is staggering, and some are much more suited to use by birders than others.

Another game changer has been the smartphone. New birding apps are launched frequently, and many are free. The Merlin Bird ID app, from the Cornell Lab of Ornithology, can instantly identify birds based on a description, photo, or sound. Powered by AI and data from eBird, Merlin can help identify thousands of bird species from around the world. Seek—an app powered by iNaturalist, another community science initiative—uses visual recognition software to identify not only birds but also non-avian animals, plants, other organisms, and even "sign," like nests, tracks, and scat. Point your phone's camera at a bird (or a wildflower, or even a slime mold), and it will suggest an ID within seconds.

Keep in mind that bird ID apps are not

Well calibrated and focused, binoculars bring telling details into view, making them an essential tool for birders eager to spot and identify species.

A hallmark of the birding community is a willingness to help others. Seasoned birders are always ready to share their knowledge.

foolproof. "Trust, but verify," as the saying goes. The artificial intelligence software that powers app-based IDs, although getting better all the time, is fallible. Confirm the output from Merlin or Seek with your own eyes and ears, and consult with human experts and, of course, this field guide. If the app makes a mistake, you may choose to report the mistake and thus help improve the performance of the app.

Apps and digital cameras, along with more traditional binoculars and print field guides, are means to an end. That end is to enjoy and identify birds in the field and, ultimately, to understand and care for them.

## THE BIRDING COMMUNITY

Today more than ever before, the study and enjoyment of wild birds in the field is a shared experience. New birders often find their way into the extensive online communities organized around eBird and iNaturalist, as well as social media groups focused on towns and cities, states and provinces. These are superb starting points, but they cannot—and they were never intended to—substitute for the in-person experience of birding with others.

Essentially everywhere in the United States and Canada, local birdwatching clubs welcome those with an interest in wild birds and their habitats. Outings and programs are often free, and annual memberships are generally affordable. Beginners and experts alike can join in their shared wonder at birds and nature, whether attending a talk given by a visitor or enjoying a walk led by a knowledgeable local. You can encounter other birders at your local park or preserve, and many will be eager to share information about how to become more involved with the local birding scene.

Community birding, whether online or in person, has the potential to make a difference for bird conservation. For starters, every upload to eBird or iNaturalist adds to databases that inform conservation science about the current state of birds in your region. Bird clubs and ornithological societies frequently sponsor projects such as nest box monitoring, migration counts, and habitat restoration. The Christmas Bird Count—a survey of bird populations in the Western Hemisphere in late December and early January, sponsored by the National Audubon Society—is perhaps the most famous long-term ecological survey in history, powered entirely by the birding community.

By sharing observations, members of a birding community can begin to envision the abundance of avian life that surrounds them.

Some may choose to travel to observe new species, but many enjoy the pleasures of backyard birding right at home.

As you spend time birding with other people, you begin to respect the importance of responsible birding. An essential resource in this regard is the American Birding Association's Code of Birding Ethics: *aba.org/ethics*.

## BIRDING TODAY AND TOMORROW

Why birds? Why do we go birding? Those are questions that almost any birder gets asked, whether they've been birding for six weeks or 60 years.

An undeniable part of the reason is to "get back to nature," to reconnect with the rhythms of the outdoors—the dawn chorus we've known since childhood, the predictable seasonal movements of hawks on fall migration and Tundra Swans in the spring, the simple pleasure of songbirds in the quiet winter woods or Sanderlings chasing the ocean waves. Experiencing nature has such intrinsic value that, for most of us, it requires no further justification. But it is also worthwhile to reflect on the other benefits of time spent outdoors.

Especially in the post-pandemic era, there has been considerable interest in the health and wellness benefits of spending time in nature and of birding in particular. Birding brings us into contact with the world around us—and with others who share our delight and wonder at the beauty of birdsong, the vibrant colors and intricate patterns of birds' feathers, and the pageantry of bird migration.

Birding challenges us. Watching and listening to birds; studying their ecology, behaviors, and populations; and pondering birds' roles in the broader environment are mentally and emotionally stimulating. Birding leads us along paths of science and conservation, activism and environmentalism, art and poetry, and more. Time spent in the field with birds keeps us sharp.

Perhaps the grandest motivation of all for the birder is the promise of discovery and new knowledge. We enjoy birdwatching because we are seekers of surprise and wonder; we go birding with open minds, receptive to the possibility of new ways of engaging with nature. *National Geographic Field Guide to the Birds of the United States and Canada—East* is a traditional field guide, comprising bird art and species accounts, bound together between two covers. But it is intended for a contemporary audience, eager to chart a new course for nature study, and dedicated to bettering the welfare of the human and nonhuman inhabitants of the world around us.

## WHISTLING-DUCKS

Two species of these mostly tropical, freshwater waterfowl occur in our area chiefly in the Southeast. By day, they lounge in and around natural and artificial waterways. At dusk and dawn, they are flighty and vocal.

### Black-bellied Whistling-Duck

*Dendrocygna autumnalis* | BBWD | L 21" (53 cm)

Sociable and successful, this species has adapted well to the modified habitats of the Southeast: lawns, canals, rice fields, catfish farms, etc. It nests in tree cavities and often roosts on snags and boughs.

■ **APPEARANCE** Large and long-necked; dark overall. Gray-faced with a staring demeanor; coral bill striking. White on wing prominent, especially in flight. Sexes similar. Bill of juvenile lead gray, unlike adult's; compare with juvenile Fulvous Whistling-Duck.

■ **VOCALIZATIONS** Calls wheezy and whistled, often in rapid series of five to seven short utterances, *pip wee-dee wee-dee wip,* frantic and rushed. Especially noisy dusk and dawn, and even right through the night.

■ **POPULATIONS** Regular vagrant well north of range, sometimes in flocks of 10 or more. Range expanding northward; flourishes at commercial fish farms, to the annoyance of some in the aquaculture industry.

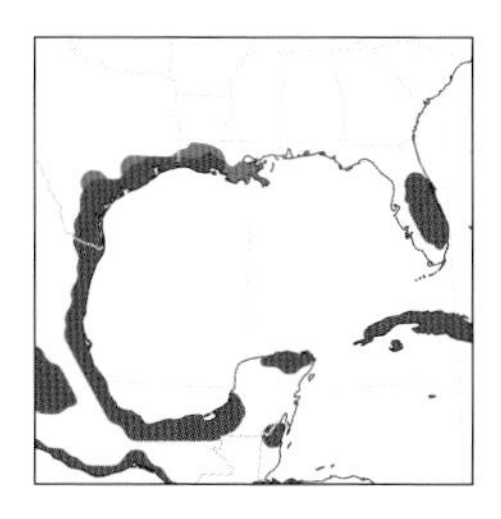

### Fulvous Whistling-Duck

*Dendrocygna bicolor* | FUWD | L 20" (51 cm)

Clad in bright earth tones, this bird is similar in behavior and ecology to the Black-bellied Whistling-Duck. However, it has not adapted as well to human-modified landscapes as that species, and it usually nests on the ground (not in trees). "Fulvous" denotes the bird's yellow-tan color, noticeable at great distances.

■ **APPEARANCE** Dark wings contrast with richly colored head and body. The two whistling-duck species are easily distinguished as adults, but compare gray-billed juveniles: Fulvous is dark-winged at all ages, with extensive pale or buff on tail; juvenile Black-bellied has all-dark tail, always flashes white in wings. Female Northern Pintail (p. 38) can resemble Fulvous Whistling-Duck but is colder and grayer.

■ **VOCALIZATIONS** Like a Black-bellied Whistling-Duck that stops after the second note: *pip wee,* often repeated slowly. Flocks chatter excitedly at dusk and well into the night; differentiating between the two whistling-ducks' calls is difficult.

■ **POPULATIONS** Global range is extensive. Numbers in the U.S. have risen and fallen in complex ways over the past century. Like Black-bellied, this species is known to wander well north of core range, sometimes in medium-size flocks.

### Why Whistling-Ducks Aren't With Other Ducks

The birds generally known as geese, swans, and ducks belong to the order Anseriformes, a word that means "gooselike" (*anser,* "goose," and *formes,* "shape"). They differ greatly in size and shape, in color and pattern, and in behavior and ecology: Compare the ornate, diminutive, tree-nesting Hooded Merganser with the hulking, pure-white, ground-nesting Trumpeter Swan. Nevertheless, those two species are both found within the order Anseriformes, and they occur in the same part of this book. Our two whistling-ducks are also in this order.

Birds within the same order share a common ancestor. Mergansers, swans, and whistling-ducks form an evolutionary assemblage distinct from, say, chachalacas, bobwhites, and turkeys. Those latter three are in a different order, Galliformes, with their own common ancestor.

So far, so good: The Anseriformes get their run of pages in this field guide, separate from the Galliformes. Within the Anseriformes, though, the birds called whistling-ducks come first, followed by the geese and swans, then other birds called ducks. But why do we separate whistling-ducks from all the other ducks?

It turns out that whistling-ducks aren't "true" ducks. Among the Anseriformes, they form a group distinct from all the other species in North America. All the others—the familiar Mallard and Canada Goose, the scaup and scoters and mergansers—have a common ancestor that is *not* shared with the whistling-ducks. The name whistling-duck is evocative, and does not seem likely to change. But whistling-ducks aren't, in an evolutionary sense, ducks like all our other birds called ducks. They keep their beautiful name in this book, though, and their placement reflects their unique evolutionary story.

## WHITE GEESE

Populations of the closely related Snow Goose and Ross's Goose have exploded in recent decades. White geese in the wild are usually one or both of these species, but be aware of escapes from captivity of non-indigenous, mostly or wholly white waterfowl.

### Snow Goose

*Anser caerulescens* | SNGO | L 26–33" (66–84 cm)

In coastal salt marshes in winter, it is stirring to see great clouds of this species rising from the back bays and cordgrass flats. Tremendous numbers likewise congregate in the colder months, often in agricultural landscapes, near the Mississippi R. and western Gulf Coast.

■ **APPEARANCE** All adults have long pink bill with peculiar "grinning patch." Typical adult plumage white except for black primaries; white feathers often suffused with pale tawny-rust, especially on head. Dark morph, called "Blue Goose," mostly brown-bodied, wings frosted pale bluish gray. Juvenile of both morphs has dusky bill.

■ **VOCALIZATIONS** Quiet while grazing or swimming, but flocks in flight are clamorous. Calls are higher, less growling than those of Canada Goose: *clee, cleek,* etc., on one pitch, monosyllabic but often wavering.

■ **POPULATIONS** Atlantic seaboard winterers ("Greater") average larger than mid-continent winterers ("Lesser"). Dark morph scarce along Atlantic coast, common inland. Spectacular numerical increase raises concern about fragile tundra breeding grounds where the birds forage voraciously.

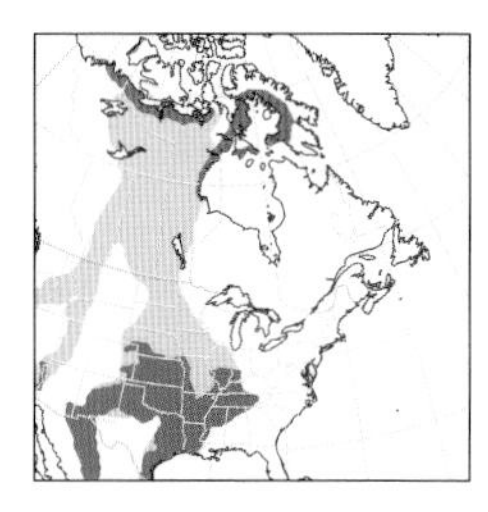

### Ross's Goose

*Anser rossii* | ROGO | L 23" (58 cm)

Uncommon but increasing, this little white goose is a dainty version of the larger Snow Goose. Singles or small groups keep to themselves in mixed-species flocks of geese.

■ **APPEARANCE** Petite; notably small-headed. Adult Ross's cleaner white than adult Snow; juvenile not nearly as dusky as juvenile Snow. Key mark is bill: short, stubby, triangular, and pink, with odd blue-green warts at base.

■ **VOCALIZATIONS** Calls softer, more nasal, than those of Snow Goose; in big flocks, drowned out by louder, richer calls of other, noisier species.

■ **POPULATIONS** Rare in winter along Atlantic seaboard; less uncommon in lower Mississippi R. drainage. Hybridizes with "Lesser Snow Goose" where breeding ranges overlap in Canadian Arctic; hybrids have intermediate bill morphology. Striking dark morph, analogous to "Blue Goose," very rare.

### Graylag Goose

*Anser anser* | GRGO | L 29–32" (74–81 cm)

Common wild goose of Eurasia, accidental to Northeast. Widely domesticated here, though, including a popular all-white or nearly white strain. These domestics often get into the wild and can be told by their tameness, portly build with drooping belly, and reluctance (or inability) to fly.

### Swan Goose

*Anser cygnoides* | CHGO | L 45" (114 cm)

Indigenous to East and Central Asia, where uncommon and declining. No known natural occurrences in N. Amer., but, like Graylag Goose, common here in captivity; very small feral populations can persist for years in city parks. Variable; whitish strains—including confusing hybrids with Graylag—are superficially similar to Snow Goose.

Snow Goose
adults
longer, more sloping head than Ross's
blue-morph adults
long bill with dark "grinning patch"
head often stained rusty
variably grayish with dark bill
variably dark below
e-morph juvenile
mostly dark brown
blue-morph juvenile
white-morph adult
faster wingbeats than Snow
adult
Ross's Goose
white-morph juvenile
much paler than juvenile Snow
short bill in the shape of a right triangle
round head
blue at base with no "grinning patch"
short neck
blue-morph adult
white-morph adult
brighter white than Snow
rare

Graylag Goose
large with heavy belly
adult
anser
domestic type
Swan Goose
domestic type
huge with long neck
variable; often all-white

## DARK GEESE

Although they nest primarily on low-lying tundra at high latitudes, these geese are known to birders chiefly as winter visitants to coastal and agricultural districts. Look for them in large flocks.

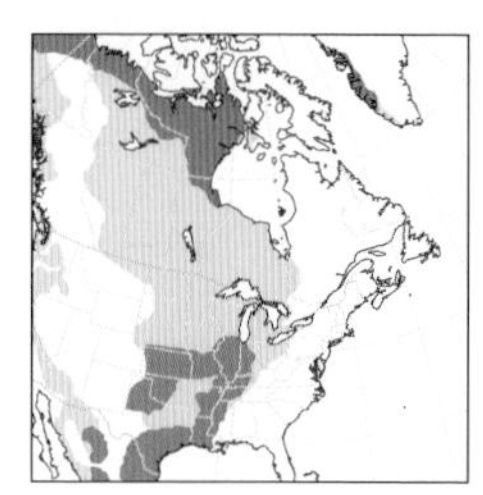

### Greater White-fronted Goose

*Anser albifrons* | GWFG | L 28" (71 cm)

Despite being one of the most widespread goose species on Earth, the Greater White-fronted is typically uncommon. The presence of a "Specklebelly" in big waterfowl flocks is often betrayed by its orange feet, or, when the birds all put up, by its distinctive call.

■ **APPEARANCE** Muddy brown overall. Adult has white feathers (the "front") at base of pink bill, black splotches on belly (hence the old name, "Specklebelly"); juvenile lacks white "front" and black splotches. All have prominent orange legs. Beware: Feral Graylag domestics (p. 22) can be similar.

■ **VOCALIZATIONS** Heard mostly in flight, a ringing *klee leek* or *klee-wee leek;* carries well, often stands out in a cacophony of other, more common geese.

■ **POPULATIONS** Relatively pale "Tundra" subspecies locally common in winter in lower Mississippi River Valley; darker "Greenland" subspecies rare in winter in Atlantic seaboard states and provinces.

### Pink-footed Goose

*Anser brachyrhynchus* | PFGO | L 26" (66 cm)

A recent addition to avifauna of East. Breeds in northern Europe, eastern Greenland; now found annually in winter along coast from mid-Atlantic northward. Currently quite rare, but future increases here may be expected. Like a juvenile Greater White-fronted but darker-headed; upperparts, especially in flight, are paler, grayer than that species.

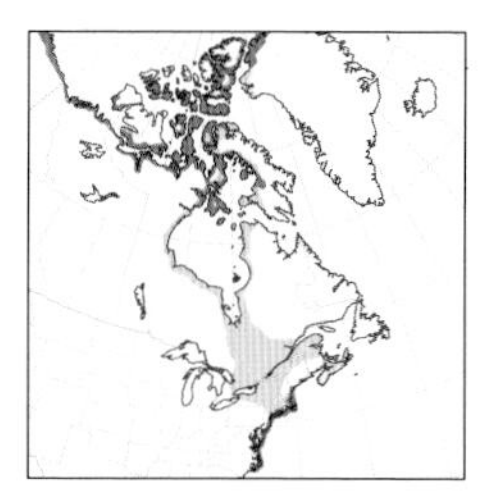

### Brant

*Branta bernicla* | BRAN | L 25" (64 cm)

Most wintering geese occur in large numbers in agricultural landscapes, but this species is highly coastal. Compact flocks feed in salt marshes and roost in sounds and inlets. Curiously, this bird's scientific name, *B. bernicla,* means "barnacle goose."

■ **APPEARANCE** Adult plumage grayscale: Head and breast black, except for elegant white band ("bow tie") across neck; belly gray; tail and vent mostly white. Young birds, seen throughout winter, are duskier and blotchier.

■ **VOCALIZATIONS** Usually heard in chorus of hundreds; low, growling *ccrruuk, rrrook,* etc. Effect is more laid-back than urgent-sounding calls of other geese.

■ **POPULATIONS** Winter distribution closely tied with availability of eelgrass (genus *Zostera*), although some enterprising flocks wander to parks and ball fields. Winter sightings guaranteed along the Belt Parkway in Brooklyn and Queens!

### Barnacle Goose

*Branta leucopsis* | BARG | L 27" (69 cm)

Like the Pink-footed Goose, this Old World species is increasing in winter along the Atlantic seaboard. But the Barnacle Goose arrived here first and started increasing earlier, and is more frequently seen. White face mask distinctive, but Cackling and Canada Geese (p. 26) on rare occasion show similar pattern.

"Greenland" adult
flavirostris
orangish bill
ightly darker
Greater White-fronted Goose
gambelii
tail white at tip and base
adult
adult
pinkish-orange bill
white border on flank
immature
unmarked warm gray
bright orange legs
juvenile
Pink-footed Goose
small, dark head
occasionally with white rim at bill base
dark bill with pink band
pale gray upperparts
paler tail than Greater White-fronted
pink legs
flight fast and low with shallow wingbeats
Brant
adult
plain darkish back
Barnacle Goose
sharply contrasting overall
strongly barred upperside
nearly white underneath
juvenile

## WHITE-CHEEKED GEESE

These familiar birds were treated as one species until 2004. Now they are classified as two: the smaller Cackling Goose and larger Canada Goose. Plumage is nearly identical between them, but differences in body shape are substantial.

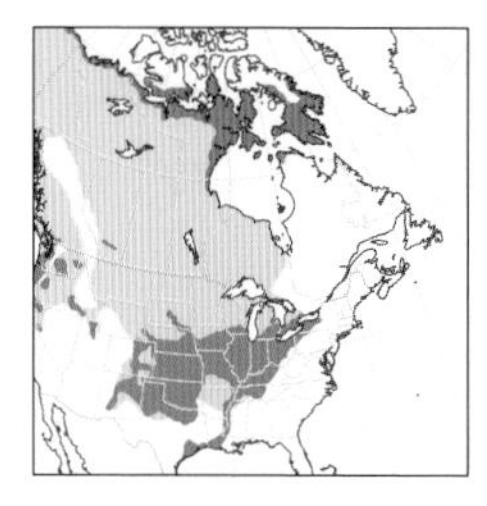

### Cackling Goose

*Branta hutchinsii* | CACG | L 23–33" (58–84 cm)

This highly migratory goose breeds in lowlands along the Arctic Ocean and Bering Sea. It winters widely across the U.S., but with a chiefly westerly distribution, often flocking with Snow (p. 22) and Canada Geese.

■ **APPEARANCE** Smaller and rounder than Canada; neck shorter, thicker; bill short and stubby; forehead steep. On many, the black neck and grayish breast are separated by a bright white band, but this mark is shown by some Canadas too. Most in East are nominate *hutchinsii* ("Richardson's"), almost as large as small Canada Geese.

■ **VOCALIZATIONS** "Cackling" is a misnomer. The birds honk like Canadas, but their calls average higher-pitched: *hink, henk, hink-a-lenk.* Much overlap, and sound spectrograms of many "Richardson's" are indistinguishable from those of Canada.

■ **POPULATIONS** Uncommon, but annual and increasing, east of Appalachians. "Richardson's" subspecies, the largest Cackling, winters in agricultural fields from the Mississippi River Valley to Tex. to the Rockies; Cackling outnumbers Canada in places in the Midwest, especially on the High Plains.

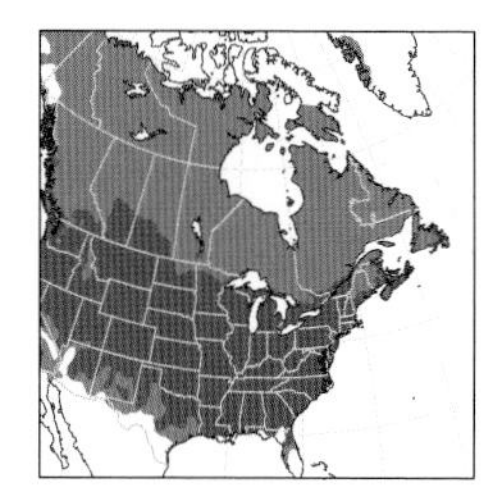

### Canada Goose

*Branta canadensis* | CANG | L 30–43" (76–109 cm)

Across much of the East, this is by far the more common of the two white-cheeked geese. It is the most numerous of all geese in most places and indeed one of the most abundant and conspicuous bird species, period.

■ **APPEARANCE** Along with Cackling, distinguished by black neck set off by white "chinstrap." Varies in size, with smaller subspecies wintering mostly in the southern Great Plains. All are long-necked, with long bills and sloping foreheads. "Lesser Canada Goose," wintering to High Plains, no larger than the biggest Cackling Geese, but body structure sets it apart.

■ **VOCALIZATIONS** Honest-to-goodness *honk,* often doubled or trebled: *ka-lonk, honk-a-lonk.* Feeding flocks murmur quietly, hatchlings wheeze softly.

■ **POPULATIONS** Until mid-20th century, famously migratory; great flocks in V-formation heralded the changing seasons. Biological introductions changed all that, and nonmigratory birds are now present year-round in many places, including parks, golf courses, and suburbia, where often considered a nuisance.

### Range Expansions, Numerical Increases

Many, although certainly not all, waterfowl species have increased in number and expanded their ranges in the past half century. The reasons are varied, complex, and complexly interwoven. Part of it is simply that numbers are rebounding from historic lows in the mid-20th century; habitat conservation, clean water legislation, and legal protections for birds have been highly effective. But another part, the complex part, boils down to the law of unintended consequences.

Wildlife biologists introduced Canada Geese from captive populations into the wild in the 1950s and 1960s, and the experiment caught on, big time. These birds and, in particular, their

offspring were mostly sedentary, and today many geese have become essentially nonmigratory. At the same time, changes in agriculture have led to the overproduction of grain on the wintering grounds; geese revel in the grain spillage, and overwintering mortality has declined.

The situation with Arctic-nesting geese, including the Cackling, is fascinating and disturbing. These "bulldozer herbivores" are devouring the tundra, causing the entire landscape to darken, which leads to warmer ground temperatures and possibly promotes the dramatic Arctic wildfires seen in recent years. In a troubling positive feedback loop, climate change is affecting birds, which are in turn contributing to climate change.

## SWANS

Huge yet graceful, these long-necked birds are closely related to geese, more distantly to ducks. The feathers of adults are entirely white, so identification depends on careful study of bill details.

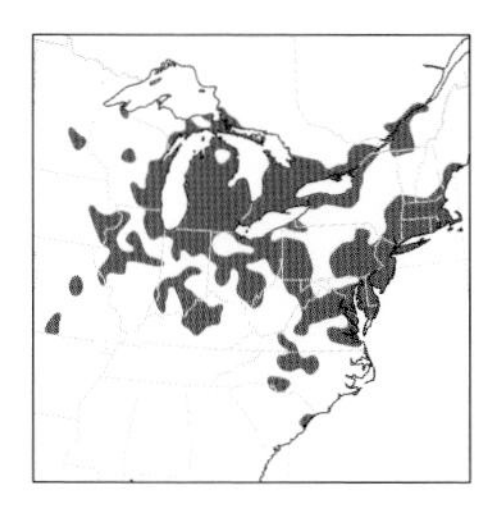

### Mute Swan

*Cygnus olor* | MUSW | L 60" (152 cm)

Indigenous to the Old World, this swan is seen in pairs, family groups, and generally small flocks around parks, millponds, and protected bays.

■ **APPEARANCE** Longer-tailed than other swans; swimming, holds neck in broad arc. Bill of adult deep orange with black knob; at rest, often raises wings above water. Juvenile dusky overall; some are rather dark, others less so. Bill of juvenile also dusky, lacks knob of adult.

■ **VOCALIZATIONS** Hisses ominously when angry; in flight, wingbeats impressively loud. Little appreciated is the species' lovely wavering whistle, audible only at close range. Despite being called "mute," this bird makes diverse sounds!

■ **POPULATIONS** Aggressive invader; beloved by many townsfolk, hated by some wildlife biologists. Local "control"—a euphemism for killing—has been widely and controversially implemented. Probably the only bird in the East capable of physically harming a human.

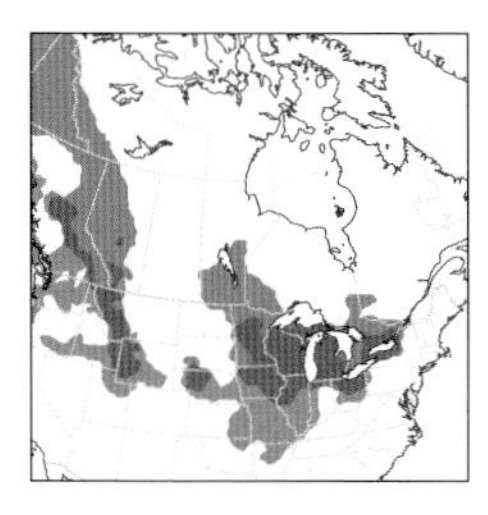

### Trumpeter Swan

*Cygnus buccinator* | TRUS | L 60" (152 cm)

A primarily northern and western species, this swan has been established via human agency in the Great Lakes region and upper Midwest.

■ **APPEARANCE** Very similar to Tundra. Note precise differences in bill shape: Base of Trumpeter's bill cuts a straight line with white feathers on sides of face; bill and feathers abut in pointy V-shape in space between eyes. Bill longer, more smoothly sloping than on Tundra. Most adults have all-black bill, lacking yellow patch of Tundra. Juvenile has dusky gray-white plumage with dusky pink bill.

■ **VOCALIZATIONS** Double honk, *hunk-unk*, muffled and nasal.

■ **POPULATIONS** Gradually expanding and increasing in the East; mosaic of migratory and sedentary flocks. Indigenous to N. Amer. but probably absent historically from much of the East; thus, introductions here are controversial.

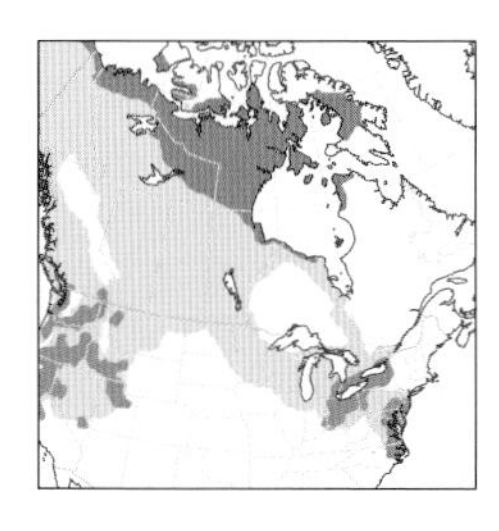

### Tundra Swan

*Cygnus columbianus* | TUSW | L 52" (132 cm)

One of the most bewitching of all migration phenomena is the spring flight of Tundra Swans over the eastern Great Lakes region; long lines power overhead, calling sonorously as they go.

■ **APPEARANCE** Base of bill subtly different from Trumpeter: Border with white facial feathers not as straight as on Trumpeter, cuts across forehead in shallower and smoother arc. Bill of Tundra also shorter overall and not as straight. Most adult Tundras have a bit of yellow at base of bill near eye. Juveniles of both species have dirty-pink bills; use bill structure, cautiously, to separate species. Also helpful is different timing in acquisition of adultlike plumage: Juvenile Tundra transitions to all-white plumage earlier in winter than juvenile Trumpeter.

■ **VOCALIZATIONS** Formerly called Whistling Swan for the baying of distant flocks in chorus.

■ **POPULATIONS** Following population losses a century ago, has largely recovered; legal protections and shift to new diets have helped.

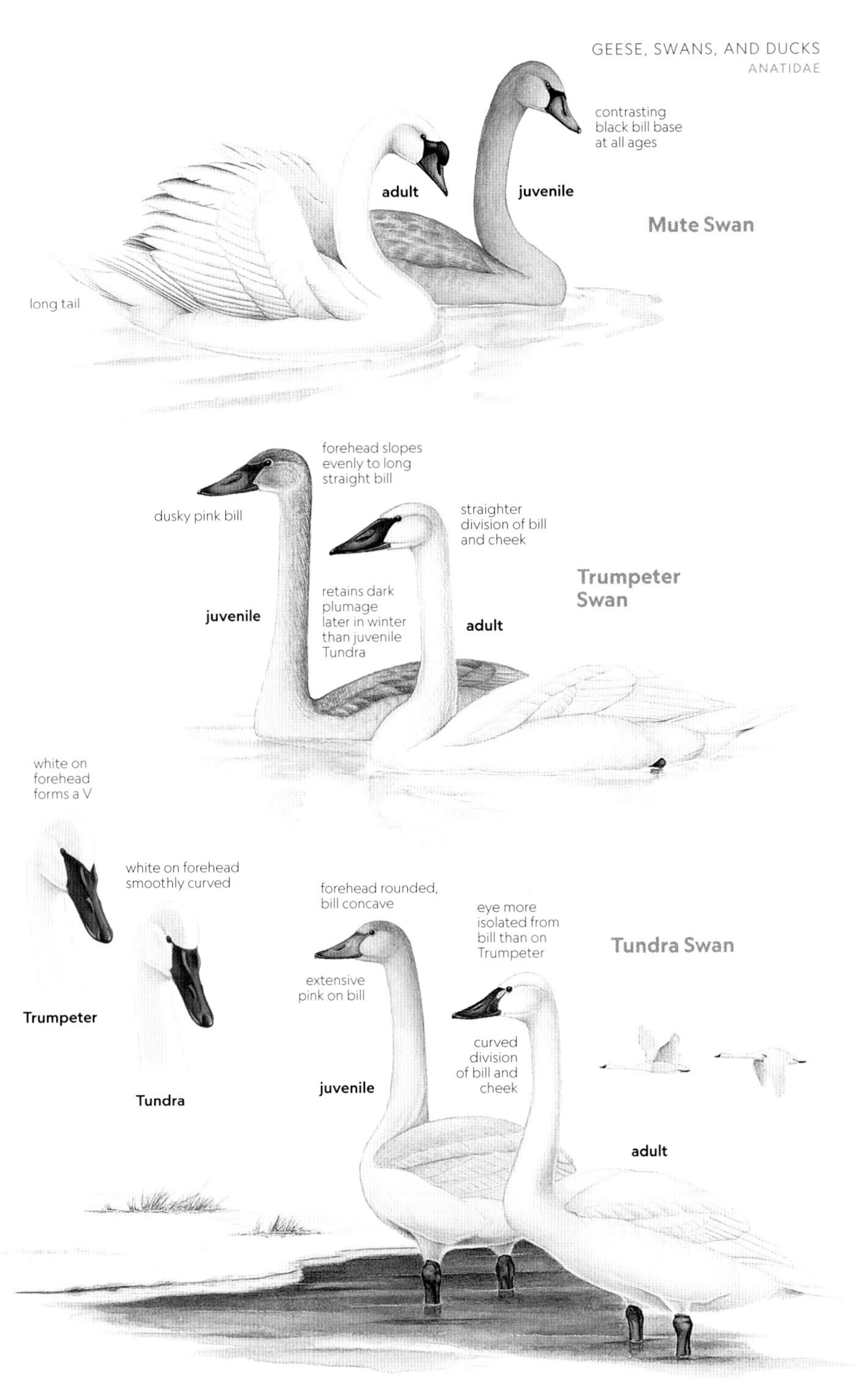
contrasting
black bill base
at all ages
adult
juvenile
Mute Swan
long tail
forehead slopes
evenly to long
straight bill
dusky pink bill
straighter
division of bill
and cheek
Trumpeter
Swan
retains dark
plumage
later in winter
than juvenile
Tundra
juvenile
adult
white on
forehead
forms a V
white on forehead
smoothly curved
forehead rounded,
bill concave
eye more
isolated from
bill than on
Trumpeter
Tundra Swan
extensive
pink on bill
Trumpeter
curved
division
of bill and
cheek
juvenile
Tundra
adult

## SOME ODD DUCKS

Within the assemblage of "true" ducks, these four form a grouping that is evolutionarily distinct from the others. There may be more garishness on this spread than anywhere else in the field guide!

### Egyptian Goose

*Alopochen aegyptiaca* | EGGO | L 25–29" (64–74 cm)

Despite the name, this bird is more of a duck than a goose; it is most closely related to Old World shelducks, widely domesticated here and sometimes seen in the wild.

■ **APPEARANCE** Large; size of a small goose. A curious mix of beige, sienna, black, and white, with reddish "lipstick" and "eye shadow." White coverts and dark remiges striking in flight, suggesting Muscovy at a distance; in good light, note dark green secondaries and orange tertials.

■ **VOCALIZATIONS** Nasal and raspy, a single note, often given in steady succession, slowly or quickly: *uunk ... uunk ... uunk ... uunk,* with little variation.

■ **POPULATIONS** Recent addition to our avifauna, expanding rapidly in Fla. and especially Tex. Not especially likely in large flocks (not yet!), often seen singly or in small groups at edges of waterways, golf courses.

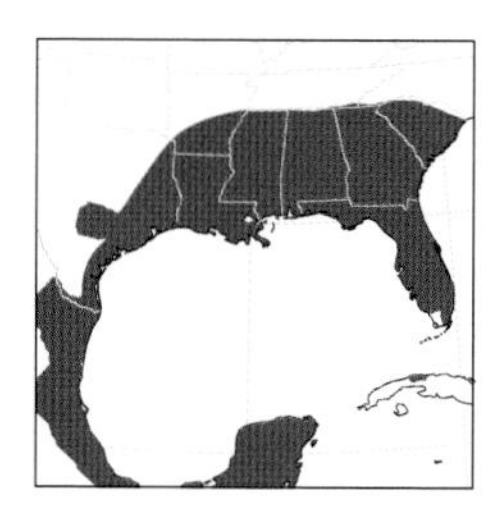

### Muscovy Duck

*Cairina moschata* | MUDU | L 26–33" (66–84 cm)

The indigenous range of this widespread tropical duck barely reaches Texas, where the species is uncommon along the Rio Grande. Muscovy Ducks of feral origin are widely noted elsewhere, especially in Florida.

■ **APPEARANCE** Birds of non-domestic origin, seen in our area only along Lower Rio Grande, all-black except for bold white wing patches (lacking in juvenile); in good light, note greenish sheen. Both escapes from captivity and their feral descendants vary greatly but typically show ample white.

■ **VOCALIZATIONS** Usually quiet in indigenous range in the tropics; listen for gruff, doglike grunts. Birds in Fla. also give Mallard-like quacks, perhaps reflecting hybridization with Mallard (p. 36).

■ **POPULATIONS** Birds of feral stock hybridize with other ducks, especially Mallard, producing "frankenducks" that may create momentary bewilderment at parks with duck ponds.

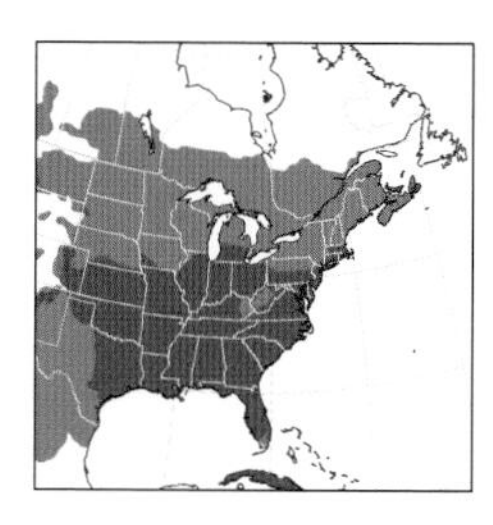

### Wood Duck

*Aix sponsa* | WODU | L 18½" (47 cm)

Despite its brilliant visage, this smallish duck has a knack for avoiding detection in the swamps and other wooded waterways it favors.

■ **APPEARANCE** All are long-tailed, long-necked, and block-headed; in flight, frequently glances down, as if to see what's going on. Female's eye surrounded by broad white patch; looks surprised. Eclipse male (summer only) like female, but eye and bill red.

■ **VOCALIZATIONS** Female, a piercing *whoo-eek,* uttered as the bird flushes; male, a drawn-out, feeble hiss, like air being expelled from a bike tire.

■ **POPULATIONS** Nest-box supplementation has aided recovery in recent decades. Does not usually aggregate in dense flocks.

### Mandarin Duck

*Aix galericulata* | MAND | L 16" (41 cm)

In same genus as Wood Duck, and breeding male is just as eye-catching. Females of the two species are similar; female Mandarin has thinner eye ring and fine "bridle" extending well behind eye. Indigenous to Asia; popular in captivity, often escaping into wild, even nesting on occasion.

Egyptian Goose
striking white wing coverts
adults
large and long-legged
juvenile ♀
white wing patches acquired in first winter
adult ♂
Muscovy Duck
adult ♂
male much larger than female
highly variable plumage, usually with knobby red face
domestic variety ♂
long tail
distinctive flight silhouette with large tail and raised head
extensive white around eye
♂
mostly dark wings
♀
juvenile ♂
♀
♂
Wood Duck
thin "bridle" extends behind eye
pale bill nail
♂
♀
Mandarin Duck

## GENUS *SPATULA*

Named for their spatulate bills, ducks in this genus are dabblers that favor shallow freshwater wetlands with emergent vegetation. They feed at or immediately below the water's surface, rarely if ever diving.

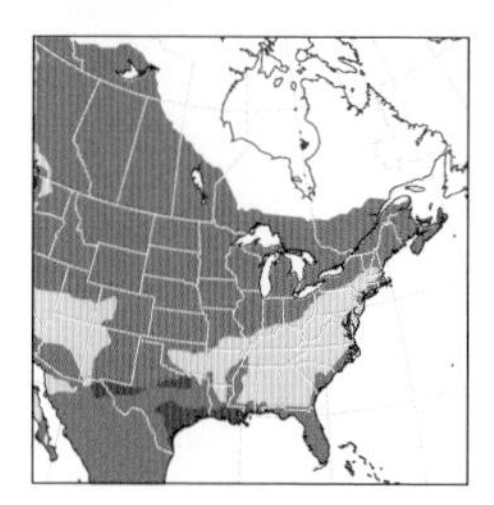

### Blue-winged Teal

*Spatula discors* | BWTE | L 15½" (39 cm)

Except along the Southeast coast, this is a warm-season duck in the East. The species returns relatively late in spring and departs early, mostly in late summer.

■ **APPEARANCE** Breeding male sports white swatch on dark purple-gray head ("crescent moon on midnight blue"). Female like Green-winged (p. 38) and especially Cinnamon: Green-winged is smaller-bodied, slimmer-billed, pale at rear, and often shows a bit of green when standing or swimming; Cinnamon is warmer brown overall and larger-billed, with blank face lacking Blue-winged's dark eyeline and white at base of bill.

■ **VOCALIZATIONS** Usually quiet; but male, especially when courting, gives steady, squeaky *chatter, peek! peek! peek!* Female's note a nasal *kvenk*.

■ **POPULATIONS** Winter range notably southern, with many reaching northern S. Amer.

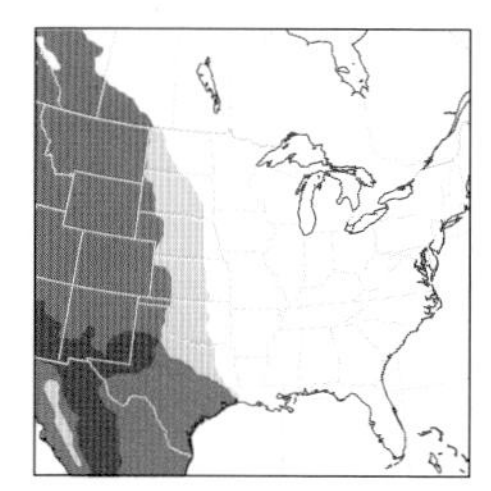

### Cinnamon Teal

*Spatula cyanoptera* | CITE | L 16" (41 cm)

Despite obvious differences in plumage between adult males, the Blue-winged and Cinnamon Teal are closely related. The name *cyanoptera* means "blue-winged," signifying the chalk-blue secondary coverts of both species.

■ **APPEARANCE** Breeding male's gingerbread plumage distinctive; glistens in bright sunlight, but appears dark and dull otherwise. Female Blue-winged quite similar but colder brown overall, with more contrastingly patterned face and smaller bill. Female Green-winged (p. 38) smaller, a bit darker, and notably smaller-billed.

■ **VOCALIZATIONS** Male call quite unlike that of male Blue-winged: a rapid rapping, *ta'ta'ta'ta'ta*. Female's nasal *kvenk* like female Blue-winged's.

■ **POPULATIONS** Unlike late-arriving Blue-winged, Cinnamon is among the earliest of all spring migrants. Hybridizes occasionally with Blue-winged; drakes intermediate in plumage, typically showing splotchy white facial crescent (Blue-winged trait) and dusky chestnut tones overall (Cinnamon trait); hybrid females may be impossible to separate.

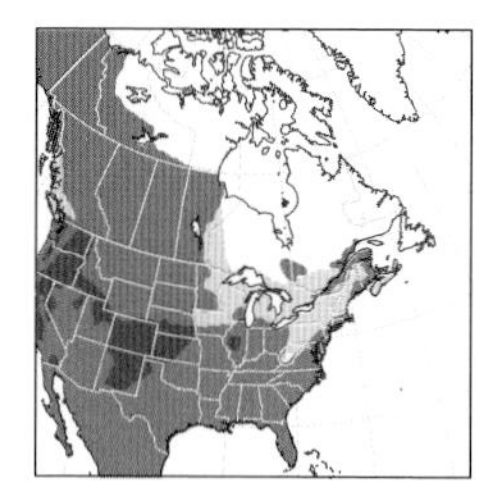

### Northern Shoveler

*Spatula clypeata* | NSHO | L 19" (48 cm)

If you see a mass of tightly packed ducks in the water rotating in the same direction, you are surely looking at a foraging flock of shovelers, which collaborate to create a whirlpool that brings food to the surface. The birds pair off in twos, as well, pinwheeling endlessly.

■ **APPEARANCE** Bill alone is sufficient for ID: broad, very long, and oddly shaped (*clypeata* means "shield"); in close view, note the odd, baleen-like structure, an adaptation for filter feeding. Female plumage similar to female Mallard's (p. 36); first-winter male often shows diffuse white facial crescent, suggesting Blue-winged Teal; but bill very different in all plumages.

■ **VOCALIZATIONS** Feeding flocks murmur constantly; listen for disyllabic note repeated slowly: *shook shook ... shook shook ... shook shook*.

■ **POPULATIONS** Breeds widely across northern reaches of Northern Hemisphere; local abundance constrained in part by availability of small crustaceans, which the shoveler strains through its amazing bill.

Blue-winged Teal
♂
♀
dark eyeline
and white
eye arcs
cool grayish
brown overall
face whitish
at front
♀
small,
hin bill
♂
white spot
at rear flank
Cinnamon Teal
♂
♂
larger, more
spatulate bill than
Blue-winged
less patterned and
warmer brown overall
than Blue-winged
♀
plain buffy
face
♂
♀
diffuse white
facial crescent
fall ♂
broad white
tail edges
boldly scalloped
with buff
♀
Northern
Shoveler
♂
swims low in water with huge
bill nearly touching the surface

## GENUS *MARECA*

These midsize dabbling ducks have slight bills that are mostly gray. Breeding males are distinctive, but females can be tricky to ID. All three species are versatile, occurring in a variety of habitats.

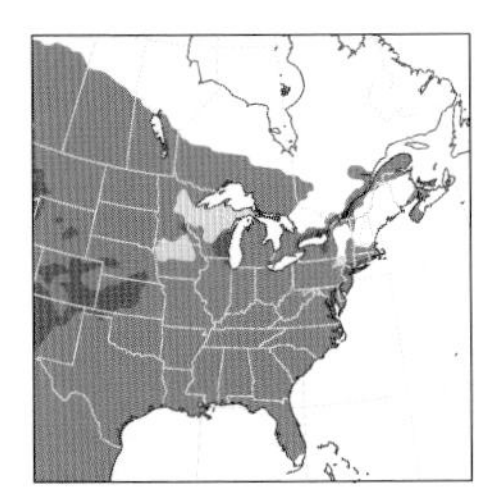

### Gadwall

*Mareca strepera* | GADW | L 20" (51 cm)

Although generally common where it occurs, this relatively unflamboyant duck is easily overlooked among showier species.

■ **APPEARANCE** Blocky, with steep forehead; both sexes flash square of white on secondaries in flight. Breeding male gray with a black "butt"; close up, the gray breast is exquisitely vermiculated; note also rich chestnut on secondary coverts. Female mostly gray-brown, but with whitish belly; good mark is bill, dark with thin orange stripe.

■ **VOCALIZATIONS** Female quack a nasal *kvunk;* courting male mixes rich whistles with endearing burps.

■ **POPULATIONS** Beneficiary of habitat protections across its N. Amer. range, including East; following steep losses earlier, numbers have been rebounding for more than 50 years.

### Eurasian Wigeon

*Mareca penelope* | EUWI | L 20" (51 cm)

Rare but regular in winter, especially mid-Atlantic coast; typically found singly in flocks of American Wigeon. Gray-and-red adult male distinctive, but female quite similar to female American; head of female Eurasian variably warmer brown. Wings of both sexes dusky gray below; American's white underwing coverts contrast with darker gray flight feathers.

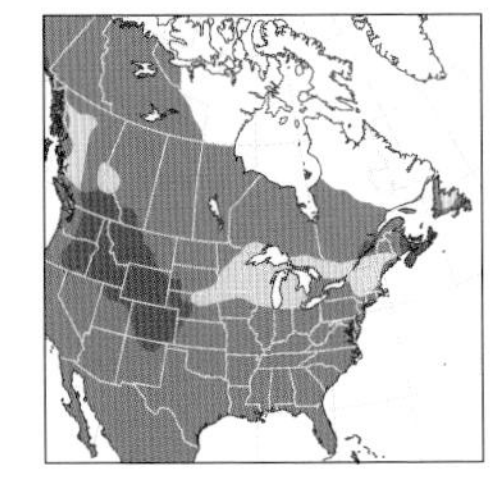

### American Wigeon

*Mareca americana* | AMWI | L 19" (48 cm)

Although perfectly capable of swimming, this common duck frequents grassy stretches, even park lawns, where it grazes in midsize flocks.

■ **APPEARANCE** Both sexes flash prominent white wing stripe in flight. Adult male has gray head, finely stippled, topped off by gleaming white crown; broad green crescent extends behind eye. Female told from female Gadwall by blue-gray bill with fine black edging. Female very similar to female of rare Eurasian Wigeon; that species is warmer-headed, with darker underwing.

■ **VOCALIZATIONS** Males exceedingly vocal; give frenzied, trisyllabic *whee WHEE whew,* often in chorus.

■ **POPULATIONS** Has recovered impressively since mid-20th century; now breeding along mid-Atlantic coast.

### Duck Genera

Scientists place closely related species in the same genus. The guiding principle here is common ancestry (see sidebar, p. 54): Ducks in the genus *Mareca,* say, share a common ancestor not shared by those in the genera *Spatula* and *Anas.* The tricky part is agreeing on how many species should be in a genus; for example, *Anas* comprises 10 species in North America, compared to four each for *Mareca* and *Spatula.*

Evolutionary relationships among birds are an area of active research right now, with frequent reassignments of genera. Typically, an old genus is dismantled when it is discovered not to be monophyletic—in other words, when it fails the test for common ancestry. That is what happened to the genus *Anas*: In 2017, the AOS reorganized the genus, assigning some species to new genera, including *Spatula* and *Mareca,* while retaining others in a pared-down *Anas.*

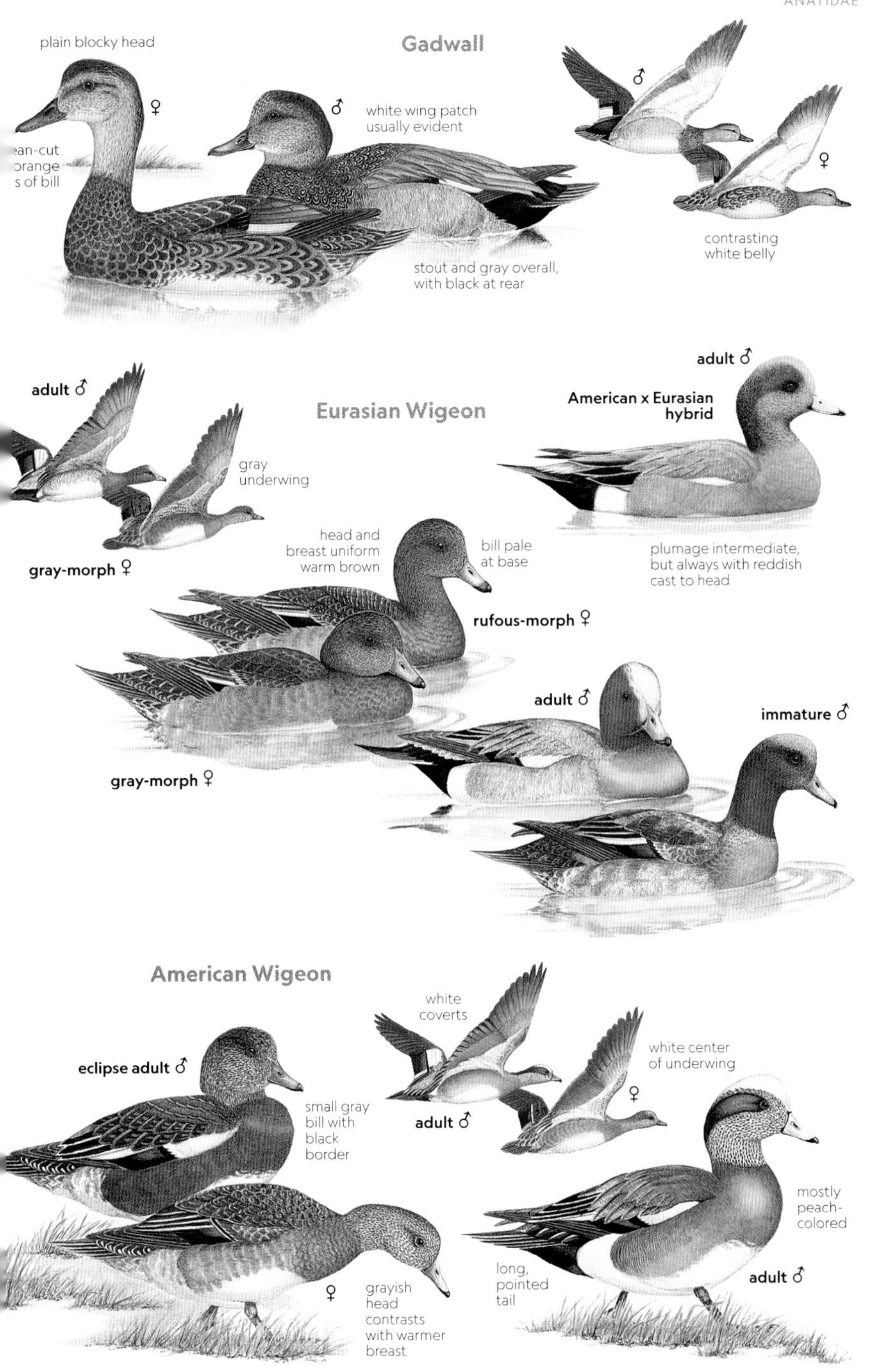
Gadwall
plain blocky head
♀
♂
white wing patch
usually evident
♂
♀
an-cut
orange
s of bill
contrasting
white belly
stout and gray overall,
with black at rear
Eurasian Wigeon
adult ♂
American x Eurasian
hybrid
adult ♂
gray
underwing
head and
breast uniform
warm brown
bill pale
at base
plumage intermediate,
but always with reddish
cast to head
gray-morph ♀
rufous-morph ♀
adult ♂
immature ♂
gray-morph ♀
American Wigeon
white
coverts
white center
of underwing
eclipse adult ♂
small gray
bill with
black
border
♀
adult ♂
mostly
peach-
colored
long,
pointed
tail
adult ♂
♀
grayish
head
contrasts
with warmer
breast

## THE MALLARD COMPLEX

This superspecies of closely related ducks comprises notably large dabblers with generally colorful bare parts. The adult male (drake) Mallard is unmistakable, but all others are dark brown overall; collectively, they are the "brown duck" group.

### Mallard

*Anas platyrhynchos* | MALL | L 23" (58 cm)

This aggressive and adaptable duck is one of the most familiar of all birds. The adult male in breeding plumage is the "Greenhead" of wildlife lore, but other plumages are pale brown.

■ **APPEARANCE** Breeding male sports glistening green head, white neck ring, chestnut breast. Bill of female marbled orange-and-black year-round. Nonbreeding (eclipse) male resembles female, but with solid yellow bill.

■ **VOCALIZATIONS** Female, not male, gives the universally recognized, straight-up *quack!* Male's utterance raspier, less exuberant. In rarely observed courtship "song," male grunts, gasps, and whistles.

■ **POPULATIONS** At home in remote mid-continent prairie marshes, as well as at duck ponds in our largest metropolises. Interbreeds with other species in Mallard complex, altering those species' genetic make-up.

### Mexican Duck

*Anas diazi* | MEDU | L 22" (56 cm)

This is the "brown duck" of our desert borderlands, recently elevated to full-species status by the AOS. It reaches our area in the Rio Grande Valley of Texas. Male is like female Mallard but with a yellow bill and darker body; tail feathers brownish (white on Mallard) and straight (curled on male Mallard). Female hard to separate from female Mallard, but note uniformly drab yellow-orange bill.

### American Black Duck

*Anas rubripes* | ABDU | L 23" (58 cm)

This generally shyer counterpart of the Mallard favors protected coastal marshes where available. Rigorously researched and written about, it is one of our best known ducks.

■ **APPEARANCE** Body quite dark, contrasting with paler, cold gray head. In flight, wing linings flash brightly; speculum not bordered with white as on Mallard. Male's bill yellow like male Mallard's; female's cold gray-green, unlike female Mallard's.

■ **VOCALIZATIONS** Array of male and female calls analogous to those of Mallard; male courtship "song," rarely observed, is longer than Mallard's.

■ **POPULATIONS** Hybridizes extensively with Mallard; adult male hybrids show variable swaths of green on crown.

### Mottled Duck

*Anas fulvigula* | MODU | L 22" (56 cm)

The Southeast's female Mallard look-alike hybridizes with Mallards and, confusingly, with Mexican Ducks. Triple hybrids occur along the Rio Grande in South Texas.

■ **APPEARANCE** Warmest and brightest of the "brown duck" complex. Head and throat buffy; body feathers edged orange-brown; gape has dark spot.

■ **VOCALIZATIONS** Only differences from Mallard involve little-known male courtship "song," with generally simple elements.

■ **POPULATIONS** Like American Black Duck, at risk from the double threat of disappearing coastal wetlands and genetic intrusion by Mallards.

Mallard
eclipse ♂
(summer only)
blue speculum
with broad
white borders
underwing contrasts
less with body than
American Black Duck
♂
♀
orange bill with
irregular black
"saddle"
mostly
white tail
♀
♂
curled tail feathers
Mexican
Duck
both sexes
with solid
yellow bill,
female
duller
darker head
pattern than
Mottled Duck,
with more
white in wing
blue-green
speculum with narrow
white borders
body
and tail
darker than
female
Mallard
♂
American Black Duck
x Mallard hybrid
♂
American
Black Duck
bill mostly
olive-gray
♀
indigo
speculum
without white
borders
underwing
contrasts strongly
with body
♂
greenish-
yellow bill
♂
blue-green speculum
with minimal white
borders
black spot
at gape
dark tail
Mottled Duck
body darker than female
Mallard, more strongly
marked buff than
American Black Duck
mainly Florida
fulvigula
yellow
bill
western Gulf Coast
maculosa
unmarked
buffy
throat
♂
paler overall than maculosa

## TWO DABBLERS

Despite obvious differences in size, shape, and plumage, these two ducks are close relatives. Both species have thin, gray bills. They round out the ducks known as dabblers, represented in the East by the genera *Spatula, Mareca,* and *Anas.*

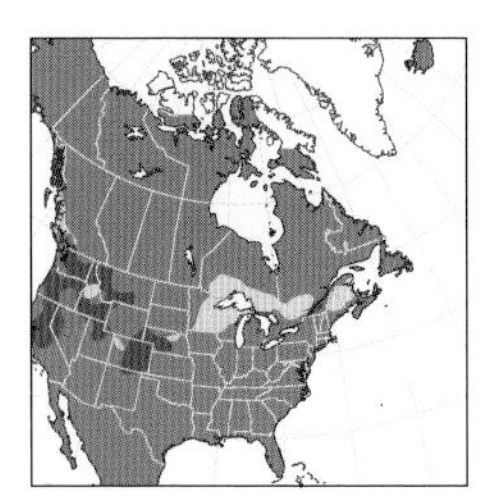

### Northern Pintail

*Anas acuta* | NOPI | L 20–26" (51–66 cm)

This is the "pointy duck." In flight, both sexes appear long-winged, long-tailed, long-necked, and pointy-billed.

■ **APPEARANCE** Central tail streamers instantly indicate breeding male; note also thin white stripe running up chocolate head, white breast, and gray body. Female uniformly gray-brown, best identified by long neck, thin bill, and relatively long tail.

■ **VOCALIZATIONS** Male in courtship gives a rising hiss transitioning to a musical hoot, then trailing off, *ssssSSLEEERPppsss;* male also gives whistled *PWEEP* year-round. Female cackles quietly.

■ **POPULATIONS** Found throughout East, but less common away from central prairies; spring migration and breeding season notably early.

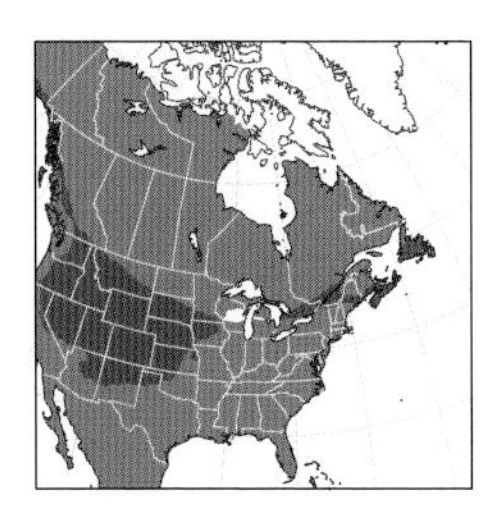

### Green-winged Teal

*Anas crecca* | GWTE | L 14½" (37 cm)

Tight flocks of this smallest dabbler probe furiously in the muddy margins of lakes and ponds. The effect is shorebird-like, and the species sometimes forages alongside dowitchers (p. 120).

■ **APPEARANCE** All have thin, gray bills, structurally distinct from those of other birds called teal; green secondaries prominent in flight. Male has chestnut head with broad green patch; sides gray with vertical white stripe. Female at rest told by small size, pale rear; often, a bit of the green wing pokes through.

■ **VOCALIZATIONS** One of our noisiest ducks, often talkative at night. Male's froglike *peee* or *pleep* enlivens marshes in early spring; females mutter nasally, *kwenk* or *kwunk*.

■ **POPULATIONS** Unlike most other dabblers, often nests in and around forested waterways. Almost all in East are American subspecies *carolinensis;* Eurasian subspecies, nominate *crecca*, treated by many as a distinct species, is very rare but annual winter to spring. Breeding male *crecca* has horizontal white stripe on sides; females of the two subspecies very similar and perhaps impossible to distinguish in the field. Most records from Nfld. south to mid-Atlantic.

### Seasonal Variation in Ducks

Many birds exhibit striking seasonal variation in plumage, most notably between the relatively bright breeding plumages of adults and the same adults' relatively dull nonbreeding plumages. Classic examples include the dissimilar spring and fall plumages of Common Loons (p. 166) and Scarlet Tanagers (p. 400). Most adult male ducks likewise exhibit pronounced seasonal variation, but the timing is peculiar: They acquire their bright plumage in *fall* and hold it through the winter; then, by late spring or early summer, they begin to molt into a comparatively drab and disheveled plumage.

For Northern Hemisphere birds with two plumages per year, the bright plumage has traditionally been termed the alternate plumage, whereas the dull, simpler plumage is termed the basic plumage. Adult birds wear their alternate plumage during the breeding season and their basic plumage otherwise; often, but not always, the differences between alternate and breeding plumages are much more pronounced in males than in females. Waterfowl biologists

all features
long and slender
in flight

k speculum
with white
ailing edge

plain gray-brown head

**Northern Pintail**

long,
slender
neck

thin gray
bill

understandably assumed that, despite the odd timing, the bright plumages of male ducks are the alternate plumage; this was especially reasonable given that courtship for many ducks begins by early winter. It turns out that this assumption was incorrect.

In the early 21st century, ornithologist Peter Pyle discovered that ducks do it in the opposite way from other birds. The bright breeding plumage in ducks is, surprisingly, the basic plumage, and the dull nonbreeding (eclipse) plumage is, oddly, the alternate plumage. The key result for birders in the field is this: Ducks, like almost all other Northern Hemisphere birds, undergo a complete feather molt in the fall; this molt results in the basic plumage, precisely as with loons, tanagers, and all other birds with two molts per year. It's just that ducks molt into their breeding plumage, which is actually their basic plumage, rather than their nonbreeding plumage, at that time of year. Finally, there is an inevitable exception to the rule: The Ruddy Duck (p. 52) does it the "normal" way, acquiring its bright breeding plumage in the spring.

## POCHARDS

The ducks on this spread and the next, in the genus *Aythya*, share the same basic color scheme: dark at both ends, and pale in the middle. All are adept at diving, although many ducks in other genera likewise are expert divers.

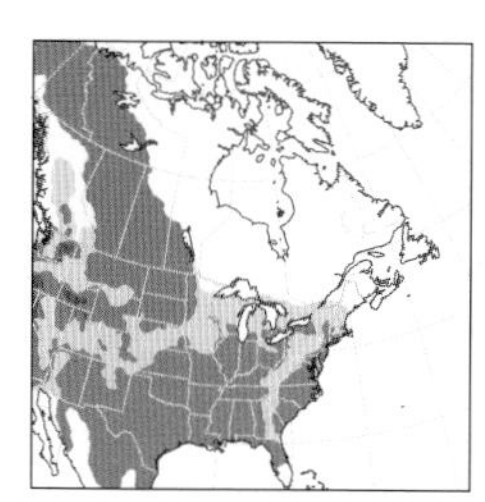

### Canvasback

*Aythya valisineria* | CANV | L 21" (53 cm)

This largest and longest of our pochards gathers in winter wherever there is open water in dense flocks called "rafts."

■ **APPEARANCE** Shape distinctive; long overall and low-slung, with a sloping forehead and long bill. Breeding male has red-brown head and gray-white (canvas-colored) back. Female separated from female Redhead by paler tones overall, all-dark bill, and especially body shape.

■ **VOCALIZATIONS** Rarely heard in winter, but the courtship display is something else: a mirthful, musical cackle, *he-lee-we-loo*.

■ **POPULATIONS** Scientific name *valisineria* means "wild celery," historically a favored food source; species has been undergoing dietary expansion in recent decades.

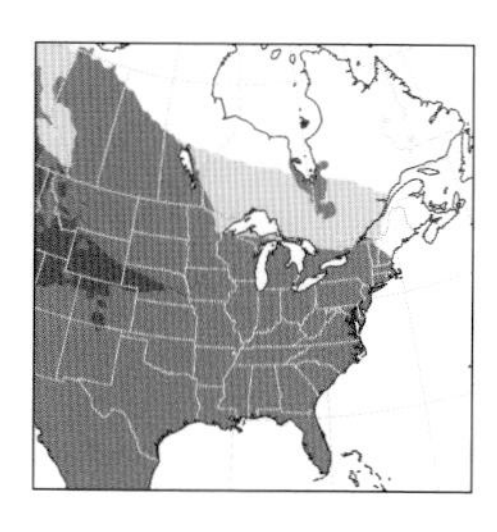

### Redhead

*Aythya americana* | REDH | L 19" (48 cm)

Ecologically speaking, this duck is intermediate between dabblers (pp. 32–39) and other pochards. Especially on the breeding grounds, it forages in shallow marshes, mixes dabbling and diving, and loafs with dabblers.

■ **APPEARANCE** More compact than Canvasback. Bill color is tripartite: A thin white stripe separates the black tip from the broad blue base; pattern more muted in female. Female darker than female Canvasback, but much variation; body shape always different. Compare also with female Ring-necked Duck.

■ **VOCALIZATIONS** Courting males in chorus astonishing: *whee* and *whew* notes, soft but far-carrying; given by many drakes at once, quietly frenzied. May be heard on calm days in late winter, even far from the breeding grounds.

■ **POPULATIONS** Numerous continentally, but can be scarce along the Atlantic seaboard. Flocks of 10,000+ routinely seen in the Laguna Madre of South Tex.

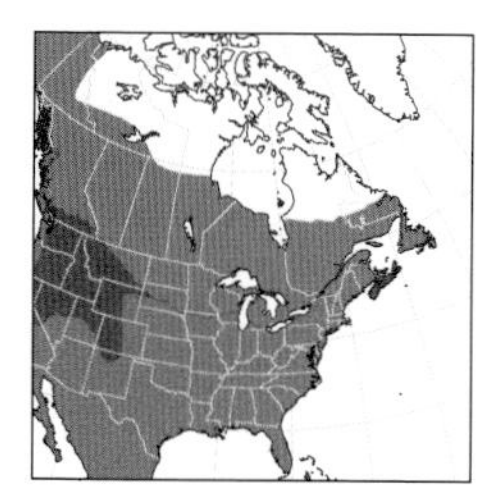

### Ring-necked Duck

*Aythya collaris* | RNDU | L 17" (43 cm)

This common duck is a perennial top-runner in the "worst name for a bird" category. The eponymous neck ring is faint on males and absent on females, but the bird's brilliant bill ring is whoppingly obvious on males and females alike.

■ **APPEARANCE** All have striking ring around the bill (not neck). Head is strongly peaked, even more than on Lesser Scaup. In good light, some males show a lovely, glistening chestnut band (the "ring") between glossy purple head and black breast; note also the white wedge at base of breast, appearing vertical on swimming male. Female shows variable white eye ring and postocular stripe ("bridle"), typically more prominent than on female Canvasback and Redhead.

■ **VOCALIZATIONS** Perhaps the most vocal pochard in winter; when flushing, gives series of annoyed, muffled quacks, *runk ... runk ... runk*.

■ **POPULATIONS** In winter, more likely found on smaller bodies of water than Canvasback or Redhead, often in and around woodlands and farm country.

long, sturdy neck
gray body contrasts with brownish breast and head
muscular and angular
♀
♂
unpatterned wings
♂
♀
long, sloping black bill in all plumages
Canvasback
uniform drab brown
Redhead
♂
♀
pale gray wing stripe
♀
♂
clean gray body
head and body more compact than Canvasback
extensive black on breast
Ring-necked Duck
adult
♂
gray wing stripe
♀
black back
head strongly peaked at rear
♂
white eye ring and "bridle" on grayish face
♀
white ring on bill
white spur divides flank and breast

## TUFTED DUCK AND SCAUP

The rare but regular Tufted Duck is intermediate in plumage between the Ring-necked Duck (p. 40) and scaup. The two species called scaup are widespread in the East, locally common, and a challenging field ID range-wide.

### Tufted Duck

*Aythya fuligula* | TUDU | L 17" (43 cm)

Common in parks in Europe's big cities, this species is much sought on this side of the pond. Adult male instantly recognized by his very long crest. Dark-headed female usually shows at least a hint of the male's crest; her dark blue bill with broad black tip is different from that of all other eastern pochards. Most occurrences here coastal, especially Newfoundland.

### Greater Scaup

*Aythya marila* | GRSC | L 18" (46 cm)

The two scaup are similar in all respects, but the ecology of this species trends more toward "sea duck" than Lesser Scaup does. Greater also breeds farther north and winters more coastally.

■ **APPEARANCE** Larger than Lesser, but size difference not always evident. In both sexes, note head shape: shallowly rounded, with peak above or in front of eye; bill broader than on Lesser with larger, more triangular black tip; white in wing extends well out into primaries. Adult male head glossed greenish; sides and back cleaner white than on Lesser. Female plumage very similar to female Lesser but averages more white on face. First-year males, commonly seen in winter, intermediate between adult males and females.

■ **VOCALIZATIONS** Rarely heard in winter; flushing birds sometimes give a rough, snarling sound. Courting male utters muffled hoots and whistles.

■ **POPULATIONS** Where both scaup species winter coastally, the Greater is more inclined to marine and brackish conditions, but there is great overlap in these proclivities. Inland, Greater shows a preference for large bodies of water.

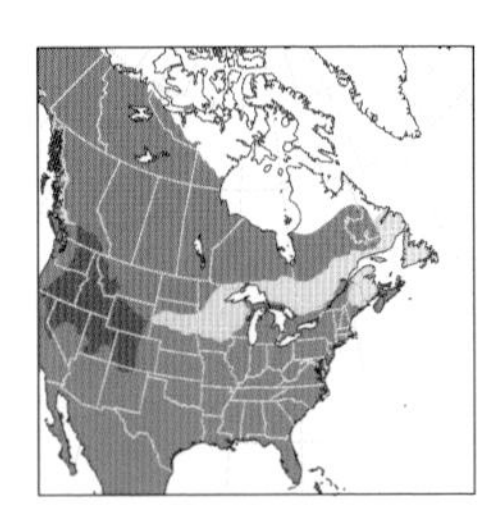

### Lesser Scaup

*Aythya affinis* | LESC | L 16½" (42 cm)

Away from the Great Lakes, this is the more expected scaup in winter in the eastern interior. The species moves north fairly late for a duck, and fall movement southward is also late.

■ **APPEARANCE** In all plumages, head is peaked or pointy, with the high point behind the eye; bill slighter than Greater's with less black at tip; male's head glossed purplish, female's face with less white than on Greater. In both scaup, head shape greatly affected by posture and behavior; white on face of female intrinsically variable and male head color heavily dependent on light. White in wing restricted mostly to secondaries, but getting a diagnostic view of the flying bird is tricky.

■ **VOCALIZATIONS** Mostly silent in winter; when flushing, sometimes gives a weak, wavering growl. Courting male gives excited popping and whooping sounds: *whoop, whoopa, whee-up,* etc.

■ **POPULATIONS** Has increased in recent winters on the Great Lakes in part because of rapidly invading nonindigenous mussels; unfortunately, wintering scaup bioaccumulate cadmium, selenium, and other toxins from the shellfish. Declining in winter in Gulf Coast region.

Tufted Duck
adult ♂
extensive and bold wing stripe
♀
occasionally shows some white around bill
♀
hint of crest
rker back
ıan scaup
♀
1st winter ♂
crest prominent
black back
broad black bill tip
adult ♂
pure white flanks
ften with
rominent
hitish ear
atches
larger, more scooped bill
♀
♀
extensive white wing stripe
adult ♂
Greater Scaup
♀
1st winter ♂
larger, rounder head than Lesser
more black on bill tip than Lesser
adult ♂
white mostly restricted to secondaries
adult ♂
brighter white midsection of body
♀
Lesser Scaup
smaller, straighter bill with limited black nail
head tall and peaked at rear, but beware of variability as posture changes
♀
adult ♂
slightly grayer back and flanks than Greater

**STUNNING SEA DUCKS**

Breeding well away from most human population centers, these three sea duck species are known to birders mostly on the wintering grounds: jetties, breakwaters, and coastal lagoons.

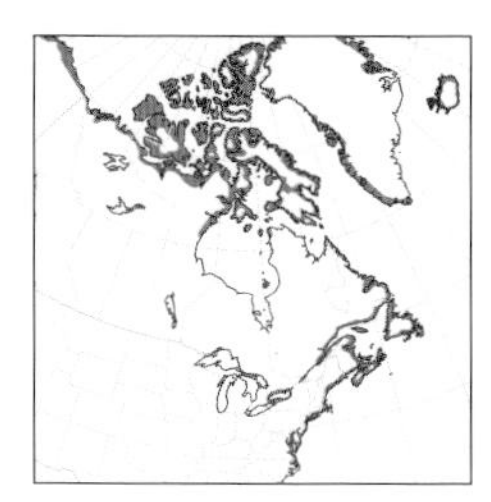

## King Eider

*Somateria spectabilis* | KIEI | L 22" (56 cm)

This is the more northerly of the two eastern eiders, with many wintering no farther south than the southern extent of the North Atlantic sea ice.

■ **APPEARANCE** Vibrantly colored adult male unique, but other plumages similar to Common Eider. Adult female King told by smaller size, blockier body shape; border between black bill and brown face not nearly as pointed as on Common. Body feathers of female King have black V marks; male King mostly black-backed. First-winter male has bill structure and overall body shape like adult female, but bill already starting to show color.

■ **VOCALIZATIONS** Adult male gives low, booming hoots, both on the breeding grounds and at sea toward the end of the winter; hard to hear over the surf, easier from boats beyond the breakers.

■ **POPULATIONS** Because distribution is so closely tied to Arctic ice conditions, demographic instability is forecast under most climate change models.

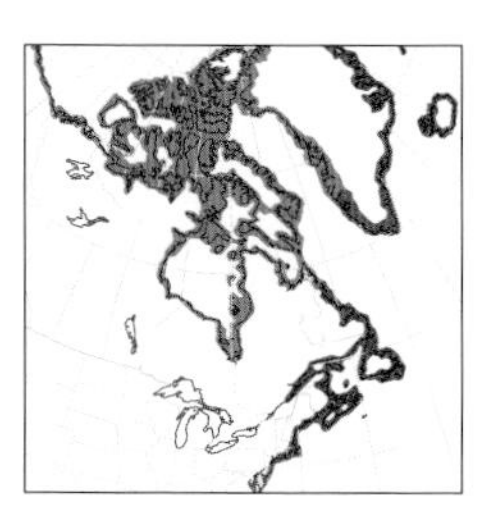

## Common Eider

*Somateria mollissima* | COEI | L 24" (61 cm)

The largest duck in the Northern Hemisphere, this is the eider of pillow and parka fame; the name *mollissima* means "maximum softness."

■ **APPEARANCE** Color of frontal lobes on adult male varies geographically from mostly green to rather bright orange-yellow; head and bill shape also variable, but always long and sloping. Male Common mostly white-backed; nape and breast suffused greenish. Females vary by subspecies in ground color from heavily rufous-tinged brown to nearly colorless gray; all are told from female King by relatively straight barring on body feathers and very pointy border between bill and facial feathers.

■ **VOCALIZATIONS** Heard mostly on breeding grounds, male gives *oooOOOooh* call, rising and falling, with human-sounding timbre.

■ **POPULATIONS** Locally increasing, with winter "rafts" of thousands sometimes seen. Unlike most other duck species in the East, which are not usually overharvested these days, Common is still hunted to excess in some places.

## Harlequin Duck

*Histrionicus histrionicus* | HADU | L 16½" (42 cm)

The habitat requirements of this plump duck may be the most unusual of any N. Amer. waterfowl species: The Harlequin breeds on ledges near turbulent streams and rivers, and it sticks close to rocky shorelines in winter.

■ **APPEARANCE** All are small, round-bodied, and short-billed. Adult male's dramatic color and pattern is spectacular. Female, brown-bodied with white spots on face, suggests female Surf Scoter (p. 46), but head shape and bill structure different.

■ **VOCALIZATIONS** Squeaky, monosyllabic utterances, *jee* and *jeep,* often given in rapid succession; when waters are calm, close-up birds often audible.

■ **POPULATIONS** Uncommon; apparently has never been a common species. Number of birds wintering in the East only in the low thousands, but these ducks gather at prominent sites, where they are easily found.

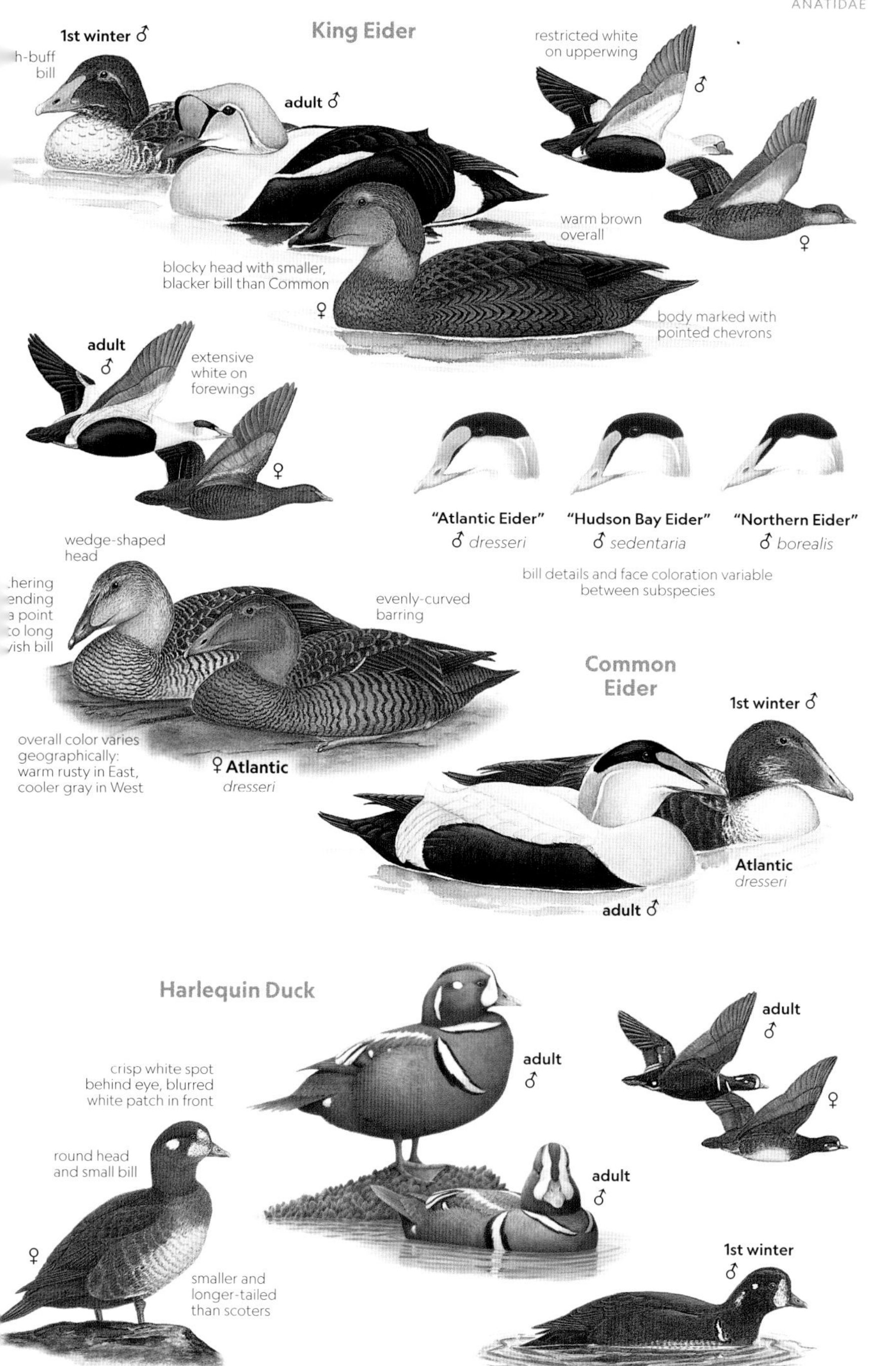
King Eider
1st winter ♂
h-buff
bill
adult ♂
restricted white
on upperwing
♂
♀
warm brown
overall
blocky head with smaller,
blacker bill than Common
♀
body marked with
pointed chevrons
adult
♂
extensive
white on
forewings
♀
"Atlantic Eider"
♂ dresseri
"Hudson Bay Eider"
♂ sedentaria
"Northern Eider"
♂ borealis
bill details and face coloration variable
between subspecies
wedge-shaped
head
evenly-curved
barring
overall color varies
geographically:
warm rusty in East,
cooler gray in West
♀ Atlantic
dresseri
Common
Eider
1st winter ♂
Atlantic
dresseri
adult ♂
Harlequin Duck
adult
♂
crisp white spot
behind eye, blurred
white patch in front
round head
and small bill
♀
adult
♂
♀
adult
♂
smaller and
longer-tailed
than scoters
1st winter
♂

## SCOTERS

Forming large to sometimes huge flocks just offshore, these sea ducks combine dark plumage with brightly colored and oddly shaped bills. The name scoter apparently relates to the word "Scotland."

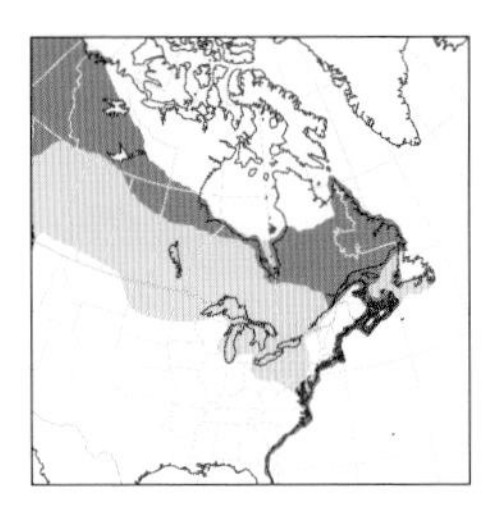

### Surf Scoter

*Melanitta perspicillata* | SUSC | L 20" (51 cm)

Along with the Black Scoter, this is one of our two dark-winged scoters. Flights well into the thousands may be seen in fall along the Atlantic coast, with smaller numbers on the Great Lakes.

■ **APPEARANCE** Appears all-dark in flight. Black-and-white head of male distinctive; "Skunkhead" is a colloquial name. Note also colorful bill. Most females show two well-formed white splotches on the face: a vertical one at the base of the bill, a rounder one behind and below the eye.

■ **VOCALIZATIONS** Relatively quiet, even on the breeding grounds; courting and otherwise excited males give rapid popping sound, *puh puh puh puh puh puh ....*

■ **POPULATIONS** Much daily movement, even in the heart of winter, to and from feeding and roosting areas; some flocks sort out by age and sex.

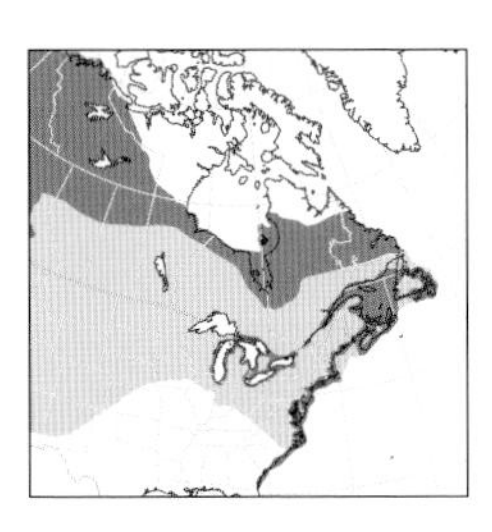

### White-winged Scoter

*Melanitta deglandi* | WWSC | L 21" (53 cm)

With its bold white wing patches and impressive heft, this scoter is often conspicuous in big winter flocks. It can be vexingly uncommon where Surf and Black Scoters form large concentrations.

■ **APPEARANCE** Entirely white secondaries contrast with otherwise dark plumage; even on birds at rest, a bit of white usually pokes out. Adult male sports broad white "teardrop" under and behind eye; two white blobs on female's dark face are more rounded than corresponding white patches on face of female Surf.

■ **VOCALIZATIONS** Like Surf Scoter, fairly quiet, even in summer. Gruff croaks and rough whistles occasionally heard in winter.

■ **POPULATIONS** Has always occurred in small numbers on Great Lakes in fall and winter; recent increases may be due to ongoing invasion of nonindigenous mussels, a favored food source.

### Black Scoter

*Melanitta americana* | BLSC | L 19" (48 cm)

This scoter winters widely on the Atlantic coast and Gulf Coast, and in smaller numbers across the Great Lakes; but nesting among the eastern population is restricted mostly to taiga lakes in a relatively small chunk of north-central Quebec.

■ **APPEARANCE** The smallest and darkest scoter. Adult male appears all-black at rest, except for bright yellow on bill; like Surf, dark-winged in flight, but remiges paler gray than coverts. Extensive white on face of adult female unlike that on other scoters; female Ruddy Duck (p. 52), similar in overall plumage, can be unexpected point of confusion.

■ **VOCALIZATIONS** Unlike the other scoters, vocal even in winter. Listen for drawn-out, sorrowful, sighing whistles, given on different pitches, often doubled: *ooooo eeeee.*

■ **POPULATIONS** Range of eastern breeders wholly disjunct from that of western breeders, mostly western Alas. Local distribution in winter varies annually in East, with proximity to shore influenced by climatic conditions.

vertical whitish oval at base of bill, diffuse patch behind eye
somewhat contrasting dark cap
white nape on adult female
adult ♀
adult ♂
uniformly dark wings
1st winter ♀
adult ♂
all immature scoters have pale bellies
sturdy triangular bill
1st winter ♀
1st winter ♂
Surf Scoter
two white ovals on face, front one extending onto bill
long, scooped bill
White-winged Scoter
striking white secondaries
adult ♂
adult ♀
1st winter ♀
1st winter ♂
adult ♂
adult ♀
white usually, but not always, visible
gray remiges contrast with black coverts
adult ♂
round head and more "standard" duck bill
unmarked pale cheeks
smaller and rounder than other scoters
adult ♂
adult ♀
adult ♀
1st winter ♂
Black Scoter

## BAY DUCKS

Named for their proclivity for gathering in large bays near the coast in winter, these four divers actually spread out quite a bit. Long-tailed Ducks often mass beyond the surf, while Buffleheads and goldeneyes winter well inland.

### Long-tailed Duck

*Clangula hyemalis* | LTDU | L 16–22" (41–56 cm)

The quip "if it quacks like a duck" utterly fails in the case of this striking duck. Flocks yodel melodiously round the clock, even in winter.

■ **APPEARANCE** Small-bodied, short-billed, and long-tailed; underwings dark. Annual variation complex. Males in winter mostly white; extensively black in summer; tail extremely long. Female lacks central tail streamers of male, but nevertheless long-tailed; note white plumage overall, becoming darker spring into summer.

■ **VOCALIZATIONS** Sonorous, far-carrying nasal hoot followed by two or three more: *ow! ... ow ow ow! (owl om-e-let);* flocks in chorus memorable.

■ **POPULATIONS** Fairly common in winter on Great Lakes, but locally so; midwinter movements away from coast tied to freezing and thawing.

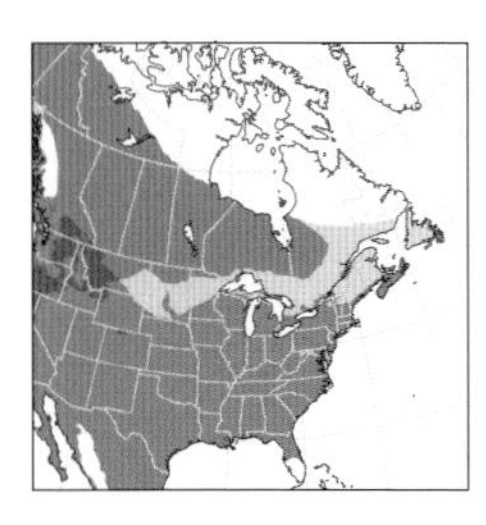

### Bufflehead

*Bucephala albeola* | BUFF | L 13½" (34 cm)

Among our smallest waterfowl, the Bufflehead is also one of the most frequently encountered diving ducks inland in winter, when it often consorts with Hooded Mergansers (p. 50).

■ **APPEARANCE** Quite small, chunky, with steep forehead; bill small and dark. Male has huge white patch on iridescent black head; female, a broad white swath across dark chocolate head. Buffleheads tilt slowly from side to side as they fly past.

■ **VOCALIZATIONS** Quiet even on the boreal breeding grounds; flushing birds in winter sometimes give harsh, flatulent quacks.

■ **POPULATIONS** Obligate cavity nester; historically depended on nest holes drilled by large woodpeckers, but now also uses nest boxes.

### Common Goldeneye

*Bucephala clangula* | COGO | L 18½" (47 cm)

Also known as "whistlers," goldeneyes make a loud, musical trilling or purring sound with their wings. A whole flock is remarkably noisy.

■ **APPEARANCE** Male's glossy green head marked with large white circle near base of bill. Female brown-headed; black bill is variably tipped with yellow. All have gently sloped forehead, distinct from Barrow's.

■ **VOCALIZATIONS** Male courtship song a long mechanical buzz followed by shrill *peent!*

■ **POPULATIONS** Common, but susceptible to human influences, for ill or for good; readily uses nest boxes, attracted to large reservoirs in winter.

### Barrow's Goldeneye

*Bucephala islandica* | BAGO | L 18" (46 cm)

Uncommon breeder in East in Quebec and Labrador, winters coastally; even on core wintering grounds from Maine northward, the rarer of the two goldeneyes. Adult male's head glossed purplish; white patch on face crescent-shaped; mostly black scapulars make swimming male appear darker-sided than Common. Female's bill mostly deep yellow. Steep forehead an excellent point of distinction from Common; angle between bill and forehead nearly 90 degrees. Hybrids with Common complicate ID.

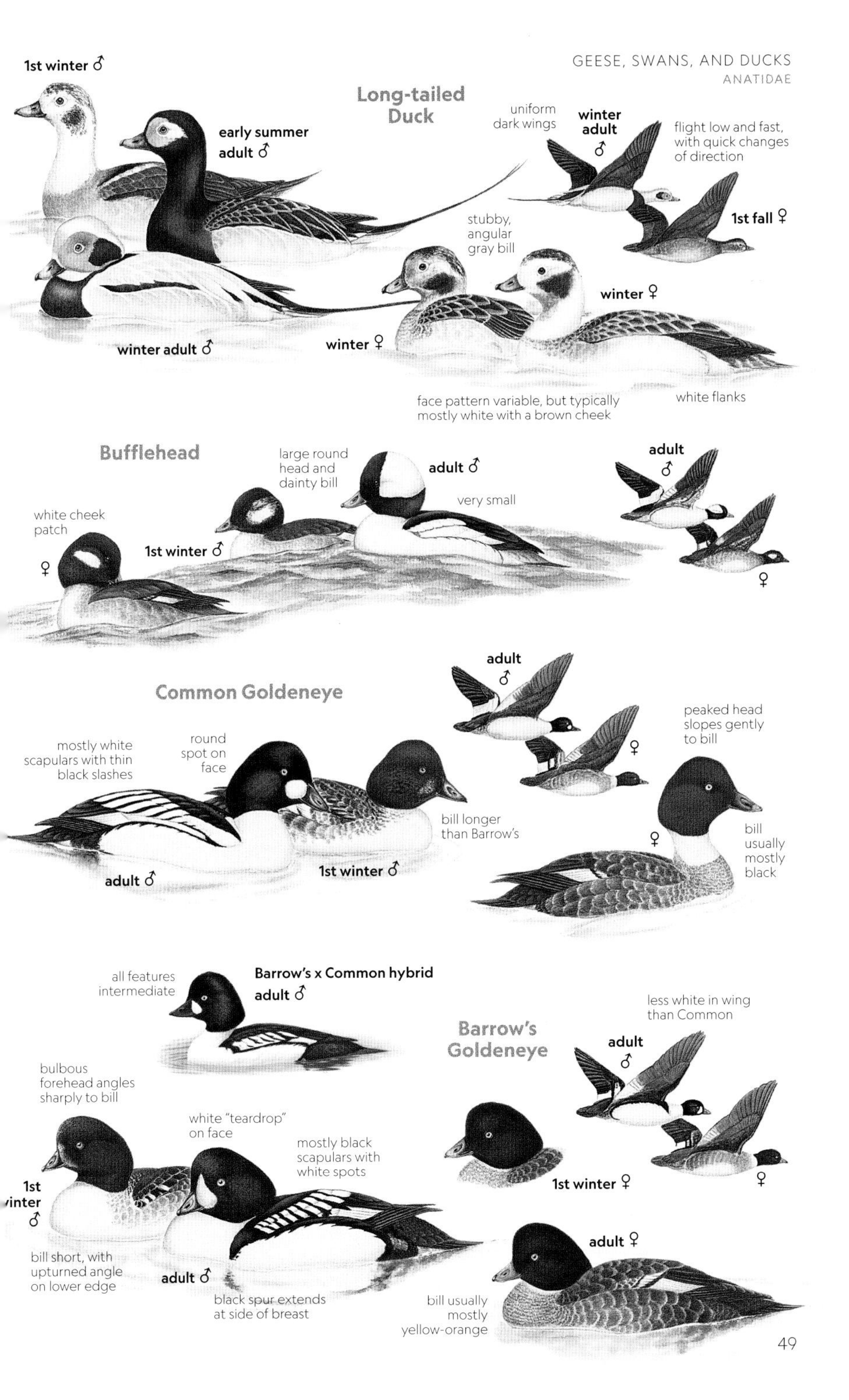
Long-tailed Duck
1st winter ♂
early summer adult ♂
uniform dark wings
winter adult ♂
flight low and fast, with quick changes of direction
1st fall ♀
stubby, angular gray bill
winter ♀
winter adult ♂
winter ♀
face pattern variable, but typically mostly white with a brown cheek
white flanks
Bufflehead
large round head and dainty bill
adult ♂
adult ♂
very small
white cheek patch
1st winter ♂
♀
♀
Common Goldeneye
adult ♂
peaked head slopes gently to bill
mostly white scapulars with thin black slashes
round spot on face
♀
bill longer than Barrow's
♀
bill usually mostly black
adult ♂
1st winter ♂
all features intermediate
Barrow's x Common hybrid adult ♂
less white in wing than Common
Barrow's Goldeneye
adult ♂
bulbous forehead angles sharply to bill
white "teardrop" on face
mostly black scapulars with white spots
1st winter ♀
♀
1st winter ♂
adult ♀
bill short, with upturned angle on lower edge
adult ♂
black spur extends at side of breast
bill usually mostly yellow-orange

## MERGANSERS

Also known as "sawbills," these adroit divers are equipped with serrated bills for capturing and manipulating crustaceans and fish. Hooded and Common Mergansers nest mostly in cavities, Red-breasted on the ground.

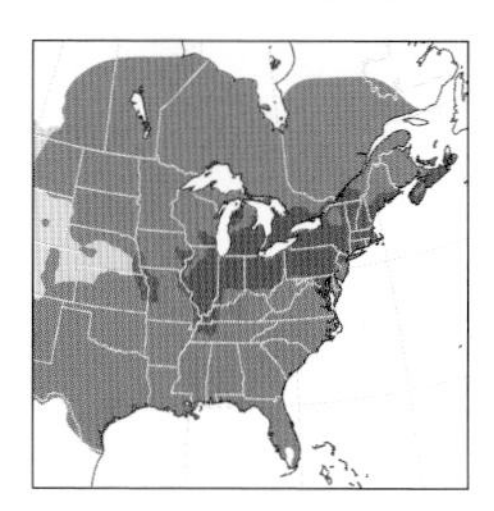

### Hooded Merganser

*Lophodytes cucullatus* | HOME | L 18" (46 cm)

This smallest and most distinctive merganser is widespread inland in winter on ponds and rivers, often with woods nearby.

■ **APPEARANCE** Male's extraordinary bonnet can be raised or lowered in less than a second; at a quick glance, could be mistaken for male Bufflehead (p. 48), with which it often consorts. Female's crest tawny, rest of body colder brown; eyes brown. First-winter male and rarely seen summer-plumage (eclipse) adult male like female, but with staring yellow eye.

■ **VOCALIZATIONS** Nonvocal wing whistle soft but distinctive: a rapid trill like a field cricket.

■ **POPULATIONS** In winter, spreads out across the landscape singly or in small flocks. Breeding range expanding, probably in response to nest box supplementation. The "Hoodie" now breeds alongside the "Woodie" (Wood Duck, p. 30) in many wooded wetlands.

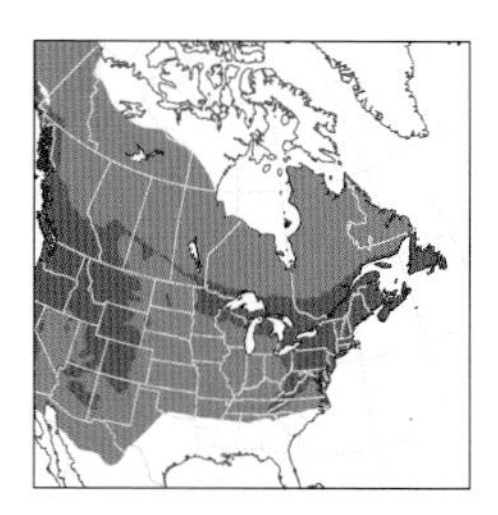

### Common Merganser

*Mergus merganser* | COME | L 25" (64 cm)

Widespread inland in winter as far north as there is open water, this largest merganser migrates north as soon as the spring thaw begins across the mostly boreal breeding grounds.

■ **APPEARANCE** Males on the wing look like flying bowling pins, with slim white bodies and all-dark heads. Breast white, sometimes tinged rosy. Female's orange-brown head sharply demarcated from pure-white breast and throat; head color on female Red-breasted blends more gradually with paler breast and throat. In all plumages, bill and forehead structure distinctive: Broad-based bill of Common slopes gradually up toward long, shallow forehead.

■ **VOCALIZATIONS** Flushing and sometimes swimming, female issues muffled quacks, in slow succession: *woohf... woohf... woohf....*

■ **POPULATIONS** Sex segregation, especially in late summer, can be pronounced; males move locally to molt while females, also molting, stick closer to nest sites. Eschews saltwater; winter reports in marine settings suspect.

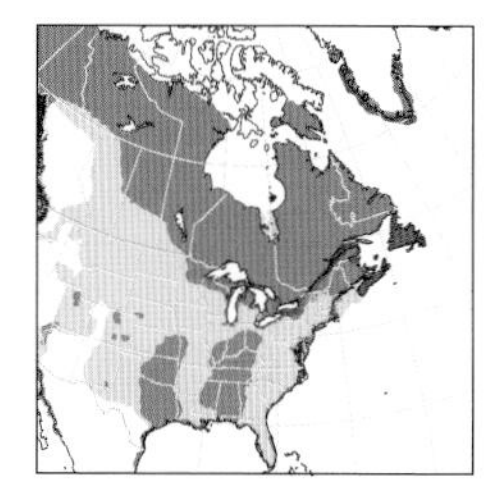

### Red-breasted Merganser

*Mergus serrator* | RBME | L 23" (58 cm)

The only seafaring merganser, this species breeds north to Arctic coasts and winters widely along the Atlantic coast and Gulf Coast. Overland fall migration creates prodigious flights in Great Lakes region.

■ **APPEARANCE** All are shaggy-headed; thin-based bill forms sharp angle with steep forehead, distinct from sloping profile of Common. Adult male's gray-red breast and green head separated by broad white band. Some females closely resemble Common, but bill and forehead structure a reliable point of distinction.

■ **VOCALIZATIONS** Female's quack a bit higher, more nasal than female Common's; heard much less frequently in winter than Common is.

■ **POPULATIONS** Heavy presence around Great Lakes has raised concerns about competition with fisheries, but most are passage transients and alleged conflicts probably minimal.

## Hooded Merganser

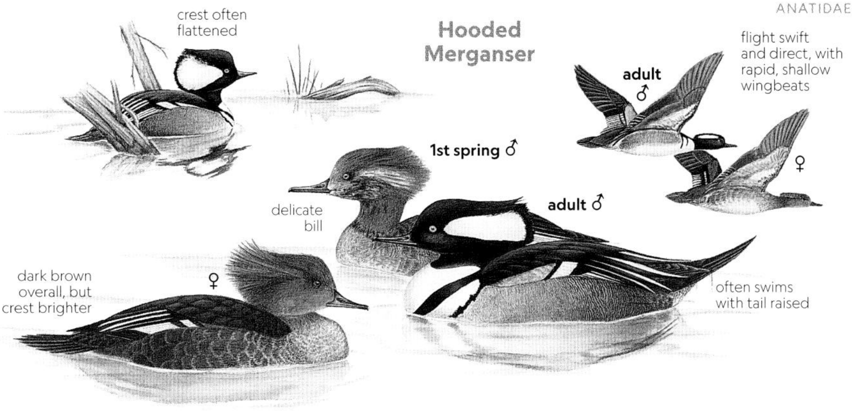

crest often flattened

adult ♂

flight swift and direct, with rapid, shallow wingbeats

♀

1st spring ♂

delicate bill

adult ♂

dark brown overall, but crest brighter

♀

often swims with tail raised

## Common Merganser

sloping forehead

1st spring ♂

adult ♂

streamlined and powerful in flight

underparts mostly white, often tinged rosy

thick-based bill tapering to thin drooped tip

♀

♀

rich rufous head and neck contrast sharply with white chin and breast

adult ♂

## Red-breasted Merganser

adult ♂

more slender than Common in flight

♀

1st winter ♂

steep forehead

long, shaggy crest

bill thin along entire length

adult ♂

dull tawny head blends into pale neck

♀

## STIFFTAILS AND A GUAN

Despite obvious differences between the two "stifftail" ducks and one cracid on this page, they and other fowl share common ancestry as members of the clade Galloanseres, described below.

### Masked Duck

*Nomonyx dominicus* | MADU | L 13½" (34 cm)

Widespread in the Neotropics, this small duck is probably regular in far southern Texas, but it is secretive and easily overlooked. Breeding male has black face; female, juvenile, and nonbreeding male show two dark horizontal lines across pale face; similar female Ruddy Duck has only one.

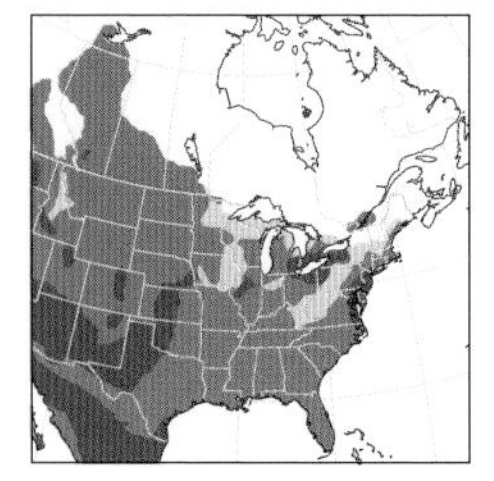

### Ruddy Duck

*Oxyura jamaicensis* | RUDU | L 15" (38 cm)

Loafing in compact "rafts" on lakes and bays in winter, these little ducks are practically inert—until summer, when courtship is highly animated.

**APPEARANCE** Short body with slightly upturned bill and stiff tail gives it "rubber ducky" look. Unlike all other regularly occurring waterfowl in East, acquires breeding plumage in spring (see sidebar, p. 38). Breeding male deep rufous with black cap, bright white face, intense blue bill; in winter, bill black and body cold brown. Female has long dark line across dusky white face.

**VOCALIZATIONS** Song of courting male a loud rapping, followed by a harsh buzz: *tik-tik-tik-tik-tik-rrrrrr.*

**POPULATIONS** Introduced to western Europe, but has since been largely eradicated because it competes with endangered White-headed Duck, *O. leucocephala.*

### Plain Chachalaca

*Ortalis vetula* | PLCH | L 22" (56 cm)

You may not see them, but you sure will hear them. Sex-segregated choruses call back and forth in treetops of bosques and even residential districts.

**APPEARANCE** Only species in Neotropical guan family in our area. All are long-bodied and long-necked, large-tailed and small-headed; male has red wattle. Naturally arboreal but eagerly tends feeding stations on the ground.

**VOCALIZATIONS** Raucous *cha-cha-la,* repeated steadily. Listen for differences in pitch, usually male vs. female. One well-known mnemonic is *Keep it up!* for males, *Cut it out!* for females.

**POPULATIONS** Apparently stable in South Tex.; has been introduced to several islands off Ga., where small numbers persist.

### Galloanseres

They have always been known as gamebirds: the waterfowl, comprising ducks, geese, and swans; and the upland grouse, turkey, quail, and relatives. Now we know that the pairing is grounded in evolutionary history. The orders Anseriformes (pp. 20–53) and Galliformes (pp. 52–63) together form a so-called parvclass—the Galloanseres (literally, "rooster geese," from *gallus* and *anser*)—separate from all the other birds in this book.

What is it that unites the Ruddy Duck, say, and the Plain Chachalaca? By the early 21st century, a broad array of independently derived results from molecular genetics all pointed to a relationship between the two groups, and the AOS conjoined them in 2003. There are similarities in morphology and life history too. Most notable are some well-known features of their developmental biology, including large clutch sizes and extreme precociality. Young quail can run, swim, and even fly within hours of hatching.

Masked Duck
♂
winter ♂
blocky bill
two dark bars on face
♀
winter ♀
body densely marked in all plumages
breeding ♂
juvenile
breeding ♀
adult male with clean white cheek in all plumages
winter ♂
female with pale cheeks crossed by single blurry line
Ruddy Duck
winter ♀
breeding ♂
long tail often raised
breeding ♂
♀
compact body with large head
seldom seen in flight; almost never on land
breeding ♀
Plain Chachalaca
♂
ample tail tipped with white

## NEW WORLD QUAIL

With their elegant crests and plumes, these upland gamebirds are popular across a broad swath of human society: hunters, homeowners, and manufacturers of holiday ornaments.

### Northern Bobwhite

*Colinus virginianus* | NOBO | L 9¾" (25 cm)

In much of its range, this iconic gamebird is the only quail—and by far the smallest upland gamebird. The bird's eponymous call is one of the most evocative sounds of the rural Southeast.

■ **APPEARANCE** Rotund and rufescent, with small head and short bill. Male's face boldly patterned in dark and white; male has the least impressive head adornment of any quail in the U.S. and Canada. White regions of face on male replaced by warm buff on female.

■ **VOCALIZATIONS** Famous *bob... white!* call usually preceded by much softer introductory note. Flocks cackle quietly among themselves.

■ **POPULATIONS** Range contracting, numbers declining. Reasons for losses unclear.

### Scaled Quail

*Callipepla squamata* | SCQU | L 10" (25 cm)

Popularly known as "Cottontop" for the male's puffy white crest, this desert quail is the only representative in our area of its genus. Like the bobwhite, it frequents feeding stations.

■ **APPEARANCE** Typically seen scooting down a dirt road or across the desert, singly or in frantic groups. Close-up, check out the breast feathers, leaden with fine black edges ("scales"). Male has white crest, reduced in female.

■ **VOCALIZATIONS** Song comprises two paired notes, delivered in rapid succession: *chipchurr-chipchurr-chipchurr ....* Also gives an explosive, startling *rreurk!*

■ **POPULATIONS** Intrinsically variable; numbers crash regionally following heavy snowfall but also rise in response to good growing seasons. Probably in long-term decline owing to overgrazing.

### Montezuma Quail

*Cyrtonyx montezumae* | MONQ | L 8¾" (22 cm)

Formerly known as the Harlequin Quail for the male's intricately patterned face, this cryptic gamebird occurs primarily in Mexico. In the East, it ranges into the limestone-and-savanna country of the western Edwards Plateau of Texas. Listen for the haunting song: a loud, long, low whistle, descending slightly and weakly modulated.

**In a Family Way**

In Carl Linnaeus's seven-step hierarchy from kingdoms down to species, families are near the middle. The question of what constitutes a family is somewhat arbitrary: A family in one avian order might be considered a subfamily or superfamily in another order. Across even broader groupings—birds vs. insects, say, or birds vs. dicots—the parallels are yet weaker.

The essential idea, as with all taxonomic groupings above the species level (see sidebar, pp. 20–21), is that of monophyly, the sharing of a common ancestor. In the case of upland gamebirds in the U.S. and Canada, the Plain Chachalaca has long been placed in the Neotropical family Cracidae. But recognition of the quail family was more recent. Analysis of skeletal structure, coupled with evidence from molecular biology, led the AOS to take the New World quail—but not the Old World quail!—out of the family Phasianidae and place them in their new family, Odontophoridae.

western ♂
taylori/texanus
white
throat and
supercilium
buffy
throat and
supercilium
paler
than other
subspecies
♂
ıdespread
virginianus
Northern
Bobwhite
♀
smaller, with
more black
below
juvenile
Florida ♂
floridanus
white-tipped
crest on
forehead
Scaled Quail
West Texas ♂
pallida
South Texas ♂
castanogastris
juvenile
male has dark
chestnut
belly patch
shaped like male,
but with indistinct
pattern
♀
juvenile
♂
Montezuma
Quail

## FOREST GAMEBIRDS

Like most other upland gamebirds, these three prefer to walk or run away when flushed. Although the most photogenic encounters are of birds on or near the ground, they spend a great deal of time well up in trees.

### Wild Turkey

*Meleagris gallopavo* | WITU | L 37–46" (94–117 cm)

Until recently, turkeys avoided humans and limited their activities to forests and forest edges. But in a spectacular example of a rapid behavioral shift, the species has adapted to urban life, occurring now in many of our largest eastern cities.

■ **APPEARANCE** Huge; our largest upland gamebird. Appears all-dark in poor light, but the plumage glistens bronze and vinaceous in the sunshine. Male has naked red head; female's unfeathered head gray. Many males and a few females sport a "beard" of feathers hanging from the breast. Typically seen in flocks of 10–25.

■ **VOCALIZATIONS** Male's well-known gobble audible at great distances; flocks cackle quietly, sometimes erupt into short shrieks.

■ **POPULATIONS** Extensive management has led to a mixture of subspecies, varying especially in tail and rump color, across much of the East.

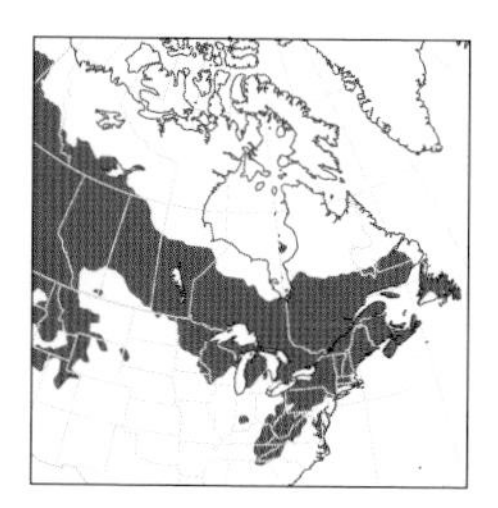

### Ruffed Grouse

*Bonasa umbellus* | RUGR | L 17" (43 cm)

Hear an old motorcycle start up deep in the forest? Chances are, it's the drumming of a Ruffed Grouse, especially if you are in an early successional broadleaf forest in early spring.

■ **APPEARANCE** Named for the black neck feathers puffed out by male in display. In most encounters, the bark-colored bird saunters slowly through the woods; broad black tail band especially prominent when the bird flushes. All sport a modest crest.

■ **VOCALIZATIONS** The drumming is nonvocal, created when the male beats his wings: *bup ... bup ... bup bup bu'bu'bu'B'B'B'B'.* Clucks quietly when walking about, especially hen with young.

■ **POPULATIONS** Two color morphs sort out geographically: Red predominates in the Appalachians, gray predominates elsewhere. Color differences most pronounced on tail.

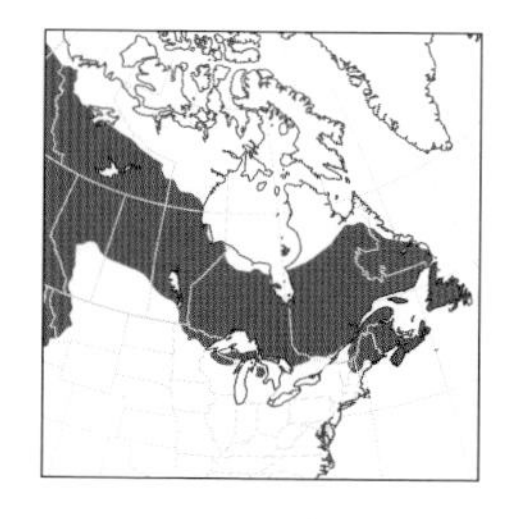

### Spruce Grouse

*Canachites canadensis* | SPGR | L 16" (41 cm)

This is the "Fool Hen" of the Northwoods, allowing one to approach almost to within arm's reach. On the flip side, the bird is so cryptic and motionless that it often escapes notice altogether.

■ **APPEARANCE** A bit smaller than Ruffed Grouse. Male Spruce Grouse shows much more black below than Ruffed; red patch above eye hard to see except in display. Both sexes distinguished from Ruffed by tail: dark with pale red-brown terminal band, the opposite of Ruffed. Lacks crest of Ruffed.

■ **VOCALIZATIONS** Male's display, like that of Ruffed, created by beating of wings; but much shorter, quite rushed.

■ **POPULATIONS** Like Ruffed Grouse, ranges in color from brown-red to brown-gray. Note microhabitat difference with Ruffed: Spruce feeds mostly on buds of spruce and other conifers, Ruffed heavily on aspens and other broadleaf trees.

Wild Turkey
eastern
silvestris
displaying ♂
western
merriami
male much larger than female
whitish tips to tail feathers
♀
rufous tips to tail feathers
solid black tail band
displaying gray-morph ♂
spiky crest
red-morph ♂
Ruffed Grouse
red-morph ♀
distinct black bars on flanks
female with broken tail band
no crest
Spruce Grouse
dark tail with rusty tips
♀
displaying ♂
crisp black-and-white underside
short dark tail

## PTARMIGAN

The three species of ptarmigan, two of which occur in the East, are fantastically adapted to life on the tundra. The name *Lagopus* means "rabbit foot." The silent *p* in the bird's English name was put there to make a perfectly good Scottish Gaelic name look respectably Greek.

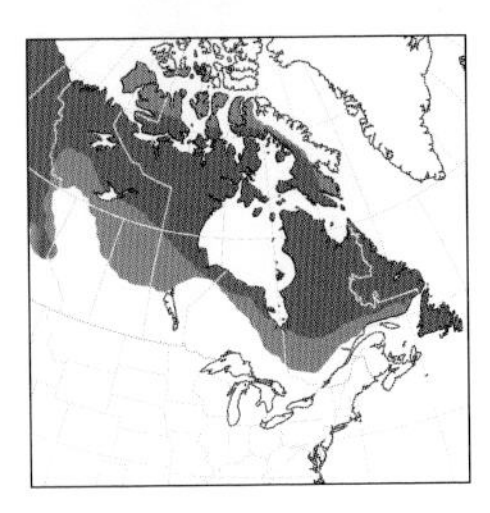

### Willow Ptarmigan

*Lagopus lagopus* | WIPT | L 15" (38 cm)

This is the less extreme of the two eastern ptarmigan, favoring relatively lusher landscapes with low shrubs, especially small willows. It breeds a bit farther south, and winters more widely—occasionally to the Canada-U.S. border.

■ **APPEARANCE** In famous winter plumage, nearly all-white; by summer, extensively rufous-brown above (whence comes "Red Grouse," the name in England). Separation from Rock Ptarmigan difficult: Willow is larger overall, with bigger bill.

■ **VOCALIZATIONS** Calls are impressively humanlike. Short utterances like *Oh? Oh!* Also longer series, *Go back! Go back! Go BACK!* Like most grouse, cackles quietly, sometimes loudly, while foraging.

■ **POPULATIONS** A textbook example of a vertebrate with natural, long-term abundance oscillations, typically 10–11 years. Causes not well understood, might even involve sunspot cycles.

### Rock Ptarmigan

*Lagopus muta* | ROPT | L 14" (36 cm)

This just might be the most ecologically extreme bird in the Northern Hemisphere. It is common in winter on all of Baffin I. and even the north shore of Greenland.

■ **APPEARANCE** Male in rarely encountered winter plumage nearly pure white; differs from Willow by black line through eye; male's summer plumage grayer overall than Willow's. Females especially hard to separate: Rock smaller overall, with smaller bill. The small bill of this species is thought to be an adaptation for minimizing heat loss.

■ **VOCALIZATIONS** Courting male gives unmusical rattle or rapping, typically in short bouts of flight.

■ **POPULATIONS** Despite superficial sameness of its high-latitude habitat, the species exhibits much geographic variation; nonbreeding plumage varies subtly within range in East, and genetic variation apparently extensive.

### Arctic Adaptations

The three species of ptarmigan—two are widespread, a third is restricted to the Rockies—are very obviously adapted to life in extreme snow and cold. They are white in the winter to blend in with the ubiquitous snow cover; in summer, they are mottled brown and gray and white, also to blend in with the tundra. Ptarmigan are round overall with small bills, maximizing their volume-to-surface-area ratio and thereby minimizing heat loss. Their feet are feathered right down to the talons, an adaptation both for preserving warmth and for walking on the snow. And when conditions are particularly harsh, they burrow in the snow.

Ptarmigan suffer essentially no migration mortality—because they are practically nonmigratory. They are prolific breeders, so, despite some hunting pressure, their numbers are quickly replenished. What's to worry about? For now, not much. But anthropogenic climate change is affecting the high Arctic more than any other environment on Earth. All models forecast heavy reduction in total tundra acreage worldwide and, on top of that, rapid transformation of what tundra habitat remains.

summer ♂
Willow Ptarmigan
male mottled golden and brown, similar to Rock Ptarmigan
summer ♀
transitional spring ♂
winter
hefty bill
summer ♂
male rich rufous
bold black eyeline
more delicate bill
winter ♂
Rock Ptarmigan
winter ♀
female very similar to Willow Ptarmigan
summer ♂
summer ♀
male finely marked cool brown
fall ♂

## DANCING CHICKENS

In these species, males perform communal displays called leks. Commercial tours for viewing the lekking males are popular—and big bucks for local economies.

### Greater Sage-Grouse

*Centrocercus urophasianus* | GRSG | L 22–28" (56–71 cm)

All stages of the life cycle of this enormous grouse are closely associated with sagebrush-steppe landscapes. Males lek in clearings in late winter, hens and chicks forage on sage in summer, and extended families flock together starting in fall.

■ **APPEARANCE** Heavier than even a pheasant. Cold gray with black belly; tail long, pointed. Flying away, appears uniformly dark above.

■ **VOCALIZATIONS** Lekking males emit vocal and nonvocal popping and whooshing sounds, many notes with liquid quality.

■ **POPULATIONS** Declining range-wide due to loss of sagebrush habitats from overgrazing; disturbance from oil field infrastructure a new threat.

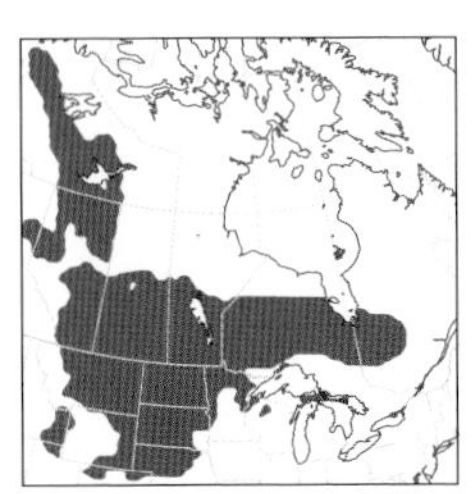

### Sharp-tailed Grouse

*Tympanuchus phasianellus* | STGR | L 17" (43 cm)

Closely related to the prairie-chickens, this grouse is found widely in small numbers across much of the northern prairie biome of N. Amer.

■ **APPEARANCE** Told from similar Greater Prairie-Chicken by spotted, not barred, plumage; pale tail is pointed. Easily confused with female Ring-necked Pheasant (p. 62); female Ring-necked's tail is longer and lacks extensive white of Sharp-tailed.

■ **VOCALIZATIONS** Males caper about on leks, leaping high; listen for loud foot stomping and tail rattling, interspersed with shrill yapping calls.

■ **POPULATIONS** Fairly common throughout range. More accepting of agricultural landscapes than its congeners; comes to feeding stations.

### Greater Prairie-Chicken

*Tympanuchus cupido* | GRPC | L 17" (43 cm)

The spooky "booming" of this midsize grouse is one of the great sounds of the tallgrass prairie—indeed of any haunt or habitat in N. Amer.

■ **APPEARANCE** Both sexes prominently barred, especially below; tail short, rounded. Sharp-tailed Grouse and female Ring-necked Pheasant (p. 62) have pointed tails, lack barring.

■ **VOCALIZATIONS** Male gives low moan, unsteady and wavering. Notably ventriloquial, especially in morning mist of the featureless prairie.

■ **POPULATIONS** Declined drastically in 19th and 20th centuries, but numbers have stabilized and may be rebounding in some places. "Heath Hen" subspecies of eastern seaboard extinct since 1932; "Attwater's" subspecies of southeastern Tex. hanging on by a thread.

### Lesser Prairie-Chicken

*Tympanuchus pallidicinctus* | LEPC | L 16" (41 cm)

This rarer, range-restricted counterpart of the Greater Prairie-Chicken is holding on in extensive tracts of shortgrass prairie, especially those under active conservation management.

■ **APPEARANCE** Very similar to Greater; smaller, slighter overall, with finer barring. Male Lesser's air sacs reddish, yellow-orange in Greater.

■ **VOCALIZATIONS** Male in courtship gives short, clipped notes, *poo* and *p'doo,* in rapid series.

■ **POPULATIONS** Hybridizes with Greater in narrow zone of range overlap.

Greater Sage-Grouse
♂
♀
displaying ♂
♀
black belly patch
displaying males on lek
large spiky tail
Sharp-tailed Grouse
lightly marked white belly
displaying ♂
prominent central tail feathers
purple air sacs
♀
intricately spotted
Greater Prairie-Chicken
short fan-shaped tail
displaying ♂
large golden air sacs
♀
strong barring over entire underside
Lesser Prairie-Chicken
displaying ♂
smaller reddish air sacs
♀
barring fades in central belly

## NONINDIGENOUS FOWL

More so than with most avian orders, species in the Galliformes have been widely transplanted by humans. All the species on this page are indigenous to the Old World.

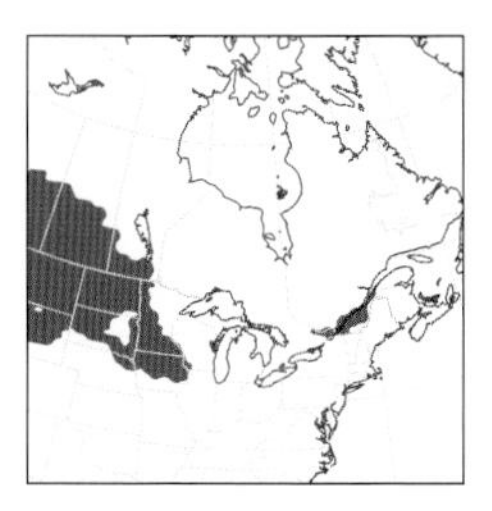

### Gray Partridge

*Perdix perdix* | GRAP | L 12½" (32 cm)

Suggesting a large quail, this plump partridge is widespread but generally uncommon in parts of southern Canada and the northern U.S. "Partridge in a pear tree" is a pun based on its French name, *perdrix* (sounds like *pear tree).*

■ **APPEARANCE** Foxy gray and russet all over; orange face contrasts with gray underparts, finely barred. Male has black belly patch; female pale-bellied.

■ **VOCALIZATIONS** Three-syllable call, the middle note shortest: *kreeh d' kriih,* like a squeaky gate swinging open. Flushing flocks boisterous.

■ **POPULATIONS** Established here in early 20th century. Flourishes around hedgerows in farm country; releases may be seen far from core range.

### Ring-necked Pheasant

*Phasianus colchicus* | RNEP | L 21–33" (53–84 cm)

Named for the extinct kingdom of Colchis, this popular gamebird was introduced to western Europe from Asia long ago. Centuries later, the descendants of those transplants were introduced to the New World.

■ **APPEARANCE** Dapper male unmistakable, but females suggest indigenous grouse. Note long tail of female, with warm buff-tan tones overall.

■ **VOCALIZATIONS** Male in display gives rasping double honk followed by soft rushing of wings. The two honking notes carry far.

■ **POPULATIONS** Constant reintroductions by game agencies cloud understanding of whether the species is truly self-sustaining here. Stock from diverse sources—a mix of domestic variants and well-differentiated subspecies—adds further complication.

### Indian Peafowl

*Pavo cristatus* | INPE | L 40" (102 cm) ♀; 60–80" (152–203 cm) ♂

The well-known peacock in zoos, aviaries, and private collections. Indigenous to Indian subcontinent. Small populations often get established for a short while in the wild, but none in the East are known to have caught on. Feeds on the ground during the day, roosts in trees at night. Male's powerful, catlike cry carries far.

### Red Junglefowl

*Gallus gallus* | REJU | L 17" (43 cm)

The domesticated "chicken," derived chiefly from Red Junglefowl, is by far the most abundant terrestrial vertebrate on Earth, with a global population exceeding 30 billion. Indigenous to Southeast Asia; apparently established on Fla. Keys. Elsewhere in East, frequently wanders from farms, petting zoos, etc., but rarely gets established. Male's famous *cock-a-doodle-doo* can be heard anywhere, including big cities in recent decades.

### Chukar

*Alectoris chukar* | CHUK | L 14" (36 cm)

Indigenous to Central Asia and the Middle East, well established in the U.S. Interior West. Escapes from game farms, typically single birds, noted widely across the East. Rotund like a quail but larger; both sexes gray overall with bold black markings. Bill and eye ring blood-red. *Chukar* is an anglicization of the Urdu word for "partridge" (*chakor*).

Gray Partridge
♀
peach-orange face, brighter in male
♂
dark belly patch
♀
lightly marked buff
barred flight feathers
Ring-necked Pheasant
♂
long, graduated tail with no white
♀
Indian Peafowl
♂
greenish gray with white belly
♀
not to scale with other species on page

Red Junglefowl
typically with cocked tail
♂
♀
Chukar
pale gray and buff with bold black markings
red bill
juvenile

## FLAMINGOS AND GREBES

In one of the most startling revelations of the ongoing revolution in our understanding of evolutionary relationships among birds, it turns out that flamingos and grebes are each other's closest relatives.

### American Flamingo

*Phoenicopterus ruber* | AMFL | L 46" (117 cm) WS 60" (152 cm)

It is easy to recognize a flamingo: long-legged, extremely long-necked, and pink. The tricky part is distinguishing between escapes from captivity and birds that got here under their own wing power. The latter occur annually in Fla. Bay; they act "wild" and are often seen at a distance. Note tricolored bill of adult American Flamingo; a good point of distinction from nonindigenous species like Greater Flamingo, *P. roseus* (also pictured opposite).

### Least Grebe

*Tachybaptus dominicus* | LEGR | L 9¾" (25 cm)

Suggests a miniature Pied-billed Grebe—with which it often associates. Bill pointy; yellow eye contrasts with gray face. Breeding adults gray with black throat; nonbreeders with whitish throat. Breeds year-round in South Texas; uncommon and generally a recluse, but small flocks sometimes loaf out in the open in marshes and resacas. Also casual in South Florida; has bred.

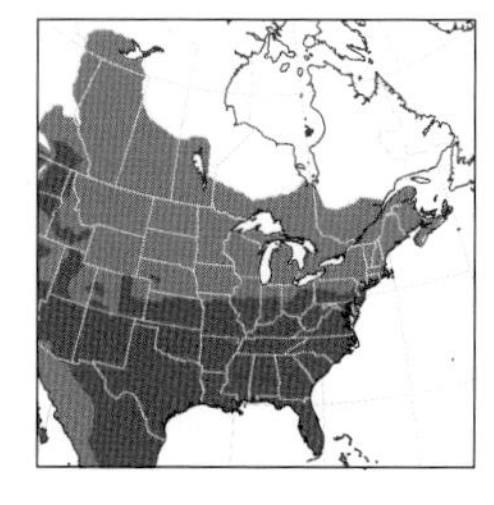

### Pied-billed Grebe

*Podilymbus podiceps* | PBGR | L 13½" (34 cm)

A small brown bird on a quiet pond suddenly sinks! Like a submarine, the Pied-billed Grebe instantly reduces buoyancy—by expelling air from the spaces among its feathers.

■ **APPEARANCE** Eponymous pied bill—pale blue-gray with a thick black ring—seen only on breeding adults. In all plumages, bill more bulbous than other grebes' bills. Nonbreeders warm gray-brown all over, with puffy white rear. Young birds transition slowly from strikingly patterned fledglings to weakly striped juveniles, the size of adults.

■ **VOCALIZATIONS** Primary song a long series of clucking or gulping notes, decelerating like a Yellow-billed Cuckoo's (p. 78); adults at any time of year issue monosyllabic grunts and shrieks; dependent young pipe urgently.

■ **POPULATIONS** Accepts smaller and shallower waterbodies than other grebes. Declining as a breeder in parts of the East; officially listed as endangered across much of New England and the mid-Atlantic.

#### Miraculous Birds

For much of the 20th century, grebes were understandably thought to be related to loons (pp. 164–167): Both are expert divers, with broadly convergent molts, morphologies, and migratory strategies. The placement of flamingos was less clear: An affinity with other long-legged waders made sense, but there was also conjecture about an evolutionary relationship with ducks. Enter molecular biology. Recent genetic evidence, derived from independent analyses, strongly supports a close relationship between grebes and flamingos. And it's not just the DNA: These two groups share numerous morphological similarities, mostly involving skeletal and muscular structure.

What are we to call this strange, almost miraculous pairing of the supremely elegant, plankton-eating *pink* flamingos with the mostly squat, deepwater, incessantly immersive grebes? The straightforward name Phoenicopterimorphae ("flamingo-shaped") has been proposed, but Dutch ornithologist George Sangster has put forward a better idea: In a 2005 paper, he called this grouping the Mirandornithes ("wondrous birds" or "miraculous birds") for the unexpected, yet eminently sensible story of their evolutionary biology.

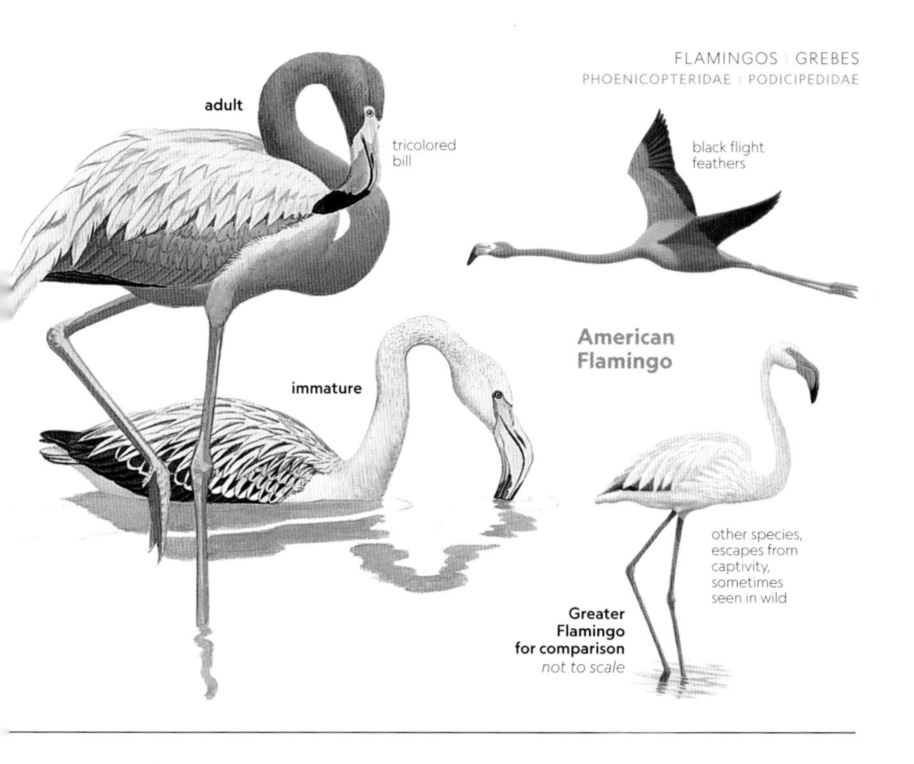
adult
tricolored bill
black flight feathers
American Flamingo
immature
other species, escapes from captivity, sometimes seen in wild
Greater Flamingo for comparison
not to scale

juvenile
Least Grebe
small, thin bill
winter
whitish throat
breeding adult

breeding adult
Pied-billed Grebe
juvenile
winter
puffy white rear
straw-colored bill
downy young

## GENUS *PODICEPS*

Three species in this genus, differing greatly in summer and winter plumages, are represented in North America. Breeders are colorful and distinctive, but nonbreeders require some scrutiny.

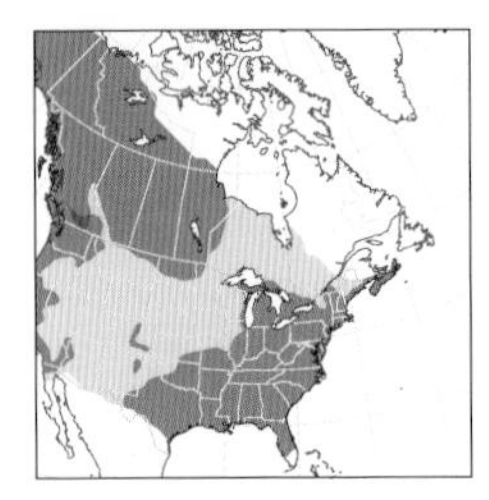

### Horned Grebe

*Podiceps auritus* | HOGR | L 13½" (34 cm)

The "horns" of the breeding adult are in fact a broad swath of yellow feathering that covers much of the face. The name is a bit off, but an alternative name, the "Slavonian Grebe," is odder yet.

■ **APPEARANCE** Larger than Eared Grebe, causing it to ride lower in the water; Horned also has a smoother, flatter head profile. Breeding adult has black head with extensive yellow behind eye; neck dark rufous. Horned in winter shows more contrast on face than Eared. An excellent field mark in all plumages is the bill: relatively stout with a white patch at the bulbous tip.

■ **VOCALIZATIONS** A quiet grebe, almost never heard in winter. On breeding grounds, listen for a wavering, gull-like squeal, *uh-reee*.

■ **POPULATIONS** Overland migrants put down suddenly in bad weather, sometimes at improbably small ponds, even on flooded streets and parking lots.

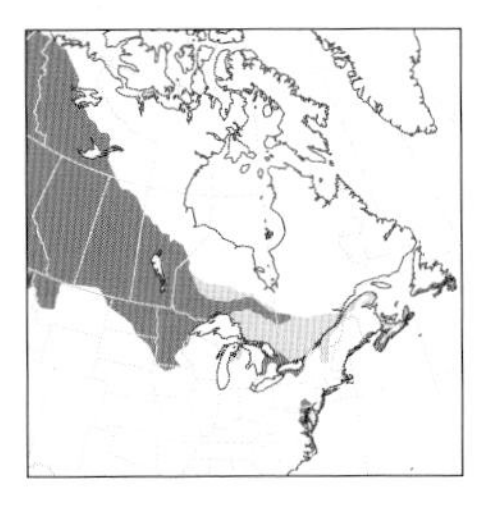

### Red-necked Grebe

*Podiceps grisegena* | RNGR | L 20" (51 cm)

Like the two swan-necked grebes (p. 68), this species has a long neck and a golden dagger of a bill. It is more closely related to the Horned and Eared Grebes, however.

■ **APPEARANCE** Breeding adult presents combo of blackish cap, gray face, and rufous neck; bill is long and straw yellow. In winter, especially when hunched, easily confused with smaller Horned and Eared Grebes; but even the dingiest Red-necked shows yellow at base of long bill.

■ **VOCALIZATIONS** Animated on the breeding grounds, with displaying birds mixing nasal wailing with prolonged chatter.

■ **POPULATIONS** Winter distribution inland strongly governed by ice cover on Great Lakes; following freeze-up in late winter, these grebes sometimes disperse across the mid-Atlantic and Ohio River Valley.

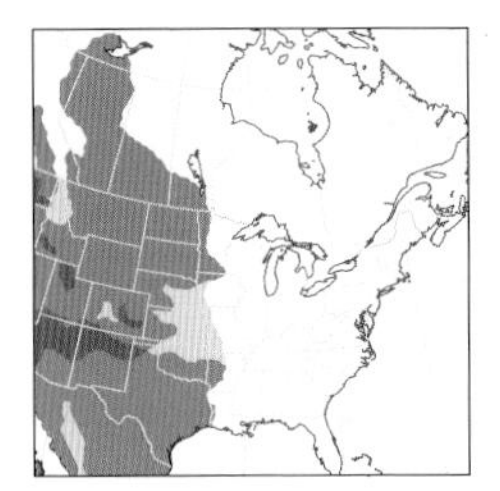

### Eared Grebe

*Podiceps nigricollis* | EAGR | L 12½" (32 cm)

This small grebe is a physiological marvel. During the course of the annual cycle, it combines hyperphagia (extreme eating) with rapid molt and drastic reduction in flight-muscle mass. Most are completely flightless more than half the year, but they also perform impressive long-distance migrations.

■ **APPEARANCE** The "ears," analogous to the Horned Grebe's "horns," are a tuft of yellow-orange feathers behind the eye. In winter, distinguished from Horned by smaller size, thinner and longer neck, and fine, all-dark, upturned bill. Markings on face blurry in winter; Horned shows sharper contrast.

■ **VOCALIZATIONS** On breeding grounds, a slightly rising whistle that ends in a stutter: *wheeeeee-ippuh,* often in slow series.

■ **POPULATIONS** Common in western N. Amer.; east of the Great Plains, however, this small grebe occurs mostly as a winter rarity. Molt timing useful for identification in spring: Eared molts later in spring than Horned; thus, by the time most Horned Grebes are in spiffy breeding plumage, many Eared Grebes are still molting and look messy.

Horned Grebe
golden "horns" raised
breeding adult
breeding adult
adult in spring molt
molting birds can be confused with Red-necked
darker winter
white cheek sharply demarcated from cap and neck
white bill tip at all times
winter

Red-necked Grebe
dingy gray neck
winter adult
1st winter
juvenile
long straw-yellow bill
breeding adult

cheeks less pronounced from behind; sits higher on water than Horned
Eared Grebe
winter
Horned
Eared
downy young
breeding adult
1st fall
paler winter
winter
gray cheeks blend into dark cap

## GENUS *AECHMOPHORUS*

For most of the 20th century, these largest of the North American grebes were considered a single species. Remarkably, one short vocal element in the two species' elaborate courtship displays appears to be enough to prevent interspecific pairing most of the time.

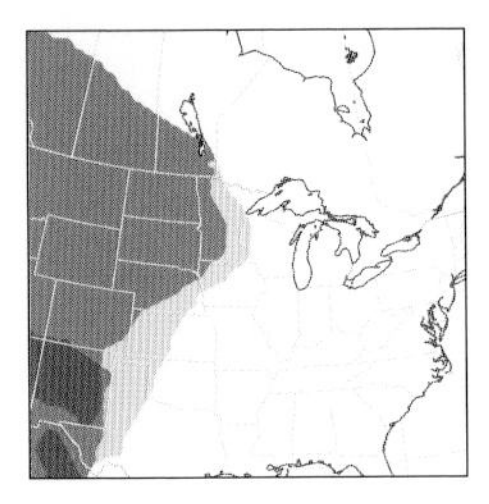

### Western Grebe

*Aechmophorus occidentalis* | WEGR | L 25" (64 cm)

Mated pairs perform an astonishing courtship ritual of synchronized swimming, involving wild rushing and lunging, and culminating in an extraordinary weed ceremony, in which mates present one another with aquatic vegetation.

■ **APPEARANCE** At all seasons, the darker of the two swan-necked grebes; darker-sided, with black on cap extending below eye. But beware winter Clark's, appearing duskier than in summer. Year-round, bill color is a good mark: always dusky yellow-olive in Western, always brighter yellow, with orange tones, in Clark's.

■ **VOCALIZATIONS** Far-carrying advertising call, heard year-round, day and night, a shrill, sudden, strongly disyllabic *reeeh-rreeeek*. Short piping notes, often in rapid series, heard most commonly on breeding grounds.

■ **POPULATIONS** Range overlaps broadly with that of Clark's, but Western more common, especially north and east; both species are rare east of range in winter, but Western more frequent.

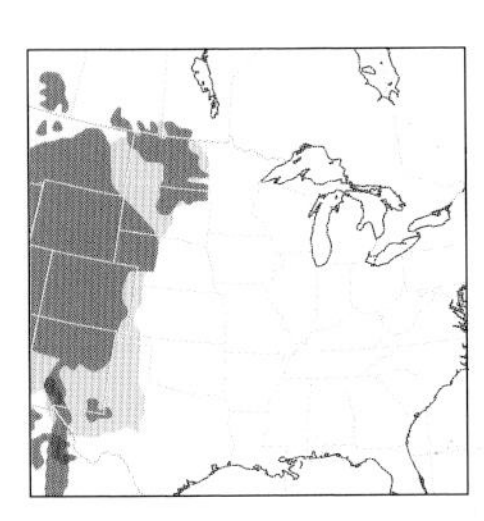

### Clark's Grebe

*Aechmophorus clarkii* | CLGR | L 25" (64 cm)

This brighter, whiter counterpart of the Western Grebe was formerly treated as a white phase or white morph of that species—if any attempt at differentiation was made at all.

■ **APPEARANCE** Garnet eye stark on white face of breeders; on Western, the black cap extends below eye. Clark's sides whiter on average than Western's, but lounging grebes of either species can show much white. Black-and-white face pattern turns to grayish mush on many Clark's in winter, so focus on the bill: bright yellow-orange, even on dingy-plumaged winter birds.

■ **VOCALIZATIONS** Key difference with Western is the advertising call, a drawn-out, monosyllabic *rreeeeek* in Clark's; sometimes slurred, *reee-eek,* but rarely as strongly disyllabic as in Western.

■ **POPULATIONS** Mixed-species pairs (Western–Clark's) uncommon but routinely noted on breeding grounds; hybrid progeny difficult to recognize, especially in winter.

#### Isolating Mechanisms

Since they were split in the 1980s, the two *Aechmophorus* grebes have attracted much notice from American birders. Bill color, the exact pattern of the face, and even the relative contrast of the flanks are all brought to bear on the field identification of these elegant grebes. Also widely known is that the two species differ in their loud, far-carrying advertising calls: disyllabic in Western, monosyllabic in Clark's.

The advertising call is one of many vocal elements in the grebes' amazing courtship display—and that is to say nothing of the astonishing choreography. Yet it is the advertising call, less than a second in total duration, that appears to make all the difference for the two species. Carefully controlled experiments have shown that grebes choose mates largely on the basis of just that one call note. Such isolating mechanisms are legion in birds and other animals and are thought to play an important role in speciation and, ultimately, biodiversity.

face paler
in winter
Western Grebe
winter
adult
dusky
yellow-olive
bill
black extending
below eye
courtship
display
breeding
adult with
downy young

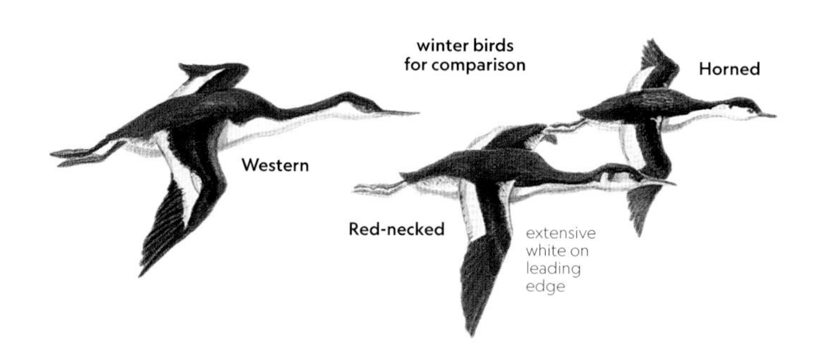
winter birds
for comparison
Horned
Western
Red-necked
extensive
white on
leading
edge

face darker
in winter
Clark's Grebe
winter
adult
yellow-orange
bill
breeding adult
with downy young

## PIGEONS

Birds in the family Columbidae are called pigeons and doves, biologically imprecise terms that apply loosely to body size. Pigeons tend to be larger, doves smaller. But there are exceptions!

### Rock Pigeon

*Columba livia* | ROPI | L 12½" (32 cm)

Few bird species have played a larger role in human history and culture than this one. For centuries, Rock Pigeons have been significant in the arts, in sports, in scientific research, and even in epochal military campaigns.

■ **APPEARANCE** The familiar city pigeon, known to all. So-called wild type is light blue-gray with black crescents across wings at rest; sides of neck iridescent green and lavender. Variants, derived from domestic stock, run the gamut from pure white to rusty to almost black, with many color intermediates. All show a tuft of white feathering at base of upper mandible (maxilla); in flight, even at great distance, the underwing linings gleam white.

■ **VOCALIZATIONS** Low-pitched cooing soft and ventriloquial, carries well. Wings clap noisily when the bird flushes.

■ **POPULATIONS** Indigenous to Eurasia, but their New World descendants are entirely naturalized; some nest on remote cliffs and highway overpasses, far from cities. Local clusters (called demes) in the urban northeastern U.S. are undergoing documented genetic evolution.

### White-crowned Pigeon

*Patagioenas leucocephala* | WCPI | L 13½" (34 cm)

This large pigeon of the Caribbean region reaches our area in South Florida, where it occurs in small numbers near the coast. Typical sightings are of birds in flight or in treetops.

■ **APPEARANCE** Bright white cap of adult contrasts with dark color overall; bill pale red, nape scalloped with white. White on head mostly lacking in juvenile. Beware variant Rock Pigeons, superficially similar to White-crowned; White-crowned is lankier, shows dark wing linings in flight.

■ **VOCALIZATIONS** Several hoots, a bit higher and more musical than those of Rock Pigeon: *hoooo ... hoo-hoo-hoooo,* syncopated.

■ **POPULATIONS** Numbers and nesting phenology closely tied to availability of fruiting trees. The extinct Passenger Pigeon (p. 425) was most closely related to pigeons in the genus *Patagioenas.*

### Red-billed Pigeon

*Patagioenas flavirostris* | RBPI | L 14½" (37 cm)

This species is the South Texas counterpart of the White-crowned Pigeon. Both are mostly arboreal, seen singly or in small numbers.

■ **APPEARANCE** Largest member of its family in the East. The only all-dark pigeon in South Tex., but reddish variant Rock Pigeons might be confused for Red-billed; in flight, Red-billed shows darker wing linings. Head and breast of adult suffused with deep red wine; ironically, bill mostly yellow with small reddish base. Juvenile grayer, paler.

■ **VOCALIZATIONS** Song like that of White-crowned, but the two species do not overlap in the East; introductory note averages longer in Red-billed, rises and falls in pitch.

■ **POPULATIONS** Uncommon year-round in U.S. range; more common in summer, when perhaps overlooked in dense vegetation and oppressive Lower Rio Grande Valley heat.

long, pointed wings mostly white underneath with two black bars above
Rock Pigeon
highly variable
color variations

White-crowned Pigeon
both *Patagioenas* pigeons have dark underwings
white eyes and crown, both darker in juveniles
dark slate gray overall
Red-billed Pigeon
dark overall, with wine-colored blush to head and breast

## DOVES LARGE AND SMALL

The body plan of North American columbids is remarkably consistent from species to species: All are plump and small-headed with slight bills and short reddish legs.

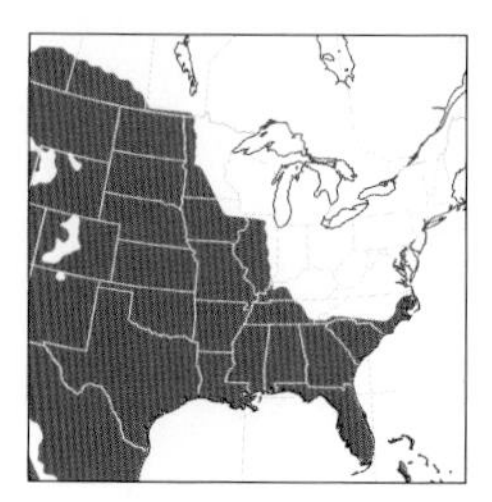

### Eurasian Collared-Dove

*Streptopelia decaocto* | EUCD | L 12½" (32 cm)

The recent takeover by Eurasian Collared-Doves is perhaps the most spectacular vertebrate biological invasion ever documented in N. Amer. Look for these big birds around roads, rail yards, farms, suburbs—anywhere with a bit of infrastructure.

■ **APPEARANCE** Impressively large; about 70 percent heftier than Mourning Dove (p. 74). Eponymous collar absent from juveniles, but all have broad white tip on tail. African Collared-Dove, *S. roseogrisea* (also pictured opposite), frequently escapes from captivity, occasionally breeds; paler overall than Eurasian, tail feathers and undertail coverts uniformly pale below.

■ **VOCALIZATIONS** Tripartite song, given year-round, has accent on second syllable: *hoo HOO hoo,* repeated slowly or quickly. The scientific name *decaocto,* for "eighteen," is onomatopoetic (*de CAOC to*). In flight, especially when landing, gives harsh growling sound, raptor-like. Rolling song and sputtering flight call of African Collared-Dove very different.

■ **POPULATIONS** Despite successes elsewhere in N. Amer., has been slow to catch on from mid-Atlantic northward, for reasons poorly understood.

### Inca Dove

*Columbina inca* | INDO | L 8¼" (21 cm)

Although generally slow-moving and inconspicuous, this bird often wanders onto lawns or perches on wires. It is the small, long-tailed dove of parks and neighborhoods across much of the southern U.S.

■ **APPEARANCE** Densely scalloped below. In flight, wings flash rufous, tail with bold white edge; even when perched, a bit of rufous may peek through. Juvenile Mourning Dove (p. 74), considerably smaller than adult and variably scalloped, can be mistaken for Inca Dove.

■ **VOCALIZATIONS** Song, once learned, is one of the characteristic sounds of suburbia in its range: a pair of weakly descending coos, the first harsher than the second, repeated endlessly, even in the hottest parts of the day: *whirlpool ... whirlpool ... whirlpool ....* Flushing, gives an enchanting wing rattle.

■ **POPULATIONS** Has expanded east, where it adapts well to human haunts: lawns, feeders, and birdbaths. However, has withdrawn in recent years from northern portions of range.

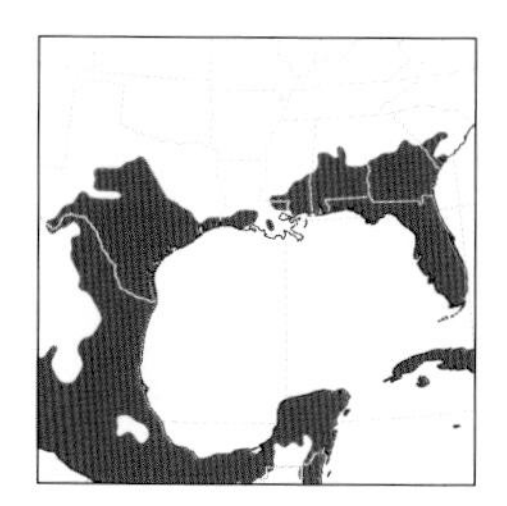

### Common Ground Dove

*Columbina passerina* | CGDO | L 6½" (17 cm)

This smallest dove in our area, plump and bobtailed, is uncommon and dwindling in number in semi-open habitats. Like the longer Inca Dove, it is easily overlooked, although it is not particularly shy.

■ **APPEARANCE** Like Inca Dove, flashes extensive rufous in wings; tail much shorter than Inca's, with limited white. Fine scaling on head and breast; Inca has coarser, more extensive scalloping. Bill reddish; dark on Inca. Gray-brown wings have black blobs, suggesting Mourning (p. 74).

■ **VOCALIZATIONS** Song a weakly rising, slightly slurred *ooo-wup,* repeated steadily. Does not give wing rattle of Inca.

■ **POPULATIONS** Tolerates disturbed, early successional habitats around farms and suburbs, but declining in many regions. A few stray well north every fall.

efty and
quare-
ailed
delicate and
attenuated
Eurasian
Collared-
African
Collared-
Mourning
Eurasian
Collared-Dove
grayish
overall
African
Collared-Dove
for comparison
contrasting
three-toned
wings
Inca Dove
rufous wing
patches
pale
overall
long tail with
white edges
very short
tail with small
white corners
♂
all-rufous
underwing
western
pallescens
♂
duller and
paler than
eastern birds
Common
Ground Dove
western
pallescens
♀
eastern
passerina
richly colored
♂

## MIDSIZE GRAY DOVES

All have the basic body plan of their family, but these three differ greatly in details of the wing and tail—especially in flight.

### White-tipped Dove

*Leptotila verreauxi* | WTDO | L 11½" (29 cm)

This is the only representative of its speciose Neotropical genus to reach the U.S. Like other "leptos," it is a forest bird that is difficult to glimpse but easily heard.

■ **APPEARANCE** Suggests a huge Common Ground Dove (p. 72): rotund and short-tailed; flashes rufous wing linings in flight. On standing or walking bird, the uniformly dark wings contrast with the paler body. The name refers to the hard-to-see bit of white on the tail corners.

■ **VOCALIZATIONS** Song a short hoot, followed immediately by a longer, slightly wavering coo: *ooh ooooOOOh*. Once learned, it is one of the characteristic and evocative sounds of South Tex. bosques.

■ **POPULATIONS** Like many Lower Rio Grande Valley specialties in Tex., has adapted well to feeding stations; otherwise, infrequently encountered but probably fairly common within limited range in our area.

### White-winged Dove

*Zenaida asiatica* | WWDO | L 11½" (29 cm)

The story of this indigenous dove's northward expansion has been overshadowed by that of the Eurasian Collared-Dove (p. 72), but it is nevertheless impressive—and apparently still in progress.

■ **APPEARANCE** White on upperwings always visible; appears as long crescent on perched bird, bright white patch in flight. In same genus as Mourning Dove, but plumage more suggestive of paler, grayer Eurasian Collared-Dove. In flight, collared-doves have the same tripartite color scheme as White-winged: primaries dark, back gray-fawn, coverts paler gray; corresponding feather tracts of White-winged contrast more strongly. Tail from below also similar to that of White-winged: black with broad white band at tip.

■ **VOCALIZATIONS** Classic song a tetrasyllabic *ooo ooo uh ooooo* (*Who cooks for you?*), heavily syncopated; often runs out to 5–10 syllables.

■ **POPULATIONS** Expanding north into southern Great Plains and east along Gulf Coast to Fla., where recently established. Regular stray, mostly fall to winter, well north of range.

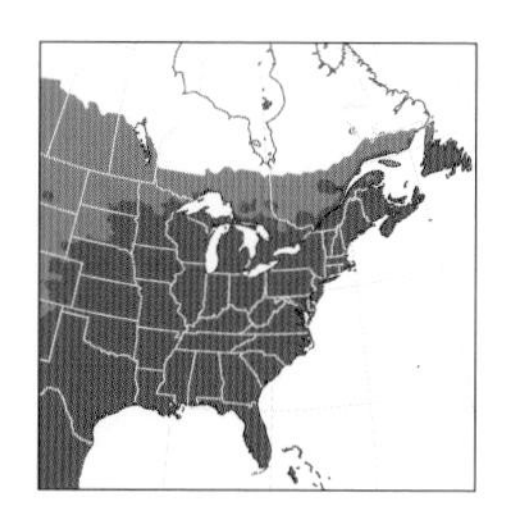

### Mourning Dove

*Zenaida macroura* | MODO | L 12" (30 cm)

Across southern Canada and much of the U.S., this is one of the very first species to greet the dawn with its exaggeratedly slow, minor-key cooing; it is also one of the last species to sing at dusk.

■ **APPEARANCE** Slender overall. Feathers of long, thin tail tipped white. Wings monochrome gray-tan, except for some irregular black blobs. Young fledge well before attaining adult size; with their scalloped plumage, can suggest smaller species like Inca Dove (p. 72).

■ **VOCALIZATIONS** Familiar "mourning" heard all day long: *ooOOOOh oooh oooh oooh*. Where resident year-round, one of the earliest songsters in the year, often singing by Jan. Airy wing-whistle distinctive.

■ **POPULATIONS** Heavily hunted but remains common across East; competition from expanding Eurasian Collared-Dove (p. 72) worries some, but there is little evidence that the two species interact significantly.

White-tipped Dove
dark wings contrast with pale body
short dark tail with small white tips

White-winged Dove
bright red eye and blue orbital ring
black flight feathers
square-cornered tail with broad white border
white wing patch visible when perched and in flight

Mourning Dove
dark eye
uniform dusty brown
long, intricately patterned tail
♂
♀
black spots
juvenile

## ANIS AND A ROADRUNNER

These birds of warm climes are associated with brushy habitats. Despite clear differences in appearance, anis and roadrunners are both in the cuckoo family.

### Smooth-billed Ani

*Crotophaga ani* | SBAN | L 14½" (37 cm)

Widespread in Caribbean region and across much of S. Amer., but occurs in our area only in South Florida. Similar Groove-billed Ani restricted in the East mostly to South Texas, but does wander a bit—sometimes to peninsular Fla. Seeing the "smooth" bill, lacking "grooves," requires a close look; a better field mark is the higher, more arched culmen. Expanded in Fla. in 20th century but has sharply declined.

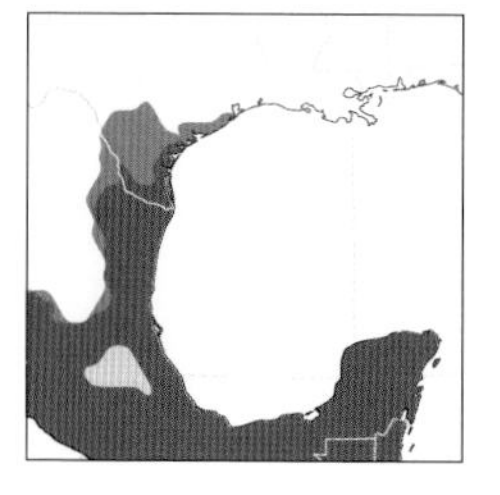

### Groove-billed Ani

*Crotophaga sulcirostris* | GBAN | L 13½" (34 cm)

Long overall with a long tail, this all-black bird superficially suggests a grackle (p. 374). On closer inspection, the loosely textured feathers, hunched posture, and peculiar bill indicate a very different story.

**APPEARANCE** Eponymous grooves in maxilla hard to see, but overall bill shape obvious at a distance: stout, decurved, and exceptionally broad-based. Anis seem to struggle to fly across clearings, landing awkwardly, then clambering through foliage.

**VOCALIZATIONS** Call sharp and squeaky, explosive, often in series: *peek! aweep! ... peek! aweep! ... peek! aweep! ....*

**POPULATIONS** Very rare but annual wanderer in fall along Gulf Coast, all the way to Fla., where now rare Smooth-billed Ani may also occur.

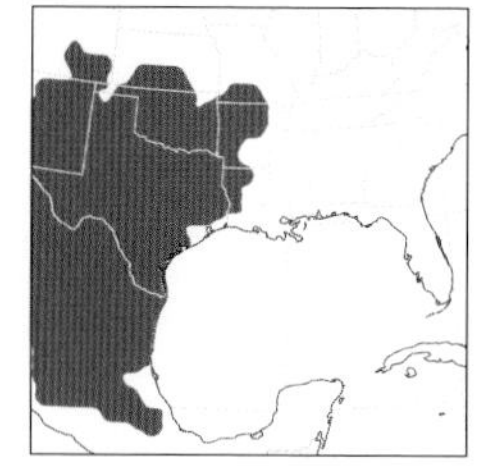

### Greater Roadrunner

*Geococcyx californianus* | GRRO | L 23" (58 cm)

This huge terrestrial cuckoo is usually seen singly. Roadrunners bask in the morning sun, but are wary and fleet of foot as soon as the day starts to warm up.

**APPEARANCE** Trots across shrublands at an impressive clip, frequently stopping to cock its tail and raise its crest. Through binoculars or a camera lens, note greenish sheen to tail and densely streaked plumage overall; red, white, and blue "racing stripe" behind eye prominent on male.

**VOCALIZATIONS** Song a series of sonorous coos, descending in pitch, faint but far-carrying, heard on still mornings, especially in late winter.

**POPULATIONS** Nonmigratory, but has been creeping north and east very slowly for many decades. At edge of range, numbers crash following harsh winters.

### Cuckoo Relations

The taxonomy of many bird families has undergone dramatic revision in recent decades, but the approximately 140 species in the family Cuculidae have not. In the U.S. and Canada, these are represented by the cuckoos (p. 78), anis, and Greater Roadrunner. Most cuckoos are long-tailed and long-bodied, with loosely layered feathers, zygodactyl toes (two front, two rear), and uncomplex vocalizations.

The family Cuculidae has emerged unscathed from the ongoing genomics revolution, but the order Cuculiformes has not. Cuculiformes still contains just the single family Cuculidae, but its position relative to other orders has been significantly revised. Formerly considered one of the "near passerines," an imprecise grouping, the order Cuculiformes is now placed in the "basal land birds," far removed from various other primarily terrestrial orders.

bill lacks grooves
Smooth-billed Ani
Groove-billed Ani
grooves visible at close range
short flights, often sailing to ground before running
Greater Roadrunner
long, iridescent tail with white tips

## CUCKOOS

These three species are not as notorious as the Old World cuckoo, which lays its eggs in the nests of other species. But two of them do this from time to time—and all three exhibit other unusual breeding behaviors.

### Yellow-billed Cuckoo

*Coccyzus americanus* | YBCU | L 12" (30 cm)

Elegantly attired but rather awkward in its movements, this cuckoo scans treetop foliage in a cautious, hesitating manner. Yet it is a powerful flier, blasting through clearings and migrating every autumn to S. Amer.

■ **APPEARANCE** Slender overall and long-tailed. Bright yellow bill of adult matches yellow orbital ring, harder to see. All are brown above and creamy white below, with prominent white spots on undertail; wings flash rufous in flight. Juvenile has dark bill like Black-billed, but usually shows extensive rufous in wings and very prominent white spots on tail.

■ **VOCALIZATIONS** Two main songs: a slow, steady series of rather harsh, slightly descending *whoo!* notes; also a rapid series of clipped *cu!* notes, decelerating to stuttering, clucking notes, *cowlp-cowlp-cowlp.*

■ **POPULATIONS** Sometimes lays eggs in other bird species' nests, a practice called brood parasitism; females occasionally help one another in a behavior known as cooperative breeding.

### Mangrove Cuckoo

*Coccyzus minor* | MACU | L 12" (30 cm)

Widespread in tropical lowlands north of the Equator, this cuckoo reaches southern peninsular Fla. The species certainly occurs in mangroves, but it is also accepting of a diversity of dense woodland types.

■ **APPEARANCE** Has yellow bill and big white tail spots like Yellow-billed; but note uniformly dark wings, black face mask, and extensive buffy wash below. Black-billed does not normally occur in the U.S. range of Mangrove Cuckoo.

■ **VOCALIZATIONS** Song is a series of identical *aahr* notes, notably harsh. In typical song, the delivery is perfectly regular, three to four notes per second; but also mixes it up, slowing down, speeding up, and pausing.

■ **POPULATIONS** Unlike Yellow-billed and Black-billed, does not appear to be a brood parasite, though sometimes forms loose aggregations, called colonies, of breeders.

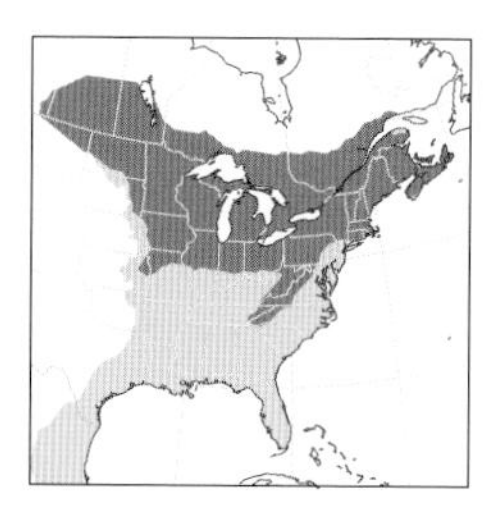

### Black-billed Cuckoo

*Coccyzus erythropthalmus* | BBCU | L 12" (30 cm)

Even more than other American cuckoos, this species is a lover of hairy caterpillars. It is particularly adept at gutting caterpillars, avoiding consumption of the most toxic part of the prey.

■ **APPEARANCE** Shaped like Yellow-billed, but even slimmer; bill thinner. Orbital ring red; bill dark. Has less white in undertail than Yellow-billed, and subtle buffy wash on throat and breast. Flushing, the adult shows little rufous in wing; but some juveniles have rufous in wing, inviting confusion with juvenile Yellow-billed.

■ **VOCALIZATIONS** Song a series of rapidly uttered *cu* notes: *cu'cu'cu'* ... *cu'cu'cu'* ... *cu'cu'cu'* .... Surprisingly, can sound like some songs of Least Bittern (p. 186).

■ **POPULATIONS** Annual, even intraseasonal, booms and busts related to caterpillar outbreaks; species is in long-term decline. Occasional brood parasite, less so than Yellow-billed Cuckoo.

adult
Yellow-billed Cuckoo
adult
juvenile
rufous primaries
large white tail spots

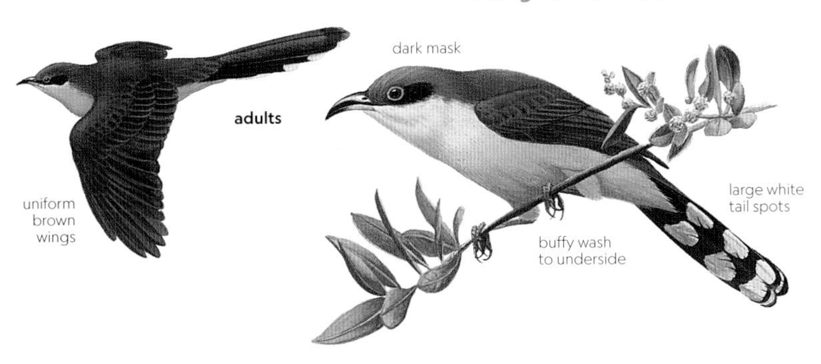
Mangrove Cuckoo
dark mask
adults
uniform brown wings
large white tail spots
buffy wash to underside

Black-billed Cuckoo
adult
uniform brown wings
light buffy wash to underside
adult
reduced white tail spots
juvenile

## NIGHTHAWKS

Despite their name, these nocturnal insectivores have never been considered close relatives of the "true" hawks in the order Accipitriformes. They and other nightjars (p. 82) are more closely related to swifts and hummingbirds.

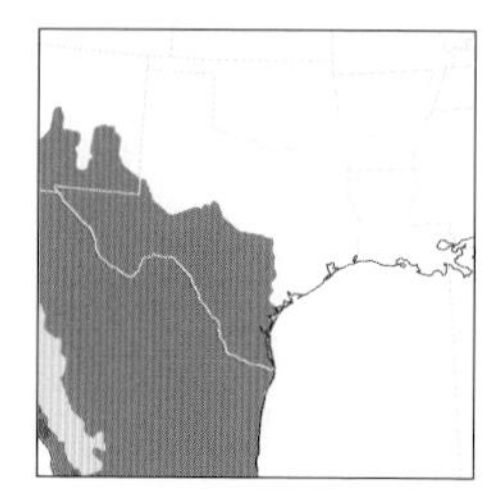

### Lesser Nighthawk

*Chordeiles acutipennis* | LENI | L 8½" (22 cm)

Where Lesser and Common Nighthawks co-occur, a good way to tell them apart is by how high they fly when feeding: Commons often way overhead, Lessers frequently just above the ground.

■ **APPEARANCE** A bit smaller than Common Nighthawk and buffier overall. Compared to Common, Lesser in flight shows shorter, more rounded wings, with the white bar closer to the wing tips. Bounded by this white bar, the dark wing tip of Lesser forms an equilateral triangle, the longer wing tip of Common an acute isosceles triangle.

■ **VOCALIZATIONS** Utterly different from Common. Song, given while perched, is a low, slow, toadlike trill, suggesting tremolo of Eastern Screech-Owl (p. 218), but running on for 10–30+ seconds. In flight, gives mirthful, maniacal yips and yaps, run together.

■ **POPULATIONS** Annual vagrant in fall and winter across Gulf Coast region, where care must be taken to distinguish this species from Common. Differences in voice, behavior, and habitat, so useful on the breeding grounds, rarely apply in the case of vagrants.

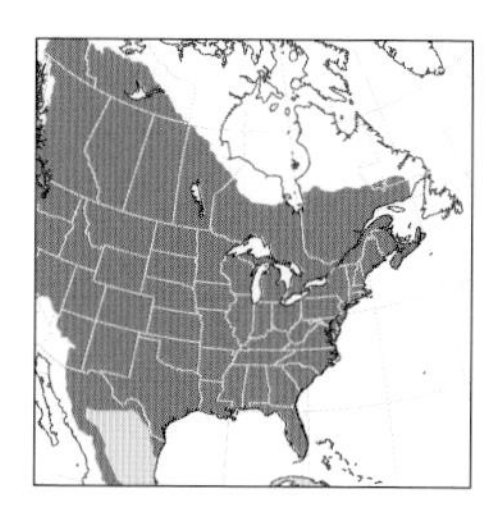

### Common Nighthawk

*Chordeiles minor* | CONI | L 9½" (24 cm)

This is the "Bullbat" that whips high above the lights of ball fields, shopping centers, and even busy downtowns continent-wide. The scientific name confusingly means "lesser nighthawk."

■ **APPEARANCE** In most of the East, the only nighthawk. Flies erratically, "shifting gears" as it goes. Both sexes have prominent white bar, larger on male, across pointed wings. Male has broad white tail band; female does not. Like all nighthawks, may be discovered roosting by day; assumes horizontal posture on ground, rooftops, and tree boughs.

■ **VOCALIZATIONS** Nasal call, typically given in flight, a startling *peeent!* or *peeezhr!* In display, plunge-dives almost to ground level, pulling up at the last second, creating a whirring "sonic boom" with the wings.

■ **POPULATIONS** Has declined worrisomely in recent years; possible causes include loss of aerial prey and nest predation by urbanizing crows.

### Antillean Nighthawk

*Chordeiles gundlachii* | ANNI | L 8" (20 cm)

Until the early 1980s, this Caribbean specialty was considered a subspecies of the Common Nighthawk. The two species differ hugely in voice.

■ **APPEARANCE** Antillean is a bit browner and smaller overall than Common, with shorter wings, but those differences are hard to assess in the field.

■ **VOCALIZATIONS** Call a series of five scratchy notes in rapid succession: *jit j' j' j' jit*. Sounds a lot like the Common True Katydid, *Pterophylla camellifolia*.

■ **POPULATIONS** "Snowbirds" to South Fla. will likely never encounter this species, for it occurs there only during the hot summer months. Found regularly in Fla. on Keys and in Miami region; almost never recorded north of there, but vagrants, mostly silent, are very difficult to detect.

juvenile
adult ♂
white bar near wing tip; forms equilateral triangle
extensive tan spotting on underwings
adult birds molting in late summer are Lessers; Commons molt in South America
molting adult ♀
Lesser Nighthawk
white bar farther from wing tip; forms isosceles triangle
Common Nighthawk
minor
♂
♂
juvenile
white throat
juveniles show distinct geographic variation; sennetti is silvery gray
eat Plains
sennetti
juvenile
Antillean Nighthawk
best distinguished from Common by range and voice
often warm-toned above, but Common can appear similarly
♂
♂

## NIGHTJARS

Along with the nighthawks (p. 80), these four are known, oddly, as goatsuckers. The word "nightjar" can apply to all the goatsuckers, but it also sometimes excludes the nighthawks, in a different subfamily.

### Common Pauraque

*Nyctidromus albicollis* | COPA | L 11" (28 cm)

As night falls in the Lower Rio Grande Valley, clearings and pastures come alive with the sounds of pauraques. Though hard to see in the gathering dusk, they have the convenient habit of putting down out in the open.

■ **APPEARANCE** Flashes white wing patch like nighthawks (p. 80), but wings are rounded. Long tail is likewise rounded, with white streaks on males, buff-white on females.

■ **VOCALIZATIONS** A harsh whistle, loudest in the middle: *puhWEEEurrr.* Call often preceded by several weaker *puh* or *pup* notes.

■ **POPULATIONS** Our only nonmigratory goatsucker. In summer, the only nightjar in subtropical woodlands in South Tex.

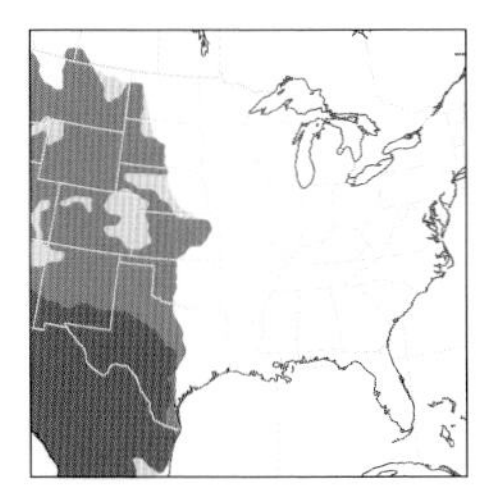

### Common Poorwill

*Phalaenoptilus nuttallii* | COPO | L 7¾" (20 cm)

This smallest "will" engages in prolonged periods of torpor, when metabolic activity nearly ceases. Other birds are capable of brief bouts of torpor, but the poorwill takes it to an entirely different level.

■ **APPEARANCE** Small and gray. Flushing, appears bobtailed with short, rounded wings. Often rests on back roads at dusk, where its orange eye-shine gleams in cars' headlights.

■ **VOCALIZATIONS** A sad, haunting *poor-will-ip,* proclaimed repeatedly from shrubby badlands and canyons. Terminal *ip* note much softer.

■ **POPULATIONS** This is the western "will," with poorly understood movements after nesting. Reclusive during winter torpor.

### Chuck-will's-widow

*Antrostomus carolinensis* | CWWI | L 12" (30 cm)

Like a beefed-up whip-poor-will, the "Chuck" is the largest goatsucker in our area. Where they overlap on the breeding grounds, "Chucks" tend to favor drier, more open habitats than "Whips" do.

■ **APPEARANCE** Warmer brown than Eastern Whip-poor-will and notably larger. Lacks black on chin and extensive white in tail of whip-poor-will, but these differences hard to see even on sleeping birds found by day.

■ **VOCALIZATIONS** The bird's peculiar name may have four syllables, but the loud song has five: *chip! widow! widow!*

■ **POPULATIONS** Probably in long-term decline, but these wary, nocturnal birds are difficult to survey.

### Eastern Whip-poor-will

*Antrostomus vociferus* | EWPW | L 9¾" (25 cm)

Although declining in number, this remains one of the most celebrated voices of the night in eastern forests. Like other nightjars, "Whips" sing from the ground—and they do so seemingly endlessly.

■ **APPEARANCE** Smaller, grayer than Chuck-will's-widow. Black and white on face and tail contrast well on male, less so on female. On roosting birds seen by day, tail projects less than on Chuck-will's-widow.

■ **VOCALIZATIONS** Famous song sneaks in a fourth syllable: *whip! poor-a-will!*

■ **POPULATIONS** Numbers dropping overall, but range has also expanded a bit southward; eats mostly large moths, in dwindling supply.

Common Pauraque
prominent buffy tips on wing coverts
♂
♀
♂
long barred gray tail
rounded wings with large white bar; buffy in females
♂
large-headed appearance
Common Poorwill
short, dark tail with white tips
adult ♂
large, flat head
Chuck-will's-widow
most are rufous
adult ♂
reduced white in tail
gray morph
♀
rufous morph
black chin with white collar on males
♂
Eastern Whip-poor-will
extensive white in outer tail feathers
♀
♂

## SWIFTS AND A HUMMER

The short-billed, mostly drab swifts glean tiny flying prey high above rooftops and rimrocks, whereas the long-billed, outrageously parti-colored hummingbirds sip nectar from flowers and feeders. Despite these huge differences, they are close relatives.

### Chimney Swift

*Chaetura pelagica* | CHSW | L 5¼" (13 cm)

The only swift in most of the East, this high-flier is a common summer sight and sound above cities and suburbs. It nests almost exclusively in human-built structures.

■ **APPEARANCE** Ordinarily seen only on the wing. Tubular body and long, arc-shaped wings, swept back, separate swifts from swallows (pp. 288–293). Entirely gray-brown, appearing simply all-dark in most views. Alternates short glides with twittering wingbeats. Feeds anywhere there is open sky, often at considerable altitude where it captures aerial plankton, whence comes the scientific name *pelagica,* meaning "ocean."

■ **VOCALIZATIONS** Sharply down-slurred notes in rapid chatter.

■ **POPULATIONS** Has declined due to diminishing prey base of airborne insects and ballooning spiders. Local distribution affected in part by availability of chimneys.

### White-throated Swift

*Aeronautes saxatalis* | WTSW | L 6½" (17 cm)

Nests mostly in foothills and canyons to our west, but migrants and wanderers are regular slightly east of the Rockies; breeds east to Badlands N.P., S. Dak. Larger than monochromatic Chimney Swift; White-throated boldly patterned in dark and white. Call, given on the wing, a loud, shrill chatter, the whole series descending slightly in pitch.

### Mexican Violetear

*Colibri thalassinus* | MEVI | L 4½" (11 cm)

A glittering green hummingbird of woodland edges in Mid. Amer., breeding north to southern Tamaulipas, Mexico. Now annual in summer in very small numbers in the Hill Country beyond San Antonio, Tex.; scattered records elsewhere to Great Lakes, mid-Atlantic. A large hummer; note dark blue splotches on face (the violet "ear") and breast.

#### "Birds Without Feet"

Hummingbirds and swifts have long been recognized as sister taxa, meaning they are more closely related to one another than to any other group (or taxon) of birds. They have tiny feet, and their wings are long, narrow, and arc-shaped. They are supreme aerialists: Swifts can stay aloft for months on end, and hummingbirds hover and fly backward. Together they form the order Apodiformes ("lacking feet"), surpassed only by the massive order Passeriformes (the passerines) in terms of the total number of extant species. And there may be more to the story.

Although the exact details remain to be worked out, there is emerging consensus that the goatsuckers (pp. 80–83) are closely related to the Apodiformes, perhaps even belonging in the same order. Like swifts and hummingbirds, the goatsuckers have reduced feet. Numerous details of wing and facial physiology are similar, and species in both groups are capable of torpor. Recent molecular work supports a strong connection among these birds, but let's give credit where credit is due: Avian anatomists before Darwin's time tentatively proposed these relationships. Their technology may have been primitive, but they were careful and insightful students of nature—and they got it substantially right.

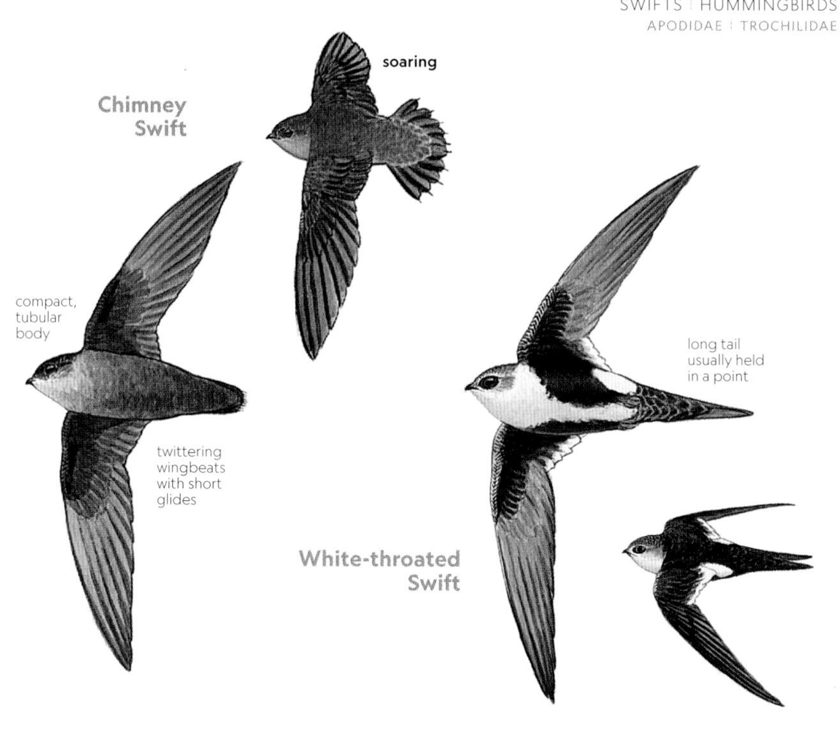
soaring
Chimney Swift
compact, tubular body
twittering wingbeats with short glides
long tail usually held in a point
White-throated Swift

immature
gently curved bill
Mexican Violetear
adult
large tail with dark subterminal band
appears mostly dark
adult

## GREEN-AND-GRAY HUMMERS

The brilliant throat feathers, or gorget, of the males attract our attention, but overall coloration is important for identifying males and females alike. These three are green and gray, with at most a hint of faint orange on the flanks.

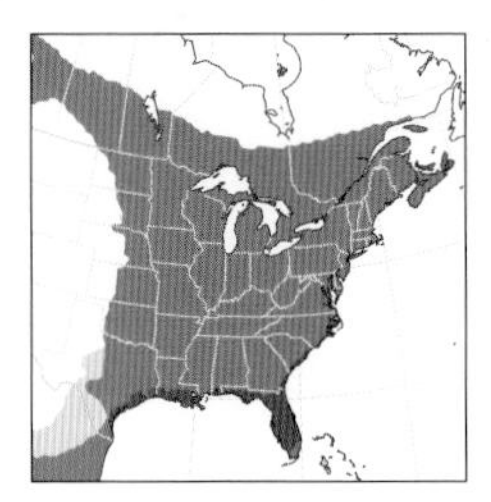

### Ruby-throated Hummingbird

*Archilochus colubris* | RTHU | L 3½" (9 cm)

This is probably the most celebrated and certainly the tiniest of all the trans-Gulf migrants. After traversing the Gulf of Mexico Mar. to Apr., the species quickly spreads north across much of the East.

■ **APPEARANCE** Except in a thin sliver of Tex. and Okla., the only hummingbird in its breeding range; overlaps more broadly with similar Black-chinned on migration and in winter. Male Ruby-throated has extensive ruby on gorget; male Black-chinned, purple. Female Ruby-throated brighter green above than female Black-chinned, especially on crown. Both sexes a bit shorter-billed than Black-chinned. In good view, primaries of perched Ruby-throated thin and pointed; primaries of Black-chinned broader and rounder. Longer tail of Ruby-throated extends beyond folded wings on perched bird, but beware the effects of the bird's posture and the observer's angle.

■ **VOCALIZATIONS** Twitters as it feeds; chases off other hummers with loud sputtering sounds. Wingbeats audible (the "hum") at close range.

■ **POPULATIONS** Increasing in winter in Southeast, especially Fla., where other hummers, including look-alike Black-chinned, may also occur.

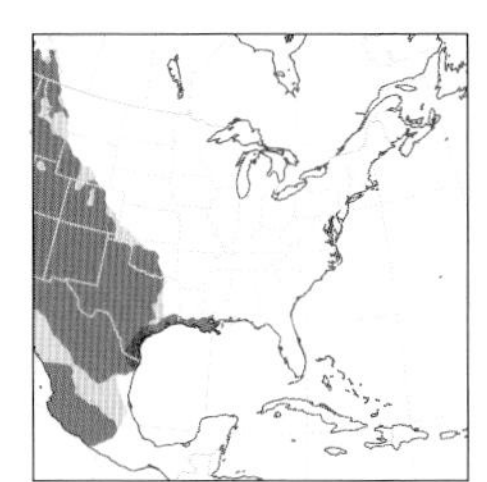

### Black-chinned Hummingbird

*Archilochus alexandri* | BCHU | L 3½" (9 cm)

The western counterpart of the Ruby-throated, this relatively demure hummingbird is widespread in canyon country, dry foothills, and residential districts.

■ **APPEARANCE** Looks washed-out in most views. Purple swatch on adult male's gorget hard to see; except when lit, gorget appears dusky gray-black. On female, note green-gray upperparts, not as bright as Ruby-throated's; crown and face, in particular, comparatively dull. Some bright female Ruby-throateds have a weak golden tint; female Black-chinned colder green-gray overall. Slightly longer-billed than Ruby-throated. Black-chinned's primaries rounder and broader, extend nearly to tail tip on perched bird; careful study of photos usually required to confirm those differences. When drinking from flowers or feeders, pumps tail up and down; Ruby-throated relatively stiff-tailed while feeding.

■ **VOCALIZATIONS** Calls given while feeding and fighting, like those of Ruby-throated. Adult male in flight gives dull wing buzz; in arcing display, bottoms out with an endearing tinkling sound, oddly suggestive of wing whistle of Mourning Dove (p. 74).

■ **POPULATIONS** Breeding range expanding north and east; hybridizing with Ruby-throated in Okla. and perhaps Tex.

### Anna's Hummingbird

*Calypte anna* | ANHU | L 3¾" (10 cm)

A western hummer, regular in very small numbers in winter in western and southern Tex., with scattered records elsewhere throughout East, typically at feeders. Magenta-headed adult male spectacular; female and immature usually show at least some color on throat. Otherwise, dusky green-gray; broad-tailed and relatively stout-billed. In good photos of perched bird, note very short greater coverts.

ult ♀
adult ♀
black chin and cheek
adult ♂
Ruby-throated
Hummingbird
er
maries
rrow
d
inted
last
primary
pointed
bright green
above,
including
crown
deeply
forked
tail
mostly white
below
adult ♀
adult ♂
tail projects
beyond wings
immature ♂
dult ♀
adult ♂
Black-chinned
Hummingbird
adult ♀
er primaries
ow, rounded
dull
green
above;
crown
grayish
slightly longer bill
than Ruby-throated
adult ♀
adult ♂
outer primaries
broad, curved
wing tips nearly
reach tail tip
immature ♂
venile ♂
immature ♀
Anna's
Hummingbird
adult ♂
tail extensively
gray
dult ♀
adult ♀
very short
greater
coverts
red
throat
patch
adult ♂
primaries
broad and
evenly spaced
adult ♀
dingy gray below
with extensive
green spots
immature ♂

## GREEN-AND-GOLD HUMMERS

The hummingbirds of the U.S. and Canada breed mostly in the West, but several westerners disperse well eastward in fall. As fall turns to winter, many find their way to feeders.

### Rufous Hummingbird

*Selasphorus rufus* | RUHU | L 3½" (9 cm)

Relative to its body length, this hummer migrates farther than any other organism on Earth. While hummingbirds are known to feud, the Rufous is notably pugilistic.

■ **APPEARANCE** Adult male Rufous fiery orange; male Allen's has mostly green back, but otherwise resembles Rufous. Males hovering at feeders can only be identified by photographic analysis of exact shape of splayed-out tail feathers: Male Rufous has broader tail feathers, with diagnostic notch in next-to-central tail feather. Females and immatures largely unidentifiable in the field in winter, but tail shape mirrors that of males: Outer tail feathers of female Rufous relatively broad, those of female Allen's narrower.

■ **VOCALIZATIONS** Aggressive at feeders, flowers, and elsewhere. Chases off other hummers—and even hawks and bears—with rapid, pentasyllabic *zee! zickety-zoo!*

■ **POPULATIONS** Until late 20th century, very rare but annual in the East; has increased greatly in recent decades, aided and abetted by the proliferation of feeders, particularly in the Southeast.

### Allen's Hummingbird

*Selasphorus sasin* | ALHU | L 3¾" (10 cm)

Breeds mostly coastal Calif., replaced by Rufous northward well into Alas. Despite Rufous's numerical dominance and capacity for long-distance migration, a fair number of green-and-gold hummers in the East are known or suspected to be Allen's. Separation from Rufous extremely difficult; many, especially females and immatures, must be rendered "Rufous/Allen's."

### Broad-tailed Hummingbird

*Selasphorus platycercus* | BTHU | L 3¾" (10 cm)

Breeds mostly southern Rockies and Great Basin; regular on fall migration in western High Plains. Along with tiny, short-billed, short-tailed Calliope Hummingbird (p. 411), annual in very small numbers to central Gulf Coast. Even dull females and immatures distinguished from Black-chinned and Ruby-throated by rufous on outer tail feathers and by usually orange-buff wash on sides; from female Rufous and Allen's by broader tail with less orange.

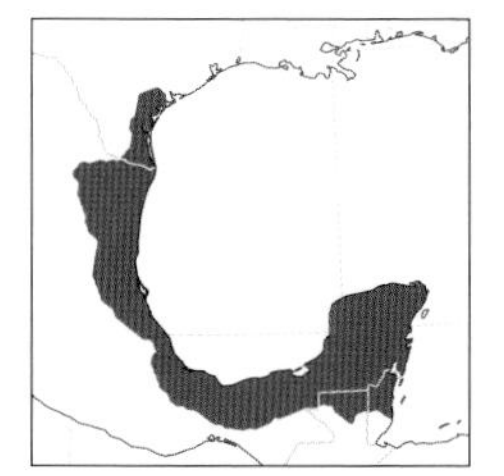

### Buff-bellied Hummingbird

*Amazilia yucatanensis* | BBEH | L 4" (10 cm)

This is the hummingbird people go to South Texas to see. It is the only U.S. representative of the Neotropical genus *Amazilia*.

■ **APPEARANCE** Relatively large; only hummingbird in East with red bill. Tail rufous; belly pale buff; rest of body green, variably flecked with buff. Male and female about the same in plumage, unlike other eastern hummers.

■ **VOCALIZATIONS** Most frequent call an abrupt *tsick*, typically given in rapid, rapping series of two or three notes: *tsick-tsick, tsick-tsick-tsick*, etc.

■ **POPULATIONS** Fairly common in summer, less so in winter, around gardens, parks, and woodlands in South Tex. Rare but regular in winter along Gulf Coast to La.; casual, but increasing in recent years, in winter to southern Atlantic coast.

ufous tails
om above
extensive rufous in tail, including central feathers
Rufous Hummingbird
♀
bright rufous contrasts with white collar
adult ♀
juvenile ♀
♂
adult ♂
some males with extensive green back
notched tip
adult ♂
Allen's Hummingbird
males always with green back
females and immatures virtually identical to Rufous
narrower tail feathers in all plumages
uvenile ♀
adult ♂
♂
Calliope Hummingbird for comparison
minimal rufous in tail
adult ♀
adult ♀
adult ♂
adult ♀
tiny and compact with short tail
evenly spotted throat
ample tail projects well beyond wing tips
smooth buff underneath
rosy gorget with thin white line under bill
Broad-tailed Hummingbird
adult ♂
bright red bill
adult ♂
slightly duller than male
adult ♀
immature
adult ♂
variable scaly throat
Buff-bellied Hummingbird
rufous tail

## MIDSIZE TO LARGE RAILS

Heard far more than seen, these chicken-shaped birds inhabit densely vegetated marshes. Glimpses are often brief, so focus on color and contrast, especially on the upperparts.

### Clapper Rail

*Rallus crepitans* | CLRA | L 14½" (37 cm)

This habitat specialist is almost never found outside the thin strip of coastal cordgrass salt marsh just inshore from the outer beaches of the Gulf of Mexico and Atlantic Ocean.

■ **APPEARANCE** Large and long-billed. Atlantic birds washed-out gray-brown overall, especially above; feathers on back dull brown with gray edges. Gulf birds brighter, especially above; back feathers are warmer brown than Atlantic birds' but nevertheless gray-edged.

■ **VOCALIZATIONS** Hesitant clicks or clucks change tempo erratically.

■ **POPULATIONS** Infrequently reported in winter from Long I., N.Y., northward; some migrate south, but others may remain and go undetected.

### King Rail

*Rallus elegans* | KIRA | L 15" (38 cm)

The freshwater and inland counterpart of the Clapper Rail, this brightly colored bird is uncommon even in favored habitat.

■ **APPEARANCE** Size and structure the same as Clapper; black feathers on back have contrasting orange edges. Much larger than Virginia Rail; King Rail has buffy face, Virginia is gray-faced.

■ **VOCALIZATIONS** Clicks and clucks identical to Clapper's; King Rail utters notes more steadily on average, but either species can vary call's tempo.

■ **POPULATIONS** Hard to study, but believed to be in decline everywhere, especially well inland. Hybridizes with Clapper where habitats overlap.

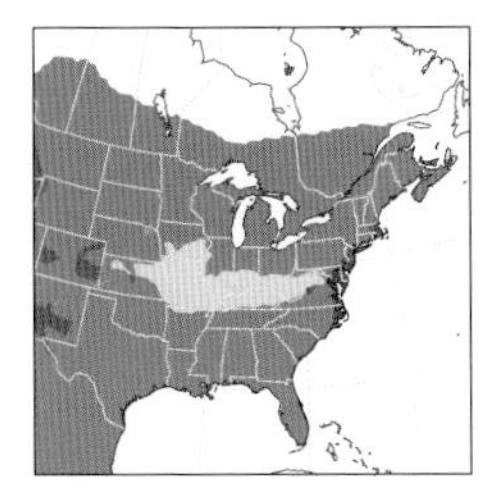

### Virginia Rail

*Rallus limicola* | VIRA | L 9½" (24 cm)

Across much of its extensive range, this is the only long-billed rail. It sometimes wanders into the open, creeping slowly or even standing still, then quickly darting back into cover.

■ **APPEARANCE** Rarer King Rail is much larger, with warm buffy cheeks; Virginia Rail's gray cheeks contrast with orangish breast.

■ **VOCALIZATIONS** Song an irregular series of *kick* and *k'dick* notes, sometimes with terminal *kjurrr*. Call a series of piglike grunts, each note descending.

■ **POPULATIONS** Highly dependent on habitat availability: Immediately abandons drained marshes, but also rapidly establishes even in unimpressive artificial wetlands.

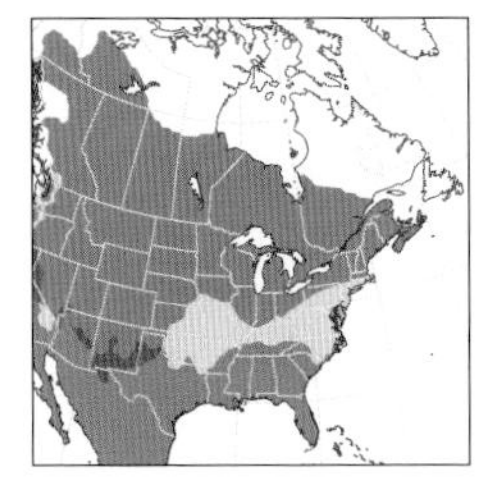

### Sora

*Porzana carolina* | SORA | L 8¾" (22 cm)

Even more than the Virginia Rail, this generalist rail tolerates a wide array of conditions. All it needs is dense vegetation and mud near water.

■ **APPEARANCE** All have short, thick-based bill. Adult has black on face and breast, contrasting with otherwise gray plumage. Juvenile buffier overall, similar to elusive and smaller Yellow Rail (p. 94) but plainer above, lacking bold gold lengthwise stripes and fine white lateral barring.

■ **VOCALIZATIONS** Song a rich, low, rising whistle, *oooh-whee!* (*so-ra!*), repeated steadily. Call is a long series of short, sharp whistles, descending in pitch after the first few notes; slows down and drags out toward end.

■ **POPULATIONS** Makes ready use of agricultural habitats, especially in winter.

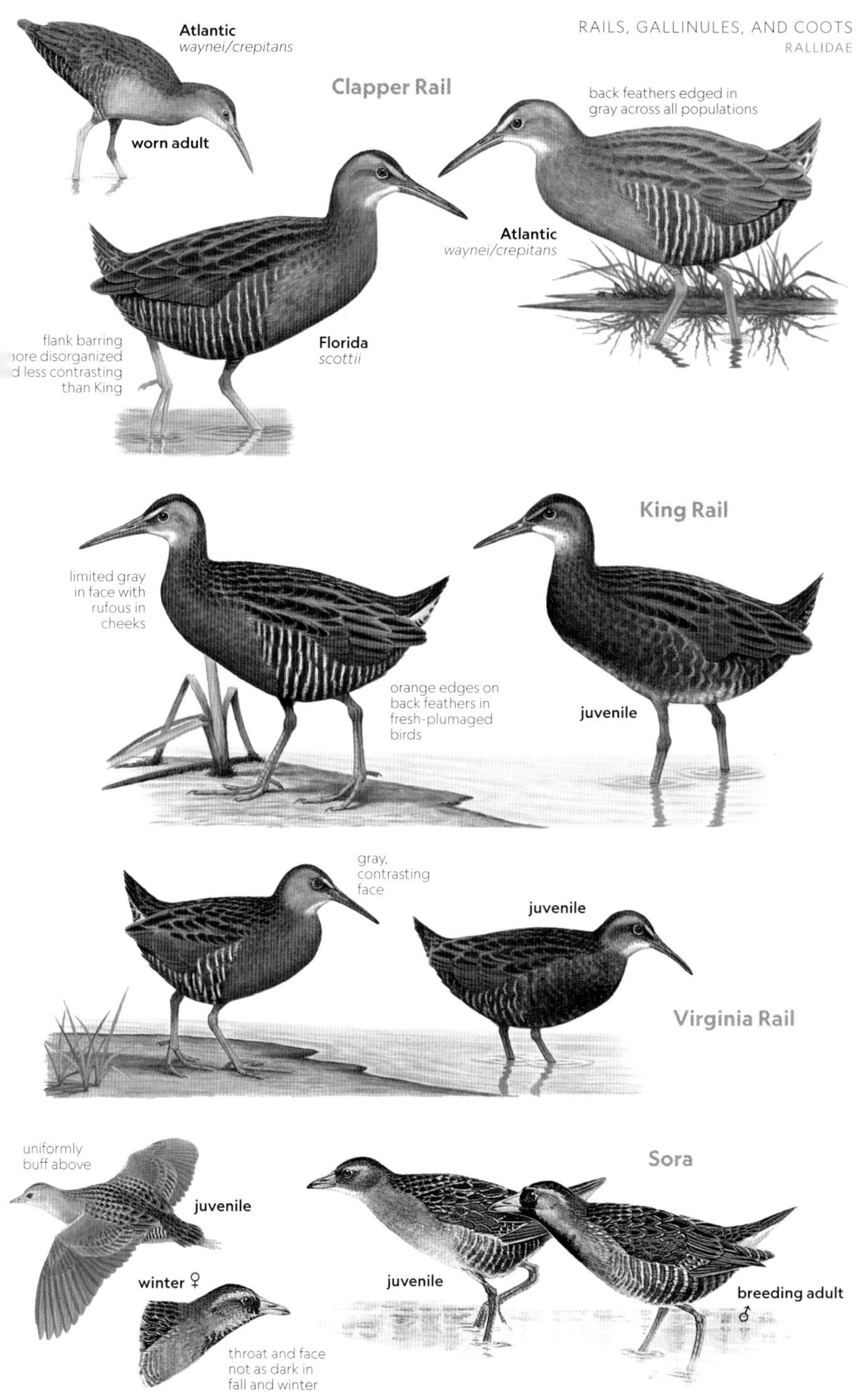
Atlantic
*waynei/crepitans*
worn adult
Clapper Rail
back feathers edged in gray across all populations
Atlantic
*waynei/crepitans*
flank barring
nore disorganized
d less contrasting
than King
Florida
*scottii*
King Rail
limited gray in face with rufous in cheeks
orange edges on back feathers in fresh-plumaged birds
juvenile
gray, contrasting face
juvenile
Virginia Rail
uniformly buff above
juvenile
Sora
winter ♀
juvenile
breeding adult
♂
throat and face not as dark in fall and winter

## DARK MUDHENS

These birds aren't called rails, but they are in the same family as the birds known by that name. Compared to our other rallids, these four are more likely to be out in the open.

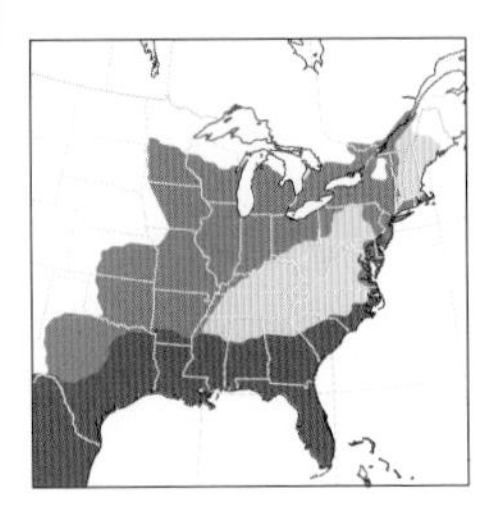

### Common Gallinule

*Gallinula galeata* | COGA | L 14" (36 cm)

Smaller and more boldly patterned than the coot, this bird favors freshwater marshes with pond lilies, duckweed, and other floating plants.

■ **APPEARANCE** All are dark slate and brown in good light, appearing black in most views. Breeding adult has red frontal shield and red bill with yellow tip; straw-yellow legs with variable red on tarsus. Nonbreeding adult and immature grayer-billed; all plumages show extensive white in tail and thin white stripe on flanks.

■ **VOCALIZATIONS** Varied monosyllabic utterances, squeaky and nasal; given singly, *penk,* or in raucous, decelerating series, *p' p' p' peh peh peh penk penk penk ....*

■ **POPULATIONS** Has undergone multiple name changes and taxonomic revisions in recent years; *G. galeata* is endemic to the Americas.

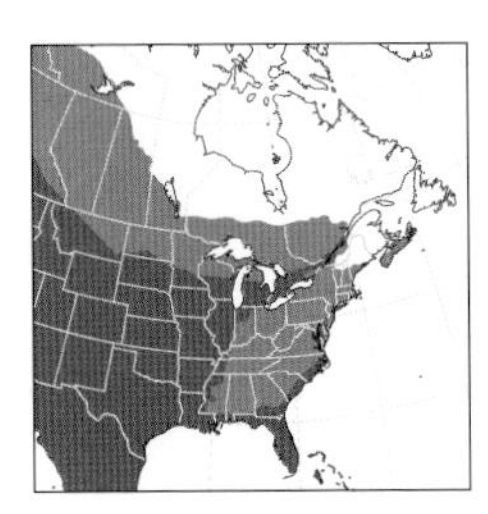

### American Coot

*Fulica americana* | AMCO | L 15½" (39 cm)

The black "duck" at the duck pond isn't a duck at all. Its bill and feet are totally different, and it bobs its head like other rails.

■ **APPEARANCE** A black ball of a bird. Adults are ivory-billed with a red knob on the white frontal shield, mostly absent in winter. Juvenile plumage, held well into fall, paler overall. Lobed toes large and gray-green.

■ **VOCALIZATIONS** Outbursts, *frup* and *frap,* with undeniable flatulent quality. Also higher squeals, rising, often in series.

■ **POPULATIONS** Adaptable and aggressive. Adept at both dabbling and diving; flocks in winter often fan out onto golf course fairways.

### Purple Gallinule

*Porphyrio martinicus* | PUGA | L 13" (33 cm)

In bright light, this lover of lily-laden waterways is as colorful as any Painted Bunting: blue and green all over, with garish yellow feet and a bright red bill.

■ **APPEARANCE** Smaller than a coot. Adult mostly rich blue with green wings and back; red bill has yellow tip; frontal shield baby blue; feet huge and bright yellow. Juvenile mostly buff, but with greenish wings and upperparts like adult. Clambers about vegetation, even climbing into shrubs.

■ **VOCALIZATIONS** Short *pik* and *pek* notes, given singly or in series, not unlike the calls of Black-necked Stilt (p. 98).

■ **POPULATIONS** Restricted in our area mostly to southeastern coastal plain, but a notorious vagrant, even to tiny ponds and swimming pools.

### Purple Swamphen

*Porphyrio porphyrio* | PUSW | L 18–20" (46–51 cm)

Looking like supersize Purple Gallinules, escapes from captivity in Broward Co., Fla., in the 1990s have become well established on the peninsula.

■ **APPEARANCE** Larger, bulkier than Purple Gallinule. Adults sky blue below, darker blue above; legs, bill, frontal shield red. Juvenile duskier.

■ **VOCALIZATIONS** Low, nasal moans and honks, *aaahhr* and *uuuurr;* also an explosive, ascending *creeEEK!*

■ **POPULATIONS** Has defeated eradication efforts; continues to expand, with records to Ga. and S.C. The eBird name is Gray-headed Swamphen.

Common Gallinule
thin white stripe along flank
breeding adult
winter adult
juvenile
often seen swimming in deep water in wetlands
frontal shield variable
adult
American Coot
often swims like a duck
juvenile
lobed toes
baby blue shield
Purple Gallinule
juvenile
absurdly long legs and feet
notably large
oversize bill
Purple Swamphen
long, sturdy legs

## SMALL, SECRETIVE RAILS

In the East, all our birds with "rail" in their name are difficult to glimpse, but these two take it to an extreme. They are small, nocturnal, and loath to emerge from dense vegetation.

### Yellow Rail

*Coturnicops noveboracensis* | YERA | L 7¼" (18 cm)

Although capable of swimming, in summer this rail is more typically encountered in wet meadows with sedges than in cattail marshes. In winter, it finds its way to paddy fields and the edges of coastal salt marshes. At all seasons, it is legendarily hard to see.

■ **APPEARANCE** Flushing, flashes white in secondaries. Sora (p. 90), much more likely to be seen, is larger, with plainer back; Yellow Rail has yellow lengthwise stripes and thin white transverse barring on back.

■ **VOCALIZATIONS** At night and on cloudy days on breeding grounds, an arhythmic clicking, *tik ... tik ... tik-tik ... tik ... tik-tik ... tik ... tik ... tik ...,* like two stones clicked together. Song of Virginia Rail (p. 90) also an arhythmic clicking, but notes reedier, more sonorous. A recently discovered call, given in winter, sounds like a distant Sandhill Crane (p. 96).

■ **POPULATIONS** Movements poorly understood: May disperse by midsummer to little-known molting grounds; breeding sites favored one year may be abandoned the next, and vice versa. Can be locally common (but still very hard to see) at favored wintering locales.

### Black Rail EN

*Laterallus jamaicensis* | BLRA | L 6" (15 cm)

This smallest rail, only the size of a large sparrow, is one of the holy grails of birding. It can be fairly common locally, but it is also absent from large swaths of seemingly appropriate habitat.

■ **APPEARANCE** Dark and stocky with a small bill; in good view—difficult to attain—note white speckles above and rich chestnut on nape. Recently fledged young of all rails are small and black; beware confusion with chicks of more common rail species, like Sora and Virginia Rail (both on p. 90).

■ **VOCALIZATIONS** Two or three piping notes in rapid succession, followed by a lower, harsher whistle, *ki'ki'JJrrrr or ki'ki'ki'JJrrrr,* sung repeatedly. Rich and far-carrying; some Virginia Rail songs suggestive of Black Rail's, but more tentative.

■ **POPULATIONS** Coastal salt marsh breeders in long-term decline due to habitat loss and rising sea levels. Newly discovered breeding colonies well inland may be climate refugees from coasts or simply previously undetected birds.

### Situational Ethics

Seeing a rail, even a relatively confiding species like a Sora or a Virginia Rail (both on p. 90), requires luck or, more often, effort. Rails can be coaxed into view with smartphone audio of their songs; to see anything more than a dark form in the sedges and bulrushes, especially at night when rails are most active, a flashlight is necessary. More extreme measures can be taken: A group of birders may drag a chain through a marsh or meadow in a technique appropriately called rail dragging. Or one can bum a ride on a huge combine making its rounds in the Louisiana rice fields, flushing rails as it lumbers along. These methods are, needless to say, stressful for the birds. Similar concerns apply in many other wildlife-watching scenarios: "taping out" nesting songbirds with audio playback; "baiting" owls with mice and "chumming" for seabirds with fish oil and popcorn; and the simple act of getting too close to a roosting hawk or shorebird for the perfect photo.

The act of birding is, to a greater or lesser extent, disruptive for birds and their habitats. Yet birds need more people, *many* more people, to know about them, to marvel at them, and, ultimately, to care for their populations and habitats. Judicious use of audio and artificial lighting, and even the occasional wild ride on a combine, can expose whole groups of students and other citizens not only to wildlife, but also to their own hidden wild side.

To maximize your chances for an encounter with a rarely seen owl, rail, or other secretive species, consider signing up for a field trip with a birding festival. Such trips typically are led by expert birders with extensive training in minimizing threats to wildlife and maximizing festival participants' chances for seeing a sought-after species. In the final analysis, even responsibly observing nature is slightly disruptive—whereas ignoring nature has much greater negative consequences for birds and their habitats.

## TALL RELATIVES OF RAILS

If you somehow reduced these species to about a quarter of their size, the result would be something rather like a rail. In their locomotion, vocalizations, overall body plan, and general habitat requirements, these three have much in common with their smaller relatives.

### Limpkin

*Aramus guarauna* | LIMP | L 26" (66 cm)

The Limpkin is fluid in its movements, not limping. Surprisingly agile, it is adept at navigating trees and tangles in freshwater swamps.

■ **APPEARANCE** Shaped like a *Rallus* rail (p. 90), but much larger. Plumage mostly brown with white spots on back and upperwing coverts, white streaks on face and neck; bill straw yellow. *Plegadis* ibises (p. 196) and juvenile White Ibis (p. 194) slimmer and a bit smaller with reddish legs.

■ **VOCALIZATIONS** Wild, wailing sounds in Fla. and Gulf Coast swamps are likely this species. Most notes monosyllabic but slurred: *rrreeee!* and *rrrooo!* Wails often preceded by soft clicking. Like cranes and French horns, Limpkins have elongated tracheae for amplifying sound.

■ **POPULATIONS** Feeds primarily on large apple snails, genus *Pomacea;* range rapidly expanding with the advance of nonindigenous snails.

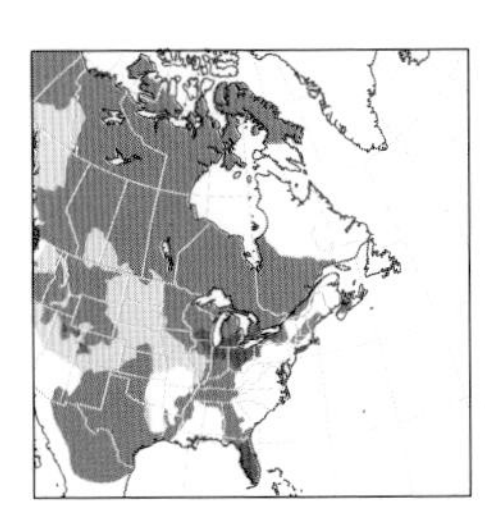

### Sandhill Crane

*Antigone canadensis* | SACR | L 41–48" (104–122 cm) WS 73–84" (185–213 cm)

Few sounds in nature are more stirring than the proclamations of cranes migrating high overhead. The late winter gathering of this species along the Platte R. in Nebr. is widely acknowledged to be one of the grandest spectacles on the continent.

■ **APPEARANCE** Nearly as tall as a Great Blue Heron (p. 186) and similarly gray overall. Cranes fly with necks outstretched, often soar; herons seldom do either. Adult Sandhill has red crown; immature plain-crowned with extensive rust in plumage. All have "bustle" of long tertials.

■ **VOCALIZATIONS** Rich, rolling *g'r'r'roo,* given in flight or by pairs capering in agricultural fields; urgent and wild-sounding, carries tremendously far. Shrill, reedy whistle given in flight, apparently only by juveniles.

■ **POPULATIONS** "Greater Sandhill" larger-bodied, relatively longer-necked and longer-legged, and deeper-voiced than "Lesser." But "Lesser" relatively longer-winged, with darker wing tips; migrates farther.

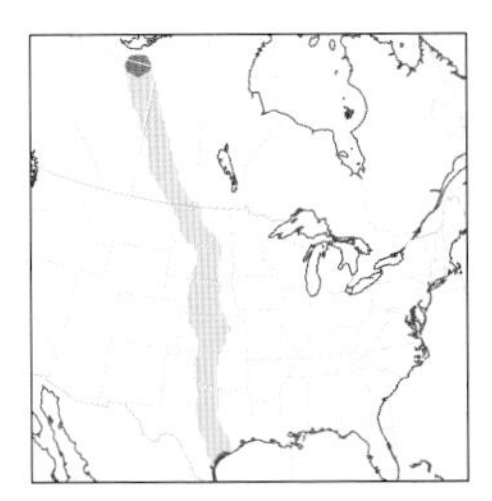

### Whooping Crane EN

*Grus americana* | WHCR | L 52" (132 cm) WS 87" (221 cm)

This "comeback kid" of avian conservation numbered in the low double digits in the mid-20th century, but counts are now well into the hundreds. It remains endangered, however, and full recovery, if it is to happen at all, will be many decades into the future.

■ **APPEARANCE** Stands taller than any other bird in our area. Adult white with black primaries; has extensive red on crown and below eye. Adult Sandhill smaller and grayer, with less red on head. First-winter Whooping shows extensive rust like Sandhill, but with white, not gray, ground color. Beware leucistic (whitish) Sandhills, sometimes misreported as Whoopings.

■ **VOCALIZATIONS** The "whooping" is more of a bright bugling, higher and purer than the Sandhill's calls.

■ **POPULATIONS** The one remaining indigenous (wild) flock breeds mostly in Wood Buffalo N.P., Alta. and N.W.T.; winters Aransas N.W.R., Tex. A single chemical spill could wipe them all out. Small introduced and captive flocks occur elsewhere in the East.

adult

extensive white streaks and spots on upperparts

## Limpkin

all-dark back and belly

juvenile

adult

## Sandhill Crane

stained adult

adult

neck extended in flight

dark red mask and crown

black primaries

adult

adult

## Whooping Crane

juvenile

## THREE STRIKING SHOREBIRDS

The term "shorebird" is applied to the members of several families in the diverse order Charadriiformes. Most trend brownish, and identification of some is difficult. But these three, looking like they're decked out in their finest evening wear, rarely present field ID challenges.

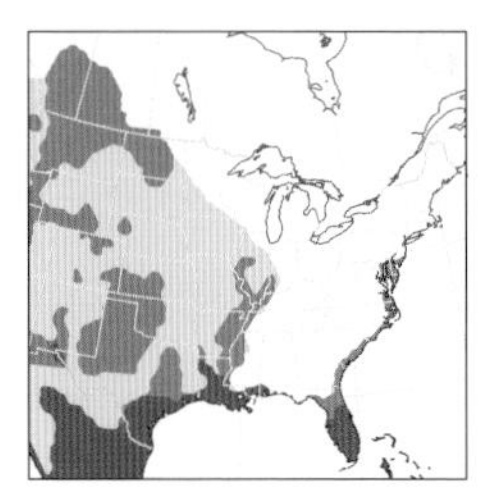

### Black-necked Stilt

*Himantopus mexicanus* | BNST | L 14" (36 cm)

For a bird so supremely elegant, this species has an odd predilection for some of the gnarliest of habitats: sewage lagoons, evaporation ponds, and naturally hypersaline wetlands.

■ **APPEARANCE** Bright pink legs are extremely long; the bird looks as if it should fall over. Perfectly straight bill is exceedingly thin. Adult male jet-black above, snow-white below. Female a bit browner above; juvenile's dark brown back feathers have buff edges.

■ **VOCALIZATIONS** An explosive *KEK*, equal parts annoyed and annoying, often repeated in long series. Birds around nests are fiercely defensive, driving off humans with constant yapping.

■ **POPULATIONS** Slowly expanding as a breeder in mid-Atlantic and Midwest. Opportunistic; found in artificial and degraded wetlands, where it can be affected by heavy metal contamination.

### American Avocet

*Recurvirostra americana* | AMAV | L 18" (46 cm)

In the same family as the Black-necked Stilt, this dapper shorebird is especially adept at handling deepwater situations. Avocets wade up to their belly, and they often swim.

■ **APPEARANCE** Head, neck, and breast of breeding adult pastel orange; wings and back with large patches of black and white; long legs a ghostly gray-blue. Variably gray in winter; some are black and pure white. Female's bill more sharply upturned than male's, corresponding to average differences in feeding behavior: Females swish more, males jab more, but much overlap.

■ **VOCALIZATIONS** Piercing *SHPLEENT*, less nasal than Black-necked Stilt's note. Avocets, like stilts, are highly aggressive around nests.

■ **POPULATIONS** Nested along East Coast long ago. Increasing in summer in mid-Atlantic, especially Delmarva region, might foretell reestablishment there. Annual vagrant to Great Lakes.

### American Oystercatcher

*Haematopus palliatus* | AMOY | L 18½" (47 cm)

The oystercatcher's brilliant bill looks as if it were about to burst into flame. It is a morphological marvel, laterally compressed for prying open bivalve shells; the stout bill can also be used to hammer right into the shells.

■ **APPEARANCE** Boldly pied like stilts and avocets, but more heavyset. Forages lumberingly by itself or in small flocks; roosts out in the open on gravel bars, often with Black Skimmers (p. 160). Thick white wing stripe contrasts with black wings in flight. Gleaming red bill of adult tipped yellowish, eye red and yellow; bill and eye duskier in first winter.

■ **VOCALIZATIONS** Flushing, gives a loud, curlew-like cry, *eeWEEEEee*, one of the evocative sounds of the seashore; in feeding squabbles, issues short notes intermediate in timbre between those of stilts and avocets.

■ **POPULATIONS** Completely tied to coastal habitats; sightings even a couple of miles from favored habitat are very rare.

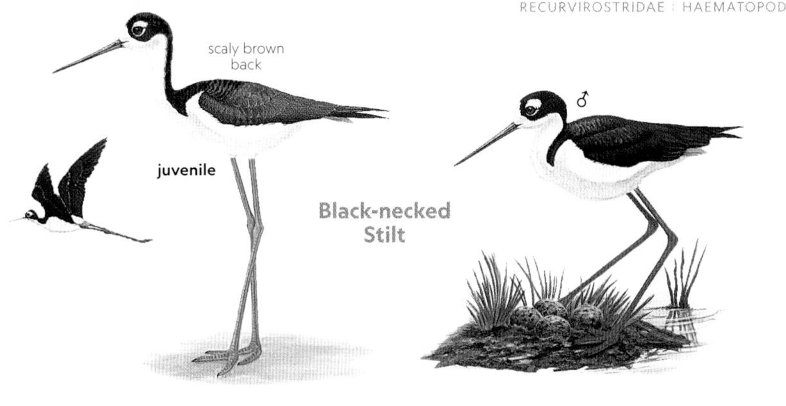
scaly brown back
juvenile
Black-necked Stilt
♂

bold pattern in flight
bill strongly upturned on female, less so on male
breeding ♀
winter ♂
juvenile ♂
American Avocet

wide white bar on wings
adult
American Oystercatcher
adult
juvenile

## LARGE PLOVERS

A shorebird trots along a mudflat, stops for a while, then keeps going. That run-stop-run sequence strongly suggests that this is a plover, a species in the second largest family of shorebirds.

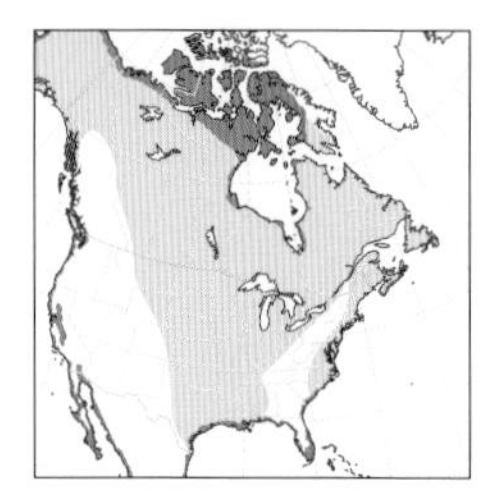

### Black-bellied Plover

*Pluvialis squatarola* | BBPL | L 11½" (29 cm)

The biggest eastern plover, this species is also one of the hardiest and most widespread. It occurs in small numbers, often singly, inland but forms larger flocks on the coast.

■ **APPEARANCE** Heavyset and block-headed, with stout bill. Breeding adult, especially male, black below and strikingly black and silver above. Nonbreeding adult and juvenile plain gray overall, with blank face; browner, smaller-billed nonbreeding American Golden-Plover shows dark cap and prominent supercilium. Adults on migration often in transitional plumage. In flight, all plumages of Black-bellied high-contrast; note white rump, bold white wing stripe, and especially black axillaries ("wingpits").

■ **VOCALIZATIONS** Call, given in flight, a rich, slurred whistle, *uhweeee-ee* or *weee-eee.* Sad-sounding; carries well, audible over the surf.

■ **POPULATIONS** Winters well north along Atlantic coast, but also migrates far south of the Equator; one of the most widely distributed bird species coastally on the planet.

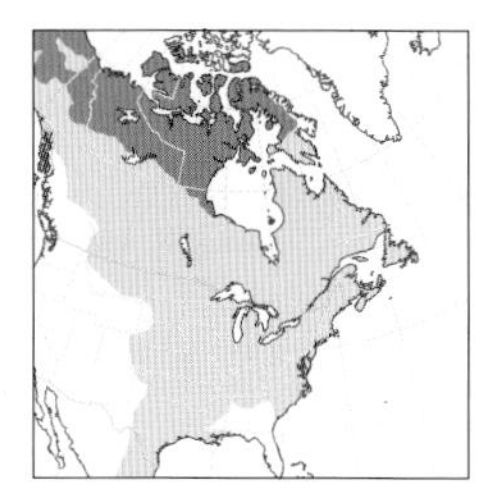

### American Golden-Plover

*Pluvialis dominica* | AMGP | L 10¼" (26 cm)

On spring migration northward, big flocks stage in disked agriculture fields in the Midwest; but on fall migration, most are out over the Atlantic, on their way from Canada to S. Amer. This plover has one of the most pronounced elliptical migrations of any bird in the East.

■ **APPEARANCE** Breeding adult spangled golden above; breeding Black-bellied whitish above. Smaller and slimmer than Black-bellied, with slighter bill. Nonbreeders and juveniles have dark cap and contrasting pale supercilium, unlike Black-bellied. In flight, all are plain-winged, smoky brown above and below. European Golden-Plover (p. 411), nearly annual to Nfld. in spring, is stockier and shorter-winged than American; underwing white, not smoky brown.

■ **VOCALIZATIONS** Flight call an abrupt *queedy-quee* or *queedy;* whistled like Black-bellied, but shorter and choppier.

■ **POPULATIONS** Declined sharply during the market hunting era, late 19th and early 20th centuries, but has recovered; a new threat is Arctic warming, forecast to eliminate more than half the species' breeding habitat.

### Killdeer

*Charadrius vociferus* | KILL | L 10½" (27 cm)

Perfectly at ease on mudflats and sandbars, but as likely to be found in pastures, on golf courses, or around parking lots and construction sites.

■ **APPEARANCE** Has two broad, black breastbands; our other *Charadrius* plovers have one, often broken, or none. Long tail has extensive rufous, conspicuous in flight. Just-fledged young much smaller than adult, with single breastband; may invite confusion with Semipalmated Plover (p. 102).

■ **VOCALIZATIONS** In agitation, varied whistles: *deee and k'deee* (*kill-deer*) and *dee d'deee.* Especially when defending young, gives a low, steady trill. Flight call, heard any time of day or night, a rising *deeeEEE.*

■ **POPULATIONS** Lingers late into fall; among first species to return in spring. Unlike most shorebirds, has adapted well to human-altered habitats.

juveniles in flight
American
Black-bellied
black axillaries
brown rump
Black-bellied Plover
silvery gray on upperparts
breeding ♀
bright juvenile
plain face
winter
white undertail coverts
breeding ♂
juvenile

American Golden-Plover
light supercilium and dark cap
breeding ♀
bright juvenile
Apr. ♂
juvenile
all-dark undertail
breeding ♂
European Golden-Plover juveniles for comparison
bright rufous rump
two breastbands
Killdeer

## LITTLE ORANGE-BILLED PLOVERS

Breeding adults of these three species are orange-billed, but their bills are darker, sometimes black, in nonbreeding and juvenile plumages. Color of the bare parts is important in shorebird ID, but be aware of the effects of staining (mud, oil, etc.) and especially lighting.

### Common Ringed Plover

*Charadrius hiaticula* | CRPL | L 7½" (19 cm)

Despite this migratory species' extensive breeding range in northern Canada, it is rarely seen by birders in the East because its annual migrations take it straight across the North Atlantic to and from the Old World.

■ **APPEARANCE** Differs from Semipalmated by broader black breastband, broader white supercilium, less colorful orbital ring, and slightly longer bill. All differences slight and subtle; good photos or audio probably necessary to clinch ID.

■ **VOCALIZATIONS** Flight call a *tooee* or *tooip*, with flat affect and little pitch change; lower-pitched than Semipalmated's, less strident.

■ **POPULATIONS** In narrow zone of overlap with Semipalmated, east-central Baffin I., may hybridize with that species. Away from Arctic breeding grounds, seen less than annually on migration, most frequently in Nfld.

### Semipalmated Plover

*Charadrius semipalmatus* | SEPL | L 7¼" (18 cm)

This bird is like a subcompact, stripped-down Killdeer: notably smaller and shorter-tailed, and not as noisy. It is the most widely encountered of the small plovers in our area.

■ **APPEARANCE** Told from Killdeer (p. 100) by single breastband and smaller size, but note that recently fledged Killdeer has single breastband. Darker-backed than similarly proportioned Snowy (p. 104) and Piping Plovers; bill much smaller than Wilson's (p. 104), and usually orangey. Separation from closely related Common Ringed Plover very difficult, probably requires photos and audio.

■ **VOCALIZATIONS** Flight call a strident *ch'wee!* or *chew-awee!* Spectrograms of call show much more frequency sweep, typically falling then rising, than in Common Ringed.

■ **POPULATIONS** Bucking the trend with other small plovers, seems to be holding its own; accepts varied microhabitats across extensive breeding grounds and is relatively tolerant of disturbance.

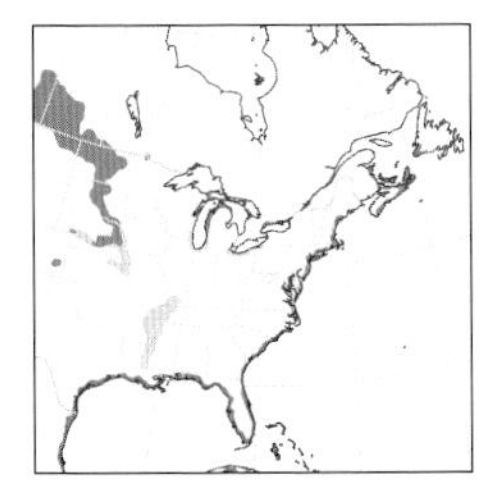

### Piping Plover

*Charadrius melodus* | PIPL | L 7¼" (18 cm)

Like human swimmers and sunbathers, this strikingly pale plover favors ocean beaches and inland lakeshores. Unfortunately, the bird is easily spooked and abandons nesting areas when disturbed.

■ **APPEARANCE** The "little gray ghost of the dunes," as pale as a Snowy Plover (p. 104). Breeding has yellow-orange legs and bill; nonbreeding adult and juvenile have black bill. Lacks dark ear patch of Snowy, has thinner breast patch.

■ **VOCALIZATIONS** Flight call richer and lower than Semipalmated's, a husky *julep* or *pieplow;* second element lower than first.

■ **POPULATIONS** Nests, roosts, and feeds above the wrack line, where dogs and dune buggies induce the birds to abandon their nests. Community education is key: For example, the 2019 film *Monty and Rose* raised awareness of Piping Plovers trying to nest at Chicago's congested Montrose Beach.

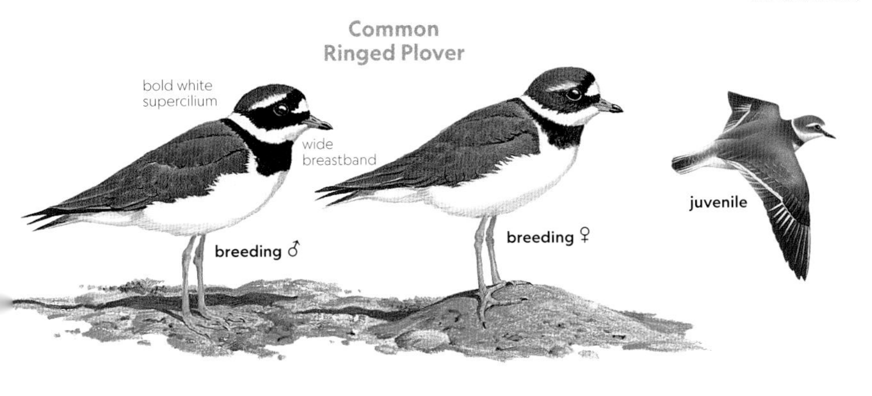
Common Ringed Plover
bold white supercilium
wide breastband
breeding ♂
breeding ♀
juvenile

Semipalmated Plover
juvenile
winter
breeding ♂
dark brown upperparts
juvenile
complete breastband
breeding ♀

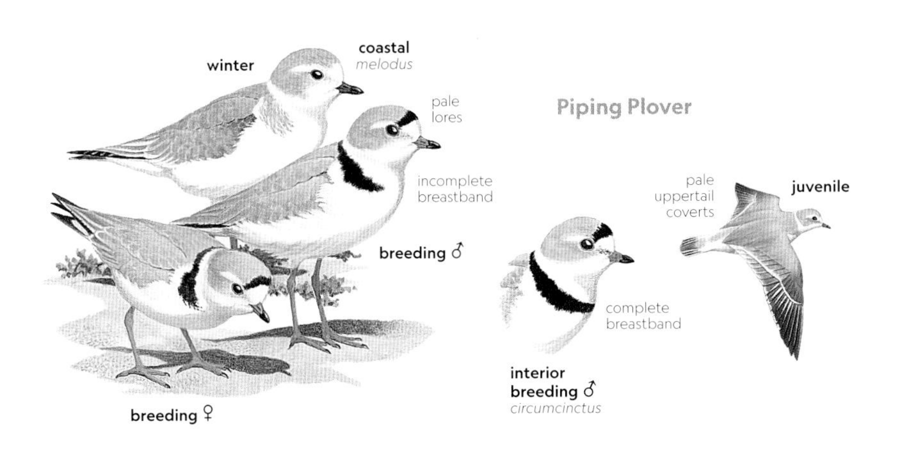
Piping Plover
winter
coastal
melodus
pale lores
incomplete breastband
breeding ♂
breeding ♀
pale uppertail coverts
juvenile
complete breastband
interior breeding ♂
circumcinctus

## DARK-BILLED PLOVERS

Like all our plovers, these three species nest in barren patches largely devoid of vegetation. And saying that they "nest" is generous, for the eggs are dropped in little scrapes right out in the open.

### Wilson's Plover

*Charadrius wilsonia* | WIPL | L 7¾" (20 cm)

More than any other plover in the East, this one is tied to the coast. Natural habitats will do, but the species also breeds on dredge spoil islands and evaporation ponds, even lightly used roads and airfields.

■ **APPEARANCE** Our largest single-banded plover; color scheme about the same as Semipalmated (p. 102), but Wilson's is larger with broader breast-band. In all plumages, bill is notably long and thick; the species is also aptly called the "Thick-billed Plover."

■ **VOCALIZATIONS** Call a sharply rising utterance, sometimes a mellow *speee,* often a sharper *speet!*

■ **POPULATIONS** Partially migratory in the East, with many departing the U.S. entirely, even breeders in South Fla. and South Tex. Range has contracted in places, and sea-level rise threatens further range contraction.

### Mountain Plover

*Charadrius montanus* | MOPL | L 9" (23 cm)

On the breeding grounds, this plover has nothing to do with aquatic habitats. It nests on disturbed shortgrass prairie far from water; in winter, it occurs on barren flats near or far from coasts and lakeshores. Despite its name, this species never occurs in mountains.

■ **APPEARANCE** Among our *Charadrius* plovers, only the Killdeer (p. 100) is larger. In all plumages, faded tan above, whitish below; essentially unbanded, although some adults show buffy smudge across breast. Black bill is long and thin; note white forehead and small black patches on face, prominent in breeding plumage, faded in winter. Confusion with other *Charadrius* plovers unlikely, but beware paler American Golden-Plover (p. 100) in nonbreeding plumage and even Buff-breasted Sandpiper (p. 118).

■ **VOCALIZATIONS** Calls harsh and descending, often short and clipped, *krr,* sometimes drawn-out, *kyeeurrr;* ternlike. Quiet as *Charadrius* plovers go; even around young, disinclined to vocalize.

■ **POPULATIONS** Faring poorly. Original habitat of bison-trampled, prairie dog-denuded, frequently burned shortgrass plains has been replaced by sterile center-pivot agriculture and, in recent years, oil-field development.

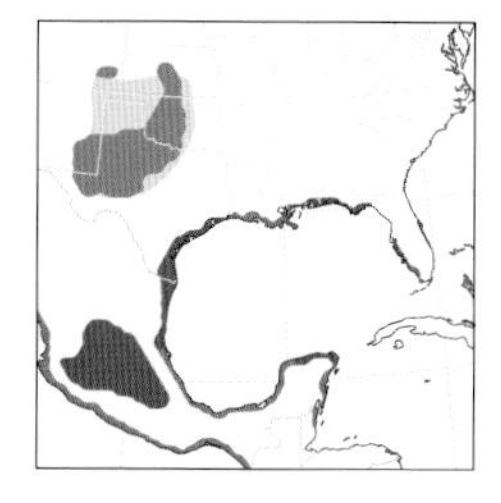

### Snowy Plover

*Charadrius nivosus* | SNPL | L 6¼" (16 cm)

In many cases, this diminutive shorebird doesn't even make a pretense to building nests. More than the other small plovers, it is inclined to alkaline flats and other hypersaline environments.

■ **APPEARANCE** A bit smaller than Piping Plover (p. 102) and similarly pale. Legs black; bill thin and black. Piping has orange legs year-round and stubbier, thicker bill, which is orangey in breeding season, mostly black in winter. In side view, adult Snowy in breeding plumage has three dark patches: atop forehead, behind eye, and along sides of upper breast.

■ **VOCALIZATIONS** Flight call a rising, wavering whistle: *churwee* or *churawee.* Also gives a gruff *grit* or *grurt* in agitation or in flight.

■ **POPULATIONS** Interior breeders average darker above than coastal breeders; inland breeders migratory, coastal birds in winter a mix of inland breeders and resident breeders. Declining across U.S. range.

Wilson's Plover
♀
large bill
thick breastband
juvenile
breeding ♂
♀
dull fleshy legs

Mountain Plover
uniform tan upperparts
plain-breasted
breeding
white underwing
winter
winter
winter
juvenile

Snowy Plover
♀
Gulf Coast ♂
tenuirostris
♀
juvenile
western
nivosus
dark patch on side
♂
dark legs

**CURLEWS**

Within the large sandpiper family Scolopacidae, these three are placed in the subfamily Numeniinae. The name derives from the Greek word for "new moon," alluding to the crescent-shaped bills of all but one species in the grouping.

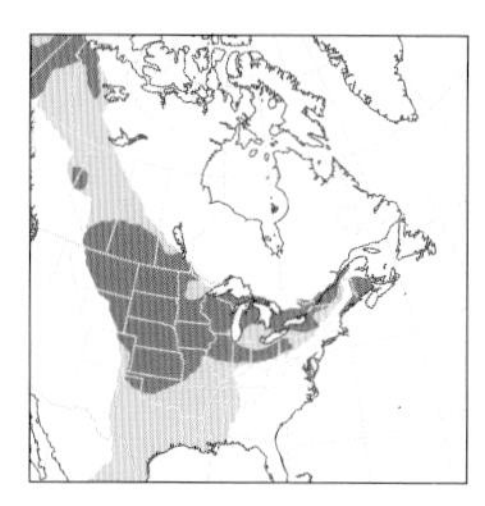

## Upland Sandpiper

*Bartramia longicauda* | UPSA | L 12" (30 cm)

The least aquatic of all sandpipers, this species flourishes best in remnant midgrass prairie. It has achieved limited success in less-ideal habitats: airfields, no-till soybean fields, even commercial blueberry farms.

■ **APPEARANCE** Cryptic plumage undifferentiated overall; brownish above, paler below. But shape distinctive: potbellied, pin-headed; note long tail and thin neck. Feet yellow; yellow bill is short and straight for a curlew. Flight direct, powerful; wings long.

■ **VOCALIZATIONS** Flight song is one of the great sounds of the prairie—a rough stutter followed by a rising then falling wail: *b'b'b'b'b'b'b'wheeEEE-wheEEEEEEeeeooo.* Flight call, often given by migrants at night, a mellow, far-carrying *quit-quit-quit* or *quiddy-quit-quit.*

■ **POPULATIONS** Formerly hunted in excess, but current threat is habitat destruction. Acceptance of novel habitats like airfields and blueberry farms has not compensated for loss of prime prairie habitat.

## Whimbrel

*Numenius phaeopus* | WHIM | L 17½" (44 cm)

This is our coastal curlew, occurring in salt marshes and on sea rocks. But like all curlews, it is at least somewhat terrestrial; flocks stage in summer in eastern Canada in dry heathlands, where they devour crowberries before migrating over the Atlantic.

■ **APPEARANCE** Quite large; the long bill droops sharply at the tip. Cold gray-brown overall; legs blue-gray. Black crown has thin white median stripe, often hard to see; black eyeline separated from crown by pale eyebrow.

■ **VOCALIZATIONS** Flight call, heard especially near noisy surf, a rich piping in rapid succession: *peep! peep! peep! peep! peep! peep!* In breeding display, wails like other curlews.

■ **POPULATIONS** Eurasian subspecies, probably annual to eastern Canada and maybe northeastern U.S., has white back and tail; may be a separate species.

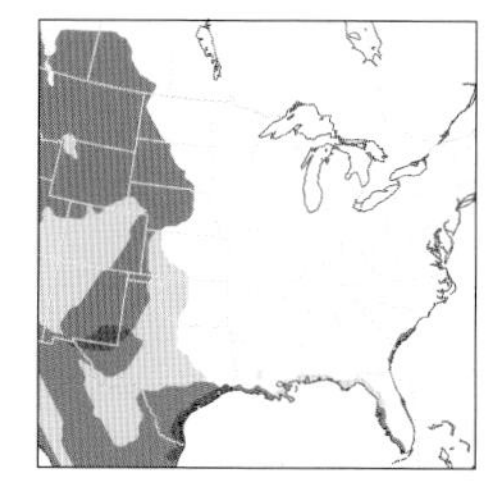

## Long-billed Curlew

*Numenius americanus* | LBCU | L 23" (58 cm)

Like the Upland Sandpiper, this curlew nests far from water. As a breeder in our area, it occurs in sandhills, shortgrass prairie, and even open juniper woodlands.

■ **APPEARANCE** A colossal sandpiper; bill astonishingly long. Legs blue-gray. Undifferentiated, low-contrast plumage is warmer and paler overall than Whimbrel's; face plain and unmarked compared to Whimbrel. In flight, flashes warm cinnamon in wings, like large-bodied Marbled Godwit (p. 108), but that species has slightly upturned bill and blacker legs.

■ **VOCALIZATIONS** Powerful flight call an urgent whistle that ends abruptly on a higher note: *eeeee-EE* or *ooeeee-EE* (*cur-lew!*). In flight display on breeding grounds, gives longer whistles interspersed with bubbly trills, similar to song of Upland Sandpiper.

■ **POPULATIONS** Formerly wintered on much of Atlantic coast; now restricted in winter in East mostly to Tex. coast.

Upland Sandpiper
juvenile
pin-headed appearance
adult
large, dark eye
plump and potbellied
long tail
juvenile
juvenile
dark crown and eye stripes
adult
Whimbrel
American
hudsonicus
white rump
light underwing
European
phaeopus
juvenile ♂
bill length highly variable; shortest in juvenile males
Long-billed Curlew
adult ♀
blue-gray legs
cinnamon underwings

## GODWITS

These four are similar in size and shape to large curlews (p. 106) and are frequently found alongside them. But they are believed to be more closely related to all the other sandpipers than to the curlews.

### Bar-tailed Godwit

*Limosa lapponica* | BTGO | L 16" (41 cm)

Of the two rare godwits in the East, this is the less rare. Occurs annually, with most sightings along the Atlantic coast. Suggests Marbled Godwit, also rare on the coast north of Fla. Marbled is larger overall, with longer legs and a slightly longer bill. Bar-tailed in flight shows gray-white underwing and rump, plus eponymous gray-barred tail.

### Black-tailed Godwit

*Limosa limosa* | BLTG | L 16½" (42 cm)

The rarest godwit in the East, but recorded almost every year. Concentration of records from eastern Canada, but sightings not as strongly coastal overall as those of Bar-tailed. Boldly pied in flight like Hudsonian Godwit, but wing linings white (black in Hudsonian). Two-toned bill straight; upturned on other godwits.

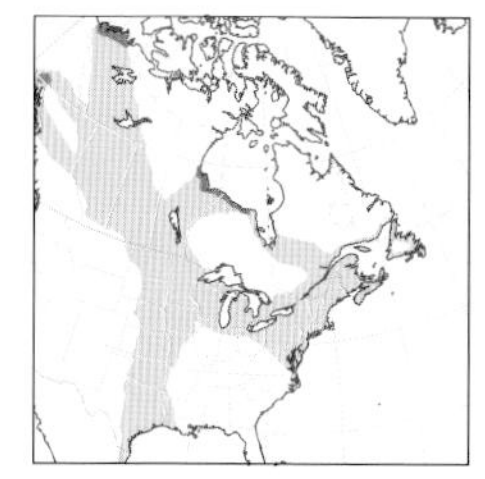

### Hudsonian Godwit

*Limosa haemastica* | HUGO | L 15½" (39 cm)

Like the American Golden-Plover (p. 100), this godwit is an elliptical migrant. The spring migration is north through the Great Plains, the fall migration mostly off the Atlantic coast.

■ **APPEARANCE** Smaller and darker than Marbled Godwit; like Marbled, has two-toned, upturned bill. Underwings black in all plumages. Breeding adult male dark red below, gray-black above. Nonbreeding adult gray-brown; suggests Willet (p. 124), but note bill structure, long and pointed wings, and black on tail. Juvenile similar to adult, but has warmer tones.

■ **VOCALIZATIONS** Call a squeaky *pip* or *pyep*, often given in flight.

■ **POPULATIONS** Locally common in spring in narrow migration band up through the Midwest; although most are offshore on fall migration, small numbers on coast and even well inland, especially when they put down in nasty weather.

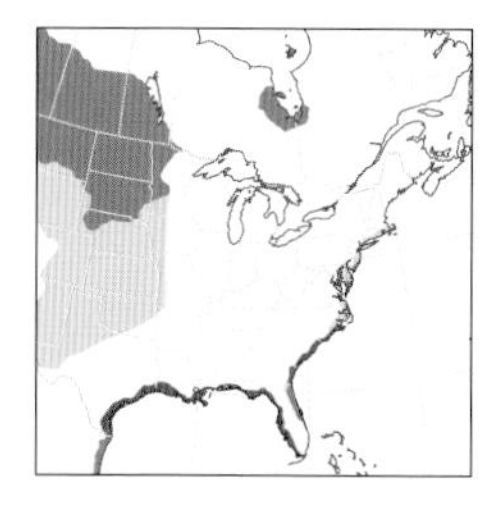

### Marbled Godwit

*Limosa fedoa* | MAGO | L 18" (46 cm)

This prairie godwit is the largest species in its genus, yet it migrates the shortest distance. The species breeds only in the U.S. and Canada and winters no farther south than northern Mid. Amer.

■ **APPEARANCE** Other than curlews, the largest sandpiper in the East. Buffy tones, especially in flight, suggest Long-billed Curlew, but very long bill of this godwit is slightly upturned. Legs black; curlew's legs blue-gray. Breeding adult shows fine dark barring all over, juvenile and nonbreeding adult more washed-out, buffier. Smaller dowitchers (p. 120), with their orangey tones and very long bills, can appear similar to Marbled Godwit; dowitchers' bills dark and nearly straight, underwing linings cold gray.

■ **VOCALIZATIONS** Call nasal and rising, strongly slurred: *oh-wee!* or *ooh-week!* Vocal around nests, typically silent on migration and in winter.

■ **POPULATIONS** Adults leave breeding grounds early; "fall" migration gets underway while it is still technically spring. Scarce migrant inland in East; winter concentrations in our area mostly Tex. and Fla.

Bar-tailed Godwit
winter
grayish underwing
winter
juvenile
shorter legs than Marbled
breeding ♂
breeding ♀
Black-tailed Godwit
juvenile
white
long straight bill
juvenile
winter
breeding ♂
Hudsonian Godwit
juvenile
black
juvenile
breeding ♂
female larger
molting fall adult ♂
patchy underparts
breeding ♀
Marbled Godwit
buffy underwing
winter
breeding ♂
two-toned bill long, upturned
black legs
winter ♀

## LARGE ARENARIINAE

The subfamily Arenariinae, made up of genera *Arenaria* (the turnstones) and *Calidris* (pp. 110–119), comprises a group of mostly small-bodied, relatively short-billed sandpipers. The three birds below are the largest of the group in the East.

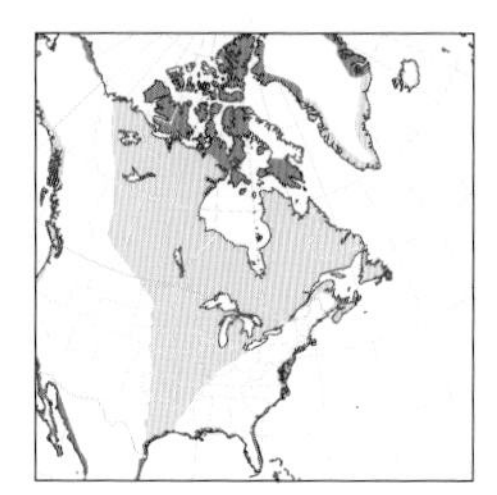

### Ruddy Turnstone

*Arenaria interpres* | RUTU | L 9½" (24 cm)

This parti-colored beachcomber occurs singly or in small flocks on sea rocks, around wrack lines, and less frequently on open sandbars and mudflats.

■ **APPEARANCE** Head and breast complexly patterned black-and-white, strikingly so in breeding plumage; nonbreeders retain general pattern. Breeders rich rufous above, nonbreeders darker and duller above. Bill stout and upturned, an adaptation for flipping debris (turning stones) in search of food; feet red-orange all year. Upperparts flash white in flight, a pattern similar to that of diverse "rockpipers" of the West Coast.

■ **VOCALIZATIONS** Dull *chut* or sharper *pep* notes, often run into a short chatter.

■ **POPULATIONS** Mostly coastal on migration, especially Delaware Bay in late May. Some become oddly habituated to humans.

### Red Knot

*Calidris canutus* | REKN | L 10½" (27 cm)

A poster child for resource specialization, the Red Knot depends critically on the flush of horseshoe crab eggs in late May and early June on protected beaches of Delaware Bay.

■ **APPEARANCE** Our largest *Calidris*, the size of a golden-plover (p. 100), but plumper; short legs and straight bill are black. Breeding adult smooth orange-pink below, scaly rufous-gray above. Nonbreeder plain gray above; juvenile prominently scalloped above. Bunches up in flocks near shoreline, often with Ruddy Turnstones.

■ **VOCALIZATIONS** Rough *chup* and sharper *wheee* sometimes heard on migration and in winter; whoops and whistles on breeding grounds.

■ **POPULATIONS** Most spring migrants along East Coast stage at Delaware Bay, where they are imperiled by overharvest of horseshoe crab eggs; an oil spill in this heavily trafficked waterway could be disastrous.

### Ruff

*Calidris pugnax* | RUFF | L 10–12" (25–30 cm)

Amazing sandpiper from Eurasia; has four genders. Annual vagrant in very small numbers in the East, mostly Atlantic seaboard and Great Lakes. All are potbellied, small-headed, and long-necked; most adults have orangey legs and at least some orange on bill. Bright buff juvenile scalloped above, plain below; similar to Buff-breasted Sandpiper (p. 118).

### Sandpiper Subfamilies

The sandpipers, family Scolopacidae (pp. 106–127), have a basic look and body plan: brown overall with long, thin bills for feeding on terrestrial and aquatic invertebrates. Most are superb migrants, with long, dark wings.

Recent molecular work in the Scolopacidae has revealed five basic subfamilies: curlews (p. 106); godwits (p. 108); turnstones and genus *Calidris* (pp. 110–119); dowitchers, snipes, and woodcocks (pp. 120–123); and genus *Tringa* and phalaropes (pp. 122–127). These relationships are relevant to identification: Knowing the subfamilies reinforces our understanding of key differences in morphology and behavior, eminently observable in the field.

## Ruddy Turnstone

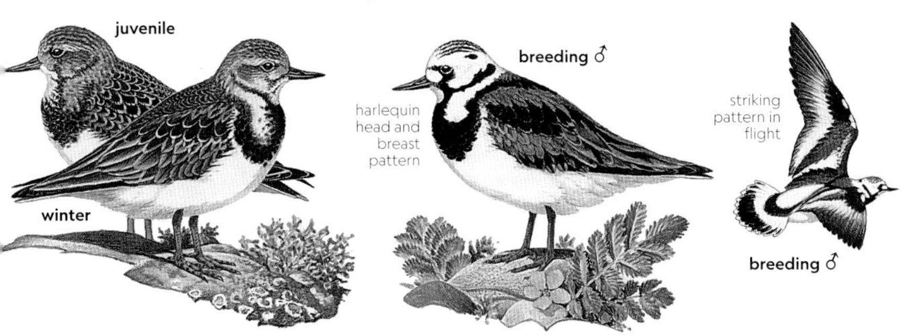


## Red Knot

## Ruff

## LONG-LEGGED *CALIDRIS* SANDPIPERS

The breeding males, dark and colorful, are among our most handsome *Calidris* sandpipers. In nonbreeding plumage, they are much plainer and mostly gray; posture and bill structure are especially important for ID in fall and winter.

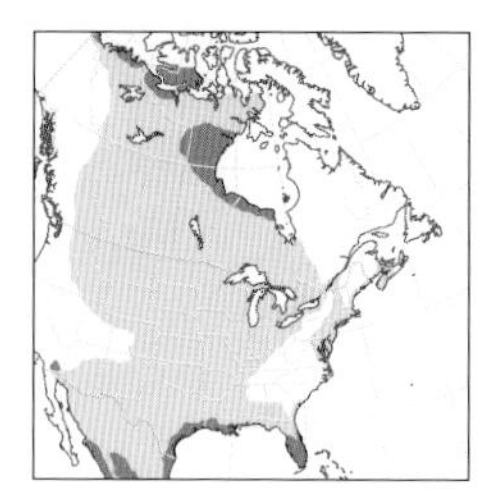

### Stilt Sandpiper

*Calidris himantopus* | STSA | L 8½" (22 cm)

With its long bill and habit of feeding in standing water, this sandpiper might be more readily mistaken for a dowitcher (p. 120) than a congener in the genus *Calidris*.

■ **APPEARANCE** Orange ear patch and extensive dark barring of breeders distinctive, but this plumage not fully attained until late spring. Juvenile and nonbreeding adult washed-out gray overall with white eyebrow. In all plumages, note long legs, usually dull yellow; bill long, drooping.

■ **VOCALIZATIONS** Flushing and socializing, gives harsh *chyurp*, descending.

■ **POPULATIONS** Migration mostly through mid-continent in fairly narrow band, especially spring; spreads out more in fall, occurring in small numbers to East Coast.

### Curlew Sandpiper

*Calidris ferruginea* | CUSA | L 8½" (22 cm)

Its breeding range is restricted to the Arctic Ocean coastal plain of north-central Russia, but this sandpiper winters widely across the Old World. Each year a few reach the eastern U.S. and Canada, where the species is a highly sought vagrant.

■ **APPEARANCE** Same size and heft as Dunlin (p. 114), but bill longer and more drooping, legs longer. Full-on breeding male unmistakable with deep-red plumage, but most sightings in East are of birds in transitional, nonbreeding, or juvenile plumage. Curlew Sandpiper has white uppertail coverts, conspicuous in flight, in all plumages; eyebrow of nonbreeding Curlew more prominent than Dunlin's.

■ **VOCALIZATIONS** Varied calls include short trills, slurred whistles, and brief bouts of squeaky chatter; calls of Dunlin harsher overall.

■ **POPULATIONS** Annual vagrant to East, especially Atlantic coast; probably annual to eastern Great Lakes, too.

### Leg Length and Bill Shape

Sandpipers in the genus *Calidris* have long been regarded as one of the thorniest field ID conundrums for birders in the East—indeed for birders across much of the world. The complaint, in its barest essence, is that "they all look the same." Even our relatively straightforward species—for example, the two on this page—are in muted grays and browns outside the breeding season, when many birders encounter them. Interspecific variation in size is considerable in this genus, but individuals within a species also vary considerably. A better approach, then, is to assess body shape, especially with regard to leg length and bill structure.

In the case of the appropriately named Stilt and Curlew Sandpipers, the birds' long legs and decurved bills, respectively, are usually obvious (even though the Stilt Sandpiper has a decurved bill and the Curlew Sandpiper has relatively long legs). But these characters are of great utility for field ID even for species with less-impressive appendages. With a bit of comparative experience, birders instantly recognize the surprisingly variable Least Sandpiper (p. 116) by its fine, slightly drooping bill; and a Purple Sandpiper (p. 114) in silhouette—the way they're often seen—is distinctively short-legged. Combine body structure with feeding behavior, microhabitat, and migration timing, and the *Calidris* sandpipers aren't so hard to sort out after all.

ong and
urved
juvenile
pale
supercilium
winter
breeding
molting
juvenile
juvenile
Stilt
Sandpiper
legs long
and dull
yellow
white rump
juvenile
Curlew
Sandpiper
long bill thin
and drooping
juvenile
breeding ♂
winter
Red Knot
Sanderling
Dunlin
Purple
Sandpiper
Baird's/
White-rumped
Sandpiper
Pectoral
Sandpiper
Stilt
Sandpiper
Western
Sandpiper
Semipalmated
Sandpiper
Least
Sandpiper

## COASTAL *CALIDRIS* SANDPIPERS

These midsize *Calidris* sandpipers are staples of winter birding along the East Coast. They roost in tightly bunched flocks, sometimes in mixed-species assemblages, and they employ distinctive foraging strategies.

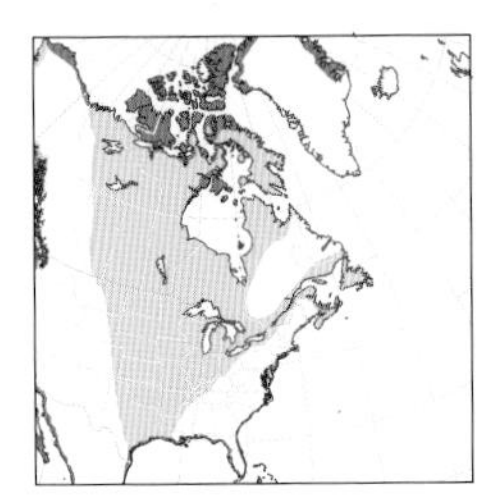

### Sanderling

*Calidris alba* | SAND | L 8" (20 cm)

Few sights are more bewitching than a flock of Sanderlings chasing the surf. The birds run down the beach as a wave retreats, then just as quickly race back up as the next wave advances.

■ **APPEARANCE** Stocky and sturdy; bill straight, black, and fairly short. Most of the year, conspicuously pale; at rest, shows black wedge at "shoulder" (leading edge of wing). Reddish breeding plumage not acquired until late on spring migration; compare with smaller Least Sandpiper (p. 116). Coordinated running alone identifies the species, but beware that individuals and even flocks often forage like other small sandpipers.

■ **VOCALIZATIONS** Industrious feeding flocks too busy to call, but listen for rough, rising *cheet?* or *chlit?* from birds in flight.

■ **POPULATIONS** Occurs coastally year-round, including nonbreeders in summer; regular in decent numbers on migration across Great Lakes, scarce but widespread inland singly and in small flocks.

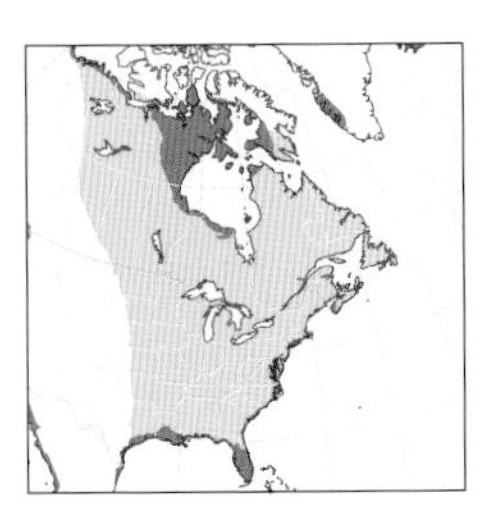

### Dunlin

*Calidris alpina* | DUNL | L 8½" (22 cm)

This curve-billed *Calidris* is notably gregarious on migration and in winter. Feeding flocks are disorganized and frenetic, but large winter flocks in flight twist and turn in perfect unison.

■ **APPEARANCE** In all plumages, bill is long, black, and drooping. Breeding adult reddish above with extensive black blotch on belly. Gray-brown above in winter; in flight, uniformly dark above with only weak wing stripe. Western Sandpiper (p. 118) in winter smaller, grayer, and not as long-billed. Compare also with rare Curlew Sandpiper (p. 112).

■ **VOCALIZATIONS** Flight call a scratchy, slightly descending *kriih* or *prrriih*.

■ **POPULATIONS** A hardy *Calidris,* wintering well north coastally; generally scarce inland away from Great Lakes, but flocks of overland migrants sometimes put down in late autumn storms.

### Purple Sandpiper

*Calidris maritima* | PUSA | L 9" (23 cm)

"Purple" is a bit of stretch, "sand" even more so. This sought-after species winters almost exclusively on jetties, breakwaters, and sea rocks.

■ **APPEARANCE** Dark and dumpy. Short legs yellow-orange in winter, duller in summer; bill slightly drooped, variably orange at base. Purple sheen, more of a cold gray, most prominent in winter; breeders browner, more splotchy. Many in winter show nearly solid purplish-gray hood, often with complete eye ring. Feeds and roosts right at the water's edge, where splashed with surf; dark plumage blends in remarkably well with sea rocks, black and slickened.

■ **VOCALIZATIONS** Winter flocks twitter softly; notes not as harsh or scratchy as those of Sanderling and Dunlin.

■ **POPULATIONS** Winters far north, but also expanding south along coast, regularly reaching Fla.; regular in late fall in very small numbers at breakwaters on Great Lakes.

winter
Sanderling
black wedge on wing
juvenile
breeding
winter
distinctive gait
lacks hind toe
Dunlin
winter
faint streaking on flanks
winter
breeding
bill long and drooping
black belly
juvenile
juvenile
winter
Purple Sandpiper
winter
dark slate gray hood
bill and legs orangey
streaked sides
breeding

## PEEPS

The moniker "peep" is applied imprecisely to the smallest and drabbest *Calidris* species. These three are exemplary peeps, feeding on open mudflats at or near the water's edge.

### Baird's Sandpiper

*Calidris bairdii* | BASA | L 7½" (19 cm)

Along with the White-rumped Sandpiper, this is one of the two large peeps—"large" being relative. It is a remarkable migrant, breeding north to Ellesmere I., wintering widely in western and southern S. Amer.

■ **APPEARANCE** Sandy gray-brown in all plumages, with black, nearly straight bill; legs black. Wings long and dark; at rest, they cross over like the blades of scissors. White-rumped Sandpiper same size and shape, but grayer overall, with eponymous white rump; bill of White-rumped droops slightly, shows pale pink-yellow at base. Least Sandpiper similar in color and pattern; Baird's larger and longer-winged, with straighter bill (thin, decurved in Least) and black legs (dull yellow in Least).

■ **VOCALIZATIONS** Flight call a short trill, rough and reedy, slightly falling in pitch: *brrrrrt* or *brriiip*. Song, given in flight on breeding grounds, consists of scratchy, gasping notes, run together in longer trills, slightly rising in pitch.

■ **POPULATIONS** Migration, both spring and fall, mostly through middle of continent. Rare east of Appalachians, but annual in small numbers.

### Least Sandpiper

*Calidris minutilla* | LESA | L 6" (15 cm)

The tiniest shorebird on Earth, this bird is more diminutive than a fair number of our sparrows. In mixed-species flocks, it tends to occur a bit farther in from water's edge than other peeps.

■ **APPEARANCE** A smidgen smaller than Western and Semipalmated Sandpipers (p. 118). All plumages muddy and dark; adults gray-brown, juveniles warmer. Bill thin, droops down, tapers to fine point. Yellowish legs distinguish Least from other peeps, but lighting and mud can make yellow legs appear dark, and vice versa. Larger Baird's Sandpiper has longer wings, straighter bill, black legs.

■ **VOCALIZATIONS** Flight call higher and sweeter than Baird's, a rising, rolling, *prreee?* Song, given by both sexes, a rapid utterance of short, rich whistles.

■ **POPULATIONS** Fall migration protracted; some linger late, even well inland.

### White-rumped Sandpiper

*Calidris fuscicollis* | WRSA | L 7½" (19 cm)

This species is an extreme migrant, surpassing even the Baird's Sandpiper. Most winter from Uruguay south, and a few reach subantarctic islands.

■ **APPEARANCE** A large peep, structurally similar to Baird's Sandpiper; both have long wings that cross over on the bird at rest. White rump distinctive in all plumages, but beware that any fast-flying sandpiper can catch bright glare as it twists and turns. Gray in breeding plumage, with rufous highlights above and gray streaks on flanks; compare with smaller Western Sandpiper. Bill slightly decurved; shows a bit of pale yellow-pink at base.

■ **VOCALIZATIONS** High-pitched flight call squeaky and scratchy, short of duration, and dropping sharply in pitch: *squit* or *tsweek*. Rarely heard song likewise scratchy.

■ **POPULATIONS** Spring migration mostly up Great Plains; fall migration likewise through mid-continent but also with heavy passage off Atlantic coast. Late spring migrant; migrants through U.S. seen into late June.

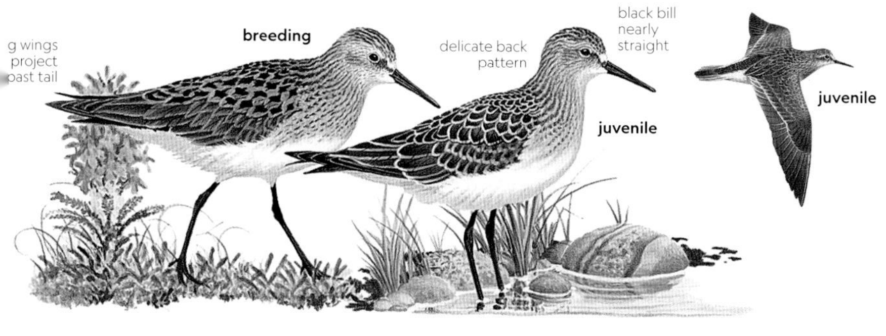
Baird's Sandpiper
g wings
project
past tail
breeding
delicate back
pattern
black bill
nearly
straight
juvenile
juvenile

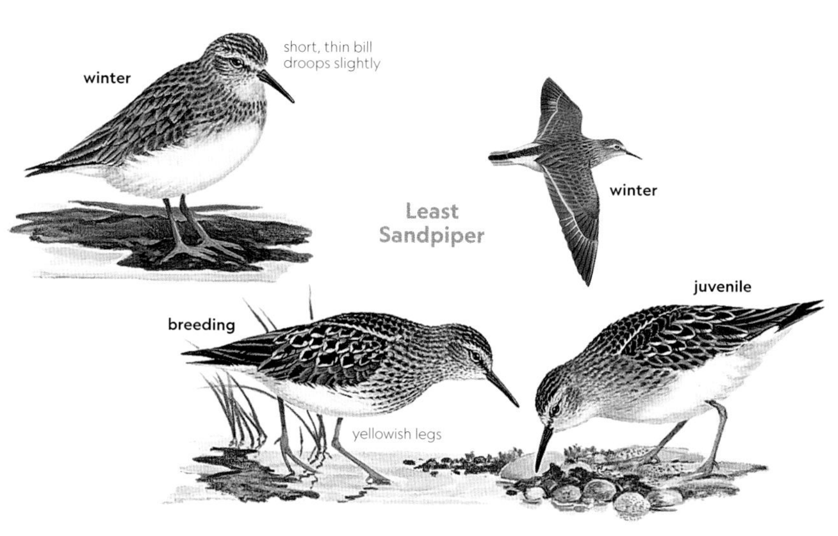
winter
short, thin bill
droops slightly
winter
Least
Sandpiper
juvenile
breeding
yellowish legs

White-rumped
Sandpiper
white
rump
juvenile
juvenile
breeding
long wings project
past tail
streaked sides
fall-molting
adult

## PEEPS, ETC.

Like most species in the diverse genus *Calidris*, these four nest in the Arctic. And as with their congeners, timing and extent of migration are useful in field ID away from the breeding grounds.

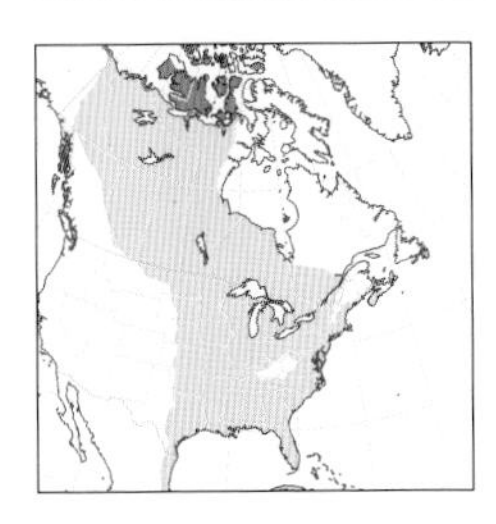

### Buff-breasted Sandpiper

*Calidris subruficollis* | BBSA | L 8¼" (21 cm)

This "grasspiper" is found during migration around sod farms and other open habitats. Like the Ruff (p. 110), it performs courtship displays at leks.

■ **APPEARANCE** Plain pale buff below; darker above with buff scaling. Erect posture, long neck, and beady black eye suggest Upland Sandpiper (p. 106). Legs yellow, bill black.

■ **VOCALIZATIONS** Soft ticking and buzzing. Rarely heard, even at leks.

■ **POPULATIONS** Migration is mainly up and down middle of continent, but wanders in fall to East Coast. Globally rare, fewer than 100,000 individuals.

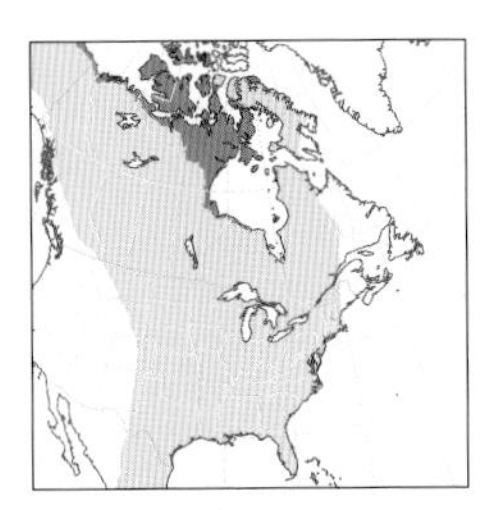

### Pectoral Sandpiper

*Calidris melanotos* | PESA | L 8¾" (22 cm)

"Pectoral" refers to air sacs inflated by the male in courtship, but it might as well denote this exceptional migrant's powerful pectoral muscles.

■ **APPEARANCE** Notably bibbed in all plumages; bib of breeding male like chain mail. Adults and especially juveniles on southbound migration more rufous than spring adults. Legs straw yellow; bill dull orange-brown, slightly decurved; wings long. Compare with Baird's and Least Sandpipers (p. 116).

■ **VOCALIZATIONS** Flight call a quick trill, *prrrrp*, not as growling as Baird's.

■ **POPULATIONS** Fairly early spring migrant; heads back south early too. Heaviest passage through mid-continent, but decent flocks to East Coast.

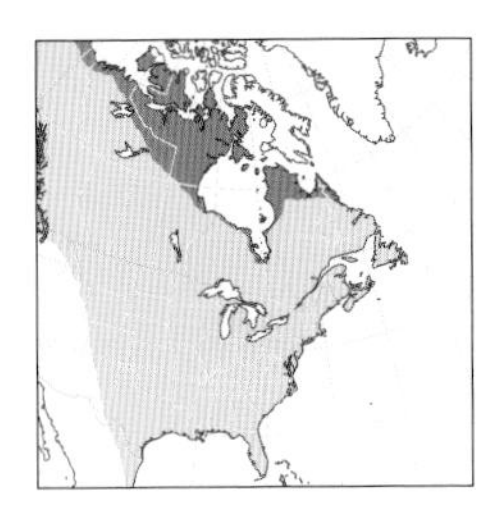

### Semipalmated Sandpiper

*Calidris pusilla* | SESA | L 6¼" (16 cm)

This small sandpiper and the similar Western constitute one of the most notorious pairs of similar species in all of birding. Body structure, migration timing, and flight calls aid in field ID.

■ **APPEARANCE** Sandy gray in most plumages. Flanks largely unspotted in spring; Western has streaked flanks in spring. Eyebrow, slightly waved, a good mark on drab fall migrants. Blunt-tipped bill short, thick, and straight.

■ **VOCALIZATIONS** Flight call a simple, descending *churk* or *churp*, rougher than Western's.

■ **POPULATIONS** Fairly common throughout East on spring migration and from midsummer to early fall. Nearly absent from U.S. in winter.

### Western Sandpiper

*Calidris mauri* | WESA | L 6½" (17 cm)

This species is the western counterpart of the look-alike Semipalmated Sandpiper. On migration in the East, it is generally much rarer than the "Semi." In some places in winter, though, it greatly outnumbers that species.

■ **APPEARANCE** Slightly larger and lankier than Semipalmated, with longer, thinner, more decurved bill, but bill structure variable. Adult in spring has rufous cap and "ears," heavy streaking down flanks; juveniles on southbound migration have rufous scapulars, showing as colorful band above.

■ **VOCALIZATIONS** Flight call high and scratchy, descending in pitch, *screet* or *jeet*, more like White-rumped (p. 116) than Semipalmated.

■ **POPULATIONS** Scarce spring migrant east to Mississippi R., rare farther east. Starts to show up in early fall after most "Semis" have moved through. Winters coastally in much of East, unlike Semipalmated.

Buff-breasted Sandpiper
displaying adult
scaly upperparts
juvenile
adult
large dark eye on blank face
juvenile
Pectoral Sandpiper
breeding ♀
breeding ♂
dense streaking on breast
juvenile
legs and bill show some color
juvenile
Semipalmated Sandpiper
winter
breeding ♀
breeding ♂
cool-toned upperparts
juvenile
short, thick bill
winter
partially webbed toes hard to see
Western Sandpiper
winter
bright rufous patches
winter
juvenile
slender decurved bill
breeding

## DOWITCHERS

Closely related and formerly treated as one species, these two forage in standing water coastally as well as inland; they often form compact flocks. The name is a loanword from the Iroquois language family.

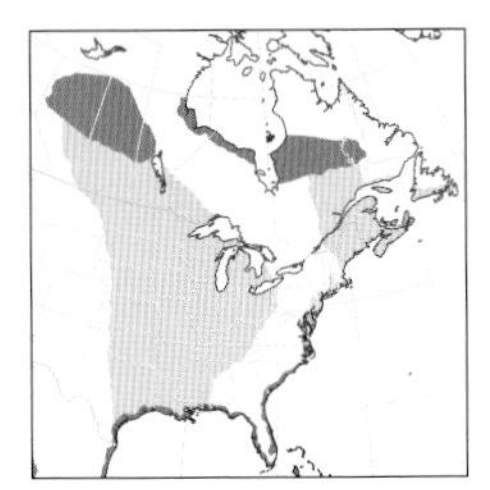

### Short-billed Dowitcher

*Limnodromus griseus* | SBDO | L 11" (28 cm)

The most striking thing about this ironically named bird is its astonishingly long bill. True, its bill is not quite as long, on average, as that of the Long-billed Dowitcher, but field ID depends on characters other than bill length.

■ **APPEARANCE** Bill long, subtly kinked or curved. White bars on tail as broad as or broader than black bars, making bird look pale-tailed in flight; Long-billed Dowitcher appears darker-tailed. Wings slightly longer than Long-billed's; eye positioned higher on face than on Long-billed. In breeding plumage, Atlantic subspecies, *griseus,* has white belly with black bars and spots; mid-continent subspecies, *hendersoni,* mostly orange-bellied in breeding plumage with small black spots.

■ **VOCALIZATIONS** Flight call a soft, descending *tyu,* typically in twos and threes, like Lesser Yellowlegs's (p. 124).

■ **POPULATIONS** Short-billed is a "warmer" migrant than Long-billed, moving on average later in spring and earlier in fall.

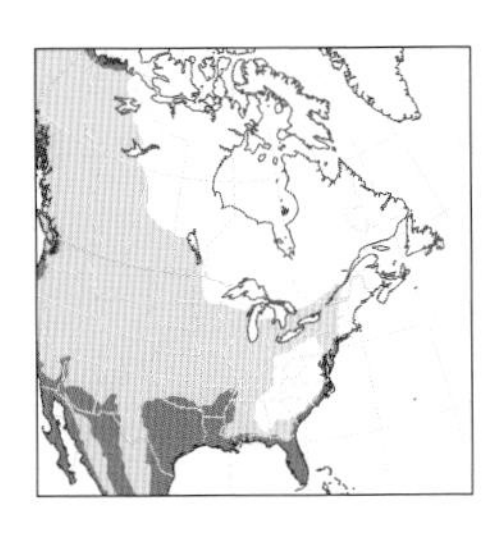

### Long-billed Dowitcher

*Limnodromus scolopaceus* | LBDO | L 11½" (29 cm)

Like the Short-billed Dowitcher, this species feeds with rapid jabs of the bill. Especially in winter, Long-billed is more likely to be found in freshwater microhabitats than Short-billed.

■ **APPEARANCE** Bill averages longer and straighter than bill of Short-billed Dowitcher, but beware of sex differences: Males of both dowitchers are shorter-billed than females, and male Long-billed and female Short-billed overlap in bill length. Compared to Short-billed, Long-billed has proportionately shorter wings, eye lower on face, and thicker dark bars on tail. Breeding adult dull brick red all over. Winter adult has smooth gray breast; breast of winter Short-billed more spotted.

■ **VOCALIZATIONS** Flight call a sharp, squeaky *pik* or *plik*. Feeding flocks twitter softly, producing notes like those of Short-billed.

■ **POPULATIONS** Rare in much of East in spring. Somewhat more common in fall, with first arrivals later than first Short-billeds. Lingers late inland where there is open water.

### Aging Shorebirds

In some sandpipers, differences between adults and juveniles are pronounced. In southbound dowitchers, for example, these differences may be greater than interspecific differences. So it is important for field ID to be able to distinguish juvenile sandpipers from adults. In general, juvenile shorebirds on "fall" migration—often during meteorological summer—appear colorful and scaly compared to adults. In the case of dowitchers, the tertials of juvenile Short-billeds have orange markings crossing the entire width of the feathers; the same feathers on juvenile Long-billeds are solidly dark-centered, with gold restricted to the margins.

Timing is important. In most shorebirds, adults migrate ahead of juveniles. (The "empty nest" metaphor for humans is reversed here.) In the mid-Atlantic region, adult Short-billeds are on the move in July; in Aug., it's juvenile Short-billeds and the first adult Long-billeds; by late Sept., juvenile Long-billeds start to appear. These differences in timing reflect different molt strategies: Adult Short-billeds don't molt until arriving on the wintering grounds, whereas adult Long-billeds molt before the main fall migration.

bold internal marks on tertials
juvenile tertials
Short-billed Dowitcher
Atlantic
griseus
worn breeding
Atlantic
griseus
breeding
white belly
spotted flanks
interior
hendersoni
winter
molting juvenile
juvenile
interior
hendersoni
winter
Atlantic
griseus
molting juvenile
winter
winter
juvenile
faint to no internal markings on tertials
juvenile tertials
Long-billed Dowitcher
white scapular fringes
worn breeding ♀
barred flanks
fresh breeding ♂

## FAMILIAR INLAND SANDPIPERS

Most sandpipers are sought at beaches, mudflats, and grasslands, but these four are found elsewhere: They occur in varied upland habitats, often wooded or otherwise densely vegetated.

### American Woodcock

*Scolopax minor* | AMWO | L 11" (28 cm)

The astonishing "sky dance" of this terrestrial sandpiper is performed on cool evenings and frigid mornings in late winter in and around early successional woodlands.

■ **APPEARANCE** Rotund and russet, with long bill and large eyes. Plain below, cryptically patterned above, with broad transverse barring on crown.

■ **VOCALIZATIONS** Flushes on twittering wings. Spring display, at dawn and dusk, has two components: abrupt, nasal *peent* notes delivered from ground, and rhythmic, pulsating chirps and twitters at great height.

■ **POPULATIONS** Migrates early in spring and fairly late in fall, almost never associating with other shorebirds on mudflats and sandbars.

### Wilson's Snipe

*Gallinago delicata* | WISN | L 10¼" (26 cm)

This is the bird of "snipe hunt" notoriety. It has a knack for flushing the instant you see it—and for putting down inconveniently out of sight.

■ **APPEARANCE** Dowitcher-like; feeds hunched over. Head striped lengthwise; back has long white streaks. Short orange tail prominent in flight, hard to see at rest.

■ **VOCALIZATIONS** Flushing, a startling *skrrrtt,* rasping and rising. Spooky winnowing of males created by splayed-out tail feathers: *oohuhuhuhuhuhu,* given at considerable altitude. Also proclaims from fence posts or other perches: a steady *wheep wheep wheep* ....

■ **POPULATIONS** Winters well inland, fairly far north, at seeps and springs.

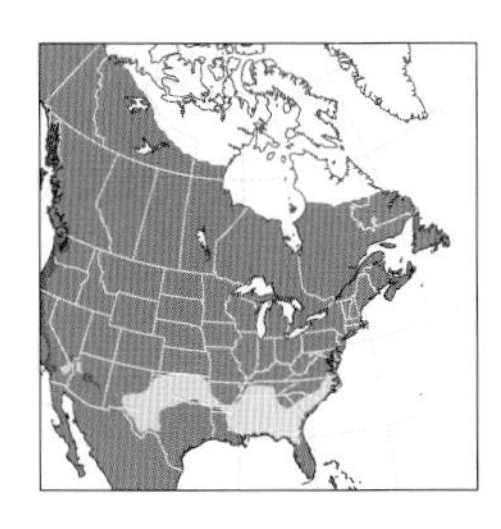

### Spotted Sandpiper

*Actitis macularius* | SPSA | L 7½" (19 cm)

This small bird teeters as it walks along a pond edge or sandbar, flying out over the water on stiff, shallow wingbeats, then returning to the shore.

■ **APPEARANCE** Small and plump; most show thin white eyebrow and equally thin dark transocular line. Boldly spotted below in breeding plumage; juvenile and nonbreeding adult plain below; note white wedge in front of wing.

■ **VOCALIZATIONS** A rising *wheee,* run together in series in song, given in twos or threes as bird flushes.

■ **POPULATIONS** Occurs singly or in small, incohesive flocks. Has largest breeding range of any N. Amer. sandpiper.

### Solitary Sandpiper

*Tringa solitaria* | SOSA | L 8½" (22 cm)

This bird is well named. Even when multiple individuals of this species find their way to the same mudflat, they still manage to keep their distance.

■ **APPEARANCE** Like a little Lesser Yellowlegs (p. 124); Solitary has shorter, greenish-yellow legs and a shorter, stouter bill. White eye ring often prominent. In flight, shows barred tail and dark underwings. Some nonbreeding Spotteds, smudgy overall with white eye ring, can suggest Solitary.

■ **VOCALIZATIONS** Flight call a strident *pee! wee!* or *pee! wee! wee!* Often heard at night. Higher, squeakier than Spotted's flight call.

■ **POPULATIONS** Nests in trees in boreal forest, threatened by logging and climate change. Migrants found widely, often around wooded ponds.

American Woodcock
dark bands on nape
underwing
displaying
Wilson's Snipe
dark lateral stripes on face
juvenile
1st winter
juvenile
Spotted Sandpiper
white wedge in front of wing
constant tail bobbing
breeding
juvenile
tail barred
breeding
white eye ring
juvenile
Solitary Sandpiper

## GENUS *TRINGA*

Rangy and leggy, these three sandpipers, along with the Solitary Sandpiper (p. 122), are the representatives of genus *Tringa* in the East. They are active and alert, walking swiftly from danger, and then flushing with strident flight calls.

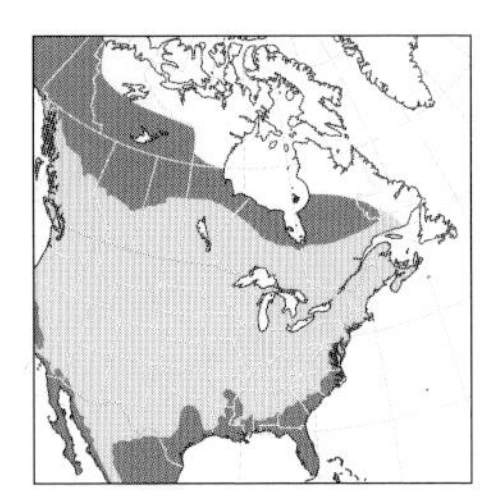

### Lesser Yellowlegs

*Tringa flavipes* | LEYE | L 10½" (27 cm)

Even among *Tringa* sandpipers, this wader is notably long-legged. It is wary and skittish, but less prone to drama than the Greater Yellowlegs.

■ **APPEARANCE** Color and pattern of feathers and bare parts almost identical to those of Greater Yellowlegs; in spring, not as boldly barred below as Greater. Differs subtly in structure: Lesser has shorter and straighter bill, longer legs, and longer wings, giving it a daintier gestalt. Beware confusion with smaller Solitary Sandpiper (p. 122) too, especially birds in nonbreeding plumage in poor light.

■ **VOCALIZATIONS** Flight call a slightly descending *tyu,* often given singly. Individual calls nearly the same as those of Greater Yellowlegs, but Lesser is not so declarative in its utterances. Similar flight call of Short-billed Dowitcher (p. 120) descends more sharply.

■ **POPULATIONS** Widespread and fairly common on migration; scarce along coasts, even to southern U.S., in dead of winter.

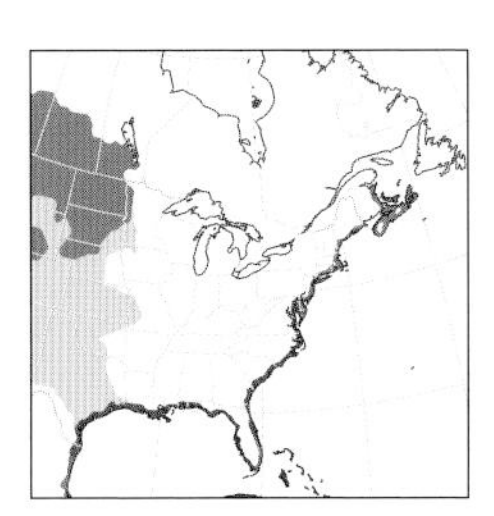

### Willet

*Tringa semipalmata* | WILL | L 15" (38 cm)

Striking in flight, this heavyset sandpiper was until recently in its own genus. Molecular data place it well within the genus *Tringa,* however, with a particularly close affinity to the Lesser Yellowlegs.

■ **APPEARANCE** Large and gray. Mostly dark bill is long and stout; legs grayish. Heavily barred in breeding plumage, plain gray in winter. High-contrast black-and-white wings striking in flight even at great distance.

■ **VOCALIZATIONS** Flight call a shrill, far-carrying *peep eeewee eewee* (*pill will willet*), notes all on the same pitch.

■ **POPULATIONS** Inland-breeding *inornata* migrates to our coasts for winter; nominate *semipalmata* breeds in coastal salt marshes, leaves our area by late summer for a coastal Brazilian vacation. They differ structurally too: *inornata* is more godwit-like, being larger overall than *semipalmata,* with longer bill, longer legs, and longer neck.

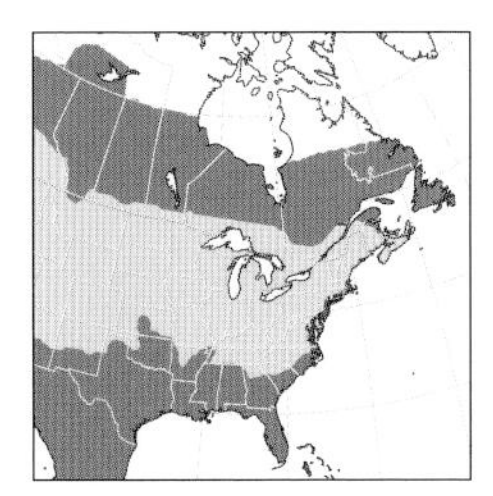

### Greater Yellowlegs

*Tringa melanoleuca* | GRYE | L 14" (36 cm)

One of the most widely distributed of our sandpipers, this large *Tringa* is equally likely to be found all by itself at a fishing pond or on mudflats teeming with other shorebirds.

■ **APPEARANCE** Nearly identical in color and pattern to smaller Lesser Yellowlegs, although in spring Greater averages more boldly marked than Lesser. In all plumages, Greater has thicker and longer bill than Lesser—at least as long as the head and subtly upturned. Greater has proportionately shorter legs and shorter wings.

■ **VOCALIZATIONS** Flight call a ringing *tyu!* like Lesser's, but Greater manages to deliver it with greater oomph. Greater often belts out four or five such notes, the first couple especially loud: *TYU!! TYU! tyu tyu tyu.*

■ **POPULATIONS** Found on migration in midlatitudes of East from late winter through late fall, with brief window of absence in mid-June. Hardy; winters well north on coast and even inland across southern tier of U.S.

slim bill shorter than Greater
breeding
Lesser Yellowlegs
winter
white rump
winter
juvenile
early Mar. in molt
breeding
Willet
contrasting scapulars
western
eastern
eastern
*semipalmata*
juvenile
winter
striking wing pattern
western Willets have longer, darker bill
worn breeding
winter
juvenile
western
*inornata*
breeding
long, slightly upturned bill
Greater Yellowlegs
breeding
winter
bill has gray base
winter
juvenile

## PHALAROPES

These swimming sandpipers spin rapidly to create vortices that pull small prey up to the surface. The ingenious birds also exploit surface tension and capillary action while feeding. And multiple individuals coordinate their actions to maximize feeding efficiency.

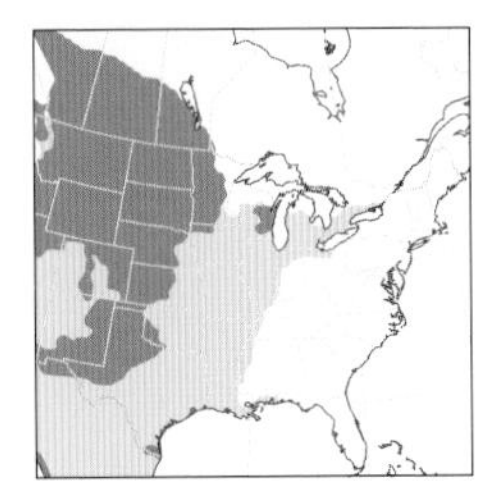

### Wilson's Phalarope

*Phalaropus tricolor* | WIPH | L 9¼" (23 cm)

This largest of the phalaropes is the species most likely to be seen inland. The species is perfectly capable of swimming, but it also feeds on shorelines—darting and dashing as it goes.

■ **APPEARANCE** Potbellied, with long neck and small head; long bill is extremely thin and straight. Bright breeding female sports tricolored neck, with long swaths of burgundy, amber, and white; breeding male duller. Juvenile notably pale; richly scalloped above. Back of winter adult uniform pale gray; overall color and graceful shape recall Lesser Yellowlegs (p. 124).

■ **VOCALIZATIONS** Muffled, nasal *uhrunk,* especially males around nests.

■ **POPULATIONS** Females, not involved in care of young, leave breeding grounds early; can be on "fall" migration by mid-June. Breeding range expanding eastward.

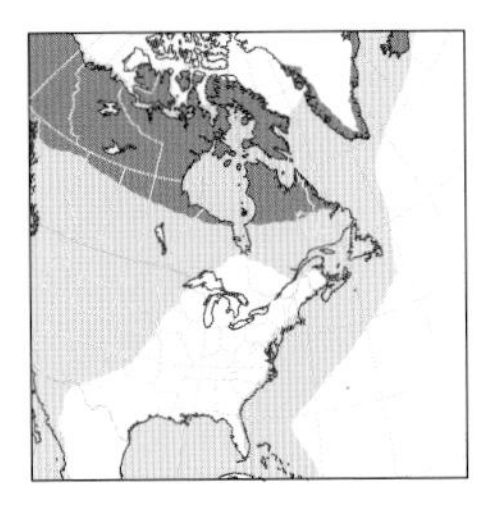

### Red-necked Phalarope

*Phalaropus lobatus* | RNPH | L 7¾" (20 cm)

A seafaring phalarope, this species is most likely to be encountered by eastern birders on offshore boat trips out to near the continental shelf edge.

■ **APPEARANCE** The smallest phalarope; bill long and thin. Red-orange neck and upper breast of breeding female contrast sharply with white cheeks; male duller. In breeding plumage, both sexes sport long golden bands, or "braces," on gray uppersides. Juvenile and winter adult show dark eye patch on mostly white head; note also dark back with white braces.

■ **VOCALIZATIONS** Call a dull, clipped *pik* or *plick,* often run together in series as bird flushes.

■ **POPULATIONS** Rare inland on migration except far western Great Plains, where more common; inland sightings typically midsummer to early fall, before sparse Red Phalarope passage overland.

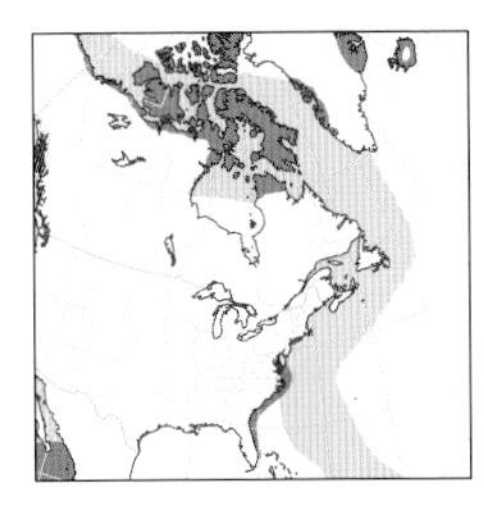

### Red Phalarope

*Phalaropus fulicarius* | REPH | L 8½" (22 cm)

Seen from offshore boat trips, this midsize phalarope often consorts with the smaller Red-necked Phalarope. But the Red Phalarope is even more pelagic, migrating farther offshore and stopping inland only rarely.

■ **APPEARANCE** Larger than Red-necked Phalarope; bill of Red much thicker than that of Red-necked and Wilson's. Breeding female deep red all over with extensive white on face; breeding male paler, more orangey. Both sexes have straw-yellow bill in breeding plumage; bill retains some color in fall and winter. Winter adult, like Red-necked, shows dark eye patch; back of winter adult Red is plain gray. Juvenile Red dark above, with buff-fringed feathers; lacks dorsal stripes of juvenile Red-necked.

■ **VOCALIZATIONS** Common call a high, sharp *peek!* or *pleek!* strangely similar to that of Hairy Woodpecker (p. 234).

■ **POPULATIONS** Rare but regular inland, especially Great Lakes eastward. Puts down on the heels of strong storms, Oct. to early Dec., then moves on as soon as the weather clears. Sometimes seen well offshore in winter.

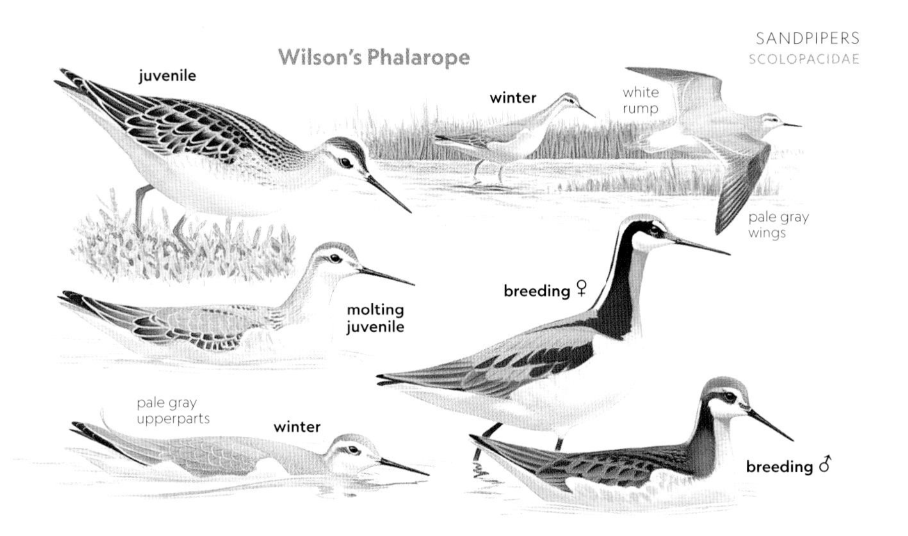
Wilson's Phalarope
juvenile
winter
white rump
pale gray wings
breeding ♀
molting juvenile
pale gray upperparts
winter
breeding ♂

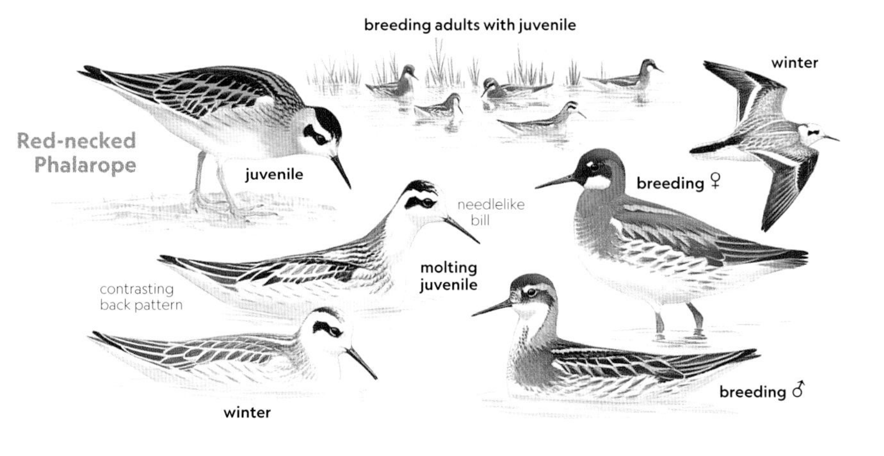
breeding adults with juvenile
winter
Red-necked Phalarope
juvenile
breeding ♀
needlelike bill
molting juvenile
contrasting back pattern
breeding ♂
winter

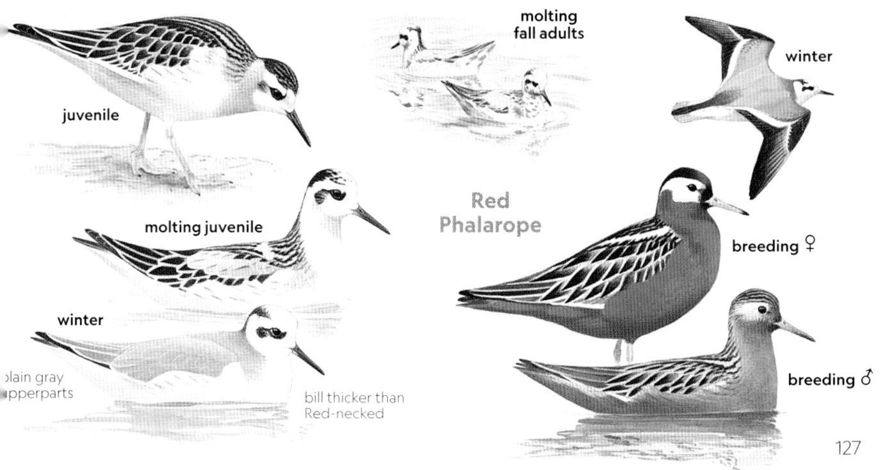
molting fall adults
winter
juvenile
Red Phalarope
molting juvenile
breeding ♀
winter
plain gray upperparts
breeding ♂
bill thicker than Red-necked

## SKUAS

Seeing one of these powerful marine predators always stops a literal boatload of birders in their metaphorical tracks. Unlike with jaegers (pp. 130-133), which migrate overland in small numbers, skua sightings are very rare even from shore—and practically unheard of inland.

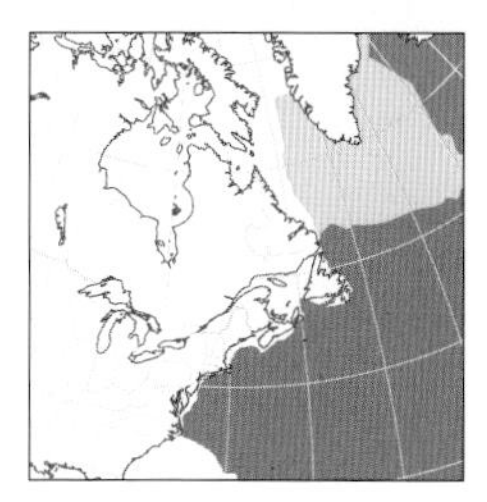

### Great Skua

*Stercorarius skua* | GRSK | L 22" (56 cm) WS 54" (137 cm)

The only skua breeding in the Northern Hemisphere, this species regularly reaches waters off eastern N. Amer. during the colder months. But the Great Skua is always uncommon, and seeing one is the highlight of a winter pelagic trip.

■ **APPEARANCE** Like South Polar Skua, flashes extensive white in broad wings, has broad tail lacking central points. Larger than South Polar, with warmer tones overall. Upperparts "messy," streaked and splotched; underparts uniform gingerbread brown. Notably larger and bulkier than jaegers, even Pomarine (p. 130). Beware similarity with young Herring Gull (p. 146), which can be dark and aggressive, flashing white in wings.

■ **VOCALIZATIONS** Usually silent away from breeding grounds, but may give descending nasal honks around other birds at sea.

■ **POPULATIONS** Off U.S. coast, probably starts to arrive in early autumn before the last South Polar Skuas have departed. Regular in summer well off Nfld. and Lab.

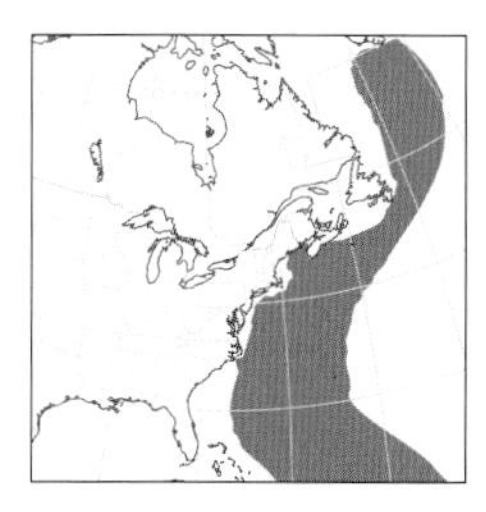

### South Polar Skua

*Stercorarius maccormicki* | SPSK | L 21" (53 cm) WS 52" (132 cm)

For birders in the East, this is a seabird sought on hot days at sea in the summer. But the South Polar Skua is a "snowbird," at the limits of its *winter* range during the northern summer.

■ **APPEARANCE** Off U.S. coast in summer, by far the more likely of the two skuas to be seen. Lacks warm tones of Great Skua; upperparts of South Polar plain and unmarked. South Polar has dark and light morphs, plus intermediates, but all are low-contrast overall. Differs from jaegers and Herring Gull as noted for Great Skua.

■ **VOCALIZATIONS** Usually silent on its northern wintering grounds, but listen for occasional barks and chatter in presence of other seabirds.

■ **POPULATIONS** Incredible migrant; breeds during its summer (our winter) almost entirely on Antarctica, one of very few birds to do so. Hybridizes with Brown Skua (*S. antarcticus*) of the Southern Ocean; how many hybrids, or even pure Browns, get to our waters is unknown.

### Genus *Stercorarius*

Birds in the family Stercorariidae (from the Latin *stercus,* meaning "excrement") spend much of their life at sea, where they steal food, including offal and other refuse, from smaller seabirds, a behavior known as kleptoparasitism. The family is a distinctive one, ecologically and morphologically. Within the Stercorariidae, however, things are much fuzzier.

The larger species, called skuas ("bonxies" in Britain), were once placed in the genus *Catharacta,* with the smaller species, called jaegers, in the genus *Stercorarius.* But that arrangement has been found to be wanting, with the Pomarine Jaeger being the crux of the problem. That species has genetic affinities with both the larger and smaller birds in this family, and biologists aren't sure why. An exotic possibility is that the Pomarine Jaeger arose in a recent hybridization event involving a large skua, perhaps the Great, and one of the smaller jaegers. One thing's for sure: These predatory seabirds get around—geographically, of course, but also genetically.

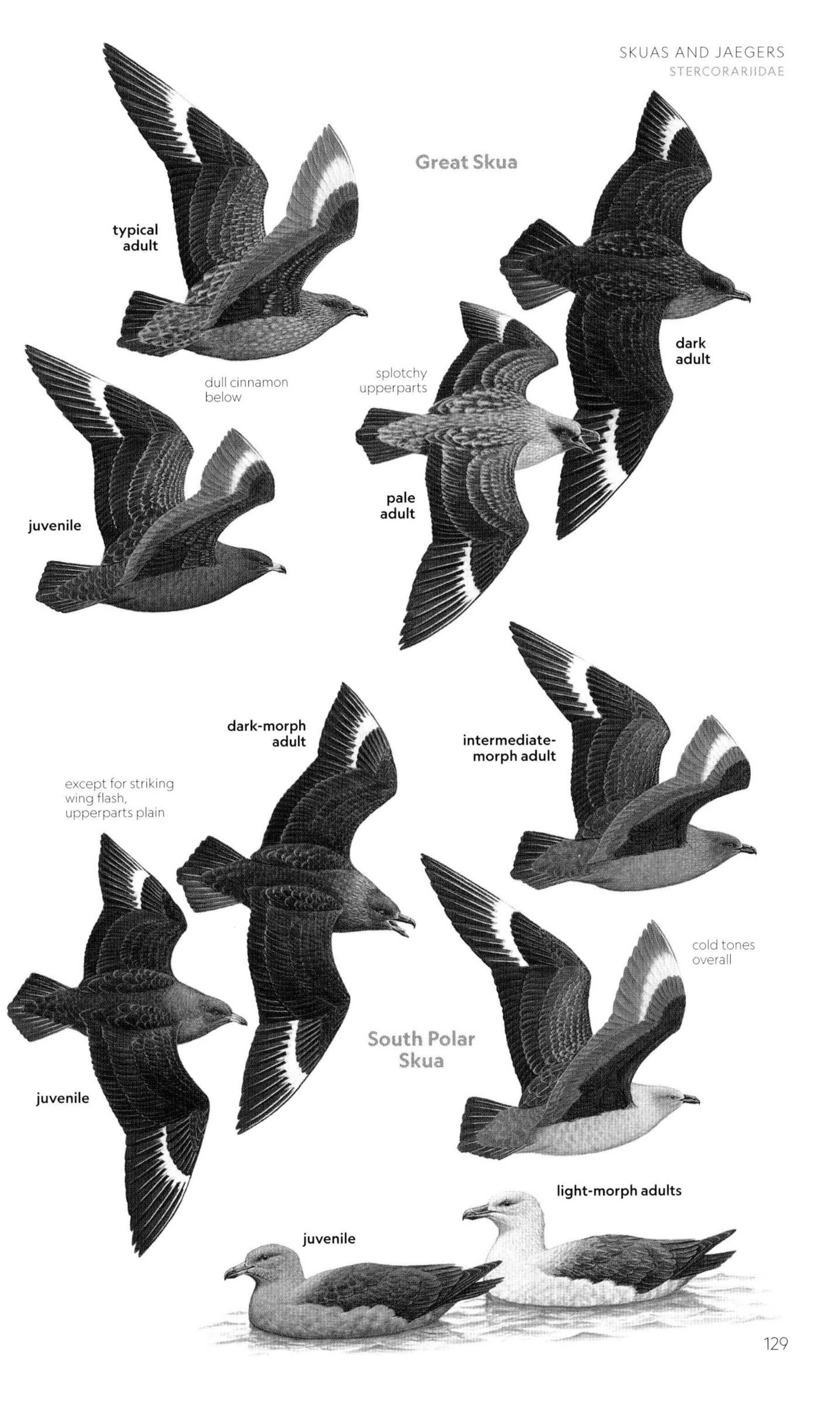
Great Skua
typical adult
dark adult
dull cinnamon below
splotchy upperparts
pale adult
juvenile
dark-morph adult
intermediate-morph adult
except for striking wing flash, upperparts plain
cold tones overall
South Polar Skua
juvenile
light-morph adults
juvenile

## LARGER JAEGERS

"Jaeger" is from the German word for "hunter." Like the hulking skuas (p. 128), these two chase gulls and other seabirds, forcing them to disgorge their prey. Both species occur inland in small numbers, mostly on fall migration.

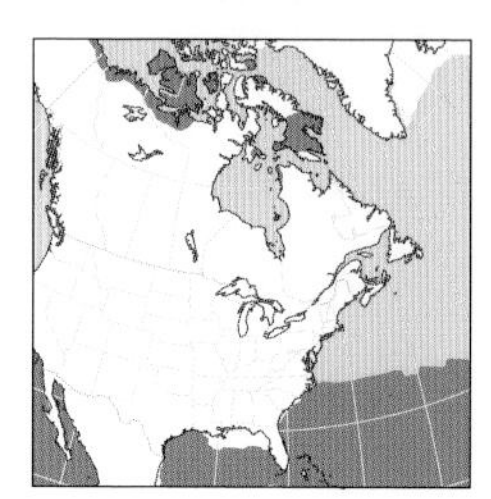

### Pomarine Jaeger

*Stercorarius pomarinus* | POJA | L 21" (53 cm) WS 48" (122 cm)

The name of this bird derives from Greek words meaning "lid" and "nose." The upper mandible of all three jaegers is covered in a lid-like sheath.

■ **APPEARANCE** Largest and heftiest jaeger; the barrel-chested "Pom" is big-billed, with prominent bulbous expansion (the gonys) of lower mandible. Light-morph adult has extensive black cap and smudgy breastband; uncommon dark morph mostly blackish. Central tail feathers of adult, often broken or missing, are rounded. Juvenile dusky all over like other jaegers; from below, note white at base of primaries *and* primary coverts, creating double wing flash.

■ **VOCALIZATIONS** Silent when hunting or resting at sea, but in presence of other seabirds, sometimes gives disyllabic yapping notes.

■ **POPULATIONS** Nonbreeders and migrants occur well offshore year-round, but bias toward summer sightings when most boat trips go out. Quite rare but regular migrant on Great Lakes in fall; most sightings later in season than those of the smaller jaeger species.

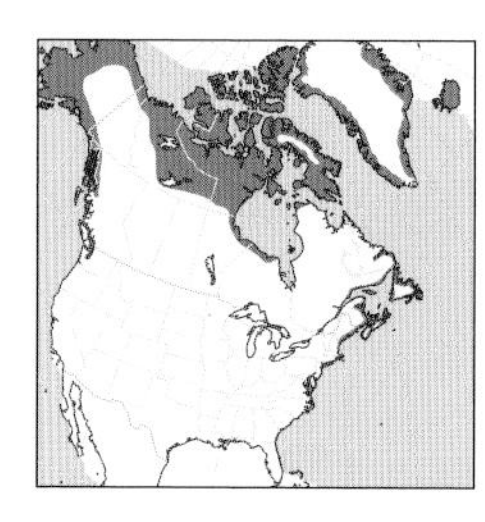

### Parasitic Jaeger

*Stercorarius parasiticus* | PAJA | L 19" (48 cm) WS 42" (107 cm)

This might be the most jaeger-like of all the jaegers, the species likeliest to be seen in close pursuit of fish-bearing gulls and terns. It is ravenous on the breeding grounds, where it feeds on rodents, birds, and even berries.

■ **APPEARANCE** Potbellied and thin-billed; intermediate in structure between Long-tailed (p. 132) and Pomarine Jaegers. Sharp central tail feathers of adult often broken or missing; light and dark morphs dramatically different, but many adults intermediate; some have subtly warm tones. Juvenile difficult to ID in the field, but good views of the wings in flight can be diagnostic: lacks double wing flash below of Pomarine, flashes more white above than Long-tailed.

■ **VOCALIZATIONS** Occasional yips and yaps in mixed-species assemblages at sea, but less inclined than Pomarine to vocalize.

■ **POPULATIONS** The jaeger most likely to be seen from beaches and headlands, especially in onshore winds. Very rare but regular on Great Lakes on fall migration; most pass through before Pomarines. Migration offshore protracted; most off Canada in summer are migrants and nonbreeders.

### Identifying Jaegers

At-sea identification of jaegers is, in a word, challenging. A typical sighting is of a distant bird in flight—and from a boat, of course. Even the trip leaders often find themselves in disagreement about the field ID of a fast-flying jaeger. Impressions of size and shape—or gestalt—are important cues, but they are not definitive. Good looks at the primaries and primary coverts, from above and below, can in fact be diagnostic; but lighting, viewing angle, and wishful thinking, not to mention distance, complicate matters.

Take pictures! Digital photos routinely confirm details of bill structure, tail shape, and wing pattern that are often indiscernible to the naked eye or, if discernible, troublingly subjective in the field. It is gratifying that many jaegers can be confidently identified from photos but sobering to realize that many cannot be.

ht-morph
breeding
adult
dark cap
extends
well below
base of bill
dark-morph
breeding adult
double
wing
flash
light-morph
breeding
adult
long, twisted
("spoon-shaped")
tail feathers
Pomarine
Jaeger
light-morph
juvenile
light-morph
1st summer
light-morph
juvenile
short, rounded
central tail points

ght-morph
breeding
adult
pale at base
of forehead
dark-morph
breeding
adult
light-morph
1st summer
ght-morph
breeding
adult
sharply pointed
tail feathers
Parasitic Jaeger
light-morph
juveniles
light-morph
juvenile
often
warm-toned
pale primary tips

## A JAEGER AND AN AUK

Other than the fact that they spend much of their lives at sea, the piratic jaegers and skuas (pp. 128–133) and bewitching alcids (pp. 132–137) would seem at first glance to have little in common. As it turns out, they are each other's closest relatives.

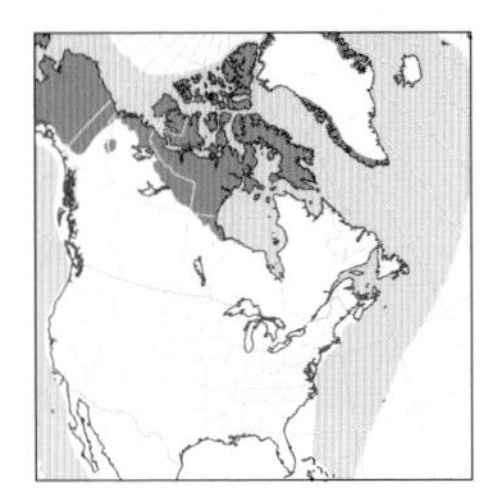

## Long-tailed Jaeger

*Stercorarius longicaudus* | LTJA | L 22" (56 cm) WS 40" (102 cm)

Discounting its long central tail streamers, this is the smallest of the jaegers. It is also the best behaved, only rarely chasing down other seabirds for their prey.

■ **APPEARANCE** Petite, with small bill, narrow wings, and long tail; upperparts of adult cold gray. In all plumages, flashes less white in wing than larger jaegers, especially above; typically shows only two or three white primary shafts above (other jaegers, five or more). Very long central tail streamers of adult are diagnostic, but often broken or not fully grown; adult's dark remiges contrast with paler upperwing coverts. Juvenile, ranging from dark brown to quite gray, extensively barred.

■ **VOCALIZATIONS** Rising, nasal honks, *aarr* and *urrnk*, given very infrequently away from breeding grounds.

■ **POPULATIONS** Leaves Arctic breeding grounds in late summer; sightings at sea and inland correspondingly early in summer and fall.

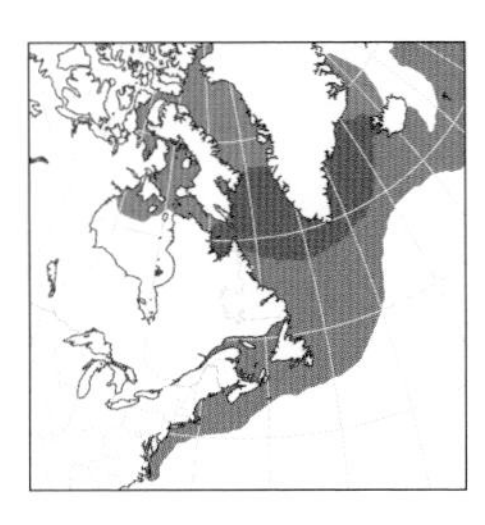

## Dovekie

*Alle alle* | DOVE | L 7¾" (20 cm)

Although it is the smallest alcid in the East, this species may be the hardiest of the lot. It breeds the farthest north, migrates the greatest distances, and winters mostly far offshore.

■ **APPEARANCE** Chubby overall and stub-billed; not even one-quarter the heft of the larger Razorbill or murres (p. 134). Almost all sightings in our area are of juveniles and nonbreeding adults, blotchy black and white above, white below.

■ **VOCALIZATIONS** Giddy wailing and chippering at the huge nesting colonies, but almost never heard at sea.

■ **POPULATIONS** Essentially a winter visitant to the East; small numbers breed on Baffin I., but most birds in our waters hail from Greenland and points east. Highly pelagic, but singles and small flocks sometimes come in as close as the surf.

### Pan-Alcidae

Watch a jaeger chasing terns, and it is natural to compare the spectacle to a young Herring Gull (p. 146) doing likewise. Indeed, the skuas and jaegers, now in the family Stercorariidae, were formerly treated as a subfamily in the family Laridae, which includes the gulls. Then watch a group of murres and Razorbills standing proud on sea rocks or propelling themselves underwater. Penguins leap to mind, and an evolutionary history with that group was in fact posited by earlier researchers.

Multiple lines of molecular evidence point to the current view: The alcids are "siblings" to the skuas and jaegers, meaning the two groups are more closely related to each other than to any other group. But it's not just DNA. The paleontological record—in the form of both fossils and paleoclimatic analysis—affirms the relationship. The two groups share a common ancestor somewhere back in the mid-Tertiary period, when Earth was warmer. They have diverged greatly in morphology, yet they share the basic strategy of breeding mostly at high latitudes and wintering at sea. But what to call them? The name Pan-Alcidae is sometimes used but hasn't really caught on. The key point is that they aren't as different as they might seem.

Long-tailed Jaeger
2–3 white primary shafts
breeding adult
gray upperparts
1st summer
pointed tail feathers
breeding adult
light-morph juvenile
dark-morph juvenile
rounded tail points
cool-toned
typical juvenile

Dovekie
winter
dark underwing
breeding adult
breeding adult
winter
short, stubby bill

## TALL BLACK-AND-WHITE ALCIDS

These three alcids stand upright on ledges and sea rocks around coastal headlands. At-sea identification in winter is tricky, requiring good looks at bill shape, tail shape, and plumage pattern.

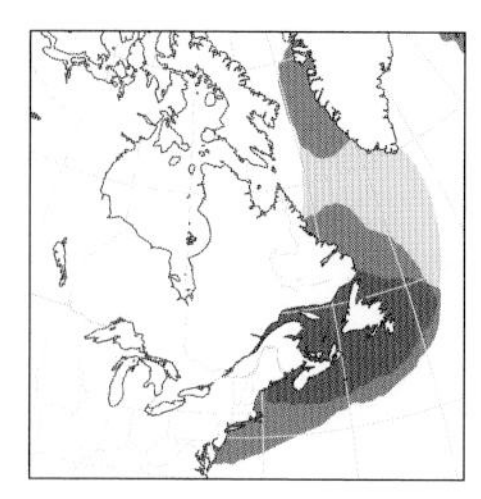

### Common Murre

*Uria aalge* | COMU | L 17½" (44 cm)

This species and the similar Thick-billed Murre are the tallest and heaviest extant alcids. Only the massive Great Auk (p. 426) was larger.

■ **APPEARANCE** All have long, thin bill; bill of first-winter birds, often seen at sea, not quite as long. In winter plumage, mostly white face marked by thin black line curving back from eye. Common Murre not as gleamingly black-and-white as Thick-billed; flanks of Common often mottled dusky. Breeding adults dark-headed, some with a "bridle": a thin white eye ring and thin line behind eye.

■ **VOCALIZATIONS** In disputes at densely crowded colonies, gives tremendous groans. Adults silent at sea, but recently fledged young solicit parents with pure *peedee* notes, often doubled or trebled.

■ **POPULATIONS** The more southerly of the murres, outnumbering Thick-billed at the famous and accessible colonies of the Atlantic provinces. Also outnumbers Thick-billed on pelagic trips off U.S. East Coast.

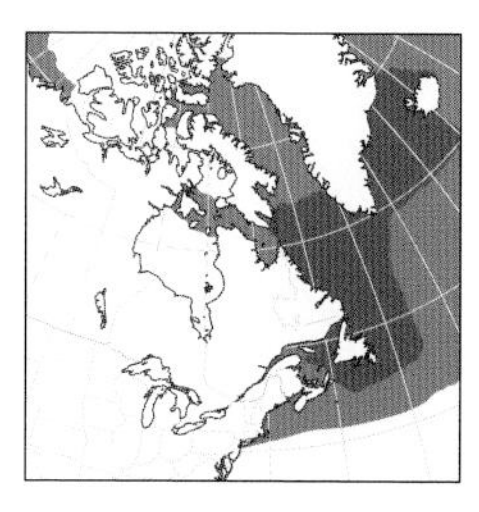

### Thick-billed Murre

*Uria lomvia* | TBMU | L 18" (46 cm)

This more northerly of the two murres generally prefers colder and deeper waters than its thin-billed congener. But the two mix freely at sea and nest side by side on sea cliffs.

■ **APPEARANCE** In good view, bill of adult shorter and thicker than that of Common; but beware age-related variation. Adult Thick-billed, especially in breeding plumage, glistens jet-black above; Common browner and paler. Flanks of Thick-billed pure white; Common often smudgy. Winter Thick-billed black-faced; winter Common extensively white-faced with thin black line behind eye.

■ **VOCALIZATIONS** Like Common, erupts in roars and groans at colonies; recently fledged young at sea whistle for parent.

■ **POPULATIONS** Away from breeding grounds seen mostly well out at sea, rarely from shore. Curiously, until early 20th century, was rare but probably regular in late fall inland to eastern Great Lakes region.

### Razorbill

*Alca torda* | RAZO | L 17" (43 cm)

Puffins (p. 136) get all the publicity for their remarkable bills, but this species deserves commendation too. Like puffins, Razorbills are able to capture and carry multiple fish.

■ **APPEARANCE** Bill of adult laterally compressed, with vertical grooves and bold white stripe. A tad smaller than murres and notably longer-tailed. Seasonal variation in plumage mirrors murres: breeding plumage clean black-and-white; winter adults more dusky-headed, with smudgy dark line behind eye.

■ **VOCALIZATIONS** Growls like murres, but less declamatorily. Solicitation whistle of recently fledged young more drawn out than murres' vocalizations.

■ **POPULATIONS** Winters farther south than murres and is considerably more likely to be seen from shore; found around jetties and breakwaters. The Razorbill is the closest living relative of the extinct Great Auk.

## Common Murre

breeding adult

matte-gray upperparts

"bridled" breeding adult

breeding adult

dark center to underwing

winter

dusky flanks

1st winter

winter adult

pure white flanks

## Thick-billed Murre

often has white line on bill

breeding adults

white underwing

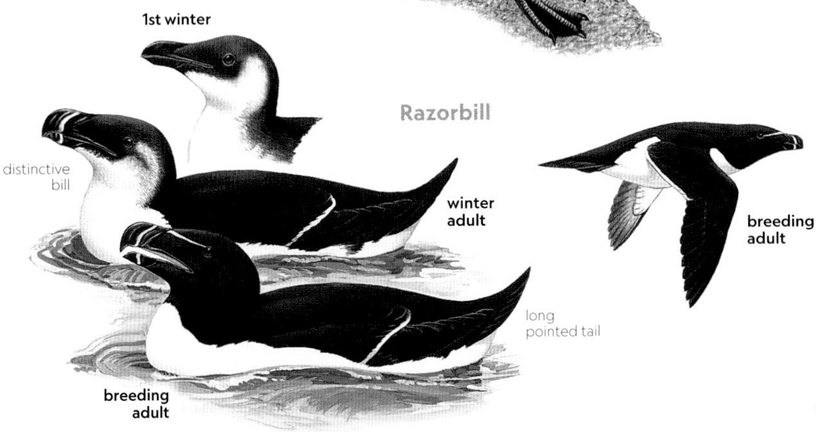

## Razorbill

1st winter

distinctive bill

winter adult

long pointed tail

breeding adult

breeding adult

## DISTINCTIVE ALCIDS

These two are our most strikingly accoutered alcids. The breeding guillemot is jet-black with scarlet feet, while the puffin may well have the most famous bill of any bird in the East.

### Black Guillemot

*Cepphus grylle* | BLGU | L 13" (33 cm)

The least pelagic of our alcids, this species is often sighted from land—sometimes *on* land—in winter. Look for it on sea rocks and even earthen embankments out of the reach of the surf.

■ **APPEARANCE** Stately, like murres (p. 134), but not as tall; bill fine and tapered. Breeding adult gleaming black with huge white oval on wing; feet brilliant red-orange. Winter adult messy, mottled black above, but white oval on wing stands out.

■ **VOCALIZATIONS** Unlike many other alcids, gives high-pitched calls, including piercing *twseet*, often in rapid series, and a long, reedy whistle.

■ **POPULATIONS** Readily seen in winter in small numbers south to eastern Mass. Almost all wintering off northeastern U.S. are the darker-backed subspecies *arcticus*. Subspecies *mandtii*, with breeding range that includes eastern Canadian Arctic, is ghostly white above in winter; very rare or absent off East Coast in winter.

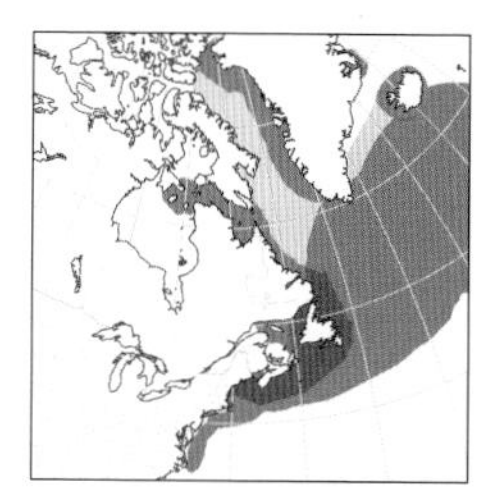

### Atlantic Puffin

*Fratercula arctica* | ATPU | L 13¾" (35 cm)

The iconic "Sea Parrot" is widely known. Trips to see puffins in summer are big business for regional economies, but the species is actually more widespread in the East in winter, when it occurs across a large swath of the North Atlantic.

■ **APPEARANCE** Bill of adult laterally compressed and brightly colored. Winter adult's bill less colorful; juvenile's bill duller still and not as huge. All are black above and white below, with orange feet. Whitish face of breeding adult becomes dusky gray in winter; juvenile also has dusky face.

■ **VOCALIZATIONS** Low-pitched; long moans and shorter snarls. More nasal and less ferocious than murres.

■ **POPULATIONS** Winters widely at sea, mostly well offshore. Despite immense popularity today, was persecuted well into 20th century; still recovering from earlier losses.

### Convergent Evolution

Take a look at a museum specimen of the extinct Great Auk (p. 426), and it is impossible not to think of a penguin. In fact, the English word "penguin" was first given to that species, only later to be applied to the "real" penguins of the Southern Hemisphere. The auk stood more than 2 ft (0.6 m) tall and weighed 12 lb (5.4 kg). It could "fly" underwater with its wings, it laid a single egg, and it hung out in dense colonies. Elegantly black and white, the auk was flightless. No wonder early ornithologists thought those flightless southern birds were the "penguins" they knew from the North Atlantic.

But penguins and alcids aren't closely related. A close affinity between the alcids and the skuas is well supported (see sidebar, p. 132), while recent research indicates a connection between the penguins and the tubenoses (pp. 168–177). What's at play here is convergent evolution, the process whereby similar environments promote development of similar traits among unrelated species or groups of species. The ancestors of both alcids and penguins inhabited cold seas and austere coasts at high latitudes, where the seas were filled with fish, and the rugged coasts are protected from predators. Given tens of millions of years of evolution, the two groups have converged on functionally identical strategies for flourishing in their respective hemispheres.

Black Guillemot
breeding adult
high Arctic
mandtii
1st winter
winter adult
white underwing
winter adult
averages darker than mandtii
winter adult
juvenile
breeding adult
eastern
arcticus
breeding adult
Atlantic Puffin
breeding adult
dusky face
juvenile
winter adult

## SEAFARING SMALL GULLS

The moniker "seagull" is a stretch for the many gull species that occur inland continent-wide. But it applies well in the case of these three, breeding near coasts and wintering at sea.

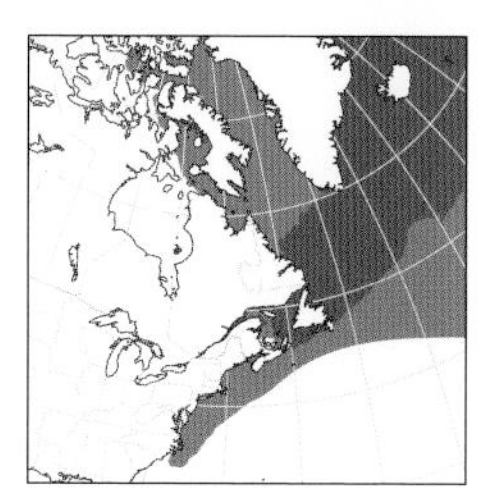

### Black-legged Kittiwake

*Rissa tridactyla* | BLKI | L 17" (43 cm) WS 36" (91 cm)

The typical kittiwake spends its entire life within earshot of the ocean. The species nests on sheer sea cliffs and winters widely offshore.

■ **APPEARANCE** Slightly smaller than Ring-billed Gull (p. 144); wingbeats stiff, ternlike. All plumages have eponymous black legs. Breeding adult has pure-white head and unmarked yellow bill; winter adult has smudgier head, typically with vertical gray smudge behind eye. Adult pale gray above; in flight, black primary tips contrast sharply with rest of wing. Distinctive first-winter bird has thick black collar and black vertical mark behind eye; in flight shows prominent M above and black-tipped tail.

■ **VOCALIZATIONS** Short nasal notes, steady or rising in pitch: *ehnk?* and *eenk?* In excitement around nests, *eenie-eenk!* ("kitti-wake"), given repeatedly. Silent except at colonies.

■ **POPULATIONS** Pelagic species; rare but regular inland continent-wide, mostly late fall, especially Great Lakes. Less rare in Midwest than in Appalachian region and Southeast.

### Ivory Gull

*Pagophila eburnea* | IVGU | L 17" (43 cm) WS 37" (94 cm)

One of the most extreme birds on the planet, normally spending its entire life around Arctic pack ice, where it feeds on offal and keeps company with polar bears. Bulky for its size; short-billed and black-legged. Adult feathers pure white, bill yellow-gray; first-winter mottled dusky above. Sightings south to U.S. in winter thrill birders but are problematic: Arctic warming imperils the species, and Ivory Gulls well out of range are climate refugees.

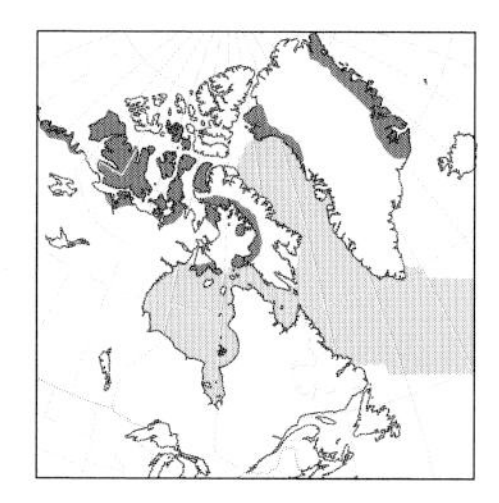

### Sabine's Gull

*Xema sabini* | SAGU | L 13½" (34 cm) WS 33" (84 cm)

This distinctively patterned gull is odd in various respects. It winters in tropical seas, its slightly forked tail is atypical for a gull, and details of its developmental biology are notable.

■ **APPEARANCE** Small; no larger than a Bonaparte's Gull (p. 140). In flight, our most distinctive gull, flashing broad swaths of dark and white, diagnostic at tremendous distances. Adult's bill dark with yellow tip; adult has slaty hood in summer, smudgy half hood by fall. Juvenile plumage, held well into fall, brown above with fine white barring. Notched tail of adult white; tail of juvenile, also notched, has black tip. Precocial young can fly before fully feathered. Juvenile and nonbreeding adult Sabine's Gulls sitting on water suggest Franklin's and Laughing Gulls (p. 142), but when the bird puts into flight, the ID is a cinch.

■ **VOCALIZATIONS** All calls grating and ternlike, some notes rising, others falling. Noisy during breeding season, calls occasionally at sea.

■ **POPULATIONS** Migrates off both coasts, but less common off Atlantic than Pacific; occurrences inland in East chiefly in fall, especially Great Lakes and Great Plains westward. Most sightings inland are of juveniles.

Black-legged Kittiwake
unmarked yellow bill
winter adult
black collar
juvenile
breeding adult
dark M pattern across wings
tip of tail black
juvenile

Ivory Gull
adult
dusky face
completely white
adult
1st winter
legs and feet black
2nd cycle adult
Sabine's Gull
breeding adult
distinct upperwing pattern
breeding adult
scaly back pattern
juvenile
juvenile

## SMALL DARK-HOODED GULLS

A very general rule of thumb with gulls is that the larger species are white-headed as adults, the smaller species dark-hooded. There are exceptions, but these three fit the model nicely.

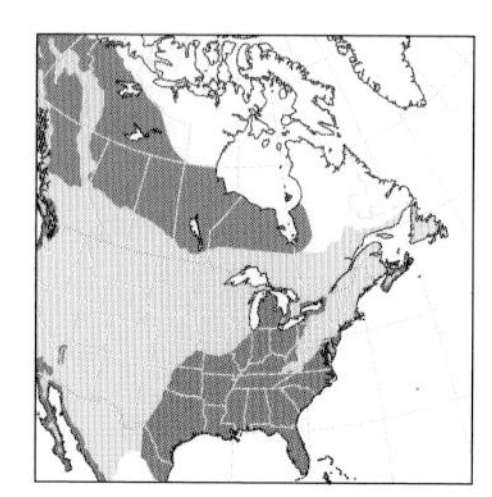

### Bonaparte's Gull

*Chroicocephalus philadelphia* | BOGU | L 13½" (34 cm) WS 33" (84 cm)

In winter, this is the widespread and well-known "bay" gull, feeding on small fish in deepwater habitats, both freshwater and marine. In summer, the species nests in forests—and actually *in* trees—in the company of warblers, jays, and thrushes!

■ **APPEARANCE** The smallest gull likely to be encountered in most birding situations in the East; appears dainty both on the water and in fluttering flight. Slight bill is black in all plumages; legs red on breeding adult, duskier on immature and winter adult. Breeding adult has black hood; mostly white-headed immature and nonbreeding adult have black blob behind eye. Wings distinctive in flight but differ greatly between adult and first-winter: Adult has flickering wedge of white on outer half of wing; first-winter has black M across upperwing like Black-legged Kittiwake (p. 138).

■ **VOCALIZATIONS** Harsh, rising snarls. "Boneys" are mostly quiet in winter but sometimes squabble at feeding frenzies.

■ **POPULATIONS** Winter distribution inland strongly tied to availability of open water and small fish, especially gizzard shad, genus *Dorosoma*.

### Black-headed Gull

*Chroicocephalus ridibundus* | BHGU | L 16" (41 cm) WS 40" (102 cm)

One of the most common gulls in Europe, this species is a scarce winter visitant along our coasts from the mid-Atlantic northward. It and the smaller Bonaparte's Gull are the only *Chroicocephalus* gulls in the East.

■ **APPEARANCE** Like an oversize Bonaparte's Gull. Long bill is red on adult, paler pink on immature; bill of Bonaparte's smaller, usually black. Ironically, hood of adult dark brown; hood of Bonaparte's black. Adult in flight shows white wedge on upperwing like Bonaparte's, but note black below, lacking on Bonaparte's. First-winter bird has black smudge behind eye like Bonaparte's, but wing is different: upperwing with broader black trailing edge and smudgy, ill-defined carpal bar; remiges dark below.

■ **VOCALIZATIONS** Ternlike squawks, higher than Bonaparte's.

■ **POPULATIONS** Reached N. Amer. and began breeding in mid-20th century; despite earlier predictions of a takeover by this adaptable gull, remains uncommon. A few nest here, mostly Nfld.

### Little Gull

*Hydrocoloeus minutus* | LIGU | L 11" (28 cm) WS 24" (61 cm)

Like the Black-headed Gull, this is a chiefly Old World species with a tenuous presence in the East. But it has penetrated farther inland, with most breeding records from the Great Lakes north to Hudson Bay.

■ **APPEARANCE** The tiniest gull on Earth, notably smaller even than a Bonaparte's. Breeding adult has dark hood and unique wings: pale gray above with white wing tips; black below, also with white wing tips. Winter adult has smudge behind eye like Bonaparte's. First-winter in flight like first-winter Black-legged Kittiwake, but smudgier overall.

■ **VOCALIZATIONS** Calls higher, more strident than those of larger Bonaparte's; closer in timbre to Laughing and Franklin's (p. 142).

■ **POPULATIONS** Nests in freshwater marshes well inland. Away from breeding grounds, to be looked for in larger gatherings of Bonaparte's.

Bonaparte's Gull
breeding adult
black bill
winter adult
winter adult
white primaries with thin black tips
1st winter
broken M pattern across wings
pale inner primaries
1st winter
juvenile

dark brown hood
breeding adult
Black-headed Gull
1st summer
red bill
dark underside to primaries
winter adult
dark inner primaries
winter adult
1st winter
1st winter

2nd winter
breeding adult
1st winter
Little Gull
winter adult
dark underwings
bold M pattern across wings
dark cap and hindcollar
breeding adult
1st winter
extensively dark coverts
juvenile

## GENUS *LEUCOPHAEUS*

This is a Pan-American genus, its five species found regularly only in the New World. The two species in our region are powerful fliers with dark backs and evocative calls. Adults have black hoods with white eye arcs, suggesting an executioner's mask.

### Laughing Gull

*Leucophaeus atricilla* | LAGU | L 16½" (42 cm) WS 40" (102 cm)

Spend any amount of time at all around public beaches in summer on the Jersey Shore, and you will absolutely make the acquaintance of this assertive, adaptable, beautiful gull.

■ **APPEARANCE** Larger than Franklin's Gull. Long, drooping bill is red-black in adult; legs black in all plumages. Eye arcs of adult not as prominent as Franklin's. All are long-winged, with primaries dark below; Franklin's primaries paler below. On adult in flight, black wing tips blend smoothly with dark gray of rest of wing. Juvenile and first-winter smudgy brown, including breast. In winter, adults and immatures alike are dusky-headed, lacking prominent half hood of Franklin's.

■ **VOCALIZATIONS** Short nasal squeals, often run together in guffawing series.

■ **POPULATIONS** Nests in salt marshes, dispersing to piers and beaches to feed. Limit of winter range apparently determined by availability of aerial plankton. Wanders well inland, especially spring and summer.

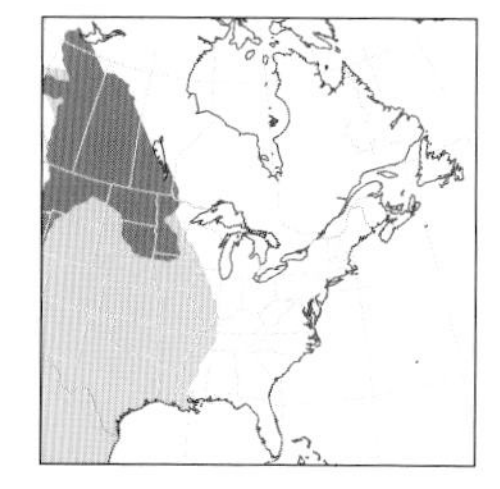

### Franklin's Gull

*Leucophaeus pipixcan* | FRGU | L 14½" (37 cm) WS 36" (91 cm)

In the U.S. and Canada, this is the prairie gull. It is often found in agricultural districts on migration, mostly around cattail marshes well inland during the breeding season.

■ **APPEARANCE** Bill slighter, wings proportionately shorter, and body smaller overall than Laughing. Bill of adult dark red; breast tinged pink in spring. Black outer primaries with white tips separated from gray mantle by white bar. Primaries from below pale in all plumages; darker in Laughing. Eye arcs of adult more prominent than on Laughing. Juvenile and first-winter mostly white on belly and breast; Laughing duskier below. All winter plumages show prominent half hood, largely absent in winter Laughing.

■ **VOCALIZATIONS** Nasal squeals like Laughing, but higher, more strident and urgent sounding.

■ **POPULATIONS** Winters along Pacific coast of S. Amer., with huge concentration around Lima, Peru. Particularly pink birds acquire color by gorging themselves at commercial shrimp farms on wintering grounds.

### Why Birds Molt

Many birds don't migrate. And some forgo reproduction for a year or more. But essentially *all* birds molt at least annually. Feathers wear out from the ardors of flight, exposure to the elements, and simple mechanical failure. So birds *have to* replace their feathers, and most birds in the Northern Hemisphere do so by means of a complete annual molt sometime from midsummer to late fall.

Gulls are no exception, and, like many other birds, gulls actually have two molts: the complete annual molt just mentioned plus a partial molt in winter. But the Franklin's Gull does things differently. Only two birds in the U.S. and Canada have two complete annual molts: the Bobolink (p. 362) and the Franklin's Gull, not closely related. Both are tremendous migrants, occurring year-round in open habitats with intense sunlight. Just one complete annual molt is evidently insufficient for their physically demanding lifestyles. Far from being arbitrary, species-specific molt strategies are in many instances logically related to a bird's life history.

Laughing Gull
winter adult
breeding adult
all-dark primaries
2nd winter
1st winter
breeding adult
2nd winter
hood indistinct
gray across breast and sides
juvenile
sludge-brown plumage
1st winter

Franklin's Gull
black-and-white wing tip pattern
1st summer
bold eye arcs
breeding adult
breeding adult
winter adult
bold half hood
1st winter
juvenile
1st winter

## SMALL WHITE-HEADED GULLS

Gulls in the genus *Larus* (pp. 144–151) are white-headed as adults and generally large. These three are the smallest of the lot but are nevertheless similar in overall morphology to the rest of the genus.

### Common Gull

*Larus canus* | COGU | L 17¼–19½" (44–50 cm) WS 46¼–52½" (117–133 cm)

Rare winter visitor from Europe to New England and Atlantic Canada; intermediate in plumage and morphology between Short-billed and Ring-billed. All are small-bodied with slight bills. Adult has dark eye, unmarked dusky-yellow bill. Adult has more black in wing tips than Short-billed; first-winter has high-contrast plumage like first-winter Ring-billed. Like Short-billed, rides higher on water than Ring-billed.

### Short-billed Gull

*Larus brachyrhynchus* | SBIG | L 16–18½" (41–47 cm) WS 43½–49½" (110–126 cm)

A gull of western N. Amer.; rare but regular in winter east to prairie states and provinces and Great Lakes. Marginally smaller than Common Gull. Like adult Common, adult has dark eye and slight bill, unmarked yellow. Adult's mantle darker than Common's; dark wing tips separated from mantle by broad white crescent. First-winter dusky overall, less contrastingly patterned than Common or Ring-billed.

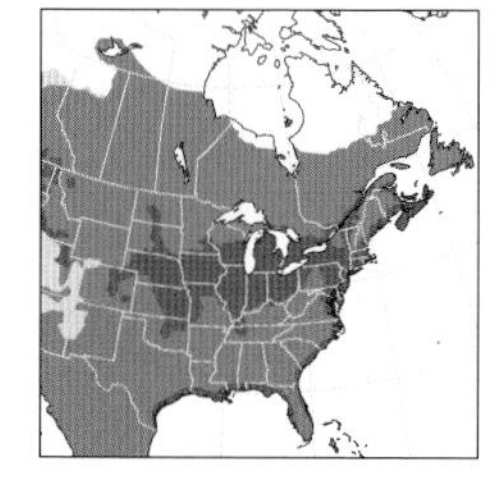

### Ring-billed Gull

*Larus delawarensis* | RBGU | L 17½" (44 cm) WS 48" (122 cm)

In many birding situations, especially inland, this is the default gull. Learn the plumages of this species and the larger Herring Gull (p. 146), and you will be well poised for identifying other species in the genus *Larus*.

■ **APPEARANCE** Adult has yellow bill with eponymous black ring; legs yellow, eyes pale, mantle gray. Juvenile uniformly brown; by late fall, gray and splotchy overall, with broad black tail band and boldly patterned wings. By second summer (one year old), plumage is still splotchy but getting grayer above. By second winter, resembles adult, except black wing tips lack prominent white spots.

■ **VOCALIZATIONS** Slurred monosyllabic squeals, nasal and distressed-sounding: *hyeee, hyee-ay,* etc. Also a lower *hyow,* given singly or in series, especially in aggression.

■ **POPULATIONS** Following severe human persecution around the turn of the 20th century, has rebounded to become one of our most abundant gulls. Nonbreeders can be seen in summer throughout East, especially coastally.

#### Vagrancy

When people speak of a rare bird, they often have in mind a vagrant, a species far out of range. Some species—and some whole families—are much more prone to vagrancy than others: e.g., gulls, which are adaptable, long-winged, and highly migratory. Accurately identifying a rare gull requires understanding when and where vagrants occur and, even more so, the variation between and within species. During the winter months, there are bound to be a few Common and Short-billed Gulls in the East, but usually not in the same place. Common Gulls are most likely in Atlantic Canada, Short-billed Gulls from the Rockies east to the Great Lakes. If you've got a candidate for one of these two species, the most critical step is ruling out the more likely Ring-billed Gull. It helps to study a presumptive Common or Short-billed in *direct comparison* with a Ring-billed; color and pattern, and especially size and shape, are difficult to evaluate on a bird by itself. And take photos—always a valuable corrective to the more subjective in-person observations.

Common Gull
winter adult
spotted pattern on head and breast
1st winter
adult
extensive black on next-to-outermost primary
brown underwings
st winter
Short-billed Gull
winter adult
2nd winter
breeding adult
dovelike head shape
breeding adult
juvenile
mantle darker than Ring-billed or Common
smooth brown plumage
1st winter
winter adult
pale eye
2nd winter
breeding adult
1st winter
2nd winter
pale underwings
juvenile
Ring-billed Gull
1st winter
tail pattern variable
breeding adult

## LARGE WHITE-HEADED GULLS

Large gulls in the genus *Larus* (pp. 144–151) are closely related. Species limits, the boundaries between species, are fuzzy, and variation *within* species is considerable.

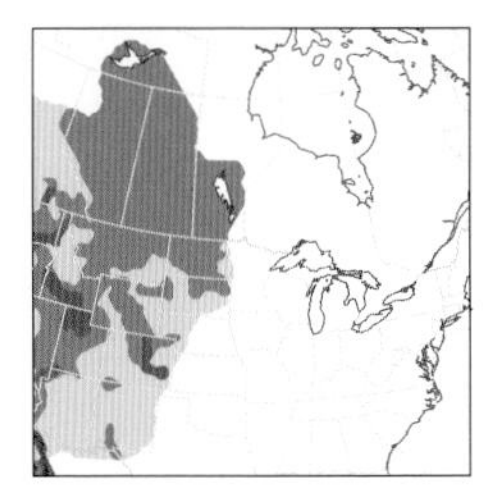

### California Gull

*Larus californicus* | CAGU | L 21" (53 cm) WS 54" (137 cm)

This is the gull that rescued Mormons from a grasshopper plague in Utah in the 19th century. "Cal Gulls" in our century still eat grasshoppers, along with anything else, animal or vegetable, alive or dead—or processed.

■ **APPEARANCE** Intermediate in heft between Short-billed (p. 144) and Herring Gulls; adult nearly as dark-mantled as Short-billed. Adult has dark eye and dull yellow-green legs; yellow bill has adjoining black and red spots. First-year birds resemble subadult Herring, Lesser Black-backed (p. 148), and Ring-billed (p. 144); assessing size and shape in direct comparison with other gulls is useful. First-year California has pale legs and pale bill with dark tip, pale head and breast, and, in flight, double dark bars across inner wing. By second and third winters, begins to acquire adult characters: dark eye, yellow-green legs, medium-gray mantle.

■ **VOCALIZATIONS** Squeals midway in timbre between squeakier Ring-billed's and gruffer Herring's squeals. Also gives a low *owl,* often doubled or trebled, a good indication of its presence amid Ring-billeds.

■ **POPULATIONS** Regular on migration across western Great Plains, where flocks numbering into the low hundreds can be found Apr. to May and late summer. Annual vagrant to East Coast.

### Herring Gull

*Larus argentatus* | HERG | L 25" (64 cm) WS 58" (147 cm)

In the mind's eye of the general public, this is *the* seagull: big and brutish, heard day and night around beaches and marinas. It also occurs well inland.

■ **APPEARANCE** Adult has pale mantle, dark wing tips, pink legs, yellow eye, and yellow bill with red spot; white-headed in breeding plumage, streaky-headed in winter. Younger Herrings easily confused with other large gulls. First-winter in flight shows pale inner primaries, an excellent mark visible at some distance; compare with first-winter Lesser Black-backed. By second winter, most show pale eye and pale gray mantle of adult.

■ **VOCALIZATIONS** Anguished squeals, given singly or in series, average lower than those of Ring-billed. Recent work suggests consistent spectrographic differences among large gull species, but more study is needed.

■ **POPULATIONS** "European Herring Gull," *argenteus,* is probably annual in winter to Nfld.; it is considered by many to be a different species from our "American Herring Gull," *smithsonianus.*

### Is it "Just" a Herring Gull?

Birds in the large white-headed gull (LWHG) complex are prone to vagrancy (see sidebar, p. 144) and thus of keen interest to birders in search of rarities. Typical adults are usually, although not always, straightforward to ID. The challenge is with notoriously variable younger birds. On seeing a possible LWHG rarity, the essential first question to ask is this: Is it "just" a Herring Gull? Or is it a rare look-alike?

First-winter Herrings can be as light as some first-winter Icelands (p. 148) and easily as dark as first-winter Lesser Black-backeds. Wear and sun bleaching, especially from late winter through summer, can make any LWHG almost as pale as a Glaucous (p. 150). Resist the temptation to identify LWHGs by plumage alone; study size and shape, especially in direct comparison with other gulls, and obtain photos of any presumptive rarity.

double dark bars
across wings
breeding
adult
1st winter
primaries
entirely
dark
California
Gull
2nd
winter
breeding
adult
winter
adult
dark smudgy
underparts
1st
winter
pale
juvenile
dark
juvenile

1st winter
European
argenteus
breeding
adult
Herring Gull
smithsonianus
pale inner
primaries
juvenile
1st winter
winter
adult
3rd
winter
often pale-
headed
2nd
winter
1st
winter
pale base
bill forms
early on
1st
winter

## HERRING GULL LOOK-ALIKES

These midsize members of the large white-headed gull (LWHG) complex are often distinctive as adults, but younger birds may closely resemble younger Herring Gulls (p. 146).

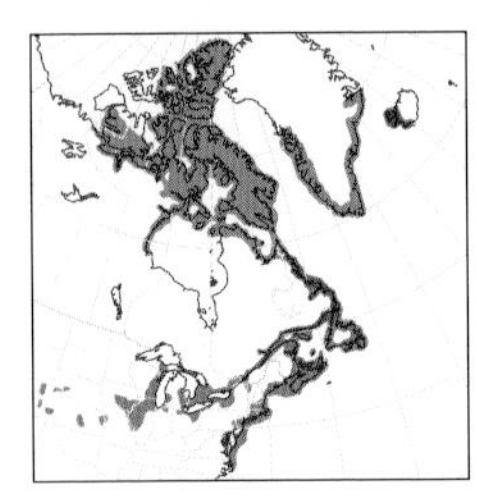

### Iceland Gull

*Larus glaucoides* | ICGU | L 22" (56 cm) WS 54" (137 cm)

This is the smaller of the two so-called white-winged gulls commonly encountered in the East, the other being the hulking Glaucous (p. 150). Despite that moniker, many Icelands show some dark in their wing tips.

■ **APPEARANCE** Like a slimmed-down Herring Gull (p. 146). Most adults in East are subspecies *kumlieni,* with pale eye, slight bill with red spot, and white wing tips with sparse gray markings. First-winters vary from fawn-brown to nearly white, inviting confusion with both Herring and Glaucous Gulls; note Iceland's smaller build and slighter bill, mostly dark. By second winter, begins to acquire paler eyes and gray mantle.

■ **VOCALIZATIONS** Squeals about the same pitch as Herring's, but not as nasal.

■ **POPULATIONS** Recently lumped with "Thayer's Gull," which is now treated as a subspecies of Iceland, *thayeri.* "Thayer's" winters mostly in the West, but is regular in winter east at least to Great Lakes. It is intermediate in size, shape, and color between Herring and other Icelands; however, adults average darker-mantled than both, and many have dark or honey-colored eyes.

### Lesser Black-backed Gull

*Larus fuscus* | LBBG | L 21" (53 cm) WS 54" (137 cm)

This species, rapidly expanding in the East, is indeed the lesser of our two black-backed gulls, the other being the titanic Great Black-backed (p. 150).

■ **APPEARANCE** Adult's mantle not quite as black as adult Great Black-backed's, but nevertheless darker than even adult Laughing's (p. 142); adult's legs yellow. Larger Great Black-backed has pink legs, huge bill. First-winter distinguished from first-winter Herring by slighter build, longer wings, and mostly white head contrasting with cold, dark upperparts; in flight, first-winter Lesser Black-backed shows mostly dark wings, lacking pale panel of first-winter Herring. By second winter, starts to acquire yellow legs and dark mantle of adult. Younger Great Black-backeds always much heftier, with stronger salt-and-pepper pattern above.

■ **VOCALIZATIONS** Like other LWHGs, gives slurred squeals and series of nasal squawks; usually more nasal and lower-pitched than Herring's calls.

■ **POPULATIONS** Breeds almost entirely in Old World, but present irregularly year-round in East, especially on coasts, south to tip of Fla. Some of the largest concentrations, curiously, are in summer.

#### Gull Subspecies

In addition to intrinsic variation within populations of gulls, many species exhibit marked geographic variation. In particular, many gulls have subspecies: geographically delineated populations with morphological characters that are diagnosable in the field or museum. The Herring Gull, a sort of baseline for LWHG identification, has multiple subspecies, several of which have occurred in the East.

Most Lesser Black-backed Gulls in the East are of the *graellsii* subspecies, and most Iceland Gulls are of the *kumlieni* subspecies—but not all. The relatively dark *thayeri* subspecies of Iceland is regular to the western Great Lakes, and the lighter nominate subspecies, *glaucoides,* may be an annual stray from Europe. One or two darker-backed Old World subspecies of the Lesser Black-backed have also occurred in the East. Knowing the differences among these is especially useful in evaluating records of vagrants.

## Iceland Gull

## Lesser Black-backed Gull

## HUGE GULLS

If you see a gull in the East that is notably larger than even a Herring Gull (p. 146), it is probably going to be one of these two species. Note especially their impressive bills.

### Glaucous Gull

*Larus hyperboreus* | GLGU | L 27" (69 cm) WS 60" (152 cm)

This gull's scientific name, *hyperboreus*, means "farthest north," and it is indeed the most Arctic of the large white-headed gulls. The "Glock" is a circumpolar breeder, nesting as far north as there is land.

■ **APPEARANCE** Second largest gull on Earth; bill huge. All ages and populations are white-winged; many Iceland Gulls (p. 148) have gray or fawn wing tips. Mantle of adult paler than that of Iceland; younger birds vary from pale fawn to nearly pure white. Adult has pink legs, pale eyes, and yellow bill with red spot. First-winter has dark eye and strongly two-toned bill; by second winter, begins to acquire pale eye and mantle of adult.

■ **VOCALIZATIONS** The "seagull squeal" of Glaucous often slurs downward. Like other large gulls, gives low, rough growl, particularly nasal in this species.

■ **POPULATIONS** Apparently quite common across vast northern breeding range. Regular in winter in U.S. to Great Lakes and coastally to Tex.; increased sightings in recent years perhaps a combination of more gulls and more attentive birders.

### Great Black-backed Gull

*Larus marinus* | GBBG | L 30" (76 cm) WS 65" (165 cm)

A huge bird—dark-bodied, with a white head and white tail—is standing on the ice. At 2½ ft (0.7 m) in length, it can only be a Bald Eagle (p. 206) or Great Black-backed Gull. That's how large this gull is!

■ **APPEARANCE** As long as an eagle, with a massive bill to match; not nearly as heavy as an eagle, though. Mantle of adult almost pure black; legs dull peach-pink, eyes yellow, bill yellow with red spot. First-winter stippled above with sharp salt-and-pepper pattern; eye dark, bill mostly or entirely black. By second winter, begins to acquire pale eye, yellow bill, and black back.

■ **VOCALIZATIONS** Befitting so large a gull, calls are lower and gruffer than those of Herring and other species in genus *Larus*.

■ **POPULATIONS** Following severe persecution a century ago, has been expanding south for decades; occurs mostly coastally and on Great Lakes, but annual inland across the East. So common in some places now that it is a threat to beach-nesting Piping Plovers (p. 102) and Least Terns (p. 154).

### Hybrid Gulls

On top of intrinsic individual variation and the effects of age, geography, and feather molt and wear, the challenge of gull ID is confounded by the proliferation of hybrids, especially among the large white-headed gulls. Hybrid gulls often attract our initial attention with their subtly odd proportions—as if their bills, heads, and overall size and shape don't quite align.

Some gull hybrid combos are so frequent that they are given names: For example, a Glaucous × Herring hybrid is referred to as a "Nelson's Gull." Such birds are predictably intermediate in appearance between their parental species and may be cautiously identified in the field. Other hybrid combos are less frequent, and their identification is most safely treated as conjectural. Photos are always valuable, and documentation of truly vexing individuals advances our understanding and appreciation of these challenging birds.

## Glaucous Gull

winter adult

breeding adult

bulkier than Iceland

winter adult

2nd winter

1st winter

bicolored bill in immature plumages

1st winter

1st winter

## TROPICAL TERNS

These four are normally encountered in the East only along the coast or well out at sea. They are lovers of warm waters, with the bulk of their ranges far from our shores.

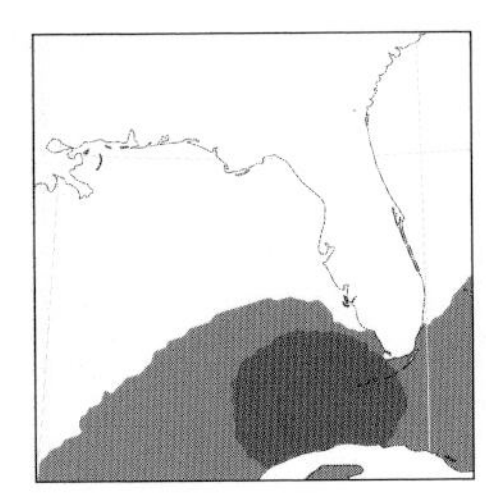

### Brown Noddy

*Anous stolidus* | BRNO | L 15½" (39 cm) WS 32" (81 cm)

Widespread in tropical oceans worldwide, this lovely seabird breeds in our area only at Bush Key, Fla., in Dry Tortugas N.P.

■ **APPEARANCE** Mostly brown; adult has frosty gray crown, reduced on immature. Black Noddy is shorter-legged and thinner-billed, with more sharply demarcated cap. Compare with adult Black Tern (p. 154) and juvenile Sooty Tern.

■ **VOCALIZATIONS** Around nesting colony at Bush Key, gives a grinding, descending *grrraaaaw.* Bowing, or "nodding," display is arresting.

■ **POPULATIONS** Seen sparingly but widely offshore to La.; sightings away from the Keys presumably a mix of U.S. breeders and visitors from elsewhere.

### Black Noddy

*Anous minutus* | BLNO | L 13½" (34 cm) WS 30" (76 cm)

Seen singly, spring and summer, at Dry Tortugas N.P. amid Brown Noddies. White cap, especially on immature, contrasts strongly with nearly black plumage overall. Smaller than Brown Noddy, with even finer bill. Like Brown Noddy, has long, dark tail, rounded at tip.

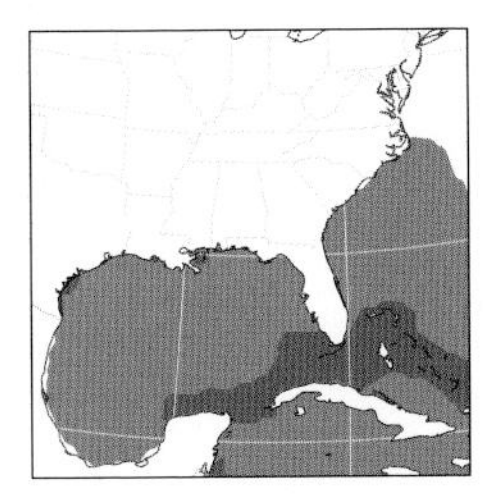

### Sooty Tern

*Onychoprion fuscatus* | SOTE | L 16" (41 cm) WS 32" (81 cm)

When Atlantic hurricanes blow warm-ocean birds off course, inland birders hope for this species. But sightings from shore or at sea are more likely.

■ **APPEARANCE** Plumage of adult contrasts well: black above, white below; black tail with bold white edges; nape black (white in Bridled). In good view, face pattern of adult diagnostic: White forehead patch ends at eye (extends behind eye on adult Bridled). Juvenile sooty dark all over; heavily spotted above and darker-bodied overall than juvenile Bridled; Black Tern more uniformly dark.

■ **VOCALIZATIONS** Gruff squawks, rising in pitch: *riiit* and *riiih.* Also shriller, rising squeaks, disyllabic or trisyllabic: *eeetie, ee-weetie,* etc. An alternative name for the species, "Wide-awake," is an allusion to this call.

■ **POPULATIONS** Hundreds of records well inland. Area in blue on map indicates range at sea in summer.

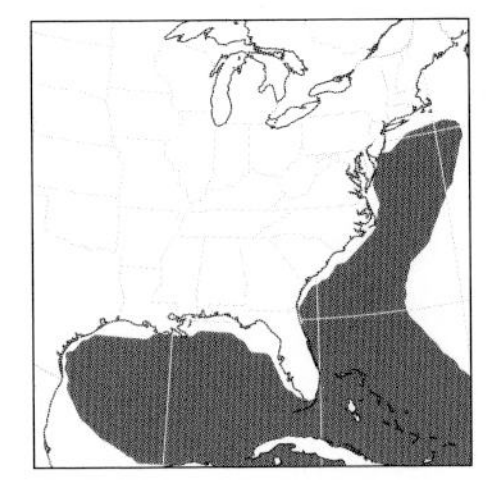

### Bridled Tern

*Onychoprion anaethetus* | BRTE | L 15" (38 cm) WS 30" (76 cm)

All terns may rest from time to time on debris at sea, but this species is especially prone to that behavior. Like the other tropical terns on this page, the Bridled feeds on small animals, especially flying fish, at and just above the water's surface.

■ **APPEARANCE** A bit smaller than Sooty Tern and not as black above. White on forehead extends behind eye, creating a black "bridle" through eye. Black cap separated from gray back by white nape; nape is black on Sooty. Young birds gray-mantled, with white head and underparts.

■ **VOCALIZATIONS** Simple whistles, sometimes pure-tone (*preeet*), sometimes rolling (*prrrrt*). On the whole, sweeter, less agitated than Sooty.

■ **POPULATIONS** Less common in East than Sooty; vagrates inland only rarely. A tiny few have nested in South Fla. amid colonies of Sooties.

frosty gray crown

## Brown Noddy

adult

limited white on forehead

adult

immature

adult

## Black Noddy

white crown sharply demarcated

immature

adult

## Sooty Tern

hite forehead es not extend past eye

Sooty

dark back

adult

juvenile

pale underwing

Bridled

white extends past eye

adult

## Bridled Tern

adult

pale back

juvenile

1st summer

## A TERN SAMPLER

These four are the sole eastern representatives of their four different genera. Despite their differences, they share the basics: pointy wings and forked tails, with bills angled down in flight.

### Least Tern

*Sternula antillarum* | LETE | L 9" (23 cm) WS 20" (51 cm)

Where there are Piping Plovers (p. 102), there are often Least Terns. These two are sort of conservation soulmates—requiring similar habitats, imperiled by similar threats, and beneficiaries of similar management strategies.

■ **APPEARANCE** Our smallest tern; wingbeats stiff, not fluid; tail relatively short. Breeding adult's bill and legs yellow, darker by end of summer; thin dark wedge on outer wing in flight. Juvenile and first-summer show dark shoulder patch.

■ **VOCALIZATIONS** Calls abrupt and irate; some are squeaky (*k'deet, kyeet*), others harsher, more nasal (*rreent, krrrp*).

■ **POPULATIONS** Disturbance and rising water threaten nesting colonies.

### Gull-billed Tern

*Gelochelidon nilotica* | GBTE | L 14" (36 cm) WS 34" (86 cm)

Equipped with the "wrong" bill, this coastal tern has a broad diet. It consumes arthropods captured in fields, in the air, and even in shrubs.

■ **APPEARANCE** Stocky, broad-winged, short-tailed. Thick, black bill has gull-like expansion (gonys) on lower mandible; legs black. Plumage mostly white, especially juvenile and winter adult. Adult bare parts simply black, unlike *Sterna* terns (pp. 156–159).

■ **VOCALIZATIONS** Call a distinctive, two-note *whit-week!* or simpler *whit*.

■ **POPULATIONS** Has wide global distribution. N. Amer. birds coastal, but species occurs well inland elsewhere.

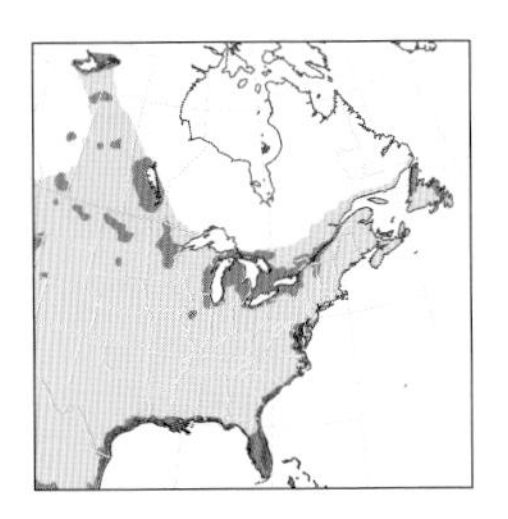

### Caspian Tern

*Hydroprogne caspia* | CATE | L 21" (53 cm) WS 50" (127 cm)

This species is the largest tern on Earth, similar in size to the California Gull (p. 146). It stands head and shoulders, literally, above other terns.

■ **APPEARANCE** Only slightly longer than Royal Tern (p. 160), but notably heftier. Bill of Caspian thick and dark orange-red; only slightly crested. In flight, shows broad dark wedge across underwing. Cap solidly black in breeding adult, mottled black and white in nonbreeding adult and juvenile.

■ **VOCALIZATIONS** Adult gives an impressive *kraawwr* or *krrock*, juvenile a thin, wavering whistle. They alternate these calls while flying in tandem.

■ **POPULATIONS** Numbers in East steady or increasing. Relatively tolerant of humans; it is common, for example, along the Chicago lakeshore.

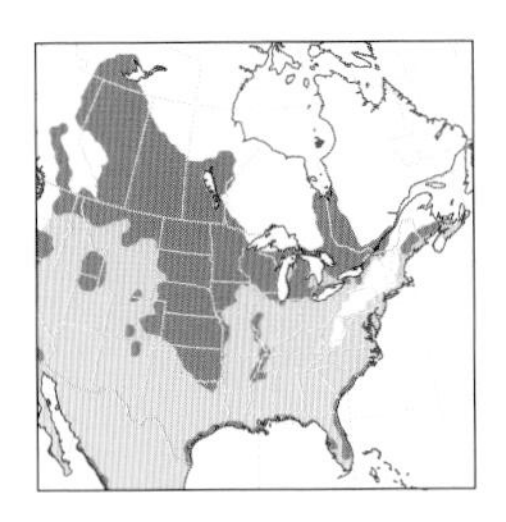

### Black Tern

*Chlidonias niger* | BLTE | L 9¾" (25 cm) WS 24" (61 cm)

In body mass and overall proportions, this tern is almost a perfect doppelganger of the Common Nighthawk (p. 80). And like the nighthawk, it feeds mostly on insects captured in flight.

■ **APPEARANCE** Quite small; the only dark tern regularly found inland in East. Breeding adult mostly jet-black, but with gray wings and white belly. Nonbreeding adult, first-summer, and juvenile paler above, with underparts largely white; head splotchy black and white.

■ **VOCALIZATIONS** Clipped *pip* or *peep*, often heard from birds on the wing.

■ **POPULATIONS** Breeds locally well inland, but disperses on migration to oceans, where confusion with tropical terns (p. 152) possible.

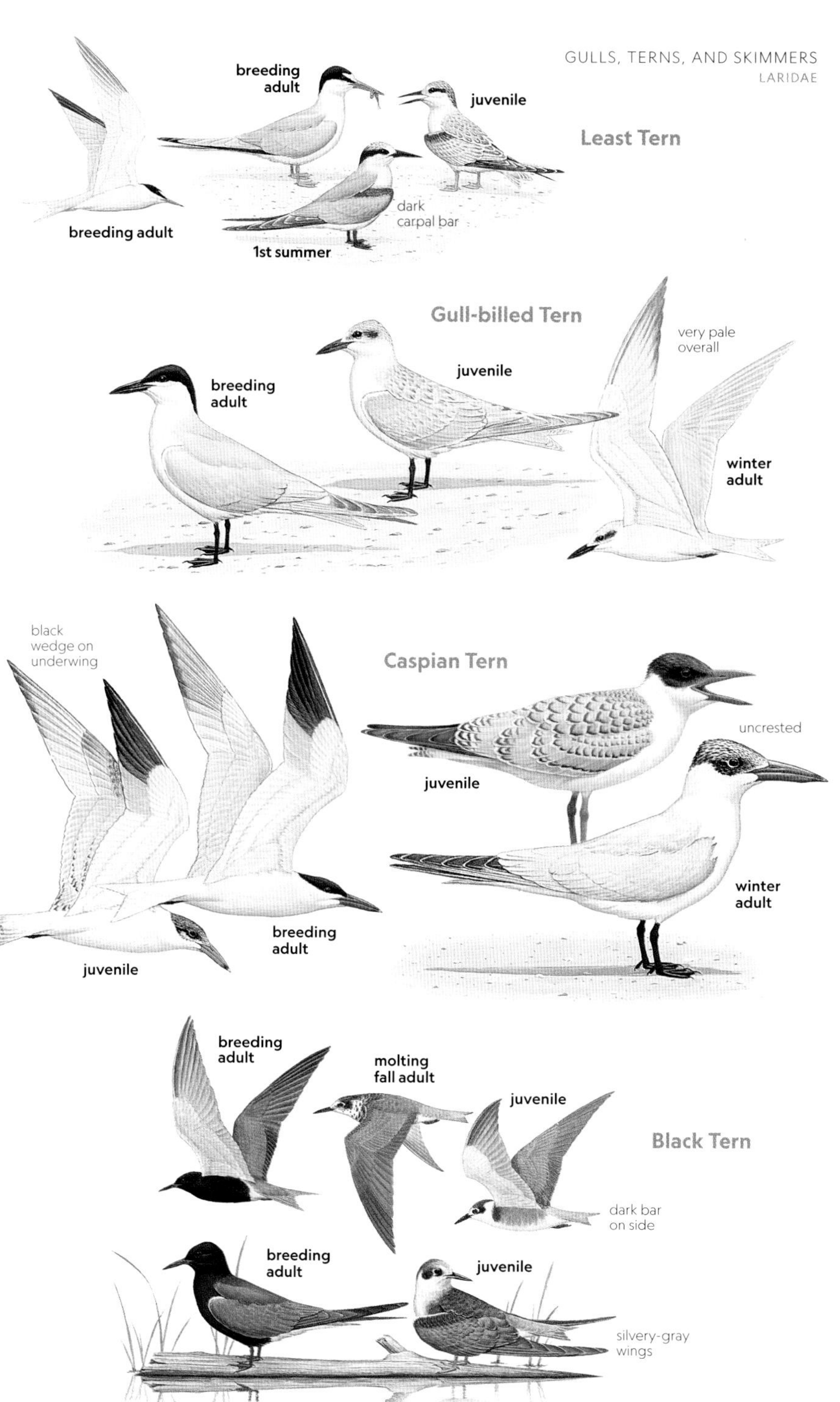
Least Tern
breeding adult
breeding adult
juvenile
dark carpal bar
1st summer
Gull-billed Tern
breeding adult
juvenile
very pale overall
winter adult
Caspian Tern
black wedge on underwing
juvenile
uncrested
winter adult
breeding adult
juvenile
Black Tern
breeding adult
molting fall adult
juvenile
dark bar on side
breeding adult
juvenile
silvery-gray wings

## GENUS *STERNA*: INLAND SPECIES

These are the two *Sterna* terns likely to be seen inland in the East from southern Canada southward. Size, shape, and flight style, if carefully assessed, are useful for field ID.

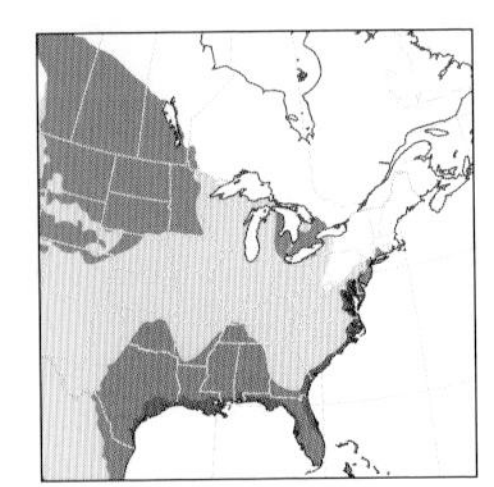

### Forster's Tern

*Sterna forsteri* | FOTE | L 14½" (37 cm) WS 31" (79 cm)

In its behavior and ecology, this widespread tern is intermediate between the Black (p. 154) and Common Terns. It is a lover of marshes but also roosts in mixed-species assemblages on piers and gravel bars.

■ **APPEARANCE** Proportionately longer-tailed and shorter-winged than Common Tern. Streamer-like tail of adult edged in white; tail of Common black-edged. Bill and feet orange in breeding plumage; more reddish in Common. Wings of breeding adult shimmering white in flight. Smooth black cap of breeding adult transitions to black face mask and white nape by late summer; bare parts darken after breeding. Juvenile sports extensive cinnamon after fledging, then shifts to plumage like that of nonbreeding adult. Juvenile and nonbreeder lack dark shoulder patch (carpal bar) of Common.

■ **VOCALIZATIONS** Calls grinding and growling, typically down-slurred: *grrreeet, grrraaah,* etc.

■ **POPULATIONS** The only tern restricted to N. Amer.; winters from southeastern U.S. to Panama; less common at sea than other *Sterna* terns.

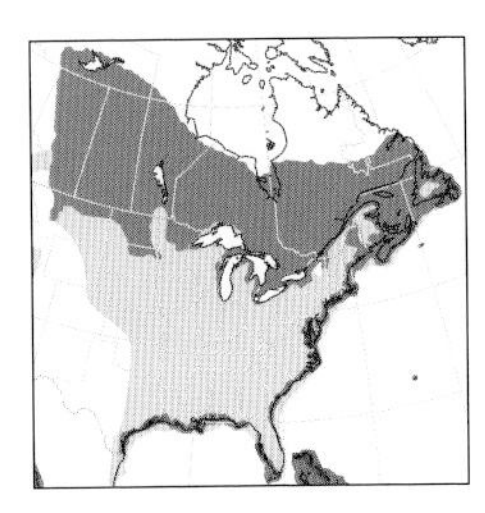

### Common Tern

*Sterna hirundo* | COTE | L 14½" (37 cm) WS 30" (76 cm)

This is the tern that has it both ways: Regularly found inland with Forster's and Black Terns, it is also at home on the open ocean amid Arctic Terns (p. 158) and tern-chasing jaegers (pp. 130–133).

■ **APPEARANCE** Slimmer overall than Forster's Tern, and longer-winged. Common Tern in flight has easier, more sweeping motions than Forster's. Tail of breeding adult black-edged; Forster's white-edged. Bill and feet of breeding adult dark red; wings plain gray above; underparts washed pale gray. Nonbreeding adult has extensive dark wedge on gray wings; note also dark shoulder patch on subadult and nonbreeding adult. Never shows dark eye patch of Forster's; instead, has black nape in nonbreeding plumages. At sea, freely mixes with Arctic Tern; see that species for comparison.

■ **VOCALIZATIONS** Calls not as grinding as those of Forster's. Many notes clipped and squeaky, but also gives a long, slightly descending *kyaaaaar.*

■ **POPULATIONS** Coastal breeders have largely recovered from terrible persecution at turn of 20th century; inland breeders seem to be doing more poorly now than their coastal counterparts.

### Shape and Proportion

Accurately assessing the build of a bird is central to field ID, but it can be problematic. A few structural characters (bill curves up vs. down, three toes vs. four, etc.) are ironclad, but others are fraught with subjectivity. It is especially valuable to gain an appreciation for how an impression of structure depends on a simultaneous consideration of multiple traits.

A Common Tern in flight arcs and bounds in a way that a Forster's Tern does not. This effect is created by the proportionately longer and narrower wings of the former, accentuated by comparison with the greater mass and proportionately longer tail of the latter. It's an important distinction, but it can also be lost in a sudden gust of wind. Even a straightforward character, like the short legs of an Arctic Tern, can be greatly affected by whether the bird being observed is crouching.

dark appears to wrap around head in some winter birds
Forster's Tern
winter adult
breeding adult
long tail projects past wings
1st winter
juvenile
white-edged tail
winter adult

Common Tern
1st summer
breeding adult
dark carpal bar
juvenile
2nd summer
dark primaries create distinctive wedge
1st fall
black-edged tail
dark gray secondaries
breeding adult

## GENUS *STERNA*: MARINE SPECIES

After nesting, these two swiftly depart their colonies. The Roseate Tern winters mostly in warm oceans at and south of the Equator, while the Arctic Tern famously winters mostly in the Southern Ocean.

### Roseate Tern

*Sterna dougallii* | ROST | L 15½" (39 cm) WS 29" (74 cm)

Like Franklin's Gulls (p. 142)—and, for that matter, flamingos (p. 64)—Roseate Terns get their pinkish hue from the food they eat, which they capture in spectacular plunge-dives well beneath the water's surface.

■ **APPEARANCE** Compared to Common (p. 156) and Arctic Terns, relatively short-winged and long-tailed; flight stiff and choppy compared to those species. Adult can be weakly tinged pink below; bill variable, reddish to black; feet red. Adult in flight shows narrow dark wedge on upperwing; underwing mostly pale gray; tail all-white. Juvenile black-capped with scalloped mantle.

■ **VOCALIZATIONS** In addition to harsh notes like other terns, gives a distinctive, two-note *kivvick* or *kivvy*, like a small plover or parakeet.

■ **POPULATIONS** A lover of the ocean; overland sightings uncommon, except in immediate vicinity of their breeding sites. The nests are almost always found within colonies of numerically dominant Common Terns.

### Arctic Tern

*Sterna paradisaea* | ARTE | L 15½" (39 cm) WS 30" (76 cm)

Although the Sooty Shearwater (p. 174) might have something to say about it, this tern is widely considered to be the greatest traveler of all animals. Many make it all the way to the Antarctic pack ice.

■ **APPEARANCE** Compared to other *Sterna* terns, short-billed and especially short-legged. Breeding adult has dark red bill and feet; underparts washed gray, wings plain gray above. Juvenile has weaker shoulder patch (carpal bar) than juvenile Common; wings in flight more translucent. Separation of Arctic and Common Terns ("Commic Terns") at sea difficult: Pale upperwing of Arctic relatively uniform in all plumages; upperwing of Common splotchier, with dark wedge on primaries (adults) and dark leading edge (young).

■ **VOCALIZATIONS** Mixes clipped, raspy *chit* notes with squeakier *kyeet* and *kyeep* calls; exceptionally aggressive around cacophonous nesting colonies.

■ **POPULATIONS** Annual but very rare on migration overland, mostly Great Lakes northward, with concentration of records in late spring in southern Ont. and Que.

### "Portlandica" Terns

Adult and hatch-year terns in the genus *Sterna* are usually identifiable with good views and adequate study. Terns in other genera are, for the most part, even more straightforward, at least in the East. But what about terns in their first full summer—when they are one year old?

One-year-old *Sterna* terns do not normally attempt to breed, with the result that many stay on the wintering grounds, which are largely outside our area. When Robert Ridgway, regarded as one of America's greatest ornithologists, encountered such a bird, he declared it a new species and called it *Sterna portlandica*—but it was actually an Arctic Tern in its first full summer. The species designation *S. portlandica* is no more, but the name "Portlandica" was retained and is still applied to one-year-old *Sterna* terns. Such birds are adultlike in morphology and plumage, resembling "winter" adults but at the wrong time of year.

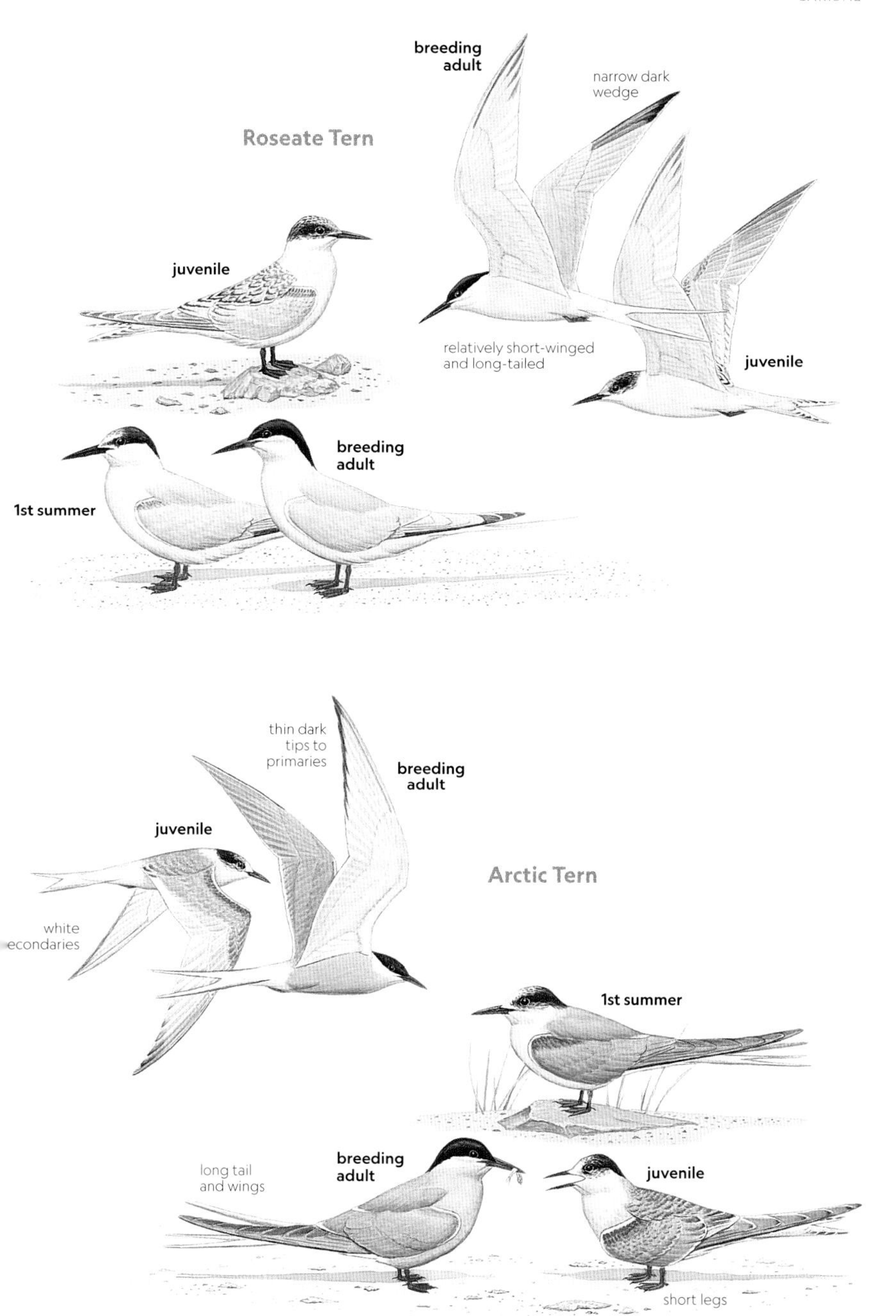
breeding
adult
narrow dark
wedge
Roseate Tern
juvenile
relatively short-winged
and long-tailed
juvenile
breeding
adult
1st summer
thin dark
tips to
primaries
breeding
adult
juvenile
Arctic Tern
white
econdaries
1st summer
long tail
and wings
breeding
adult
juvenile
short legs

## BEACH-LOVING LARIDS

Where there are white sand beaches within their ranges, these three are likely to be found. The large-bodied *Thalasseus* terns are nested deep within the tern lineage, but the status of the skimmer is unclear.

### Royal Tern

*Thalasseus maximus* | ROYT | L 20" (51 cm) WS 41" (104 cm)

This species is nearly the length of the mighty Caspian Tern (p. 154), yet it comports itself with the grace and nimbleness of smaller species.

■ **APPEARANCE** In same genus as Sandwich Tern, but more likely to be confused with Caspian. Bill of Royal thinner, more orangey than that of Caspian. Royal in full breeding plumage has solid black cap with shaggy crest; much of the year, mostly white-capped. In flight, shows largely white underwing, whereas Caspian shows more dark on underwing; tail of Royal more deeply forked than that of Caspian. Juvenile like nonbreeding adult, but with even less black on cap.

■ **VOCALIZATIONS** A rolling, wavering *keer-reet* or *keery-reet*, less harsh than *Sterna* terns' calls.

■ **POPULATIONS** Mostly coastal, but gets inland in Southeast, especially Fla. peninsula; regularly disperses north along coast following breeding. Readily nests on human-made dredge spoil islands.

### Sandwich Tern

*Thalasseus sandvicensis* | SATE | L 15" (38 cm) WS 34" (86 cm)

Along with the Royal Tern, this is one of the two crested terns in the East. The name of their genus, *Thalasseus,* derives from a Greek word for "sea," an apt moniker for these mostly marine terns.

■ **APPEARANCE** Same general body plan as Royal, but smaller; like Royal, keeps full black cap only briefly at beginning of breeding season. Bill of adult black with fine yellow tip, unlike any other tern in East; beware that lighting can create such an effect on terns with all-black bills. Adults and young paler above than corresponding plumages of Royal.

■ **VOCALIZATIONS** Call rolling and wavering like that of Royal, but higher, more grating, not as musical: *grrrp-gritt, giddy-it.*

■ **POPULATIONS** Tends to occur with Royals, especially at nesting colonies; breeding range expanding north along coast. Like Royal, wanders north a bit after nesting.

### Black Skimmer

*Rynchops niger* | BLSK | L 18" (46 cm) WS 44" (112 cm)

Of all the eastern birds in the large order Charadriiformes (pp. 98–161), this species may be the most distinctive. Whether skimmers are more closely related to gulls or terns is controversial.

■ **APPEARANCE** Only skimmers have lower mandibles much longer than upper; like Razorbill (p. 134) and puffins (p. 136), Black Skimmer has laterally compressed bill. Feeding behavior unique: It barely skims the water's surface with lower mandible; small flocks coordinate mostly nocturnal foraging with mesmerizing synchronicity. Even without the astonishing behavior and morphology, dark above and white below are distinctive.

■ **VOCALIZATIONS** A plain, nasal *enk,* often given among flockmates when skimming.

■ **POPULATIONS** Requires undisturbed beaches for nesting and roosting; feeds around inlets, canals, and marinas. In East, normally restricted to coasts, but sometimes pushed inland by hurricanes.

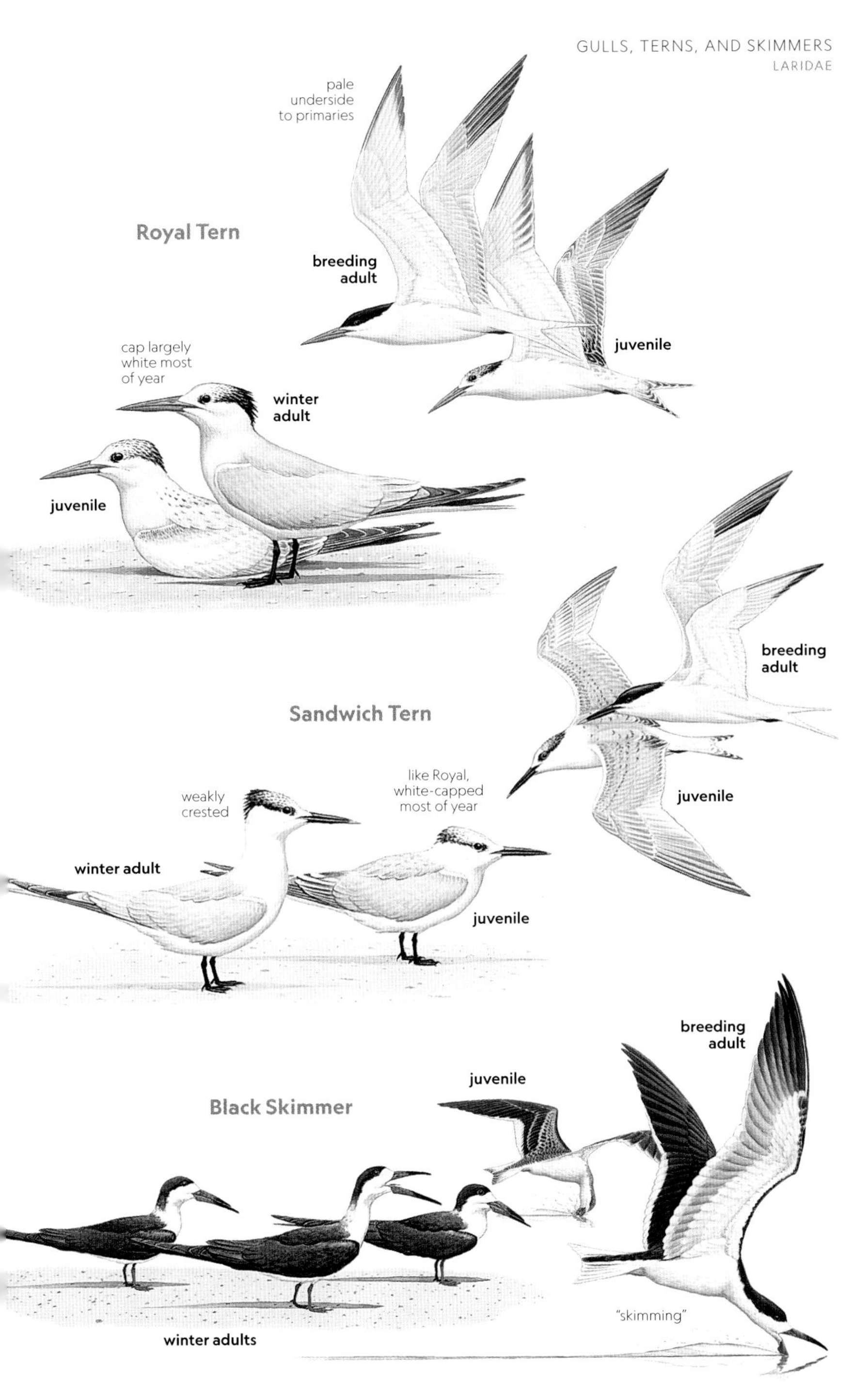
pale
underside
to primaries
Royal Tern
breeding
adult
juvenile
cap largely
white most
of year
winter
adult
juvenile
breeding
adult
Sandwich Tern
like Royal,
white-capped
most of year
weakly
crested
juvenile
winter adult
juvenile
breeding
adult
juvenile
Black Skimmer
"skimming"
winter adults

## TROPICBIRDS

In plumage, habits, and general body plan, these pelagic aerialists are suggestive of terns. They are habitual divers, and their fragile-looking central tail feathers somehow survive the plunge.

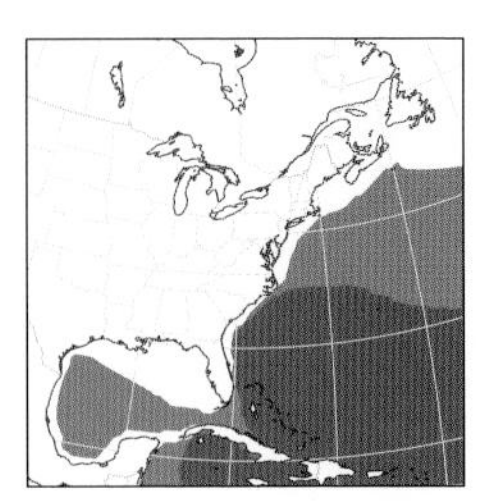

### White-tailed Tropicbird

*Phaethon lepturus* | WTTR | L 30" (76 cm) WS 37" (94 cm)

This is the smaller and more commonly encountered of the two tropicbirds in the East. Sightings in our area are typically of lone birds far from shore.

■ **APPEARANCE** Not quite as hefty as Royal Tern (p. 160). Central tail streamers of orange-billed adult longer than folded tail of any tern. In flight, adult has gleaming white upperparts, with bold black patches across inner and outer wings. Yellow-billed juvenile lacks tail streamers. Upperparts of juvenile barred gray and white; black on outer wing restricted to a few outer primaries.

■ **VOCALIZATIONS** Birds at sea sometimes utter an abrupt *rah* or *wah*, wimpy and nasal.

■ **POPULATIONS** Probably wanders widely off Gulf Coast and Atlantic coast, but sightings are concentrated off N.C. Areas in blue on map indicate presence from late spring through summer.

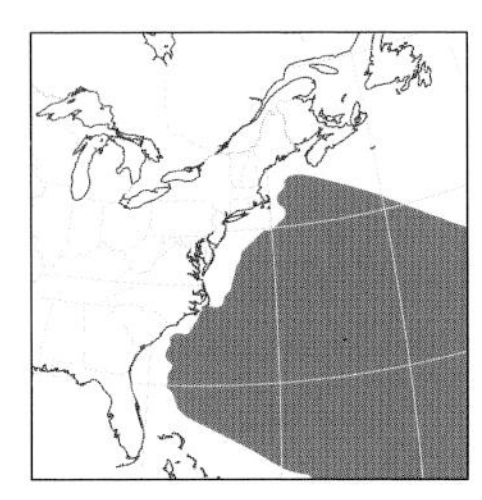

### Red-billed Tropicbird

*Phaethon aethereus* | RBTR | L 40" (102 cm) WS 44" (112 cm)

This species has one of the greatest scientific names. In Greek mythology, Phaethon drove his shining chariot through the sky; and *aethereus* means "ethereal." An emblem of rarity and wanderlust, it is found in the American Birding Association's logo.

■ **APPEARANCE** Large-billed and big-bodied; heftier than Caspian Tern (p. 154). Extensive black on outer wing of streamer-tailed adult visible on flying bird at great distance; at closer range, note bright red bill, thick black line behind eye, and finely barred upperparts. Juvenile plumages of Red-billed and White-tailed more similar: On Red-billed, note black eyeline extending to nape, finely barred upperparts, and more extensive black in outer wing.

■ **VOCALIZATIONS** Like White-tailed, rarely calls at sea; gives occasional yips and yaps (*reh, rah,* etc.), sometimes in quick chatter.

■ **POPULATIONS** Rarer of the two tropicbirds off East Coast; annual off N.C., especially June. Areas in blue on map for this species indicate range from late spring through summer.

### Of Terns and Tropicbirds

Because Tropicbirds follow terns (pp. 152–161) in this book, one might reasonably infer that these similar-seeming groups are close relatives. They are not. The three tropicbirds (two species occur in the East) are the only extant members of the order Phaethontiformes, whereas terns belong to the diverse order Charadriiformes, comprising more than 300 extant species—including many in the East.

The Phaethontiformes fall within a broader group, or clade, that does *not* include the Charadriiformes. It does, however, admit a diversity of other aquatic species, from penguins to albatrosses, from loons to herons. Where do the tropicbirds fit in exactly? This is where things get wild. Their closest relatives are apparently the Sunbittern, *Eurypyga helias,* of rivers and freshwater swamps in the Neotropics, and, even more astonishingly, the Kagu, *Rhynochetos jubatus,* restricted to upland habitats of New Caledonia. The usual caveat: The dust has yet to settle on tropicbird taxonomy, and some revisions may be in store. In any event, tropicbirds are emphatically not terns, despite striking convergent evolution between the two taxa.

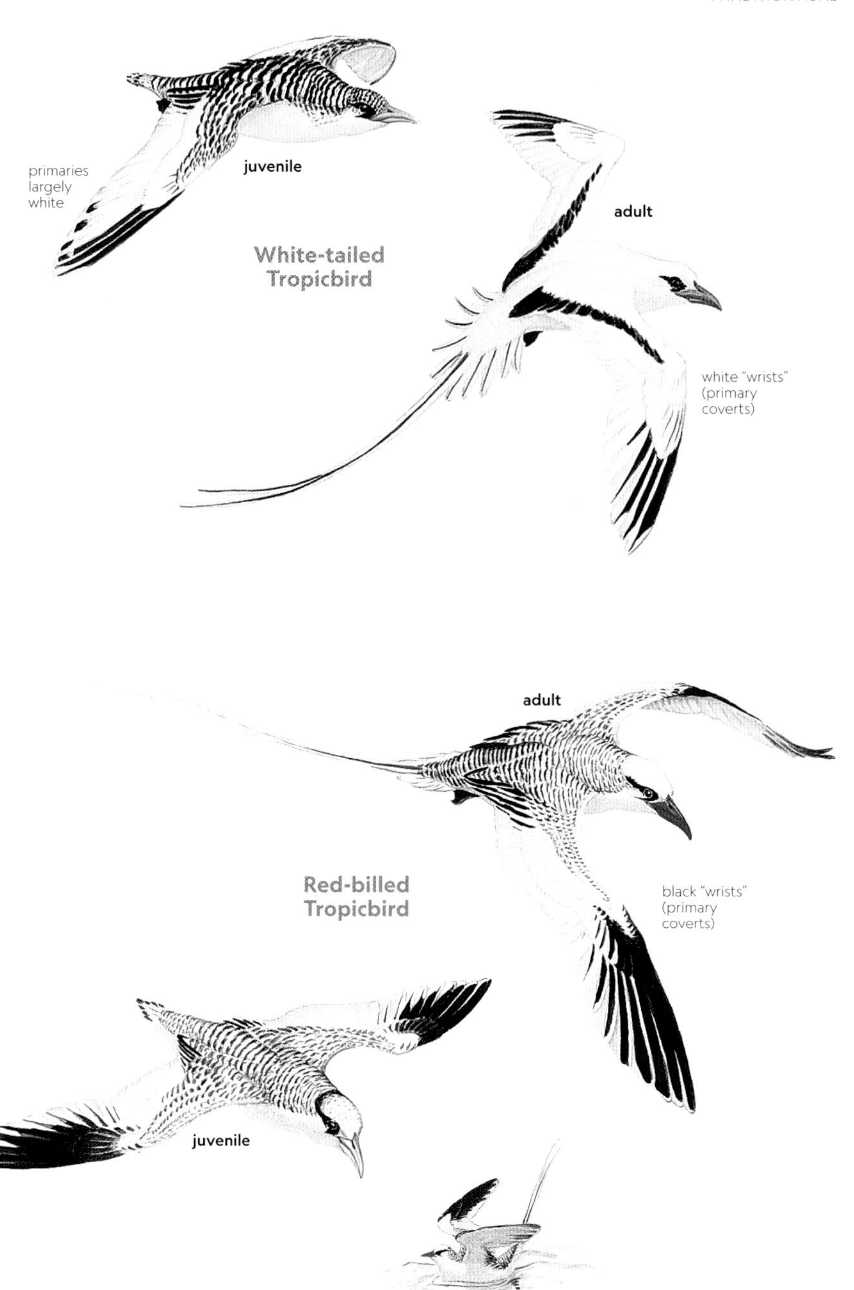
primaries
largely
white
juvenile
adult
White-tailed
Tropicbird
white "wrists"
(primary
coverts)
adult
Red-billed
Tropicbird
black "wrists"
(primary
coverts)
juvenile

## SMALLER LOONS

They're called "divers" in England because of their underwater agility. While the American name "loon" has come to be associated with their wild calls, it probably derives from a word that simply denoted the bird, not its vocalizations.

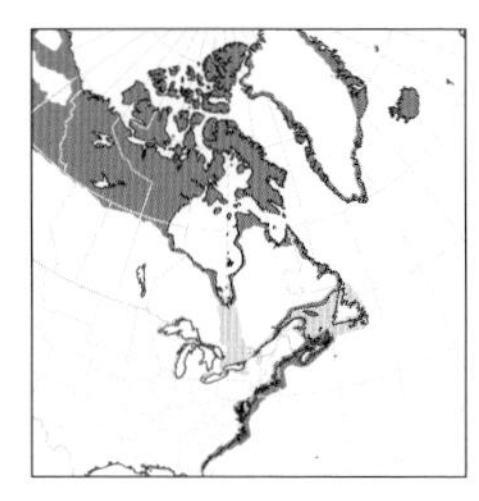

### Red-throated Loon

*Gavia stellata* | RTLO | L 25" (64 cm)

Of the two loons found commonly in the East (Common, p. 166, is the other), this one is more coastal. It often rests out on the ocean beyond the breakers, but sometimes it comes right into the surf.

■ **APPEARANCE** Barely a third the mass of Common Loon; in all plumages, note thin bill, angled up. Breeding adult has gray head with maroon foreneck. Most adults seen away from breeding grounds are in winter plumage, strikingly white-faced. Juvenile, commonly seen throughout winter, like nonbreeding adult but duskier and darker-faced. Fresh juveniles and adults are speckled white above in winter; the name *stellata* (star) indicates the starry appearance of the bird's back. Flight feather molt takes place in fall, earlier than mid-winter molt of other loons.

■ **VOCALIZATIONS** Silent in winter when roosting or feeding, but birds in flight, especially on migration, give duck-like quacks: *rruck* and *wraack*. Yodel on breeding grounds shorter, harsher, and more nasal than Common's famous wailing.

■ **POPULATIONS** Most widely distributed globally of all the loons. Away from breeding grounds, seen in East chiefly along coasts and on eastern Great Lakes. Some migrate overland from Great Lakes to coasts, but most migrate coastally in impressive daytime flights.

### Pacific Loon

*Gavia pacifica* | PALO | L 26" (66 cm)

Typically thought of as a western species, yet breeds east almost to northern Lab. Rare but annual in winter across much of East, with preponderance of records toward Rockies. Smaller than Common Loon but has the same leaden hue in winter. Also smaller-billed than Common, with smoothly rounded "cobra" head, smooth and straight border between dark hindneck and white foreneck. Many, but not all, Pacifics show "chinstrap," thin and dusky, in winter.

### Loons in Winter

Loons in breeding plumage are distinctive, but most encounters with loons in the East are of juveniles and nonbreeding adults. Two species occur regularly in winter in the East: the widespread Common (p. 166) and the primarily coastal Red-throated. The Yellow-billed (p. 166) and Pacific Loons are rare but annual, and there are a handful of records of the Arctic (p. 420). Red-throated and Common Loons in winter are normally separated by their considerable differences in bulk, bill structure, and plumage. The problem, then, boils down to identifying one of the rarer loons—especially with reference to the widespread and somewhat variable Common.

Considered in tandem, general paleness, pattern, and contrast are useful, but be aware that juveniles of all species are relatively paler-backed and often spotted and scaly above. The exact pattern of dark and light on the head and neck is particularly useful, but pay attention to lighting and viewer angle. Bill shape and color are diagnostic in some comparisons, especially on birds viewed through a scope or when analyzing photos after you get home. Differences in body size are also appreciable. Fortunately, rare loons found inland have the convenient habit of staying put, affording observers the opportunity to study them for hours—perhaps even days and weeks.

breeding
adult
winter
adult
Red-throated
Loon
pale
cheeks
1st spring
upward-
angled bill
juvenile
breeding
adult
winter
adult

breeding
adult
pale head
winter
adult
Pacific Loon
winter
adult
juveniles
variable
"chinstrap"
breeding adult
snakelike
neck
small head
winter adult

## LARGER LOONS

Notably larger than other loons, these two are similar in plumage, calls, and overall body structure. In winter, bill color and shape are the best way to tell them apart.

### Common Loon

*Gavia immer* | COLO | L 32" (81 cm)

Along with wolves and starry nights, this species is an emblem of the Northwoods. Common Loons were made famous in the 1981 film *On Golden Pond,* and today they draw tourists and locals alike to loon festivals and other events.

■ **APPEARANCE** Size of a goose; appears hulking on the water. In flight, like other loons, has gangly, floppy look, as if the head and feet are about to fall off. Bill large; head blocky and angular. Breeding plumage crisp and high-contrast; in winter, both juvenile and adult plumage smudgy, with uneven border between dark and white on neck. Bill is black in summer, gray in winter with black tip and black ridge (culmen) of upper mandible. When making IDs in winter, note that Smaller Red-throated Loon (p. 164) is white-faced with thinner, upturned bill; rare Pacific Loon (p. 164) has sharp border on neck, rounded head, and smaller bill; larger Yellow-billed Loon has plain, pale, upturned bill.

■ **VOCALIZATIONS** The pure-tone wailing (or "yodel") of this species is one of the loudest of all animal sounds, audible at well over a mile (1.6 km); it is normally heard only on the breeding grounds. A shorter, mirthful cackle (tremolo) is sometimes given away from the breeding grounds, especially by migrants in flight.

■ **POPULATIONS** The most widely encountered loon inland away from the breeding grounds. Often seen singly on large lakes and bays; also forms loose flocks, with individuals widely scattered, rarely densely packed together. Common coastally. In Britain, evocatively called the "Great Northern Diver."

### Yellow-billed Loon

*Gavia adamsii* | YBLO | L 34" (86 cm)

The largest loon, but close in size to the widespread Common. Winter Yellow-billed differs from Common by relatively pale plumage, with face especially washed-out. Bill, if seen well, is diagnostic: mostly pale, including tip; upper mandible perfectly straight (convex on Common), making bill appear upturned. Breeds east to Hudson Bay, but most in winter disperse southwest; very rare but annual in East.

#### Bioindicators

Birders quickly learn that gross habitat features (marsh, forest, prairie, etc.) are dependable predictors of avian occurrence and hence useful in field ID. The converse is also true, although typically at finer resolutions of habitat type. Like the canaries in the coal mines of yesteryear, many birds are precisely attuned to environmental quality. Louisiana Waterthrushes (p. 376) in the Appalachians and White-throated Dippers, *Cinclus cinclus,* in Britain have been employed as bioindicators of stream health, and loons are analogous proxies for water quality in large lakes.

In many cases, loons are simply absent from lakes with high levels of contaminants. But even when present, they are valuable sentinels of ecosystem health. Voracious consumers of fish, loons are top predators, so they bioaccumulate toxins. Tissue samples, typically obtained from corpses, are easily and inexpensively assayed. The results are impressively accurate—so much so that a recent international treaty aimed at mitigating mercury poisoning in humans depends on loons as credible indicators of water quality.

swimming juveniles for comparison

## STORM-PETRELS

Smallest of the tubenoses, order Procellariiformes (pp. 168-177), these flutter and zigzag right above the ocean's surface. Differences in flight style and feeding behavior are important for field ID.

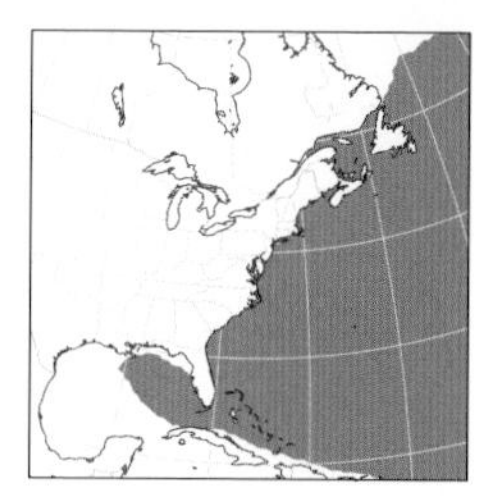

### Wilson's Storm-Petrel

*Oceanites oceanicus* | WISP | L 7¼" (18 cm) WS 16" (41 cm)

One of the world's great seafarers, this species disperses from large breeding colonies off Antarctica to northern oceans during the austral winter (our summer).

■ **APPEARANCE** Slightly smaller than other eastern storm-petrels; wings short and rounded; legs extend beyond gently rounded tail. Contrasting white rump prominent at all angles. On foraging bouts, suggests a butterfly: wanders, gently shifting course, then patters feet on the water while plucking food from the surface with bill. Compare with other storm-petrels, but also with unrelated Purple Martin (p. 290), which can get out over the ocean.

■ **VOCALIZATIONS** Rarely heard in our waters, but sometimes peeps quietly while feeding in groups.

■ **POPULATIONS** Off East Coast, generally favors cooler waters than Band-rumped does. Most sightings at sea, less often from shore.

### White-faced Storm-Petrel

*Pelagodroma marina* | WFSP | L 7½" (19 cm) WS 17" (43 cm)

Spotted annually on pelagic trips mostly from N.C. to Mass., often late summer. Distinctive in both plumage and behavior. White below, gray-black above; white face marked with black cap and eye patch, suggesting winter-plumage phalaropes (p. 126). Foraging, the bird seems to bounce off the water on long legs angled forward.

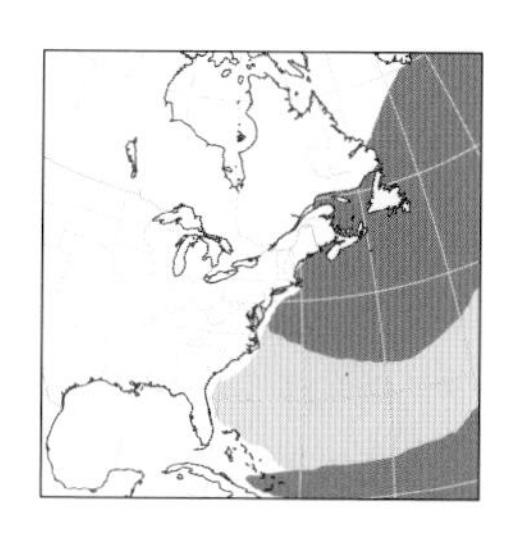

### Leach's Storm-Petrel

*Hydrobates leucorhous* | LESP | L 7-8" (18-20 cm) WS 17-19" (43-48 cm)

This is the only storm-petrel that breeds in the East. Most encounters are of birds at sea foraging in broad daylight.

■ **APPEARANCE** Wings long and sharp-tipped, often angled; tail forked. Dark overall, with white rump like Wilson's and Band-rumped; however, white on rump less prominent, split down the middle by smudgy central line. Flight powerful, zooms about in sudden bursts; relatively long wings and erratic flight suggest Common Nighthawk (p. 80).

■ **VOCALIZATIONS** At breeding colonies, mellow purring and giddy chippering.

■ **POPULATIONS** Most commonly seen off East Coast in summer from Mass. northward; on migration through our area to equatorial oceans, occurs more widely but also more sparingly.

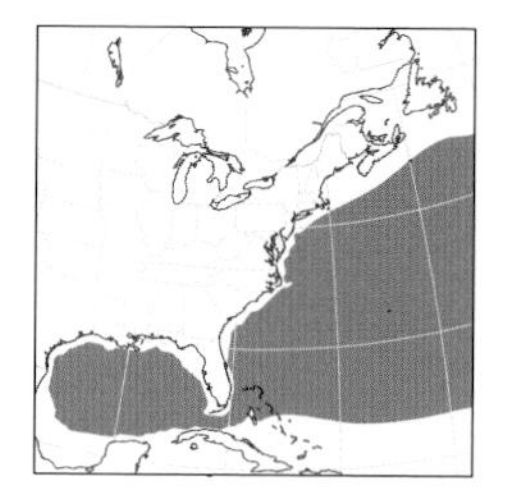

### Band-rumped Storm-Petrel

*Hydrobates castro* | BSTP | L 9" (23 cm) WS 17" (43 cm)

Our "southern" storm-petrel, this warmwater species alternates short glides with fluttering wingbeats in the manner of a shearwater.

■ **APPEARANCE** Like an oversize Wilson's Storm-Petrel, but legs shorter; white rump not as prominent as on Wilson's. Wing structure midway between that of round-winged Wilson's and pointy-winged Leach's. Shearwater-like gestalt is amplified by thicker bill.

■ **VOCALIZATIONS** In interactions while feeding, gives squeaky note.

■ **POPULATIONS** Birds in our waters, most often sighted from N.C. south, spring and summer, are nonbreeders from islands off Europe and Africa.

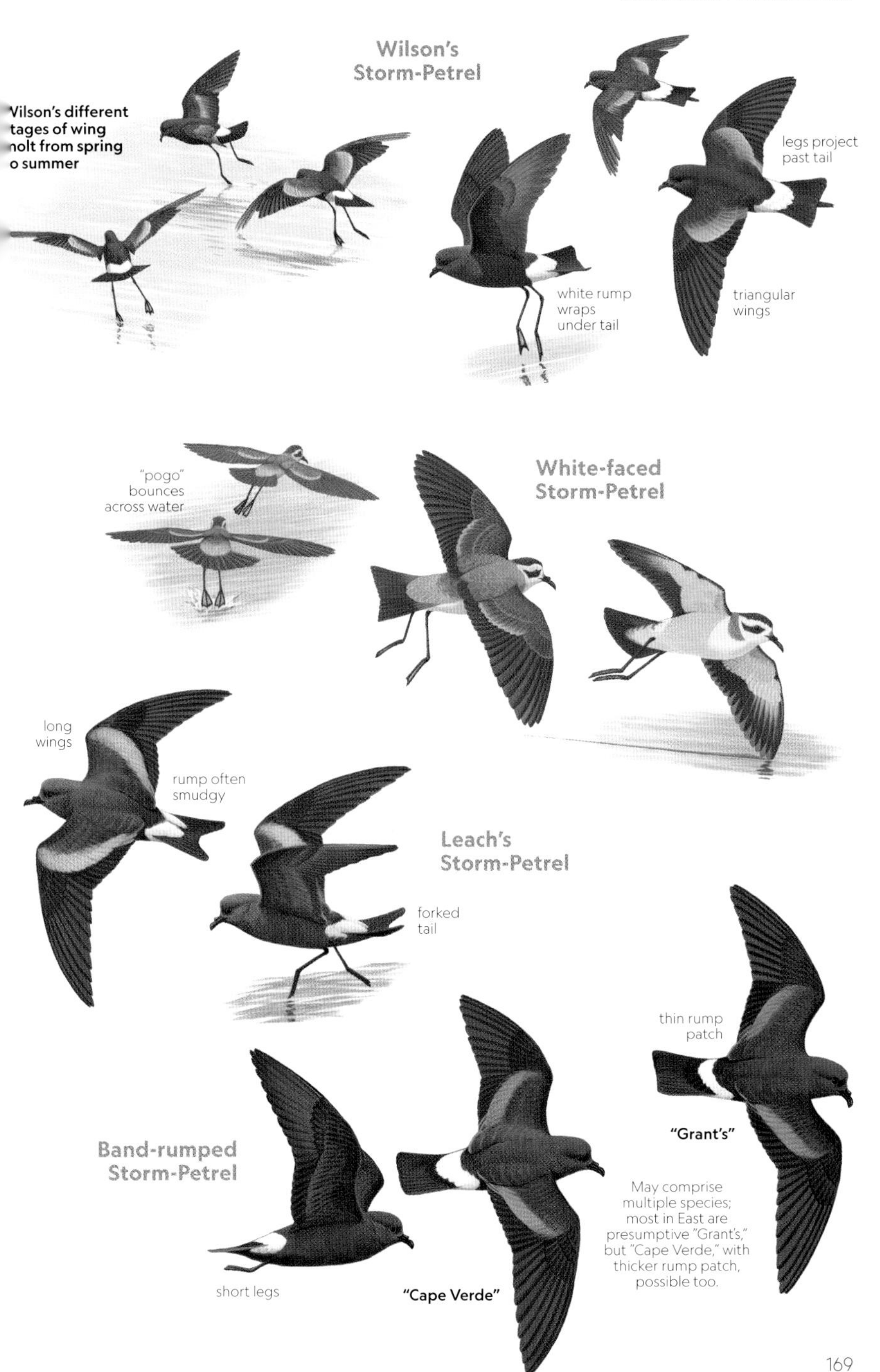
Wilson's Storm-Petrel
Vilson's different
tages of wing
nolt from spring
o summer
legs project past tail
white rump wraps under tail
triangular wings
"pogo" bounces across water
White-faced Storm-Petrel
long wings
rump often smudgy
Leach's Storm-Petrel
forked tail
thin rump patch
"Grant's"
Band-rumped Storm-Petrel
May comprise multiple species; most in East are presumptive "Grant's," but "Cape Verde," with thicker rump patch, possible too.
short legs
"Cape Verde"

## A DISTINCTIVE PROCELLARIID

The beefy Northern Fulmar is the only Northern Hemisphere species in the "fulmarine petrel" clan of southern oceans. Like other procellariids, represented in the East by the gadfly petrels (p. 172) and shearwaters (pp. 174-177), the fulmar is a medium-stature tubenose that comes to shore only when nesting.

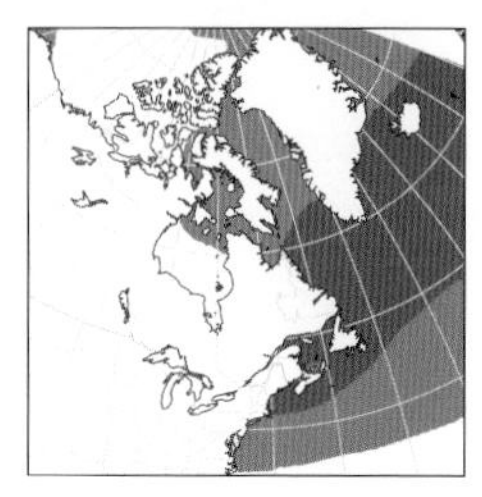

### Northern Fulmar

*Fulmarus glacialis* | NOFU | L 16-18" (41-46 cm) WS 37-45" (94-114 cm)

Although gull-like in general color and pattern, this northern tubenose flies like a shearwater, alternating stiff flaps with short glides. Unlike most tubenoses, adult fulmars exhibit considerable plumage variation.

■ **APPEARANCE** The size of one of the large shearwaters (p. 174). Stocky bill, pale with prominent nostril tubes, is discernible at some distance. Light-morph adult has gray mantle and otherwise white body, suggesting a white-headed gull (pp. 144-151), but flight style and bill structure betray the fulmar's taxonomic affinity with the tubenoses. Lovely dark-morph adult is gray all over, but lighter and softer than Sooty Shearwater (p. 174). Intermediate morphs sometimes seen. All flash pale inner primaries in flight.

■ **VOCALIZATIONS** More than most other tubenoses, pretty excitable around trawlers and other vessels at sea; listen for harsh cackles and nasal whines, audible above the drone of ship engines.

■ **POPULATIONS** Breeds on sea cliffs south in our area at least to Nfld. and eastern Que., often in company of superficially similar kittiwakes (p. 138). Sightings at sea off U.S. coast mostly in winter from mid-Atlantic northward. An adaptable species; unlike almost all the rest of the world's tubenoses, is increasing in number and expanding in range: routinely tends offshore fishing operations; has started nesting on buildings and other structures. Geographic basis for color polymorphism in the North Atlantic is complex; the most southerly breeders are usually light, with an increase in dark-morph birds farther north and especially west. They mix freely at sea in the nonbreeding season.

### Tubenoses

Few groups of birds are as challenging for field ID as the tubenoses (order Procellariiformes): the storm-petrels (p. 168), procellariids (pp. 170-177), and albatrosses (not of regular occurrence in our area). The species in our area are clad in black and white, gray and brown, without so much as a hint of color except, in limited instances, on the bill and feet. But the real problem is that they are found mostly well out at sea: Views are typically at a distance, often of birds flying fast, and from a boat that is pitching and swaying.

Assessing flight style is an excellent first step in tubenose ID. Shearwaters glide near the surface, gadfly petrels whip about in broad arcs, and storm-petrels flutter and patter at the water's edge. Overall contrast—between face, crown, and nape, between tail and upperparts, and between upperparts and underparts—is often ascertainable even from afar. But bear in mind that quite unrelated birds, such as distant gulls (pp. 138-151), gannets (p. 180), and even martins (p. 290), can be misidentified as tubenoses.

Flight style, body structure, and contrast can tell us if we're in the right ballpark, but nothing beats close comparative study of birds at sea. Fortunately, most tubenoses are readily drawn to the sterns of larger boats, where the birds may be observed on the water and in direct comparison with one another. Keep with the birds as they take flight, making note of body structure and wing pattern.

It should be noted that watching from shore is not completely out of the question with tubenoses. Easterlies, while miserable for most beachgoers, are favorable for dramatic inshore flights of tubenoses, especially shearwaters. If conditions are right, hundreds of tubenoses, representing multiple species, may be seen from sites like Montauk Point, at the eastern tip of Long Island.

Northern
Fulmar
light morphs
dark morph
sitting
on water,
suggests
a gull
pale inner
primaries
light morph
Black-capped
Petrel
Manx
Shearwater
Great
Shearwater
Northern
Fulmar
Wilson's
Storm-Petrel

## GADFLY PETRELS

They're named for their bounding flight, as if trying to get out of the way of gadflies. And their scientific name combines the words for wing (*ptero*) and run (*droma*), conveying the same idea.

### Trindade Petrel

*Pterodroma arminjoniana* | TRPE | L 15½" (39 cm) WS 37½" (95 cm)

Breeds on islands in South Atlantic off Brazil, wandering north into the Gulf Stream, especially off N.C., for the austral winter (our summer). Most in our area are dark morphs, flashing white underwings. Differs from Sooty Shearwater (p. 174) by thicker bill, pointier wings, and flight style. Light morph, even rarer off East Coast, is gray-white on breast and belly.

### Bermuda Petrel EN

*Pterodroma cahow* | BEPE | L 14-15" (36-38 cm) WS 34-36" (86-91 cm)

Formerly thought extinct, but rediscovered and slowly recovering; about 200 breed on Bermuda, with a few dispersing to warm waters off East Coast, mostly N.C., late spring into summer. Black-capped Petrel a bit larger, with thicker bill and relatively shorter wings. Black-capped more gleamingly dark and light above; Bermuda Petrel, mostly gray-brown above, has much reduced white across uppertail coverts and little if any white on nape. Compare also with similar Fea's Petrel.

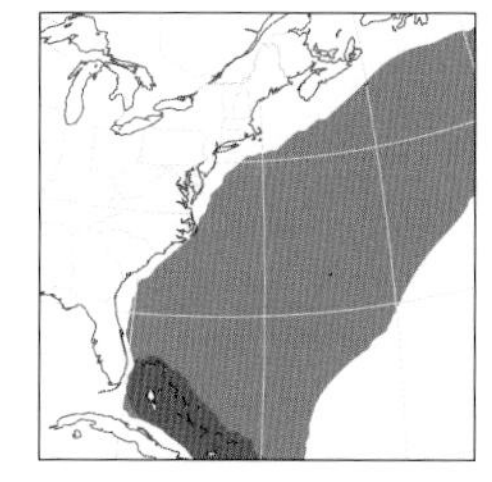

### Black-capped Petrel EN

*Pterodroma hasitata* | BCPE | L 15-18" (38-46 cm) WS 39-41" (99-104 cm)

By far the most common gadfly petrel off the East Coast, this magnificently black-and-white tubenose is routinely seen on pelagic trips starting around late May and extending throughout the summer. Tallies into the triple digits have been recorded.

■ **APPEARANCE** Larger than other East Coast gadfly petrels. Prominent black cap set off from white face and nape; white patch across uppertail coverts is huge; wings below extensively white with relatively thin bar at leading edge. In addition to separation from rarer gadfly petrels, be sure to consider similarity in plumage to Great Shearwater (p. 174), frequent off East Coast. The shearwater is larger and duskier overall, with limited white on nape and uppertail coverts; most Greats have smudgy brown on belly. Flight style and bill structure differ appreciably between gadfly petrels and shearwaters.

■ **VOCALIZATIONS** In squabbles for food at sea, sometimes emits weak, down-slurred chirps. Gives spooky yaps and yelps around nests, but apparently never at sea.

■ **POPULATIONS** Breeds only on a few islands in Caribbean; believed to number around 5,000 individuals in total. Remarkably, nests a mile (1.6 km) or more above sea level in forests on steep mountains.

### Fea's Petrel

*Pterodroma feae* | FEPE | L 14-15" (36-38 cm) WS 34-38" (86-97 cm)

Breeds on islands in eastern Atlantic Ocean, disperses west to warm waters off East Coast, especially N.C., in summer. Intermediate in plumage between rare Bermuda and light-morph Trindade Petrels. Gray-backed, with wings darker than back; tail uniformly gray. Variable, but most sightings in our area are of birds with mostly dark underwings, unlike Bermuda and Black-capped Petrels.

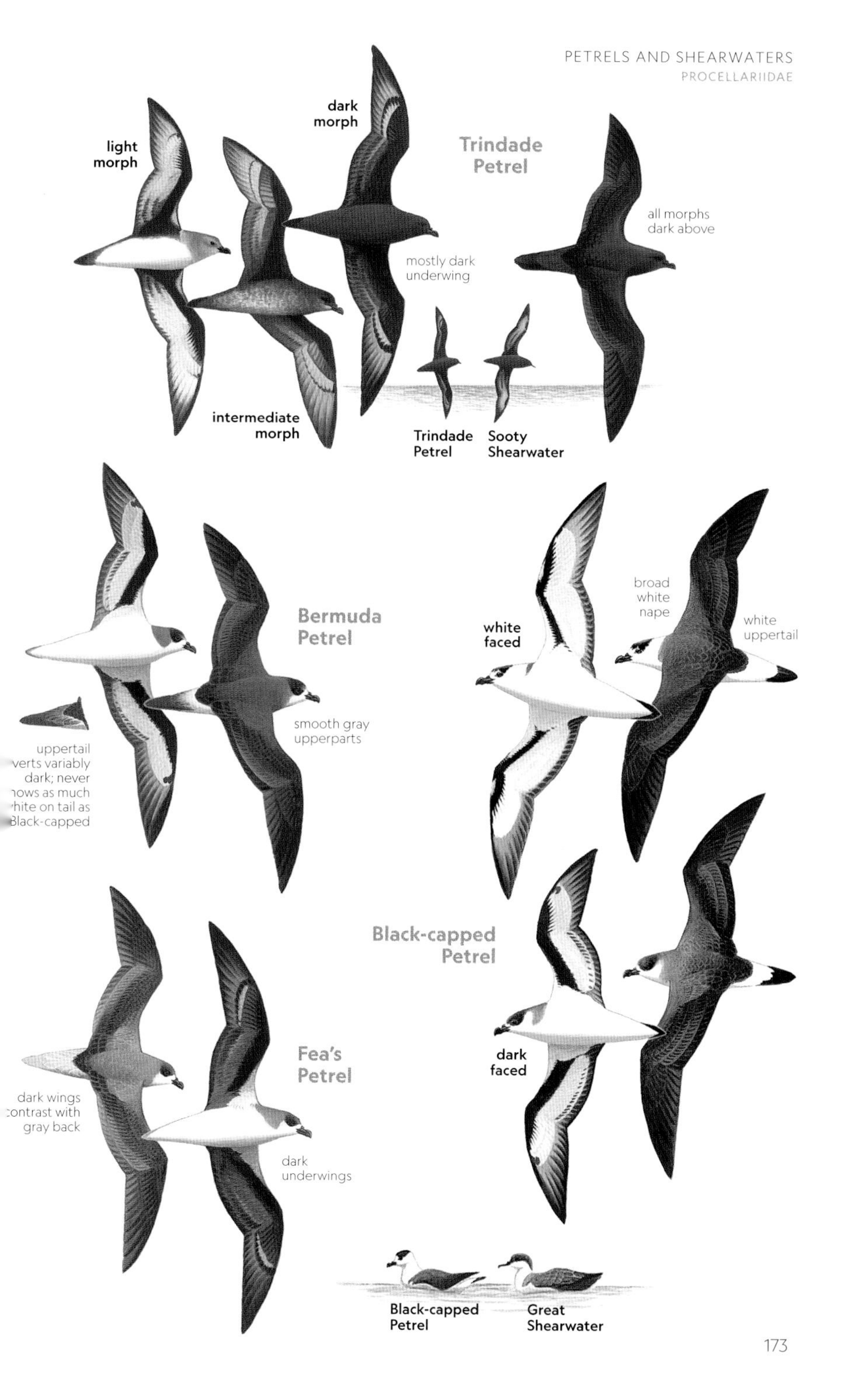
light morph
intermediate morph
dark morph
Trindade Petrel
mostly dark underwing
all morphs dark above
Trindade Petrel
Sooty Shearwater
Bermuda Petrel
uppertail verts variably dark; never ows as much hite on tail as Black-capped
smooth gray upperparts
white faced
broad white nape
white uppertail
Black-capped Petrel
dark faced
Fea's Petrel
dark wings contrast with gray back
dark underwings
Black-capped Petrel
Great Shearwater

## LARGER SHEARWATERS

With stiff wings held like scimitars, these large tubenoses fly so close to the surface that they "shear" the water with their wing tips. Along with the Northern Fulmar (p. 170), these three are the largest members of their family in our area.

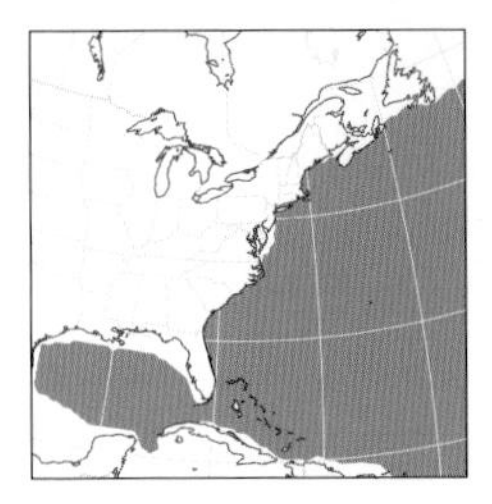

### Cory's Shearwater

*Calonectris diomedea* | CORS | L 18" (46 cm) WS 46" (117 cm)

Majestic and gratifyingly common, this species can be the most frequently spotted tubenose off the mid-Atlantic. The first part of its scientific name is from Greek *kalos* (beautiful), and *diomedea* is the name given to the albatrosses.

■ **APPEARANCE** As large as Sooty and Great, but wings broader, flight more leisurely; yellow bill, visible at some distance, distinguishes it from other shearwaters. Warm sandy brown above in good light, but appears all-dark above in either low light or bright glare. Underparts and underwing linings clean white, unlike on gadfly petrels (p. 172) and larger shearwaters.

■ **VOCALIZATIONS** Bickering over food, gives whiny notes, often doubled or trebled: *ehhh-eeehn, ehh-eh-eeehh,* etc.

■ **POPULATIONS** Two groups, perhaps representing separate species, differ in underwing pattern. "Scopoli's" (nominate *diomedea*) shows white extending almost to primary tips; *borealis,* more common in our area, has mostly dark primaries. Both occur here late spring through early fall.

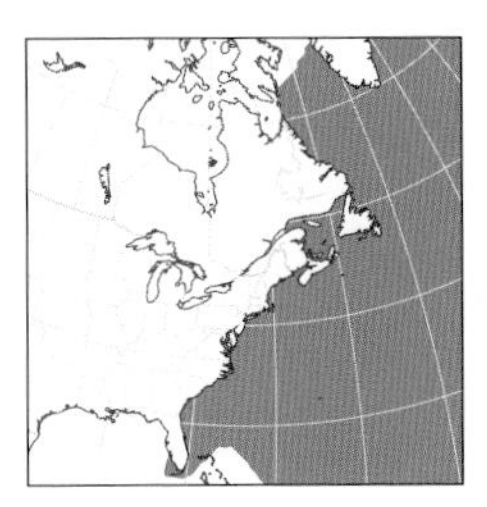

### Sooty Shearwater

*Ardenna grisea* | SOSH | L 18" (46 cm) WS 40" (102 cm)

The Arctic Tern (p. 158) is more famous, but this might be the greatest flier on Earth. Some Sooty Shearwaters, tracing figure eights around the oceans, may traverse 40,000 miles (64,370 km) in a single year.

■ **APPEARANCE** An all-dark procellariid off the East Coast, spring to fall, is likely to be this species. All flash some amount of white under the wing, suggesting rare Trindade Petrel (p. 172), jaegers (pp. 130–133), or even first-winter Herring Gull (p. 146); flight style and, in close view, bill structure seal the deal.

■ **VOCALIZATIONS** Calls like those of other large shearwaters, wimpy and nasal.

■ **POPULATIONS** Found off Atlantic coast late spring through early fall, favoring cooler waters, especially early summer.

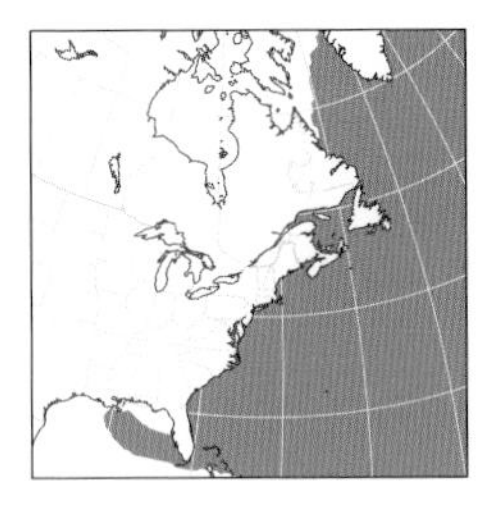

### Great Shearwater

*Ardenna gravis* | GRSH | L 18" (46 cm) WS 44" (112 cm)

It used to be called the "Greater Shearwater," but now it is simply the unqualified Great Shearwater. Like the similarly proportioned Sooty Shearwater, this species is partial to cooler waters.

■ **APPEARANCE** A well-marked shearwater. Black cap set off from white face; black tail separated from dark upperparts by narrow white band. From below, note dark belly patch and irregular black bars on underwings. Smaller Manx and Audubon's Shearwaters (p. 176) all-dark above, cleaner white below. Black-capped Petrel (p. 172) similarly patterned, but more cleanly black-and-white than Great Shearwater.

■ **VOCALIZATIONS** In rows over food, gives nasal whines like other larger shearwaters; not known whether vocalizations differ at sea.

■ **POPULATIONS** Always on the move in the North Atlantic: Birds off the East Coast (spring to fall) are mostly moving north; the movement back south is chiefly off the coasts of Europe and Africa.

dusky-gray upperparts
Cory's Shearwater
borealis
"Scopoli's"
diomedea
yellow bill visible at a distance
extensive white to underside of primaries
variably pale underwings
Sooty Shearwater
stiff-winged flight style
Great Shearwater
dark smudgy belly
mottled underwings
U-shaped band on uppertail
Cory's
Great

## SMALLER SHEARWATERS

Some like it hot: The Audubon's ranges north in summer from West Indies breeding grounds. But the Manx breeds mostly to our north, wandering widely along the East Coast even in winter.

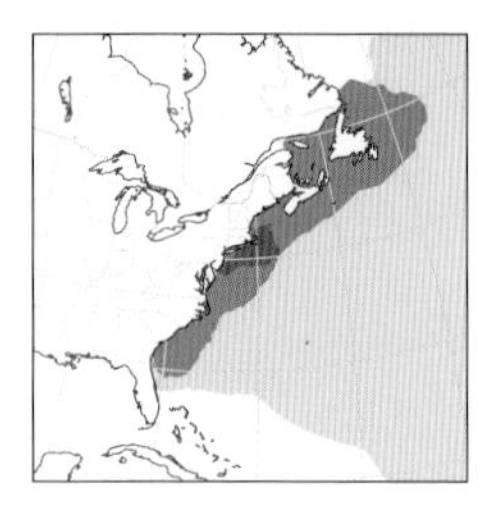

### Manx Shearwater

*Puffinus puffinus* | MASH | L 13½" (34 cm) WS 33" (84 cm)

Opportunistic and somewhat unpredictable, this is the only shearwater that breeds in the East and the only one found regularly in winter in our waters. It is generally uncommon, seen in ones and twos.

■ **APPEARANCE** Intermediate in heft between Great (p. 174) and Audubon's Shearwaters. Upperparts uniformly dark, practically pure black; underparts white. Great Shearwater has white on nape and tail, is smudgy below; "flappier" Audubon's has dark undertail coverts. In close view, note face pattern: Manx has thin white wedge behind eye, lacks white spot above eye.

■ **VOCALIZATIONS** Around nesting colonies at night, gives rich whistles and cackles. Usually silent at sea; issues occasional squeals, not as whiny as those of larger shearwaters.

■ **POPULATIONS** Gradually increasing in East. Since 1970s, has nested in small numbers in New England and Atlantic Canada, but most breed in Europe.

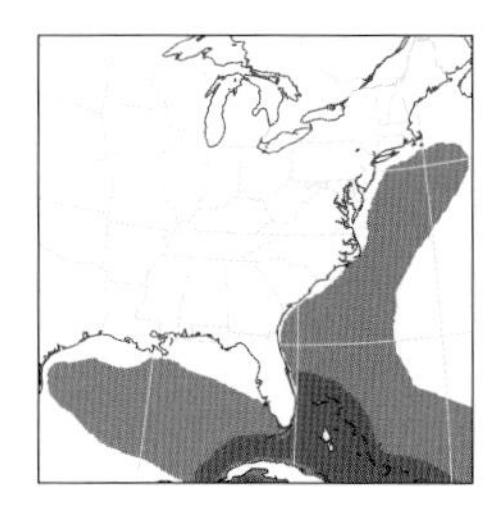

### Audubon's Shearwater

*Puffinus lherminieri* | AUSH | L 12" (30 cm) WS 27" (69 cm)

With breeding grounds as close as the Bahamas, this species ranges widely at sea in our area in the Gulf of Mexico and off the Atlantic coast north to southern New England.

■ **APPEARANCE** Pattern similar to larger Manx Shearwater; Audubon's has relatively rounded wings, shorter than those of Manx, giving it a flappier flight. Audubon's has small white spot above eye, lacking in Manx; thin white wedge behind eye of Manx ill-formed or absent on Audubon's. Undertail coverts black on long-tailed Audubon's, white on relatively short-tailed Manx.

■ **VOCALIZATIONS** Calls of birds away from breeding grounds apparently undocumented.

■ **POPULATIONS** Disperses into our waters, late spring and summer, after breeding season (winter and spring). Annual fluctuations influenced by sea surface temperature; minor irruptions north into waters off southern New England linked to warmwater events. Increasing in Gulf of Mexico.

### Flight Style

A distant shearwater—blackish above and light below—crosses the boat's path, and the trip leader instantly proclaims it to be the first Manx of the trip. How did the leader know it wasn't yet another Great Shearwater, a species the outing has already recorded a dozen of? The two species differ in details of plumage, and size differences are clear in direct comparison. But a snap ID like that of a fast-flying bird?

The trip leader had keyed in on the bird's flight style. Great Shearwaters fly in easy arcs, alternating stiff wingbeats with short glides; Manx Shearwaters, in contrast, fly straight and steady, low above the water, beating their wings quickly. The smaller, shorter-winged Audubon's flies with even faster wingbeats, rarely arcing and gliding like larger shearwaters.

Nothing beats at-sea experience for learning the differences between tubenoses, and organized boat trips led by regional experts are the best. But there's a new resource for today's birders—the Macaulay Library's crowdsourced compendium of videos of tubenoses in flight—and it's getting better every day. Go to *ebird.org/media* and search by species. Doing so before you embark is the best preparation for an encounter with an unexpected seabird.

Manx
Shearwater
white "spur"
on neck
white wraps
behind "ear"
white
undertail
coverts
Audubon's
Shearwater
long tail
dark
undertail
small white
spot above eye

## HUGE, OMINOUS SOUTHERNERS

The only two regularly occurring representatives of their respective families in the East, these are notably large—and just a tad ominous. They occur mostly in the Gulf Coast region, but both wander widely.

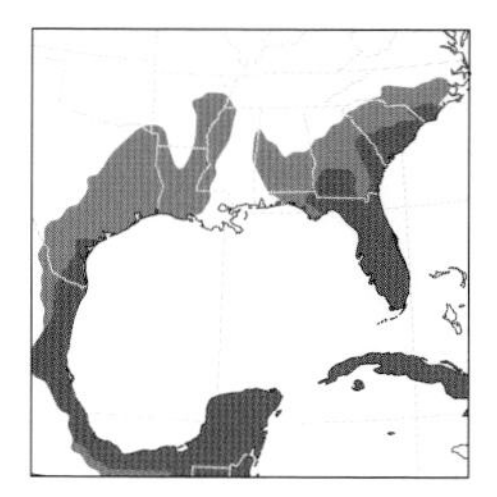

### Wood Stork

*Mycteria americana* | WOST | L 40" (102 cm) WS 61" (155 cm)

Hulking and often hunched over, this bird is right at home in the spooky bayous and brackish basins of the Southeast. Although ungainly when perched, storks are magnificent aerialists and soar masterfully.

■ **APPEARANCE** About the same length as Great Egret (p. 188) but bulkier. Head and neck of adult unfeathered; massive, antler-colored bill, very broad at base, droops downward. Perched, appears mostly white; in flight, black flight feathers of tail and wing are conspicuous.

■ **VOCALIZATIONS** Most commonly heard sound, around nests, is a dry clappering: a nonvocal sound created by rapid snapping of the bill.

■ **POPULATIONS** Routinely wanders well north and inland of core range in our area, typically following breeding season. Fla. birds in long-term decline due to habitat conversion, but the species appears to be compensating with expansion northward of the breeding range.

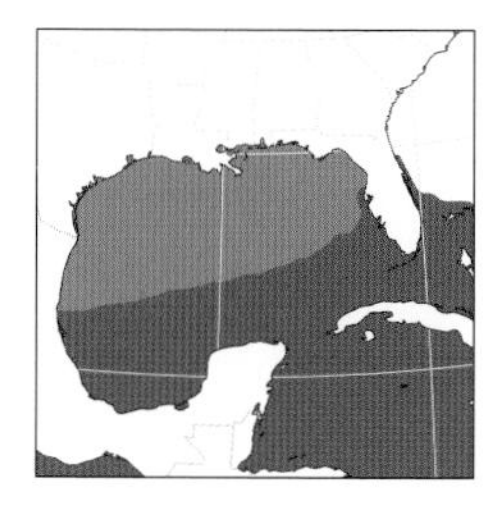

### Magnificent Frigatebird

*Fregata magnificens* | MAFR | L 40" (102 cm) WS 90" (229 cm)

The pterodactyl-like frigatebird has the longest wings, relative to body length, of all birds. Although it is the same length as the Wood Stork, the frigatebird's wingspan is 50 percent greater.

■ **APPEARANCE** Long-tailed and exceedingly long-winged. Stays aloft indefinitely, soaring and wheeling most of the time; occasional wingbeats deep and powerful, but fluid and graceful. The bird is mesmerizing, and a bit unnerving, to behold. Adult male all-black except for red throat patch, inflated garishly in display. Adult female and young variably white below, but overall aspect nevertheless dark and foreboding. Frigatebirds forage only over the ocean, but they never get wet; the bird either plucks food from the surface or catches it in midair.

■ **VOCALIZATIONS** Insect-like chatter and resonant rattle, heard around nesting colonies, are mostly nonvocal sounds created by bill snapping. Silent at sea.

■ **POPULATIONS** Most common in our area along Fla. coast from mid-peninsula southward; breeds at Dry Tortugas N.P. Occurs annually well inland following storms.

### Core Waterbirds

Most of the Earth's surface is water, and many of the world's birds have adapted successfully to aquatic lifestyles. Some groups of waterbirds—the ducks and the grebes, for example—are evolutionarily distinct from one another and from other waterbirds. But recent molecular evidence points to a "core waterbird" cohort of diverse groups sharing a common ancestor. This cohort encompasses all the species in this field guide running from the tropicbirds (p. 162) through the spoonbills and ibises (p. 197).

Within this core waterbird cohort, a large but cohesive group stands out. It includes the orders Ciconiiformes (storks), Suliformes (frigatebirds, boobies, anhingas, and cormorants), and Pelecaniformes (pelicans, herons, and ibises and spoonbills). The group has been provisionally named the Pelecanimorphae—not a bad name since, like pelicans, almost all the birds in this group are large overall and long-billed, eat mostly animal matter, pack into dense colonies, and are competent to supremely gifted fliers.

juvenile
adult
Wood Stork
all-dark flight feathers
adult

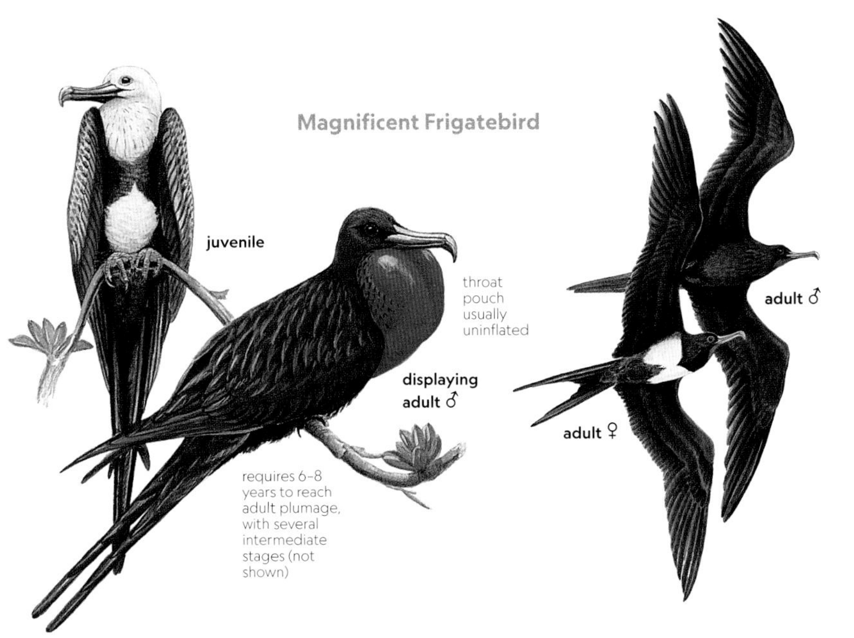
Magnificent Frigatebird
juvenile
throat pouch usually uninflated
displaying adult ♂
requires 6–8 years to reach adult plumage, with several intermediate stages (not shown)
adult ♂
adult ♀

## FAMILY SULIDAE

The boobies (genus *Sula*) and gannets (genus *Morus*) are closely related seabirds that fall to the sea like missiles in their pursuit of large fish just below the water's surface.

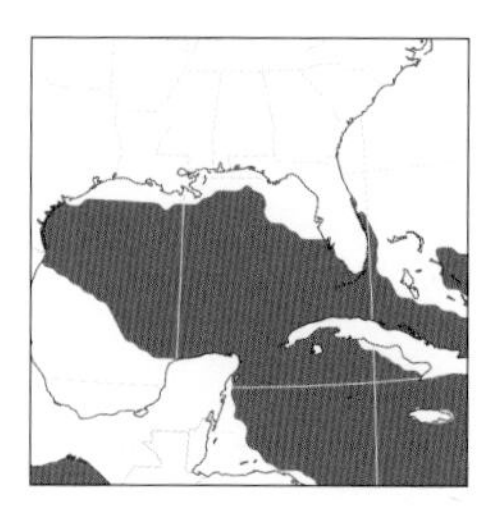

### Masked Booby

*Sula dactylatra* | MABO | L 32" (81 cm) WS 62" (157 cm)

Widespread in the "blue water" realm worldwide, this largest booby is annual in very small numbers off the Gulf and southern Atlantic coasts. Look for it roosting on structures or feeding where there are dolphin and tuna.

■ **APPEARANCE** "Mask" of adult comprises black feathering at base of bill; black tail and white underwing. Mostly brown juvenile distinguished by white underwing and white on belly extending all the way up breast.

■ **VOCALIZATIONS** Endearing honks: *uh, unk,* etc., heard primarily around breeding grounds, rarely at sea.

■ **POPULATIONS** Has bred erratically at Dry Tortugas N.P., Fla., since late 20th century. Seen widely off Gulf Coast and sparingly along Atlantic coast to N.C., with majority of sightings spring to fall.

### Brown Booby

*Sula leucogaster* | BRBO | L 30" (76 cm) WS 57" (145 cm)

Sightings of this species have proliferated in the East in recent years, presumably the result of rising sea surface temperatures. Brown Boobies are more likely than other boobies to be seen from land—and less likely to plunge-dive from dizzying heights.

■ **APPEARANCE** Adult completely brown above; note demarcation between brown hood and white belly; bill and feet dull yellow. Juvenile dark all over; subadult intermediate in appearance between juvenile and adult.

■ **VOCALIZATIONS** Like Masked Booby, emits weak honks, rarely heard at sea.

■ **POPULATIONS** May have bred on Fla. Keys, and range shift northward may portend nesting attempts in our area in the years ahead. Despite recent increase in sightings, the species has declined sharply in the past century.

### Red-footed Booby

*Sula sula* | RFBO | L 28" (71 cm) WS 60" (152 cm)

Rare visitor from tropical oceans to our area, mostly Fla. Variable adult may be brown-backed or white-backed, and dark-tailed or white-tailed; all adults have red feet. Dusky juvenile distinguished by smudgy breastband, orangey feet, gray and relatively slight bill, smaller body size overall.

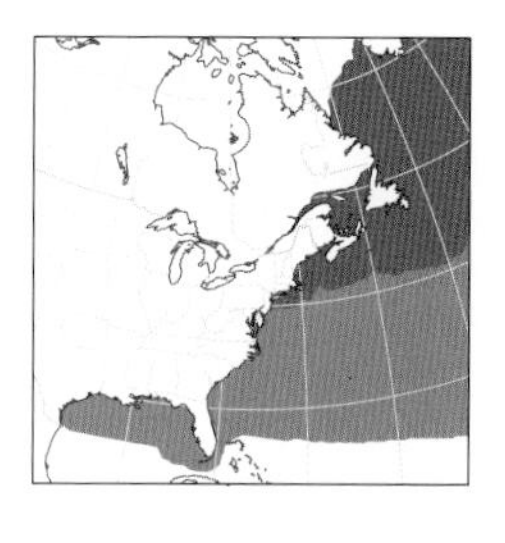

### Northern Gannet

*Morus bassanus* | NOGA | L 37" (94 cm) WS 72" (183 cm)

Notably larger than other sulids, this impressive seabird nests in immense colonies on the steep sea cliffs of Atlantic Canada. Gannets in winter are easily seen from shore, although they are usually well beyond the surf.

■ **APPEARANCE** A "flying cross," with long neck, long tail, and long wings held out straight. Adult mostly white, but note black wing tips and golden wash on head; plumage pattern recalls that of white-morph adult Red-footed Booby. Mostly dark juvenile distinguished from smaller boobies by blacker plumage with fine white spotting.

■ **VOCALIZATIONS** At colonies, rolling *rrrurr* and throatier *rraaah,* given by hundreds or even thousands of birds all at once.

■ **POPULATIONS** By far the most common sulid in the East. Even in summer, nonbreeding gannets, especially subadults, are more likely than boobies along our coasts.

Masked Booby
subadult
adult
all-dark flight feathers
adult
subadult
adult
juvenile
juvenile
white underwing
Red-footed Booby
juvenile
smudgy breast
adult
juvenile
dusky underwing
ult ♂
subadult ♀
Brown Booby
adult ♀
2nd year
juvenile
3rd year
speckled plumage
Northern Gannet
1st summer
adult
juvenile
adult
adult courtship display

**DEVIL BIRDS**
The Anhinga's name derives from the Tupi word for "devil," and the cormorant is the "devil's bird" in North American lore—for their gluttony and supposed ill omen.

## Anhinga

*Anhinga anhinga* | ANHI | L 35" (89 cm) WS 45" (114 cm)
This is the only New World representative of the darter family Anhingidae, closely related to the cormorants in the family Phalacrocoracidae.
■ **APPEARANCE** Known as the "Snake Bird" for habit of swimming with only neck and head above surface. All have spear-like bills; adult has unique corrugated tail. Adult male black with silvery wings; head and breast of female pale brown. Roosts in trees like cormorants; often soars with long tail spread.
■ **VOCALIZATIONS** Flat, quiet nasal muttering when perched, *urr urr urr*.
■ **POPULATIONS** Occurs mostly in freshwater swamps of southeastern coastal plain, but regularly vagrates well north; recently started breeding in Ill.

## Great Cormorant

*Phalacrocorax carbo* | GRCO | L 36" (91 cm) WS 63" (160 cm)
Like all cormorants, this species has a hooked bill and a colorful, but variable, patch of bare skin at the base of the bill.
■ **APPEARANCE** Bulkier than Double-crested and proportionately shorter-tailed. Adult has yellow facial skin bordered by white feathering; in breeding plumage, sports white flank patch, prominent on bird in flight. Juvenile is pale-bellied and dark-breasted, the inverse of juvenile Double-crested.
■ **VOCALIZATIONS** Around nests and winter roosts, gives soft, throaty chuckles and occasional louder squawks.
■ **POPULATIONS** Restricted in our area mostly to marine habitats, but visits freshwater sites to roost and bathe.

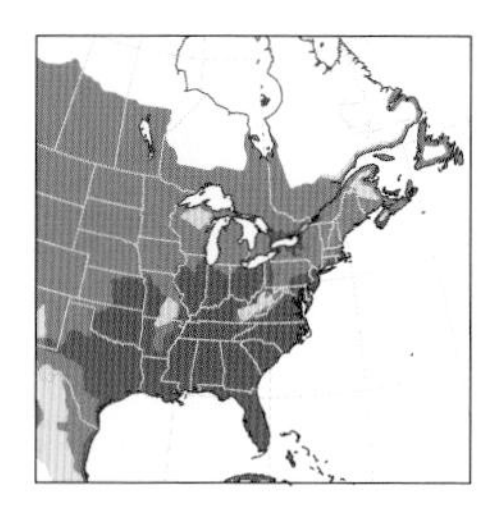

## Double-crested Cormorant

*Nannopterum auritum* | DCCO | L 32" (81 cm) WS 52" (132 cm)
In much of the East, this is the only cormorant. Even where the two other species are expected, the Double-crested can be more common.
■ **APPEARANCE** Smaller and longer-tailed than Great Cormorant, larger and shorter-tailed than Neotropic. Double-crested has yellow-orange throat patch, smoothly rounded and lacking white border; bare skin extends to front of eyes. Adult otherwise all-black; juvenile has dark belly and variably pale breast.
■ **VOCALIZATIONS** Low croaks and grunts; one call like human belching.
■ **POPULATIONS** Following sharp losses in 20th century, has rebounded remarkably; the species is now considered by some to be a menace to commercial fisheries, an exaggerated assessment.

## Neotropic Cormorant

*Nannopterum brasilianum* | NECO | L 26" (66 cm) WS 40" (102 cm)
Found mostly in Tex. and La. in our area, this species favors smaller waterbodies, typically freshwater, more than the look-alike Double-crested.
■ **APPEARANCE** Smaller, more lightly built than Double-crested, with longer tail and slighter bill. Yellowish throat patch pinches back to a sharp point, is thinly bordered by white feathering. Juvenile brown-breasted; juvenile Double-crested paler and grayer on breast.
■ **VOCALIZATIONS** Soft croaking and oinking, like wood frogs or distant pigs.
■ **POPULATIONS** Range expanding north and east. Vagrants routine to Midwest; recent breeder north to Ill.

Anhinga
♀
often soars on thermals
♂
♀
♀
nonbreeding adult
juvenile
Great Cormorant
adult Double-crested
pale belly
adult Neotropic
adult Great
short tail
white flank patch
yellow lores
Neotropic Cormorant
eeding adult
triangular gular pouch bordered with thin white stripe
juvenile
juvenile
winter adult
nonbreeding adult
Double-crested Cormorant
long tail

## PELICANS

"A wonderful bird is the pelican. His beak will hold more than his belly can." So begins the famous limerick. The key adaptation is a distensible throat pouch used as a dip net for snatching up fish.

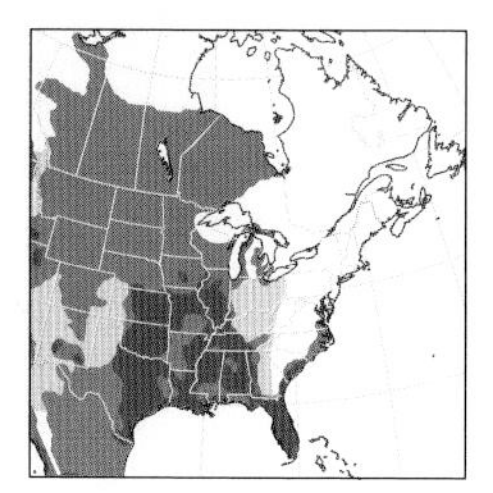

### American White Pelican

*Pelecanus erythrorhynchos* | AWPE | L 62" (157 cm) WS 108" (274 cm)

Measuring 5 ft (1.5 m) from stem to stern, with a 9-ft (2.7 m) wingspan, this is the biggest bird in the East. It lumbers on land, glides across waterways, and soars with impossible grace.

■ **APPEARANCE** Size alone is usually sufficient for field ID; it dwarfs the Great Blue Herons (p. 186), with which it often occurs. Appears all-white on land or water. Black flight feathers of wing prominent; soaring Wood Stork (p. 178) has black tail. Bill in breeding season has laterally compressed plate; in any season, bill may be stained with mud or oil.

■ **VOCALIZATIONS** Rarely vocalizes; whooshing of wings on birds in flight quite audible at close range.

■ **POPULATIONS** Breeds far inland, unlike the marine Brown Pelican; winters mostly coastally, but typically in lagoons and back bays. Vagrants widely noted far from core range.

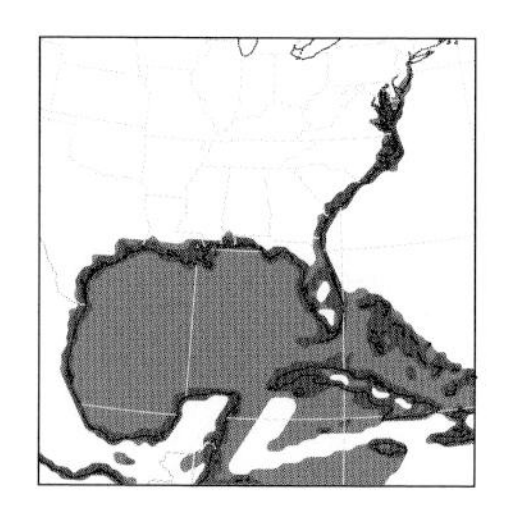

### Brown Pelican

*Pelecanus occidentalis* | BRPE | L 48" (122 cm) WS 84" (213 cm)

At "only" 4 ft (1.2 m) in length and with "only" a 7-ft (2.1 m) wingspan, this is clearly the smaller of our two pelicans. It roosts in docks and marinas, foraging just offshore.

■ **APPEARANCE** Body dark gray-brown at all ages. Breeding adult has chestnut neck and nape, variable yellow-white crown; nonbreeding adult mostly white-headed and white-necked. Juvenile dusky all over; bill and plumage always darker than that of juvenile American White Pelican.

■ **VOCALIZATIONS** Like American White Pelican, does not usually vocalize. One study has shown that the timbre of wing-whooshing in flight may be related to molt.

■ **POPULATIONS** Continues to recover impressively from terrible crash in DDT era of mid-20th century. After breeding, some move north along coast a ways before heading back south again; this strategy is much more pronounced in Brown Pelicans off the West Coast. Chiefly coastal but sometimes wanders far inland; has recently started breeding inland in Fla.

### Feeding Strategies

Even without any cues of color or pattern, our two pelican species are instantly distinguished by their utterly different foraging behaviors. Brown Pelicans fly in single-file undulations just offshore, then bank sharply and plunge into the ocean to catch fish. American White Pelicans never do that. Instead, they swim in shoulder-to-shoulder flotillas on calm waters, then immerse their bills—but not the rest of their bodies—in amazing synchrony.

The behavioral differences between these closely related species (they are in the same genus) are unusually great, but they point to a much broader strategy for the identification of birds. In example after example, consistent and easily observed feeding behaviors can help to distinguish similar species in the field. The actual method of food capture—say, sallying out for an insect vs. gleaning one from the underside of a leaf—is important to note, but so are details of microhabitat: in the shady understory vs. up in the sunlit canopy vs. down on the ground, and so on. In terms of aesthetics, few sights are more pleasing than a bird in flight or sitting on a perch; but in terms of learning bird ID, few behaviors are more useful to note than foraging.

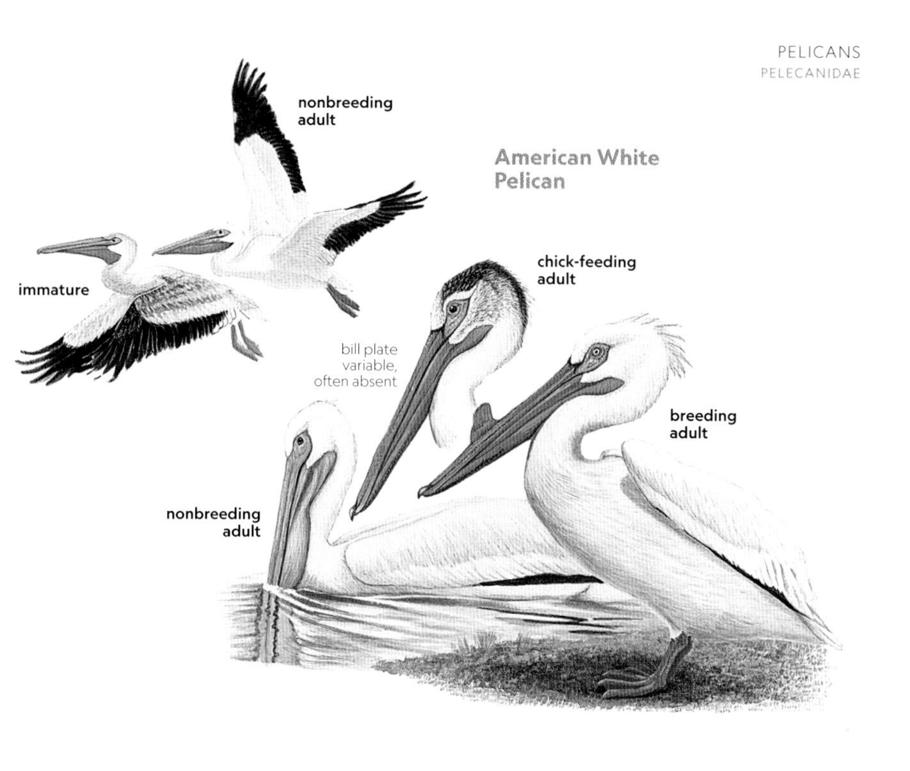
nonbreeding adult
American White Pelican
immature
chick-feeding adult
bill plate variable, often absent
breeding adult
nonbreeding adult

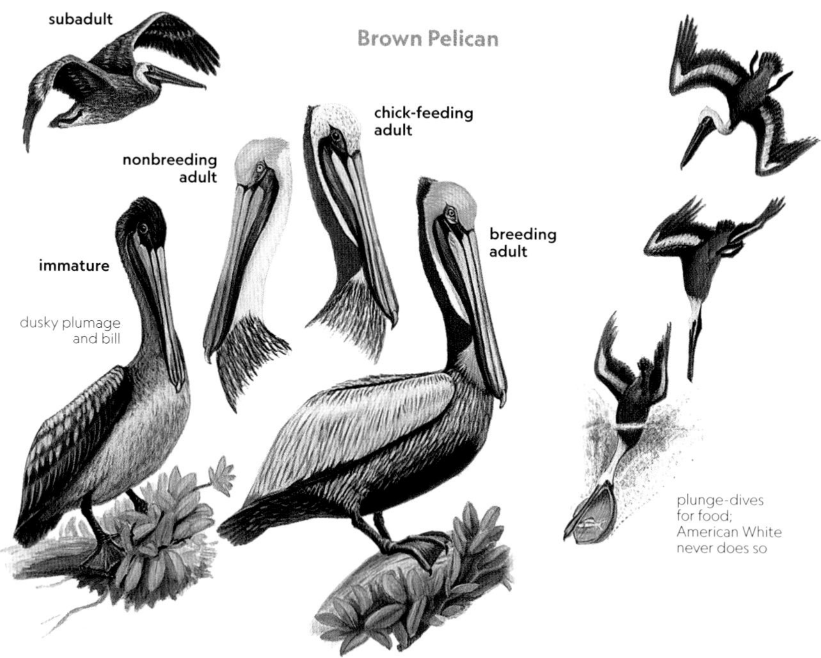
subadult
Brown Pelican
chick-feeding adult
nonbreeding adult
breeding adult
immature
dusky plumage and bill
plunge-dives for food; American White never does so

## DIVERSE ARDEIDS

Bitterns, herons, and egrets in the family Ardeidae (pp. 186–193) are well represented in the East. Ardeids are varied in color, pattern, and size, but all have daggerlike bills for spearing fish and other prey.

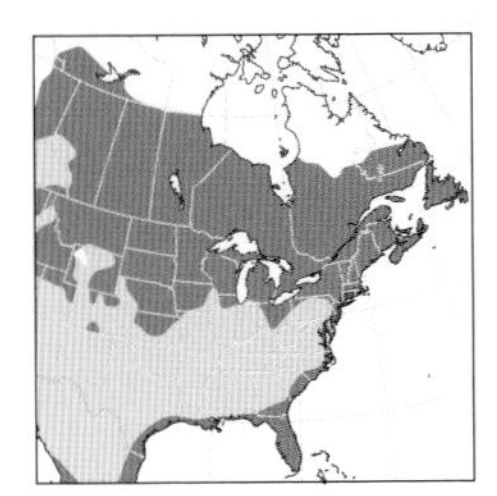

### American Bittern

*Botaurus lentiginosus* | AMBI | L 28" (71 cm) WS 42" (107 cm)

Our two bitterns are solitary and secretive. By day, they hide in bulrushes; at night, they are somewhat more active. This much larger of the two species in the East oddly freezes in place when spotted, with bill pointing straight up.

■ **APPEARANCE** Medium height but bulky. Muddy brown above; has long, thick, brown stripes below. Bill and feet dull yellow. Adult has long black stripe on face and neck, absent on juvenile. In flight, distinguished from juvenile night-herons (p. 192) by contrastingly pale brown wing coverts.

■ **VOCALIZATIONS** Called the "Thunder Pumper" for its low-pitched song, like someone swallowing air: *oonk-k'loonk,* repeated. Flight call an abrupt, monosyllabic *rrak,* similar to night-herons.

■ **POPULATIONS** Thought to be declining, but monitoring difficult owing to the species' reclusiveness. Especially secretive in winter, and northern limit of nonbreeding range imperfectly known.

### Least Bittern

*Ixobrychus exilis* | LEBI | L 13" (33 cm) WS 17" (43 cm)

Barely the size of a robin (p. 318), this smallest of all the world's herons is even more reclusive than the much larger American Bittern. The name bittern is related to the word "bull," alluding to their strange songs.

■ **APPEARANCE** Bright and buffy, but appears dark when glimpsed in dense cattails. All have blurry streaks below and huge buffy oval on folded wing. Adult male black-backed, female and juvenile not as dark-backed.

■ **VOCALIZATIONS** Song a steady series of three to five clucking or cooing notes: *rook-rook-rook-rook*. Flight call a dull *ruck*.

■ **POPULATIONS** Like American Bittern, believed to be declining; loss of marshlands the prime suspect in both species' ill fortunes. Dark morph, called "Cory's Least Bittern," unreported in recent years.

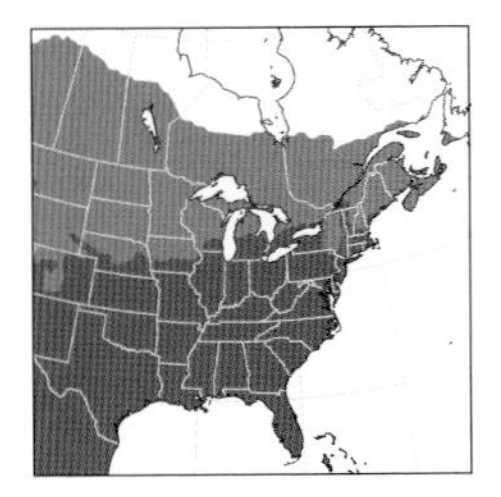

### Great Blue Heron

*Ardea herodias* | GBHE | L 46" (117 cm) WS 72" (183 cm)

A lone wader at the water's edge often proves to be this species. The bird stands like a statue—until it strikes with incredible speed at a fish or frog beneath the water's surface.

■ **APPEARANCE** Our largest heron; mostly gray-blue, with impressive yellow bill. Adult has thick black stripe on pale head; juvenile dark-capped. Dark legs of adult feathered at thigh with burgundy "trousers." Distinctive in flight, with neck tucked in, legs trailing, and dark remiges contrasting with paler wing coverts. Larger white subspecies in South Fla. has straw-yellow legs and thick bill; compare with Great Egret (p. 188). Intermediate subspecies, also Fla. only, has white head and gray-blue body.

■ **VOCALIZATIONS** When flushing, a monstrous eruption of white noise: *rrRRRAAAHH!* Young at nest solicit food with weird, machinelike ticking.

■ **POPULATIONS** Across much of the East, the most familiar and most frequently spotted heron—and indeed one of the most familiar of all waterbirds. White subspecies, of mostly Caribbean distribution, considered by some a separate species, "Great White Heron." Intermediate subspecies called "Wurdemann's Heron."

adult

**American Bittern**

light brown coverts

juvenile

adult

adult ♂

juvenile

buffy oval on wing

unreported in recent decades

high breeding adult ♂

"Cory's" adult ♂

juvenile

adult ♂

**Least Bittern**

adult ♀

adult

breeding adult

breeding adult

juvenile

yellow legs

"[...]eat White Heron" *occidentalis*

"Wurdemann's Heron" breeding adult

**Great Blue Heron** *herodias*

rufous thighs

## MOSTLY WHITE EGRETS

The name egret has more to do with color than taxonomy. The Great, Cattle, and Snowy Egrets are in three different genera, with correspondingly varied body plans and feeding behaviors.

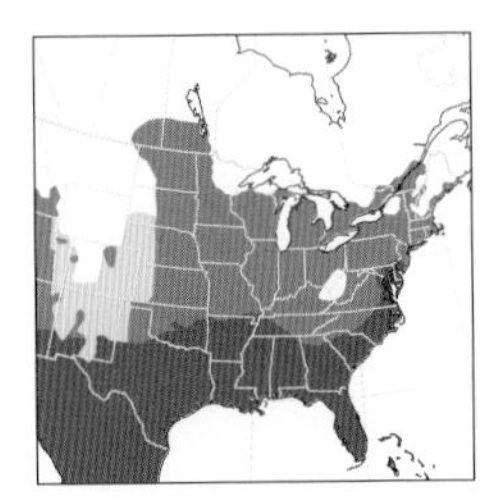

### Great Egret

*Ardea alba* | GREG | L 39" (99 cm) WS 51" (130 cm)

Although called an egret, this stately wader is more closely related to the Great Blue Heron. Because of its significance in bird conservation history, the Great Egret has long been the logo of the National Audubon Society.

■ **APPEARANCE** All-white like Snowy Egret, but proportioned like a slender Great Blue Heron. Bill yellow and legs black, the same color combo as Cattle Egret, but Great's body plan and feeding behavior are very different. At beginning of breeding season, acquires bright green facial skin and extremely thin plumes, called aigrettes, at rear. Like Great Blue Heron, stands stock-still for long periods of time.

■ **VOCALIZATIONS** Especially when courting, emits loud rumbles like a mower or buzz saw starting up.

■ **POPULATIONS** Brought to the precipice of extirpation from U.S. in early 20th century by plume hunters, but has substantially recovered. Today enjoys wide global distribution, with healthy numbers in both freshwater and saltwater habitats.

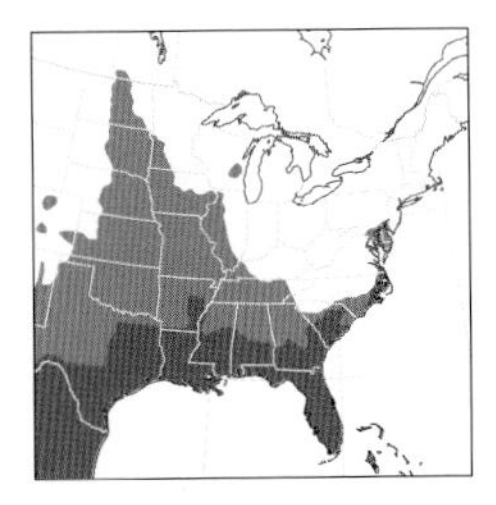

### Cattle Egret

*Bubulcus ibis* | CAEG | L 20" (51 cm) WS 36" (91 cm)

This most terrestrial ardeid is often found in pastures with cattle and horses. Cattle Egrets feed right at the feet of livestock, pumping their heads as they go. Often they feed while *on* livestock, "surfing" as the beasts lumber along.

■ **APPEARANCE** Compact and relatively small; bill stout, legs short. Adult dark-legged and yellow-billed like Great Egret, but much smaller. During courtship, acquires orange wash, bare parts become more colorful. Dark-billed juvenile superficially resembles Snowy Egret, but the latter has "golden slippers" and usually forages in standing water.

■ **VOCALIZATIONS** Call a short, nasal grunt, sometimes given in interactions while feeding: *unt, unk*, etc.

■ **POPULATIONS** Colonist from Old World via S. Amer.; opportunistic, still establishing in grazed landscapes that aren't too dry. Bulk of range in East in southeastern U.S., but routinely wanders north to southern Canada.

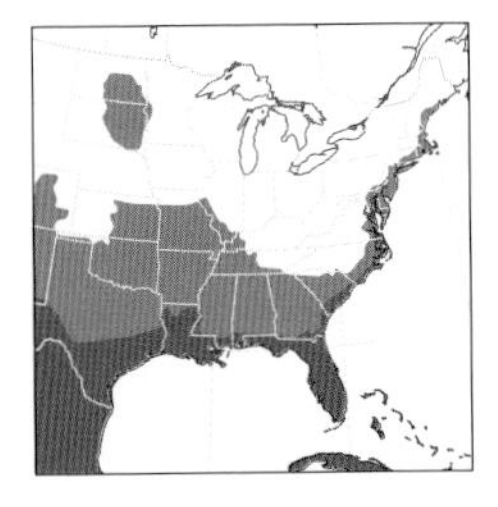

### Snowy Egret

*Egretta thula* | SNEG | L 24" (61 cm) WS 41" (104 cm)

Along with the larger Great Egret, this elegant heron was foundational to modern wildlife conservation. It is one of the "true" egrets, genus *Egretta*.

■ **APPEARANCE** Slim and dainty. Adult readily separated from larger Great Egret by thin black bill and black legs with yellow feet ("golden slippers"). Like Great Egret, undergoes color intensification for a short while at beginning of breeding season; lores and feet of Snowy become blood orange. Juvenile, with paler bill and legs, is trickier; compare especially with juvenile Little Blue Heron. Prances and capers while feeding; compare with white-morph Reddish Egret (p. 190), which likewise engages in this behavior.

■ **VOCALIZATIONS** Nasal groan, *raaa-aaah*, slightly wavering.

■ **POPULATIONS** Nearly extirpated from U.S. in early 20th century by plume hunters, but recovered quickly following legal protections and changing public attitudes. Disperses north in small numbers following breeding.

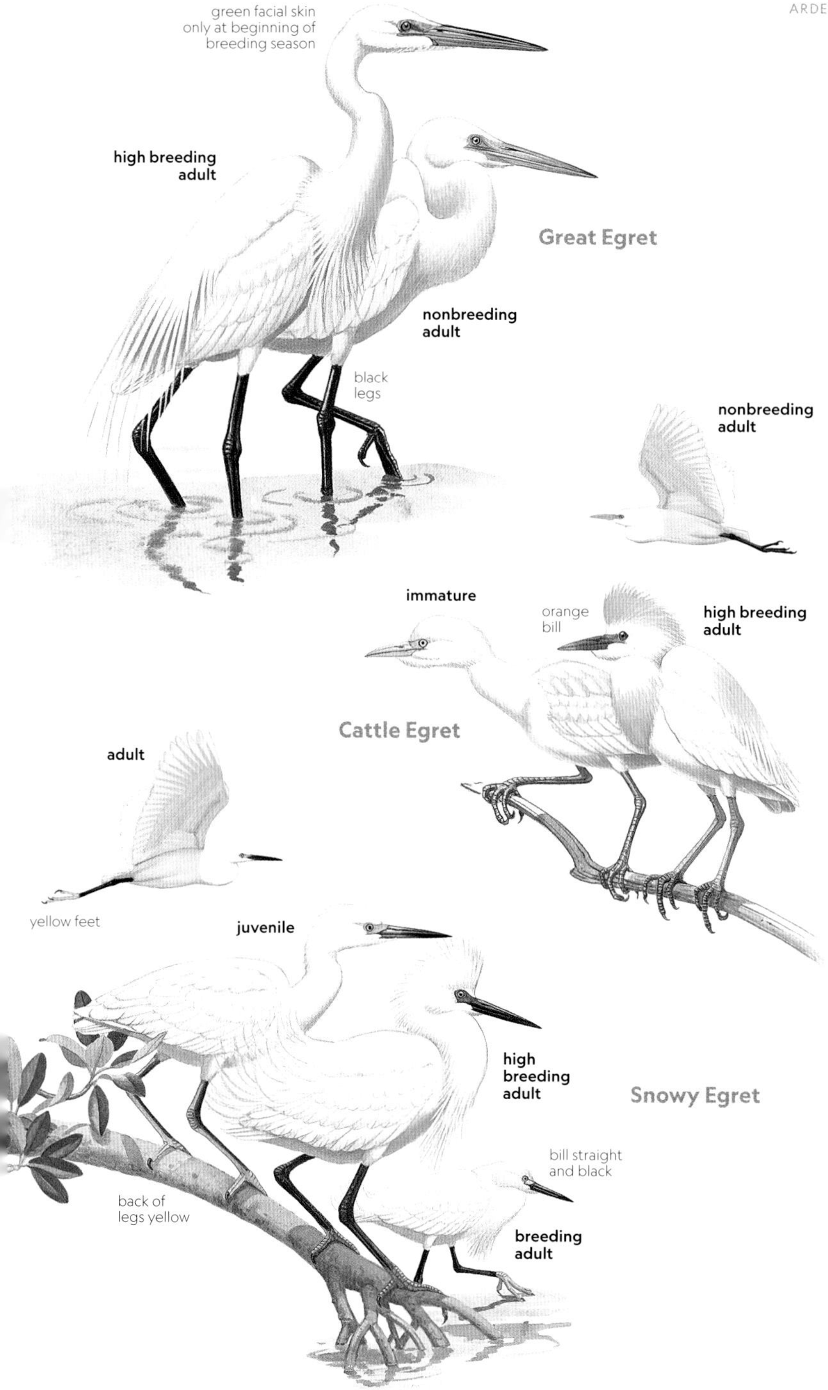

green facial skin only at beginning of breeding season

**high breeding adult**

## Great Egret

**nonbreeding adult**

black legs

**nonbreeding adult**

**immature**

orange bill

**high breeding adult**

## Cattle Egret

**adult**

yellow feet

**juvenile**

**high breeding adult**

## Snowy Egret

bill straight and black

back of legs yellow

**breeding adult**

## OFTEN-DARK EGRETS

Along with the Snowy Egret (p. 188), these three are "true" egrets in the genus *Egretta*. Snowy Egrets are normally all-white, but color variation in our other *Egretta* herons is extreme.

### Little Blue Heron

*Egretta caerulea* | LBHE | L 24" (61 cm) WS 40" (102 cm)

All herons show age-related variation in plumage, but this species takes it to the next level: Adults are dark, appearing nearly black in low light; juveniles are snow-white.

■ **APPEARANCE** Adult mostly dark blue with deep-plum neck and head; legs greenish, bill blue-gray. Color scheme of adult roughly similar to that of dark-morph adult Reddish Egret, but the latter is brighter and more contrasting, with strikingly two-toned bill. Juvenile Little Blue almost identical in plumage to juvenile Snowy Egret, so focus on bare parts: Little Blue has relatively thick bill, drooping slightly, pale blue-gray at base; bill of Snowy thinner and straighter, with yellow at base extending to lores. Legs of juvenile Little Blue dull yellow-green; legs of juvenile Snowy bicolored, black on "shins," deeper yellow on "calves." As juvenile Little Blue matures, begins to show gray splotches on otherwise white plumage; note gray-tipped primaries of older juvenile Little Blue.

■ **VOCALIZATIONS** Nasal groans like Snowy's, averaging higher-pitched.

■ **POPULATIONS** Fairly common within range, centered in the East on the Gulf and Atlantic coastal plains; routinely wanders far inland. Not as inclined to flocking as Snowy.

### Tricolored Heron

*Egretta tricolor* | TRHE | L 26" (66 cm) WS 36" (91 cm)

This species is partial to saltwater habitats—more so than the Snowy Egret and Little Blue Heron, but less so than the Reddish Egret. Relatively antisocial, the "Trike" is typically seen singly.

■ **APPEARANCE** Long-necked and long-billed, even for a heron. Adult mostly slate blue above, juvenile extensively rusty above; white underparts contrast with colorful upperparts. Bill yellow most of the year, briefly becoming baby blue during courtship. There's no general consensus on which three colors are indicated by the bird's name!

■ **VOCALIZATIONS** Call like other *Egretta* herons, but more honking.

■ **POPULATIONS** Range expanded north in second half of 20th century; has since stabilized or withdrawn slightly.

### Reddish Egret

*Egretta rufescens* | REEG | L 30" (76 cm) WS 46" (117 cm)

This is the largest, least common, and most frenetic egret. Feeding, the bird lurches about, stamps its feet, and raises its wings to create an umbrella that lures fish to the surface. This canopy feeding also makes it easier for the egret to see fish.

■ **APPEARANCE** Has two color morphs. Head and neck of adult shaggy; long bill is bicolored in breeding season. Dark-morph adult slate gray overall with red-brown head and neck; white-morph adult distinguished from other egrets by large size, long bill, dark gray legs, and shaggy neck plumes. The animated feeding behavior alone often indicates this species, but other *Egretta* herons, especially Snowy Egret, also caper about.

■ **VOCALIZATIONS** Nasal honks like Tricolored Heron's, but shorter and lower.

■ **POPULATIONS** Uncommon permanent resident along Gulf Coast. Feeds right out in the open, in shallow saltwater lagoons with little if any cover.

Little Blue Heron
juvenile
adult
gray facial skin
juvenile
bill, pale at base, droops slightly
breeding adult
nonbreeding adult
molting immature
greenish legs
Tricolored Heron
reeding adult
pink legs
long-necked
juvenile
breeding adult
Reddish Egret
bicolored bill
dark-morph breeding adult
dark-morph breeding adult
dark-morph juvenile
white-morph breeding adult shows bicolored bill
white-morph winter adult

## COMPACT MEDIUM TO SMALL HERONS

These three are the sole representatives in the East of their three respective genera. All are rather stocky, with rounded wings and relatively short legs and bills. They are often seen sitting quietly—the Green Heron alertly, the night-herons sluggishly—in brush at the water's edge.

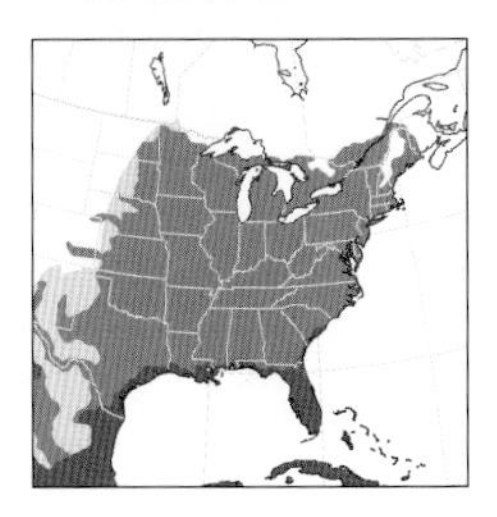

### Green Heron

*Butorides virescens* | GRHE | L 18" (46 cm) WS 26" (66 cm)

"Chalk Line," "Shitepoke," and "Poke" are among the many vernacular names—some of them unprintable—given to this species for its habit of defecating explosively when flushed.

■ **APPEARANCE** Smallish but sturdily built; bullnecked with short legs. Adult dark and glossy above, appearing more bluish than green in normal ambient light. Chestnut neck and face contrast with dark blue-green cap. Feet yellow, becoming reddish in breeding season. Juvenile brownish, heavily streaked; Least Bittern (p. 186) smaller and more colorful.

■ **VOCALIZATIONS** On flushing, a piercing *kyark!*

■ **POPULATIONS** Fairly common but typically found singly. Favors pond edges and creek banks, especially around woods.

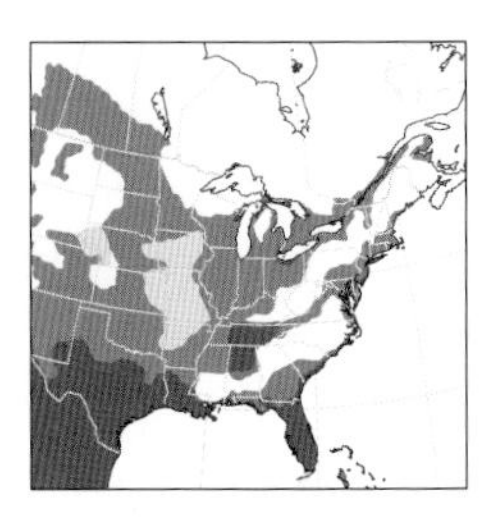

### Black-crowned Night-Heron

*Nycticorax nycticorax* | BCNH | L 25" (64 cm) WS 44" (112 cm)

By day, it's a hunched form in dense cover at water's edge. By night, it's a startling utterance in the dark. The species is equally at home in freshwater and saltwater habitats.

■ **APPEARANCE** Larger and stockier of the two night-herons; bullnecked and short-legged. Adult gray-bodied with black crown and black back. Juvenile brown with white speckling above, white streaking below; distinguished from juvenile Yellow-crowned by coarser speckling and streaking, shorter legs, and smaller, spikier, mostly yellow bill. Juvenile also recalls more warmly colored American Bittern (p. 186), which has stronger streaking below; wings of bittern in flight more obviously two-toned.

■ **VOCALIZATIONS** In flight, a sudden *quok!*, often given after dark.

■ **POPULATIONS** Stable, even increasing, across vast global range; in East, local and declining in Appalachia and interior New England.

### Yellow-crowned Night-Heron

*Nyctanassa violacea* | YCNH | L 24" (61 cm) WS 42" (107 cm)

More of a Southerner than the Black-crowned, this night-heron haunts mangroves and freshwater swamps, where it feasts on crustaceans. The Yellow-crowned often finds its way to residential districts, roosting and nesting in yards and on rooftops.

■ **APPEARANCE** Proportionately longer-necked and longer-legged than Black-crowned, but a tad smaller overall. Adult has slaty body; bold black and white markings on head. Juvenile darker than juvenile Black-crowned, with finer white speckling and streaking; black bill of juvenile Yellow-crowned is thick and blunter. Juveniles of both night-herons slowly transition to adult plumage through second calendar year; as with Black-crowned, younger juveniles can be confused with American Bittern.

■ **VOCALIZATIONS** Descending *kyaa* in flight, typically after dark.

■ **POPULATIONS** Range in East expanded north in 20th century; has since stabilized or contracted a bit.

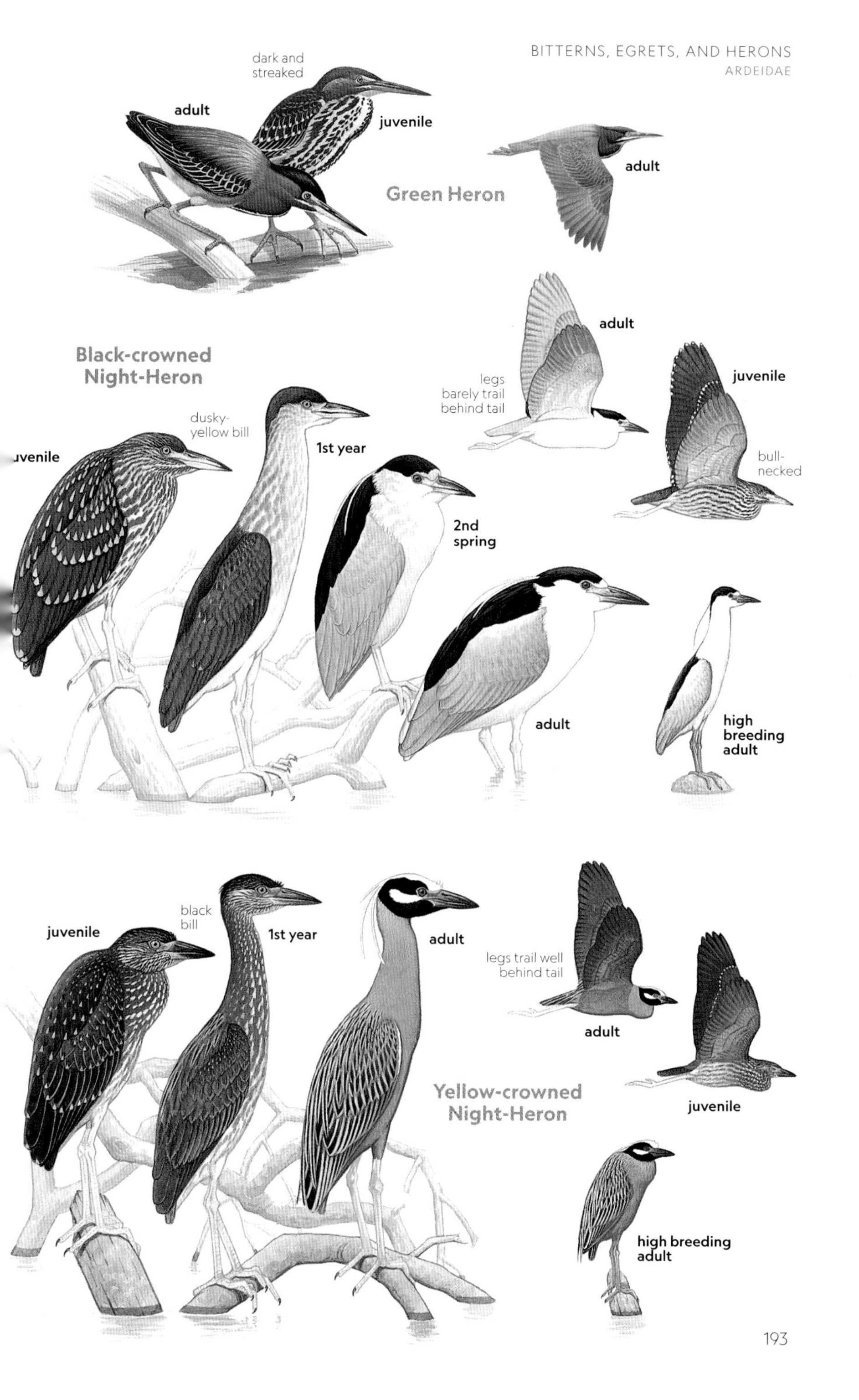
dark and
streaked
adult
juvenile
adult
Green Heron
adult
Black-crowned
Night-Heron
legs
barely trail
behind tail
juvenile
dusky-
yellow bill
uvenile
1st year
bull-
necked
2nd
spring
adult
high
breeding
adult
black
bill
juvenile
1st year
adult
legs trail well
behind tail
adult
Yellow-crowned
Night-Heron
juvenile
high breeding
adult

## SPOONBILLS AND IBISES

Together they constitute the family Threskiornithidae—from the Greek for "sacred birds"—because some were venerated in ancient religions. They are notable for their impressive bills, with special adaptations for tactile foraging.

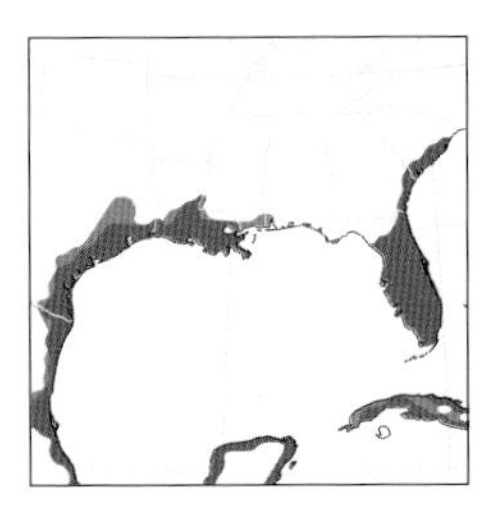

### Roseate Spoonbill

*Platalea ajaja* | ROSP | L 32" (81 cm) WS 50" (127 cm)

Except for the rare American Flamingo (p. 64), this is the only extensively pink wading bird in the East. Many locals and some wildlife biologists refer to them simply as "pinks."

■ **APPEARANCE** Heavyset wader with a remarkable bill. Variably pink: Juvenile has just a pinkish blush; older immatures more heavily imbued with pink; adults have scarlet highlights on uppertail coverts and leading edge of wing, plus broad orange tail tip. Extent of pink affected by diet and feather wear.

■ **VOCALIZATIONS** Short grunts in rapid series: *uhr-uhr-uhr-uhr-uhr* .... Also a nasal, belching *aaauuurrr.*

■ **POPULATIONS** Widespread in coastal lowlands, but with predilection for swamps and brackish or freshwater marshes. A few wander well inland each year; occasionally stages small irruptions northward, as in summer 2021.

### White Ibis

*Eudocimus albus* | WHIB | L 25" (64 cm) WS 38" (97 cm)

A characteristic and charismatic species of the southeastern coastal plain, this largest of our ibises occurs widely—not only in wetlands, but also around parks and golf courses, even condos and landfills.

■ **APPEARANCE** A bit larger than *Plegadis* ibises (p. 196), but much smaller than Roseate Spoonbill. All have long, decurved bill. Adult plumage nearly pure white; just a few primary tips are black. Bill, facial skin, and legs red. Juvenile dusky gray-brown, suggesting genus *Plegadis,* but belly and underwing white. Older immatures splotchy brown and white.

■ **VOCALIZATIONS** Short, gruff barks when flushing or squabbling over food; sometimes more resonant *hooooon,* nasal and trumpeting.

■ **POPULATIONS** Numbers increasing on central and western Gulf Coast. Breeding range expanding north along Atlantic coast.

### Bill Adaptations

Modern birds lack teeth and foreclaws, so feeding is all about bill structure and mechanics. Every bird species' feeding ecology is more or less obviously reflected in bill morphology, and the core waterbirds (pp. 162–197) are no exception. Spoonbills sweep their spatulate bills slowly through standing water, and ibises probe deep in the mud with their decurved bills—which raises a question: How do they see their prey?

Spoonbills and ibises rely heavily on tactolocation, the ability to sense the slightest of movements and even electrical activity. Birds' bills aren't like human hair or fingernail tips; instead, they are coursing with blood vessels and nerve fibers. Many birds can bend their bills, and some sandpipers and hummingbirds are particularly accomplished in this regard. And many others—including species in the family Threskiornithidae—have motion detectors, typically in nerve endings at the tips of their bills. Some birds can even detect magnetism with their bills, an aid in navigation.

Finding prey using tactolocation isn't an all-or-nothing proposition for waders like spoonbills and ibises: They are perfectly capable of sight. In fact, their eyes are set in such a way as to focus on objects immediately beyond the tip of the bill. As in so many other aspects of bird biology, multiple adaptations are brought to bear simultaneously on the essential activity of finding food.

juvenile
all-pink wings
Roseate Spoonbill
adult
extent of pink variable
adult
juvenile
white back
juvenile
White Ibis
adult
juvenile
adult
1st spring

## GENUS *PLEGADIS*

Both of these dark ibises are glossy above in good light, and both appear white-faced in many views. They may be cautiously separated by the color of their bare parts, but they nevertheless remain among the trickiest field IDs in the East.

### Glossy Ibis

*Plegadis falcinellus* | GLIB | L 23" (58 cm) WS 36" (91 cm)

On most of the East Coast, this is the default dark ibis. Flocks fly single file over coastal districts, then put down into marshes and pond edges where they feed in tight bunches.

■ **APPEARANCE** Dark overall with stocky build; long-legged, long-necked, and very long-billed. Breeding adult quite similar to White-faced Ibis, but duller overall: plumage less glossy; eyes and facial skin dark; legs gray-brown with reddish "ankles"; bill gray-brown. Two bluish lines on face are thin and pale. Nonbreeding adult and juvenile duskier, largely unglossed.

■ **VOCALIZATIONS** Short grunts and groans, low-pitched and somewhat nasal.

■ **POPULATIONS** One of the less heralded invaders of the East: Reached U.S. in early 19th century, was rare until early 20th century; has been expanding range northward and westward coastally ever since.

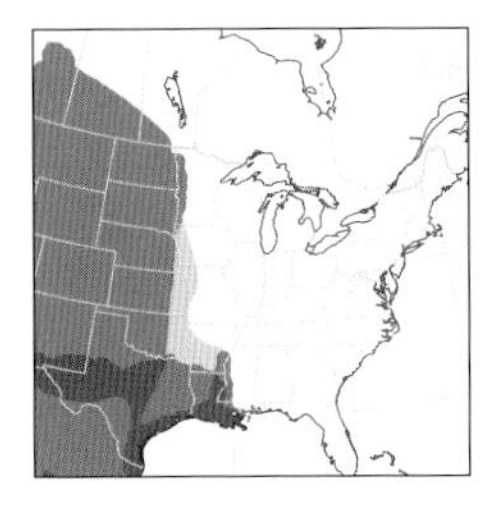

### White-faced Ibis

*Plegadis chihi* | WFIB | L 23" (58 cm) WS 36" (91 cm)

The western interior counterpart of the Glossy Ibis, this species is a frequent sight around marshes with standing water and emergent vegetation. Where the two species' ranges overlap, they occupy the same habitats.

■ **APPEARANCE** Identical in size and shape to Glossy Ibis. Breeding adult White-faced has eponymous white feathering completely encircling reddish facial skin; bill grayer, legs brighter red, and plumage a bit shinier than on Glossy. Nonbreeding adult and juvenile dusky overall, nearly identical to Glossy; but eye of adult White-faced red, and eye of adult Glossy and young of both species dark brown.

■ **VOCALIZATIONS** Monosyllabic grunts like Glossy; on average a bit higher and more rising than Glossy, but much overlap.

■ **POPULATIONS** Numbers dropped and range contracted in mid-20th century, but species has recovered well with better management of wetlands and restrictions on pesticides; vagrants increasingly noted to East Coast, partly reflecting enhanced searching by birders.

### The Dark Ibises

Even in breeding plumage, Glossy and White-faced Ibises appear identical except in close view. They were treated by some as one species, the Glossy Ibis, until well into the 20th century, with species limits (where one species ends genetically and another begins) in the genus *Plegadis* not fully worked out at the present time. And when our *Plegadis* ibises were split into two species, birders soon recognized that they had a novel problem on their hands: Glossy and White-faced Ibises were starting to hybridize.

Expanding populations of both ibises have recently come into secondary contact: After a long period of geographic isolation, they and their genes are again "in contact" with one another. The species co-occur as breeders now in the Gulf Coast region and apparently farther inland, and interbreeding may be extensive. Hybrid adults in breeding plumage show traits intermediate between the two parental species, but hybrids in other plumages are not safely identifiable in the field. Birders may call such birds "*Plegadis* sp.," which is never wrong, and is often the only responsible designation.

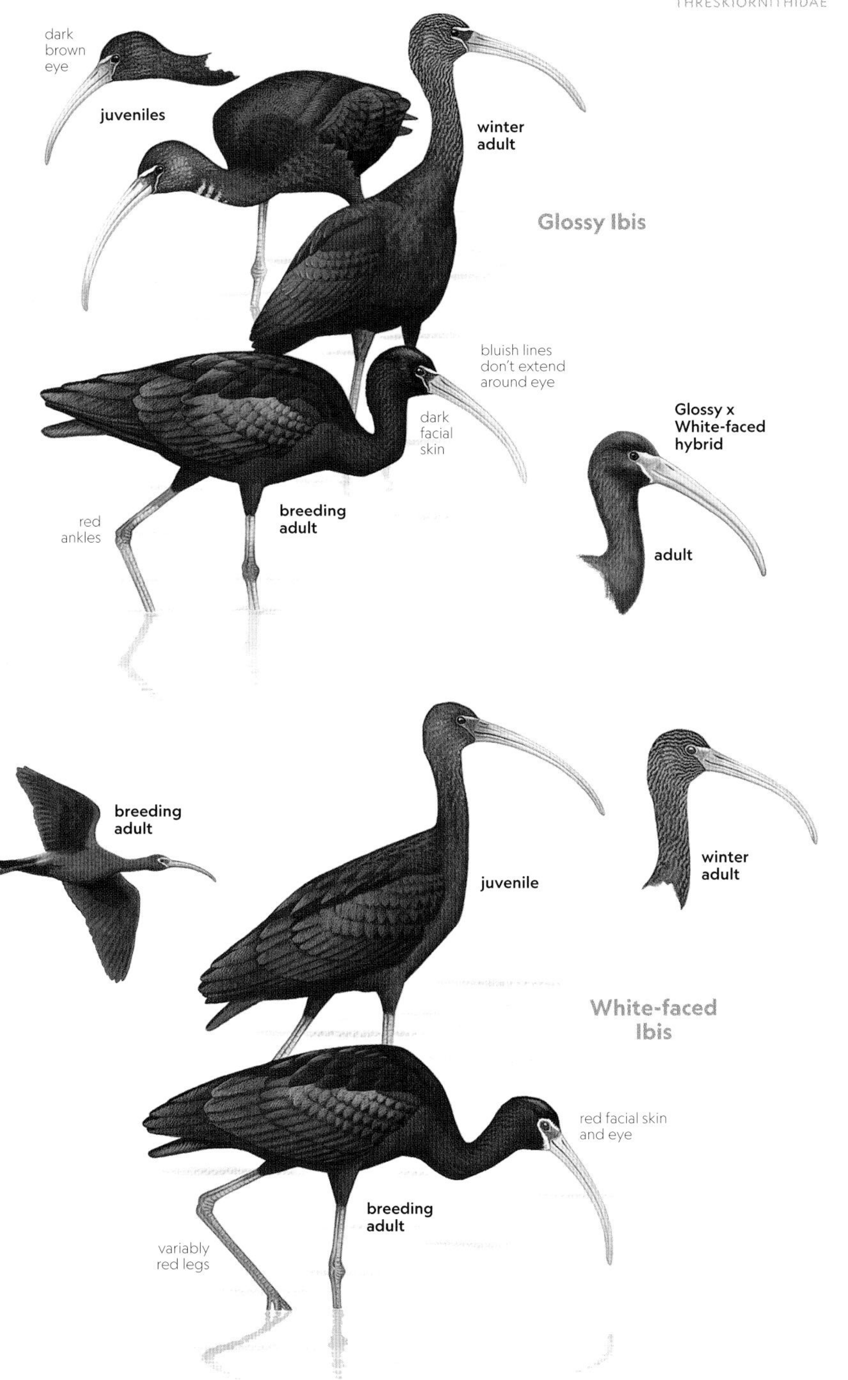
dark
brown
eye
juveniles
winter
adult
Glossy Ibis
bluish lines
don't extend
around eye
dark
facial
skin
Glossy x
White-faced
hybrid
red
ankles
breeding
adult
adult
breeding
adult
juvenile
winter
adult
White-faced
Ibis
red facial skin
and eye
breeding
adult
variably
red legs

## DIET-SPECIALIST RAPTORS

The Osprey and New World vultures are taxonomic outliers among the diurnal raptors. New World vultures, placed in their own order, eat carrion and aren't considered birds of prey. The Osprey, which eats only fresh-caught fish, is the sole representative of its family, Pandionidae.

### Black Vulture

*Coragyps atratus* | BLVU | L 25" (64 cm) WS 57" (145 cm)

The more assertive and adventurous of our two vultures, this species is often found around dumpsters and landfills. It lacks the keen sense of smell of the Turkey Vulture, and often associates with that olfactorily superior species in the quest for carrion.

■ **APPEARANCE** A bit shorter in body length and quite a bit shorter in wingspan than Turkey Vulture. Head of adult unfeathered year-round. Tail short; legs whitish, long, and sturdy. On bird in flight, whitish outer primaries contrast with otherwise dark wings; long legs trail behind short tail. Flaps stiffly, more than Turkey Vulture.

■ **VOCALIZATIONS** New World vultures lack a syrinx, the sound-producing apparatus in most birds, so they are left with breathy grunts and hissing. Feeding at carrion, Black Vultures snap at each other with a gruff *woof*; they defend young with long, drawn-out *HHHUUUUOOOOOHH*.

■ **POPULATIONS** Range expanding north. Attracted recent notoriety for preying on calves, but such reports proved exaggerated.

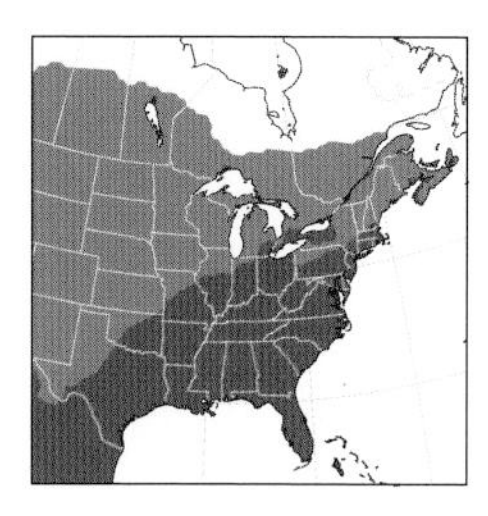

### Turkey Vulture

*Cathartes aura* | TUVU | L 27" (69 cm) WS 69" (175 cm)

Don't be put off by the name. Few sights are more serene than a kettle of these long-winged scavengers soaring effortlessly in the summer sky.

■ **APPEARANCE** Adult has unfeathered red head, gray and bristly in juvenile; longer-tailed and longer-winged than Black Vulture. In flight, tilts and rocks, wings locked in shallow V. Wings two-toned from below: wing linings dark, flight feathers paler. Typically seen in flocks wheeling about in search of carrion on the ground below.

■ **VOCALIZATIONS** Like Black Vulture, lacks a syrinx. Rarely makes sounds when feeding, unlike Black Vulture, but defends young with unpleasant groaning and hissing.

■ **POPULATIONS** One of the most frequently spotted birds across much of range. Like Black Vulture, slowly expanding north. Highly migratory; exceedingly rare in winter north of mapped range.

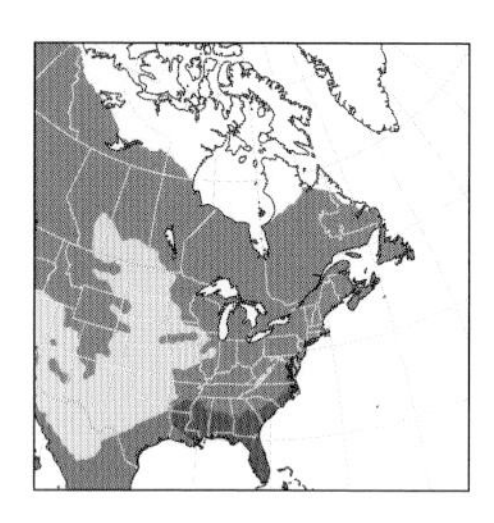

### Osprey

*Pandion haliaetus* | OSPR | L 22–25" (56–64 cm) WS 58–72" (147–183 cm)

It's not just that the Osprey catches fish, it's *how* it does it: in messy feetfirst cannonball dives from considerable altitudes. And more often than not, the drenched bird reemerges with an impressively large fish!

■ **APPEARANCE** Medium to large raptor, relatively long-winged; wings kinked in flight. Adult uniform chocolate above, mostly white below, with variable "necklace" of dark feathering, especially on female; whitish head has messy crest and thick brown eyeline. Juvenile like adult, but upperparts scaled with white.

■ **VOCALIZATIONS** Powerful down-slurred whistle, often given in slow series: *kyew ... kyew ... kyew! KYEW!*

■ **POPULATIONS** Suffered terribly during DDT era but has recovered impressively. Readily takes to artificial nest platforms, prominent along roadsides, lakeshores, waterways, etc.

Black Vulture
short tail
pale wing tips
two-toned underwing
adult
dark head
juvenile
wings held in shallow V
Turkey Vulture
long, angled wings
dark eye stripe
Osprey
carolinensis
occasionally recorded in Florida
juvenile
adult
Caribbean
ridgwayi

**THREE KITES**

The five kites in the East (pp. 200, 206) are smallish raptors with gray and black plumages. Their taxonomy is in flux, but one consistent trend is emerging: They are an artificial assemblage. Our five species represent five different genera, not particularly closely related.

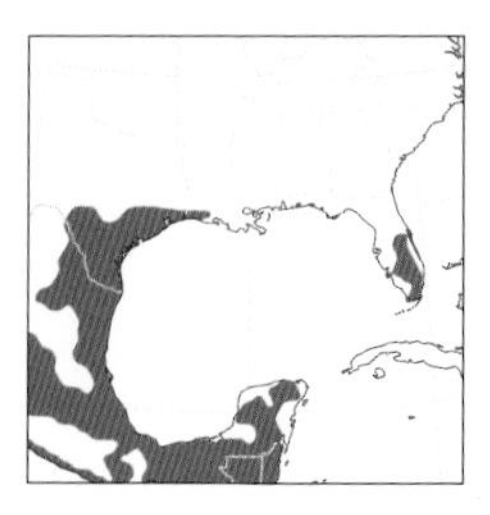

## White-tailed Kite

*Elanus leucurus* | WTKI | L 16" (41 cm) WS 42" (107 cm)

A long-tailed and long-winged bird is hovering on beating wings. It is mostly gray and white, with a black shoulder patch. The bird suggests a tern (pp. 152–161), but it is a White-tailed Kite, more inclined to hover than any other raptor in the East.

■ **APPEARANCE** A small gray hawk of open country. Black "shoulder" (upperwing coverts) and long white tail distinctive at all ages. Adult gray-backed; white face marked by black eye patch; underparts white. Juvenile washed cinnamon, but already shows black wing coverts and white tail of adult.

■ **VOCALIZATIONS** Most calls rather short with clipped quality: *cleep* and *cleet,* whistled and descending. Also gives somewhat ternlike calls, grating and stuttering: *chree-ick, chreee-iiiih,* etc.

■ **POPULATIONS** Following range expansion in early 20th century, has stabilized or withdrawn slightly. Most common in East along western Gulf Coast; scarcer in South Fla.; strays well north of core range annually, especially to Midwest. Usually seen singly, but sometimes forms wintertime roosts.

## Hook-billed Kite

*Chondrohierax uncinatus* | HBKI | L 18" (46 cm) WS 36" (91 cm)

Uncommon and wary permanent resident of dense bosques along Rio Grande of South Texas. In rare encounters with perched bird, note sharply hooked bill and staring pale eyes. Adult dark; male barred gray and white below or, rarely, all-black; female barred rufous and white below. Juvenile whitish below with brown barring on breast. Most are seen in flight, and wing shape is distinctive: pinched in at the base, bulging outward. Tail in flight often fanned, showing wide dark and light bands.

## Swallow-tailed Kite

*Elanoides forficatus* | STKI | L 23" (58 cm) WS 48" (122 cm)

An oversize Barn Swallow (p. 292) comes to mind, as does a diminutive frigatebird (p. 178), but the unique Swallow-tailed Kite is considered by many to be the most graceful raptor in the East.

■ **APPEARANCE** Slender and elegant, with strongly contrasting black-and-white plumage. Adult essentially jet-black and snow-white; juvenile, like many other raptors, tinged buffy. Typically seen in flight in loose flocks above forest clearings or open country. Deeply forked, very long tail and striking plumage unlike any other raptor's; alternates soaring with easy flapping, changing direction suddenly but smoothly. Feeds on the wing, snagging dragonflies in midair; even gleans insects from the water's surface, in the manner of frigatebirds.

■ **VOCALIZATIONS** Piercing *wheep,* often run together in rapid series, suggesting flight call of Solitary Sandpiper (p. 122).

■ **POPULATIONS** Widespread Neotropical raptor at the northern limit of its range in our area. Uncommon here, but widely noted owing to its highly aerial lifestyle. Wanders well north of core range, especially late spring and summer.

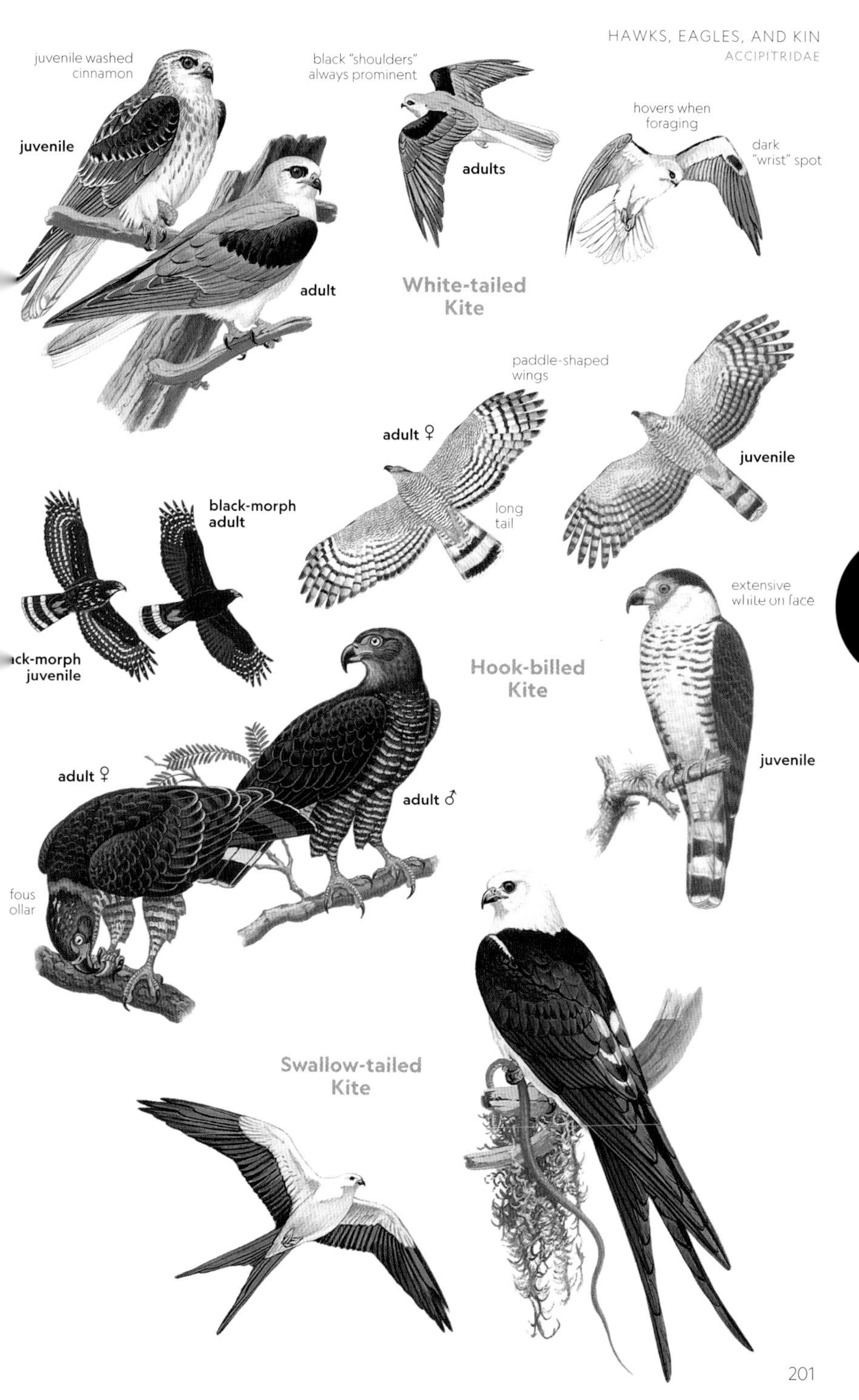
juvenile washed cinnamon
juvenile
adult
black "shoulders" always prominent
adults
hovers when foraging
dark "wrist" spot
White-tailed Kite
paddle-shaped wings
adult ♀
long tail
juvenile
black-morph adult
ack-morph juvenile
extensive white on face
Hook-billed Kite
juvenile
adult ♀
adult ♂
fous ollar
Swallow-tailed Kite
Hawks

## AN EAGLE AND A HARRIER

These two are the only representatives in the Americas of their respective genera, widespread in the Old World. The Golden Eagle is not particularly closely related to the Bald Eagle (p. 206), and the harrier has evolved certain owl-like traits.

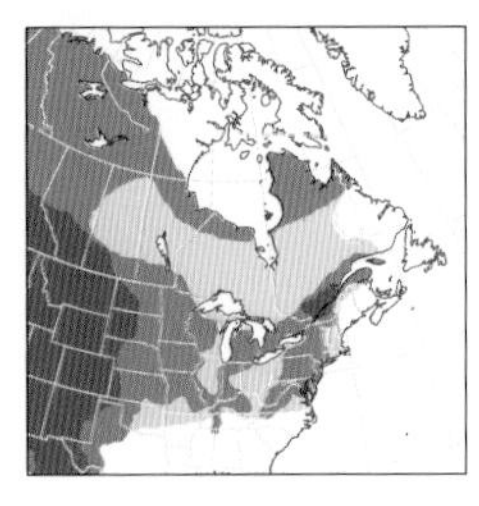

### Golden Eagle

*Aquila chrysaetos* | GOEA | L 30–40" (76–102 cm) WS 80–88" (203–224 cm)

An emblem of royalty and majesty in many cultures, this huge raptor is scarce throughout its vast range. It preys on mammals, especially rabbits and hares.

■ **APPEARANCE** Same size as Bald Eagle, but proportioned differently: relatively small-headed and long-tailed; more likely to be confused at a distance with Turkey Vulture (p. 198). All plumages have golden wash on hindneck. Adult mostly dark; juvenile has striking white patches at base of tail and on center of wing, visible from above and below.

■ **VOCALIZATIONS** Largely silent. Away from nest, gives short chirps, often repeated, typically down-slurred: *kyerp kyerp kyerp.*

■ **POPULATIONS** Widespread in U.S. and Canada, but much more common in the West. Sightings in the eastern U.S. mostly at hawk watches in late fall and sparingly in winter.

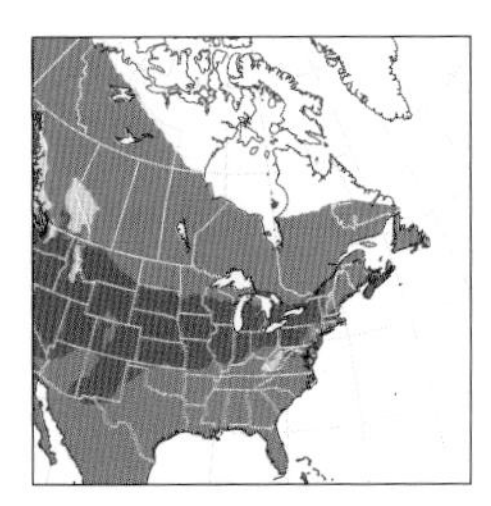

### Northern Harrier

*Circus hudsonius* | NOHA | L 16–20" (41–51 cm) WS 38–48" (97–122 cm)

With its owl-like facial disc and habit of flying just above ground level, this bird is among our most distinctive raptors. Harriers hunt over open terrain with dense cover, often in low light and relying heavily on auditory cues.

■ **APPEARANCE** Usually seen in flight. The long wings are held in a shallow V. Tail long; uppertail coverts white. Plumage variation extensive: Older males, called "Gray Ghosts," are gray above and white below with inky black wing tips; adult female dark brown with dense streaking below; juvenile warm brown-orange below, mostly unstreaked.

■ **VOCALIZATIONS** Mostly silent, but gets excited around nest, especially when chasing off intruders; gives scratchy yaps in series and harsh whistles.

■ **POPULATIONS** Distribution tightly coupled with annual variation in numbers of microtine rodents. Co-occurs with Short-eared Owls (p. 224), dependent on the same prey.

### What Is a Raptor?

Certain groups of birds—hummingbirds, parrots, and woodpeckers come to mind—are widely known. So it is with raptors, frequently called birds of prey. But whereas the terms "hummingbird," "parrot," and "woodpecker" denote evolutionarily discrete units, that is not the case with "raptor."

Take the Peregrine Falcon (p. 240), on any birder's short list of the most iconic of raptors. Yet the Peregrine is a closer relative of the parrots than of the Golden Eagle, another iconic raptor. What about Mississippi and Snail Kites (both on p. 206), which eat insects and gastropods, respectively? What about shrikes (p. 272), songbirds that prey on other songbirds? And where do owls, mostly nocturnal predators, fit in? Are any of those raptors?

The prevailing view, rooted much more in tradition than biology, is that the raptors comprise the unrelated orders Accipitriformes (represented in the East by hawks, eagles, kites, harriers, vultures, and the Osprey) and Falconiformes (falcons and caracaras). Many eat birds and mammals, but some do not. Meanwhile, many non-raptors are fiercely raptorial. Vultures and owls are considered raptors by some, non-raptors by others. The word "raptor" isn't going to be retired anytime soon, but the biological construct has been found to be wanting.

maller bill
han Bald
agle
Golden Eagle
adult
juvenile
all plumages show golden wash on nape
white wing patches on juvenile
adult
dark or faintly barred tail
juvenile
adult ♀
owl-like face
streaked belly
adult ♂
adult ♂
adult ♂
juveniles
Northern Harrier
white uppertail coverts

## GENUS *ACCIPITER*

With their short wings and long tails, these three are adroit at hunting in woodlands. They're often called accipiters, and field ID is challenging.

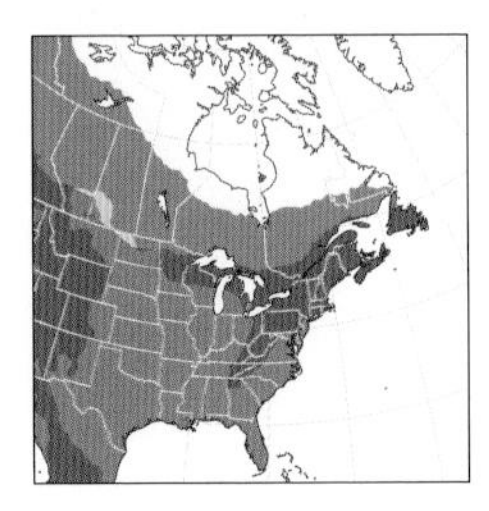

### Sharp-shinned Hawk

*Accipiter striatus* | SSHA | L 10-14" (25-36 cm) WS 20-28" (51-71 cm)

"Flap, flap, sail, with a short, square tail." That little ditty has some utility in separating this smallest accipiter from the larger, look-alike Cooper's, but there is more to it than that.

■ **APPEARANCE** Smallest hawk in the East; females over 50 percent larger than males. Adult bluish above, barred reddish below; adult Cooper's nearly identical, but has more discrete dark cap. Juvenile dark brown above, with smudgy red-brown steaks below; juvenile Cooper's has sharper and sparser streaking below. The "Sharpie" is smaller-headed and shorter-tailed (but still relatively long-tailed for a raptor) than the "Coop." In leisurely flight, Sharp-shinned has a pointy-edged tail; Cooper's is rounded.

■ **VOCALIZATIONS** Shrill whistles, *pee* or *pyew*, in rapid succession, more similar in timbre to those of goshawk than Cooper's.

■ **POPULATIONS** Formerly persecuted for being a "chicken hawk," but enjoys legal protection today. Secrėtive; breeding range probably expanding south in U.S.

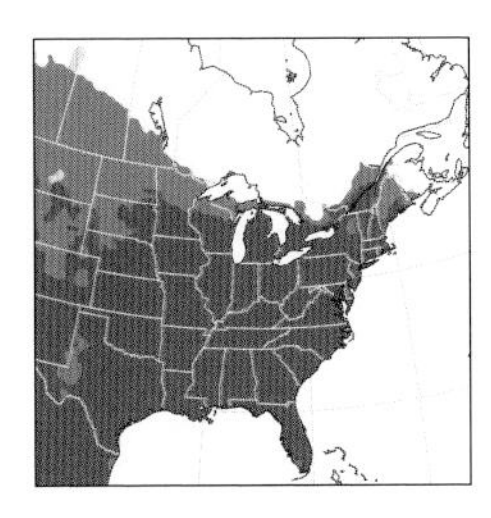

### Cooper's Hawk

*Accipiter cooperii* | COHA | L 14-20" (36-51 cm) WS 29-37" (74-94 cm)

Until the late 20th century, this accipiter was a woodland recluse. It has since urbanized, and is now fairly common in suburbs and even cities.

■ **APPEARANCE** Similar in all plumages to smaller Sharp-shinned Hawk. Dark cap of adult Cooper's, especially male, more prominent than on Sharp-shinned. Juvenile Cooper's has crisp streaking below, often shows splotchy white on back; juvenile Sharp-shinned has reddish-brown blobs and streaks below, averages less white on back. Cooper's notably long tail is flipped about, rudder-like, with ease; tail appears smoothly rounded in most views. Large female "Coops" are obviously larger than small male "Sharpies," but male Coops and female Sharpies can overlap in size.

■ **VOCALIZATIONS** Rich, nasal *kek*, given singly, in slow succession, or in rapid series. Vocal year-round on large territories, including urban neighborhoods.

■ **POPULATIONS** Shift to cities aided by greenways and acceptance of new foodstuffs, including invading Eurasian Collared-Doves (p. 72).

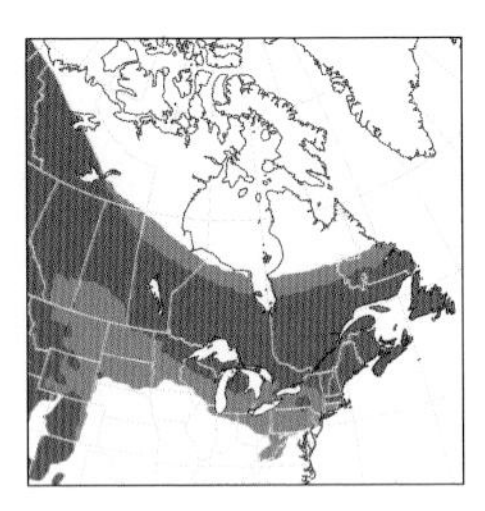

### American Goshawk

*Accipiter atricapillus* | AGOS | L 21-26" (53-66 cm) WS 40-46" (102-117 cm)

Although widespread, this sought-after forest raptor is nowhere common. An encounter with a "Gos" is often fleeting, and birds around the nest are ferocious, driving away all comers—including humans.

■ **APPEARANCE** More like a buteo (pp. 210-217), especially in soaring flight, than the smaller accipiters are. Adult blue-gray all over, finely barred below; white eyebrow contrasts with black patch behind eye. Juvenile separated from juvenile Cooper's Hawk by larger size, flared eyebrow, undertail coverts with brown splotches, and wavy tail bands.

■ **VOCALIZATIONS** Sharp, slurred whistles, often given in series; timbre like that of Sharp-shinned, but goshawk is louder and far-carrying.

■ **POPULATIONS** Breeds in dense forests, often on north-facing slopes with conifers. Numbers boom and bust with 10- to 11-year prey cycles.

Sharp-shinned Hawk
adult ♂
juvenile ♀
juvenile
small head projection
square tail
Cooper's Hawk
juvenile ♀
pale nape
adult ♂
juvenile
head projection prominent
rounded tail
American Goshawk
juvenile
pale bar on greater coverts
pale eyebrow
juvenile
adult ♂
undertail coverts streaked
wavy tail bands

**RAPTOR GRAB BAG**

They have familiar names, but these three raptors aren't what they seem. The Mississippi and Snail Kites are genetically closer to the genera *Accipiter* (p. 204) and *Buteo* (pp. 210–217) than to our other kites. The Bald Eagle is a sea-eagle, a grouping distinct from most other birds called eagle.

## Mississippi Kite

*Ictinia mississippiensis* | MIKI | L 14½" (37 cm) WS 35" (89 cm)

Like the White-tailed and Swallow-tailed Kites (both on p. 200), this species is on the wing much of the time. But it strikes a different profile in flight, suggesting a falcon (pp. 238–241) more than another kite or hawk.

■ **APPEARANCE** About the same size and shape as White-tailed Kite, but breeding ranges in East largely nonoverlapping. Adult gray-bodied and pale-headed; tail black (white in White-tailed); wings contrastingly patterned with whitish secondaries and red-shafted primaries. Juvenile coarsely streaked red-brown below, with banded tail.

■ **VOCALIZATIONS** Two-note whistle, the second note long and descending: *pee-peeeee.*

■ **POPULATIONS** Forages in loose aggregates of 10+ birds; catches dragonflies on the wing, deftly plucks katydids and cicadas from treetops. Range expanding north and west; regularly wanders north of core range, especially late spring and early summer.

## Snail Kite

*Rostrhamus sociabilis* | SNKI | L 17" (43 cm) WS 46" (117 cm)

Few birds in the East are more specialized than this one, whose diet is restricted to apple snails, genus *Pomacea.*

■ **APPEARANCE** With its hooked bill and dark colors overall, calls to mind Hook-billed Kite (p. 200), but the species do not overlap in the East. Adult male dark gray with bright red feet and bill; vent and base of tail white. Adult female dark brown with variable coarse streaking below; head contrastingly patterned. Juvenile similar to adult female, more buffy overall. Long bill, thin and sharply decurved, distinctive at all ages. Flight style likewise distinctive: low to the ground, mixes deep wingbeats with frequent stalls. Compare with white-rumped Northern Harrier (p. 202).

■ **VOCALIZATIONS** Unmusical rapping, *kuh-kuh-kuh-kuh-kuh.*

■ **POPULATIONS** Restricted in our area to freshwater habitats in Fla. Bill shape has evolved rapidly in response to invading nonindigenous apple snails, larger than indigenous apple snails.

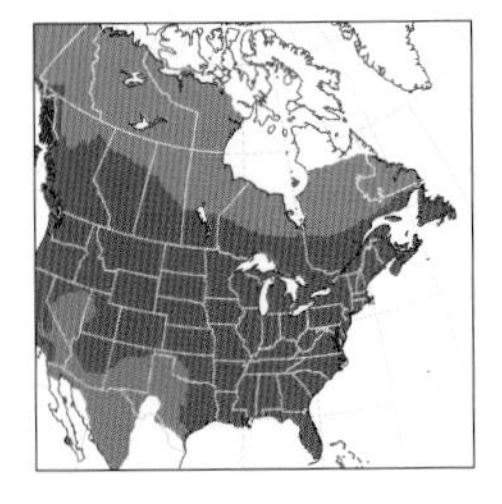

## Bald Eagle

*Haliaeetus leucocephalus* | BAEA | L 31–37" (79–94 cm) WS 70–90" (178–229 cm)

A celebrated conservation success story, this species was endangered in the late 20th century, but recovered thanks to strong legal protections.

■ **APPEARANCE** White-headed adult known to everyone, but younger birds trickier. All are long-bodied and large overall, with notably big bills; compared to Turkey Vulture (p. 198) and Golden Eagle (p. 202), Bald Eagle is large-headed and short-tailed. Dark-billed juvenile is dark brown overall with variable white splotching on tail and wings, not as sharply delineated as on Golden Eagle. Older immatures, up to four years of age, transition to adult plumage; many in third year have dark mask on smudgy white head.

■ **VOCALIZATIONS** A mirthful cackle, descending in pitch.

■ **POPULATIONS** Versatile and wide-ranging, but with strong affinity for aquatic habitats. Nests, near water, are titanic; usually placed out in the open, visible at great distances.

Mississippi Kite

juvenile

adult

adult

white secondaries

immature

adult

Snail Kite

paddle-shaped wings

white tail base

adult ♂

adult ♂

adult females

juvenile

Bald Eagle

2nd year

juvenile

whitish underwing and "wingpits"

juvenile

larger bill than Golden

adult

Osprey-like face pattern

3rd year

adult

## TEXAS RAPTORS I

Five species of hawks occur regularly in the East only in Texas, and all are stocky, medium-large, broad-winged hawks. These three are "near-buteos," not in the genus *Buteo* (pp. 210-217)—but close.

### Common Black Hawk

*Buteogallus anthracinus* | COBH | L 21" (53 cm) WS 50" (127 cm)

Rare but annual in summer to West and South Texas. Inhabits woods near rivers and ponds, where it hunts frogs and crayfish. Shape distinctive: legs long; tail broad and short; wings likewise broad and fairly short. Adult black with broad white band at base of tail. Juvenile streaked below; tail barred. Whistled call, sharp and piercing, often the best cue to this bird's presence.

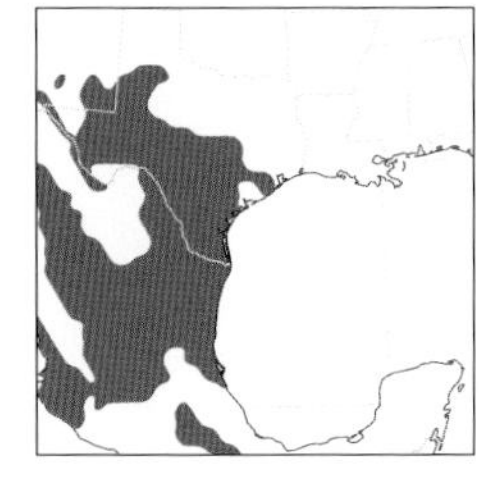

### Harris's Hawk

*Parabuteo unicinctus* | HASH | L 21" (53 cm) WS 46" (117 cm)

This brightly colored denizen of shrublands challenges the stereotype of raptors as solitary predators. Harris's Hawks hunt cooperatively, and they live in highly social assemblages with complex dominance hierarchies.

■ **APPEARANCE** Perched or flying, shows bright chestnut on wings; is called "Bay-winged Hawk" by some authorities. Long-legged; tail long and dark, with white band at tip and white patch at base. All are dark overall: Adult clean brown below; juvenile streaky below, but shows telltale bay wings and well-marked tail of adult.

■ **VOCALIZATIONS** Noisy in group settings. One frequently heard call is a low roar, trailing off: *RRAAaaauh.*

■ **POPULATIONS** Permanent resident within U.S. range, but some wander around Tex. in winter; rare vagrant north to southern Great Plains.

### White-tailed Hawk

*Geranoaetus albicaudatus* | WTHA | L 20" (51 cm) WS 51" (130 cm)

Where there are wildfires—controlled burns, usually—in the Tex. coastal plain, this raptor is sure to be found. Many raptors opportunistically scarf up prey at burns, but the White-tailed does it habitually.

■ **APPEARANCE** Short tail accentuated by long wings. Adult plumage distinctive: Upperparts gray with chestnut wash on wings; underparts unmarked white; short white tail crossed by single black band. Juvenile dark, but note messy white intrusions below and white splotches on dark head; tail of juvenile gray, not white.

■ **VOCALIZATIONS** Gull-like squeals and braying, often in series.

■ **POPULATIONS** Restricted in East mostly to open country near Tex. coast; a few wander well inland in Tex. and to coastal La.

### Raptor Behavioral Ecology

Although they are large and often easy to see, the raptors in the family Accipitridae (pp. 200-217) can be challenging to identify by morphology alone. In many instances, quick consideration of what a hawk is doing—and where it's doing it—may aid the field ID process. This interaction between behavior and ecology is termed behavioral ecology.

Among the Texas specialty raptors, Gray Hawks (p. 210) and Common Black Hawks are typically encountered in dense groves, whereas Harris's and White-tailed Hawks favor open habitat with scrubs and widely scattered trees. The behaviorally fascinating Zone-tailed Hawk (p. 210) is accepting of a wide variety of habitats, but with a twist: Zone-tails associate with look-alike Turkey Vultures (p. 198), apparently for the purpose of deceiving prey into imagining the Zone-tails are just harmless vultures.

Common
Black Hawk
juvenile
adult
juvenile
short,
broad
wings
adult
short tail with
broad white
band
juvenile
adult
Harris's
Hawk
adult
hite tail
and tip
bold white
breast
patch
2nd year
adult
White-tailed
Hawk
juvenile
all-dark
back
dark flight
feathers
adult
dark
juvenile
juvenile

## TEXAS RAPTORS II

Although placed in the genus *Buteo*, these distinctive Texas raptors break from the mold in several respects. The Gray Hawk suggests an accipiter (p. 204), and the Zone-tailed Hawk resembles the Turkey Vulture (p. 198). Adults exhibit little plumage variation, unlike our more widespread buteos.

### Gray Hawk

*Buteo plagiatus* | GRHA | L 17" (43 cm) WS 35" (89 cm)

Widespread in Mid. Amer., this dapper buteo is established and increasing in South Texas. The Gray Hawk has a well-known affinity for riparian woods, but it appears to be moving into wooded, often residential, districts away from the Rio Grande.

■ **APPEARANCE** Small for a buteo; a bit larger than, and not as slim as, Broad-winged Hawk (p. 214). Relatively short wings and long tail, combined with predilection for woodlands, call to mind an accipiter. Adult gray all over, smoothly colored on upperparts, finely barred below; long tail barred black and white. Juvenile brown above, streaked and spotted brown below; face boldly marked with white and dark chocolate; long tail finely barred with white at base.

■ **VOCALIZATIONS** Call a long, descending whistle; call of Broad-winged stays on one pitch, has short stutter at beginning.

■ **POPULATIONS** Taxonomy vexed; has been in and out of the genus *Buteo*, and variously lumped and split.

### Zone-tailed Hawk

*Buteo albonotatus* | ZTHA | L 20" (51 cm) WS 51" (130 cm)

Many animals imitate, or mimic, predatory or venomous "models" (hawks, snakes, wasps, etc.), but Zone-tailed Hawks do the opposite: They resemble innocuous Turkey Vultures (p. 198), which small prey know are harmless, and gather with the vultures in soaring flocks called kettles, the better to blend in and deceive their prey.

■ **APPEARANCE** Impressively similar to Turkey Vulture; holds long wings in shallow V like that species, rocking unsteadily as it soars in shallow circles. Also dark overall with long tail like Turkey Vulture, and wings two-toned from below; but all plumages have feathered head and yellow legs and bill. Adult's tail has broad black and gray bands, juvenile's tail barred gray and black. Breast of juvenile often flecked white.

■ **VOCALIZATIONS** Call a long, plaintive whistle. Not as steady as Broad-winged Hawk, but not as strongly descending as Gray Hawk.

■ **POPULATIONS** Rare in summer in semi-arid country in West and Central Tex., but prone to vagrancy; multiple records in recent years along East Coast from mid-Atlantic northward.

### Plumage Variation in Buteos

In the East, all species in the family Accipitridae (pp. 200–217) exhibit plumage variation. The Swallow-tailed Kite (p. 200) is relatively invariant, with only weak differences between juveniles and adults, but that species is an exception to the rule. All the other regularly occurring accipitrid raptors in the East differ greatly between juvenile and adult plumages. Adult males and females also differ markedly in several species in the East—for example, in the Northern Harrier (p. 202) and in the Hook-billed (p. 200) and Snail (p. 206) Kites. The situation takes on an additional layer of complexity with raptors in the genus *Buteo* (pp. 210–217): Most, but not all, species in the genus, collectively referred to as buteos or buzzards, exhibit well-marked color morphs.

Color morphs are genetic, they are generally stable, and they often have a geographic basis. Is a color morph, then, the same

biological phenomenon as a subspecies (see sidebars, p. 148)? It's complicated, but the short answer is no. For example, dark-morph individuals can show up within most subspecies of the widespread and variable Red-tailed Hawk (p. 212), although they are more likely in western North America than in the East. The particularly distinctive "Harlan's" subspecies of the Red-tailed Hawk itself comes in well-differentiated dark-morph and light-morph individuals—color morphs within a subspecies!

In some bird species, different color morphs exhibit different behaviors; the White-throated Sparrow (p. 346) is a notable example. But color morphs in buteos do not generally signify ecological or behavioral specialization. They don't sound different, they don't hunt differently, they don't seem to favor one microhabitat or another. They just *are*, for reasons that are not, at present, well understood.

## TWO VARIABLE BUTEOS

The species in this genus are perhaps the most hawklike of the hawks: large, well-built sit-and-wait predators seen from roadsides, along woodland edges, and in open country.

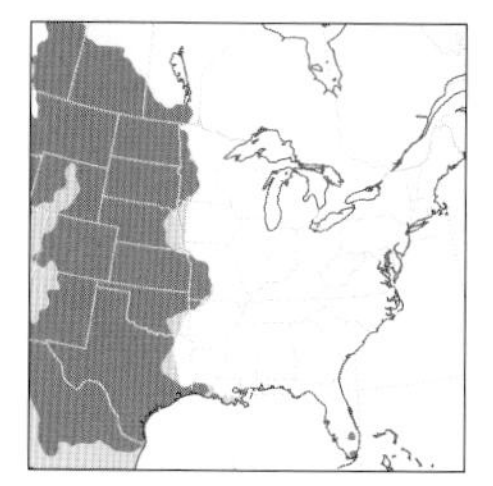

### Swainson's Hawk

*Buteo swainsoni* | SWHA | L 21" (53 cm) WS 52" (132 cm)

No other hawk in N. Amer. is more migratory than this one. It breeds far north in Canada and even Alas., and winters mostly south of the Amazon Basin on the Pampas of Argentina.

■ **APPEARANCE** Wing tips are long and dark, powering the species on its mighty migrations; flies with long wings locked in a shallow V. Most sightings in summer are of light-morph adults: uniformly gray-brown above, white-cheeked and chestnut-bibbed below, with wing linings pale and remiges dark; long tail mostly dark with a bit of white at base. Uncommon intermediate-morph adult deep chestnut below, with splotchy chestnut underwing linings. Even rarer dark-morph adult nearly all-dark above and below. Light-morph juvenile, commonly seen in late summer, quite different from adult: head and breast suffused in caramel; lacks breastband, but shows broad dark bulges on sides of breast; back and upperwing coverts scalloped warm buff.

■ **VOCALIZATIONS** An anguished squeal, wavering but mostly on the same pitch: *eeeeeuh-eeeee.* Vocal around nest, notably in parent-offspring interactions.

■ **POPULATIONS** Most in our area are breeders and migrants in the Midwest; very rare but annual on migration, chiefly fall, much farther east. Breeds on grassland with scattered trees, but also accepts agricultural landscapes. In recent years, small numbers have been wintering South Tex. and South Fla.

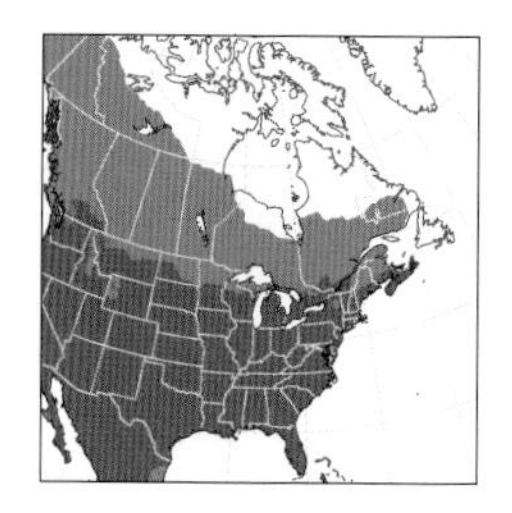

### Red-tailed Hawk

*Buteo jamaicensis* | RTHA | L 22" (56 cm) WS 50" (127 cm)

In many settings in the East, this is the default raptor: a large compact hawk watching from a utility pole on the interstate, soaring over farm country, or roosting on a bridge or building in the big city.

■ **APPEARANCE** Light-morph adult, with bright red tail, unmistakable, but plumage variation extreme. Perched or flying, appears solid and stocky; wings broad and rounded; tail fairly short; bill largish. Wings mostly pale from below, marked by darker patagial bar at leading edge of inner wing. Light-morph adult has brown back with white splotches, white underparts with smudgy band on belly, and eponymous red tail. Paler adults like poorly defined subspecies *kriderii* ("Krider's Hawk") can be confused with Ferruginous Hawk (p. 216). Subspecies *fuertesi* ("Fuertes's Hawk") mostly dark above, uniformly pale below. Subspecies *harlani* ("Harlan's Hawk") blackish overall, with whitish tail, gray remiges, and breast splotched with white. Most juvenile Red-tails in East are light morph, with tail finely barred brown and black; note belly band and, in flight, patagial bar.

■ **VOCALIZATIONS** Harsh, descending squeal, *RRReeeeeerr,* ubiquitous in movie soundtracks and television commercials, often attributed to the wrong species of raptor.

■ **POPULATIONS** Typically seen singly or in pairs; occurs in almost any habitat, from unbroken prairie to surprisingly dense forest. Although resident year-round throughout eastern U.S., it is one of the most numerous migrants at many hawk watches.

light-morph
adult
narrow,
pointed
wings
dark-morph
adult
Swainson's
Hawk
smaller bill
than Red-tailed
light-morph
juvenile
juvenile
termediate-
morph adult
light-morph
adult
dark-morph
adult
"Harlan's"
harlani
Red-tailed
Hawk
juvenile
eastern
borealis
adult
"Krider's"
kriderii
dark-
morph
adult
western
calurus
adult
"Krider's"
kriderii
dark-morph
adult
"Harlan's"
harlani
light-
morph
adult
"Harlan's"
harlani
pale
scapulars
juvenile
western
calurus
patagial bar
present on all
subspecies
belly
band
adult
eastern
borealis
adult
Texas
fuertesi
adult
western
calurus
adult
eastern
borealis

**THREE WOODLAND BUTEOS**

These midsize to smallish buteos nest within woods and even deep forest, where they are hard to see. Field ID is therefore often of birds in flight, sometimes far from wooded habitats; vocalizations are important too.

## Red-shouldered Hawk

*Buteo lineatus* | RSHA | L 15–19" (38–48 cm) WS 37–42" (94–107 cm)

Many hawks are disinclined to vocalize, but this showy species bucks the trend. The Red-shouldered seems to call constantly in and around the bottomland broadleaf forests it calls home.

■ **APPEARANCE** Relatively long-tailed like an accipiter (p. 204). In flight, all plumages show pale crescent across tip of upperwing; but be aware that other buteo species in molt may be similarly patterned. Adult of widespread nominate subspecies *lineatus* extensively orange-rufous, especially on "shoulders" and breast; tail and remiges barred black and white. Juvenile browner; distinguished from juvenile Broad-winged by longer tail and pale window across wing.

■ **VOCALIZATIONS** Short, sharp *kyeeah,* descending; often given in series, gull-like.

■ **POPULATIONS** Smaller South Fla. subspecies *extimus* pale overall; deep orange-rufous of adult *lineatus* replaced by faded orange-gray on *extimus.*

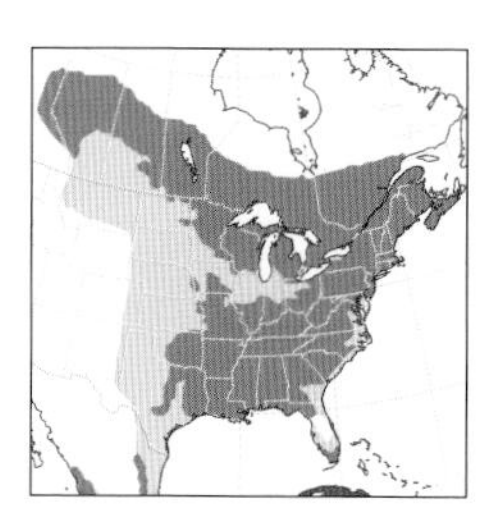

## Broad-winged Hawk

*Buteo platypterus* | BWHA | L 16" (41 cm) WS 34" (86 cm)

Regarding its migratory strategy, this buteo is the eastern counterpart of the chiefly western Swainson's (p. 212). Both engage in impressive hemispheric migrations, with most Broad-wingeds wintering in northern S. Amer.

■ **APPEARANCE** Small woodland buteo. Although the wings are broad, they are more distinctively pointed. Adult dark brown above, barred rufous-brown below; white underwings bordered in black; tail barred black and white. Juvenile resembles juvenile Red-shouldered, but note shorter tail and unmarked upperwing of Broad-winged.

■ **VOCALIZATIONS** Long monotone whistle after short introductory note: *pit-eeeeeeee.* Like Mississippi Kite's whistle (p. 206), but doesn't descend.

■ **POPULATIONS** Uncommon dark morph (not a subspecies) restricted to western fringes of range; sightings of both morphs increasing recently at western edge of range, especially on spring migration.

## Short-tailed Hawk

*Buteo brachyurus* | STHA | L 15½" (39 cm) WS 35" (89 cm)

Getting up early to see this Fla. specialty isn't worth the effort. The species isn't usually spotted until late morning, soaring well above woods and savannas—and even in the skies of urban Miami. In impressive dives, it catches prey in the treetops.

■ **APPEARANCE** Small buteo, similar in build to Broad-winged Hawk. Dark-morph adult the only mostly dark buteo likely to be seen in South Fla.; on bird in flight, note dark wing linings contrasting with paler remiges. Light-morph adult in flight patterned like Swainson's Hawk, uncommon in South Fla.; but Swainson's has prominent chestnut bib. Juveniles of both morphs resemble corresponding adults, but have more and narrower tail bands.

■ **VOCALIZATIONS** Long whistle like Broad-winged's, but no introductory stutter.

■ **POPULATIONS** Widespread in the tropics, but barely reaches our area: fewer than 1,000 individuals in Fla.; occasional strays to South Tex.

juvenile
eastern
lineatus
Red-shouldered
Hawk
lineatus
juvenile
adult
pale
crescent
adult
Florida
extimus
juvenile
barred
daries;
mpare
Broad-
winged
lineatus
pale
crescent
adult
eastern
lineatus
lineatus
wing tips
pointed
Broad-winged
Hawk
juvenile
adult
juveniles
marked
ndaries;
ompare
ith Red-
uldered
adult
juvenile
dark-morph
adult (rare)
"crossbow"
shape in
flight
light-morph
adult
Short-tailed
Hawk
secondaries
darker than inner
primaries
light-morph
adult
dark-morph
adult

## TWO PRAIRIE BUTEOS

Most of our buteos spend a fair bit of time in wide-open country, but these two are notable because they prefer particularly desolate, nearly featureless terrain. Both have feathered tarsi, an adaptation for cold temperatures.

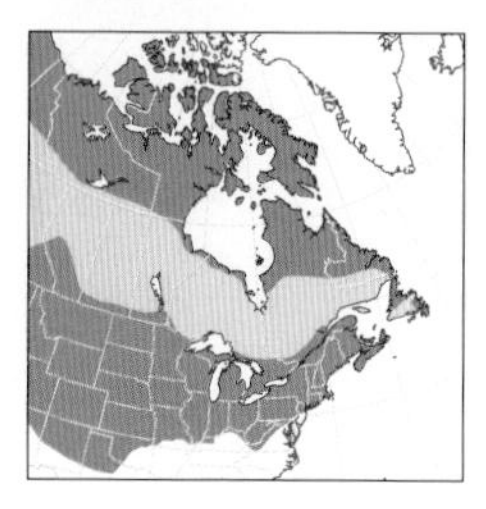

### Rough-legged Hawk

*Buteo lagopus* | RLHA | L 21" (53 cm) WS 53" (135 cm)

For most birders in the East, this dashing buteo occurs only in winter. It is our most Arctic buteo by far, breeding extensively across the Canadian Arctic Archipelago. The Rough-legged hovers habitually while feeding.

■ **APPEARANCE** A lanky buteo with small feet and a small bill. When not hunting, perches on surprisingly flimsy shrubs or twigs in small trees. Plumage variation complex: All show pale tail below with broad black band (adult female), black barring (adult male), or gray-black band (juvenile); note also black blob on "wrist" (carpal patch) seen on underwing. All light morphs have considerable pale on head: Juvenile has black on belly; light-morph adult female similar, but has solid black tail band; light-morph adult male has black belly with frosty white flecks. Dark morph nearly black when perched; in flight from below, blackish wing linings contrast with whitish remiges. Compare especially with dark-morph "Harlan's Red-tailed Hawk" where they overlap in our area (Great Plains) in winter; variable "Harlan's" is stockier, with blotchy white on black breast.

■ **VOCALIZATIONS** Usually quiet in winter, but occasionally gives a long squeal, fairly pure in tone and slowly dropping in pitch.

■ **POPULATIONS** Numbers fluctuate from year to year in response to availability of winter prey and summer success on the Arctic breeding grounds. May be undergoing long-term shift on wintering grounds away from mid-continent and toward coasts.

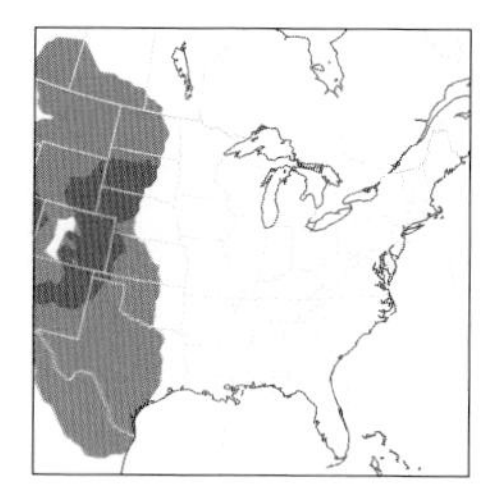

### Ferruginous Hawk

*Buteo regalis* | FEHA | L 23" (58 cm) WS 56" (142 cm)

Our largest buteo, this majestic hawk (*regalis* means "regal") is likened by many to a small eagle. Adding to its mystique, the Ferruginous Hawk inhabits lonely steppes and badlands, and it is decidedly solitary.

■ **APPEARANCE** Long-winged and long-tailed, with a big bill and large feet; gape colorful and conspicuous (but note that juveniles of any buteo species may show prominent gape). In flight, all show huge white panel on outer wings. Light-morph adult rusty above and on belly; feathered legs also rusty. Tail of adult varies from nearly white to extensively pale red, suggesting Red-tailed Hawk (p. 212). Light-morph adult shows splotchy rufous on underwing. Light-morph juvenile in flight mostly white below; note white in wings from above. Uncommon dark morph deep brown with a hint of chestnut in adult plumage; in flight from below, dark wing linings contrast with pale remiges. Adult dark morph pale-tailed above and below; juvenile dark morph darker-tailed above, but white wing panel stands out.

■ **VOCALIZATIONS** Descending whistle, richer than similar calls of other buteos. Heard mostly on breeding territory, but some in nonbreeding season, defending winter feeding grounds, do vocalize.

■ **POPULATIONS** Short-distance migrant; in winter, only rarely strays east of central and southern Great Plains, where often found in prairie dog towns. Rapid development of energy infrastructure (wind, fracking, oil field) an emerging threat for this easily disturbed raptor.

dark carpal patch
Rough-legged Hawk
small head and bill
adult ♂
light-morph adult ♀
juvenile
dark-morph adult ♂
hovers while hunting
pale primary windows
Ferruginous Hawk
rusty leg feathering
juvenile
adults
juvenile
pronounced gape
adult
dark-morph adult
feathered legs

## FAMILIAR OWLS

Owls are familiar to everyone, and these three species are particularly well represented in the cultures of both Indigenous and non-Indigenous Americans.

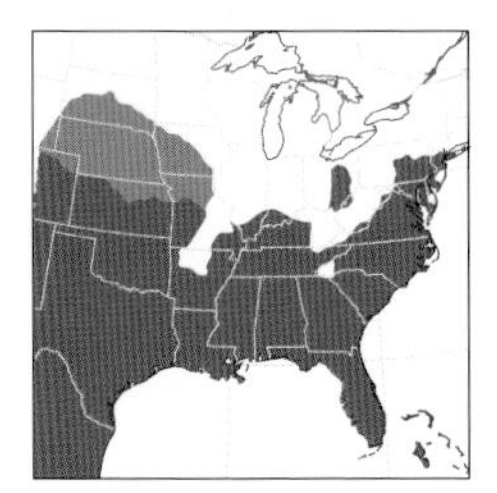

### Barn Owl

*Tyto alba* | BANO | L 16" (41 cm)

This wraithlike inhabitant of outbuildings, cliff faces, and copses with standing deadwood is the only representative in the East of the family Tytonidae. All of our other owls are in the family Strigidae.

■ **APPEARANCE** Pale overall; washed butterscotch and leaden above, mostly white below. The heart-shaped face is outlined in a red-orange frame; the facial disc of this and other owls creates a parabolic dish, an adaptation for acute hearing. Legs long and dark; eyes dark and soulful. Unmistakable when spotted in daylight (often in a silo or other structure), but be aware that any owl in headlights or moonlight can appear pale at night.

■ **VOCALIZATIONS** A sudden shriek of white noise: *HHSSSSSSHHHH*. Juvenile Great Horned Owl's is similar, but call of that species shorter, more nasal, less explosive.

■ **POPULATIONS** Voracious devourer of mice and rats; rodenticides sicken the owls, protecting vermin from predation—a lose-lose proposition.

### Eastern Screech-Owl

*Megascops asio* | EASO | L 8½" (22 cm)

Owls famously hoot, but this species whistles and gargles expertly. The bird readily responds to whistled imitations of its two songs.

■ **APPEARANCE** The only small owl in the East with "ear tufts" (not actually ears). Perched in silhouette at night, appears chunky and compact; wings broad and rounded in flight. Varies in color from orangey to gray; all are complexly striped and barred below.

■ **VOCALIZATIONS** One song is a wavering, descending, pure-tone whinny; the other song is a long, low-pitched, monotone warble. The songs are lovely, anything but screeching.

■ **POPULATIONS** Fairly common in woodlands, even in cities, but strictly nocturnal. Readily accepts nest boxes.

### Great Horned Owl

*Bubo virginianus* | GHOW | L 22" (56 cm)

You may know this owl's iconic hooting from television or a movie, but it's even better in real life! Both males and females are vocal, and easily heard on still nights in winter.

■ **APPEARANCE** A massive predator; weighs well in excess of the fluff-and-feathers Great Gray Owl (p. 222). "Ear tufts" large. Variably gray-brown with heavy barring below; white bib prominent especially when hooting. Mostly nocturnal, but readily spotted roosting by day, often in tall broadleaf trees in woodlots.

■ **VOCALIZATIONS** Deep rhythmic hooting, heard most often at dawn and dusk. Male gives three to five slow, pure hoots (*hoo HOO... hoo... hoo*), larger female four to eight faster, slightly higher, more nasal hoots (*hoo hu-hu HOO hu HOO hoo*); mated pairs often duet, especially fall and winter. Female may add weird, nasal barks; rasping, wavering screech of juvenile can suggest Barn Owl.

■ **POPULATIONS** Versatile; nests in rugged wilderness but also in large cities. Faring well, but many are hit by cars. Varied prey includes threatened species of raptors and other birds.

♂

## Barn Owl

♀

pale rounded wings lack dark markings as on Short-eared Owl

"ear tufts" often indistinct

**rufous morph**

**gray morph**

## Eastern Screech-Owl
*asio*

**juvenile**

**western Great Plains**
*maxwelliae*

color of facial disc varies geographically

## Great Horned Owl
*virginianus*

**northern interior**
*subarcticus*

## TWO NORTHERN OWLS

Their ranges practically circle the globe at high latitudes of the Northern Hemisphere, but these two are known to most birders only as rare winter visitants. Both are easy to spot by day, roosting out in the open, unlike most other owls.

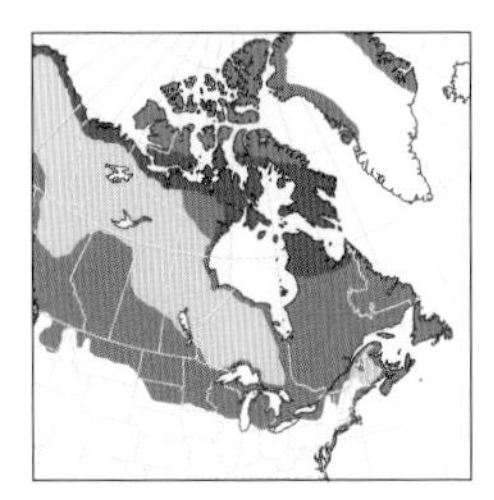

### Snowy Owl

*Bubo scandiacus* | SNOW | L 23" (58 cm)

Even before the species was immortalized in the highest-selling children's fantasy books in history and the blockbuster films that followed, this huge white owl of the tundra was beloved and much sought by birders.

■ **APPEARANCE** In the same genus as the Great Horned Owl (p. 218), but even larger. Adult, especially male, mostly white; lacks "ear tufts"; yellow eyes stare from white face. First-winter, especially female, densely barred black above and below, but nevertheless whitish overall.

■ **VOCALIZATIONS** Male hoot is a short, gruff lowing, normally heard only on breeding grounds. In interactions with other predators in winter, gives short yaps and a longer, gull-like squeal.

■ **POPULATIONS** Winters south to northern U.S., but numbers vary greatly from year to year. Many at southern edge of winter range are first-winters, roosting by day on dunes, barns, water towers, etc. They often stay put for a week or more at a time. Snowy Owls wintering in and around midlatitude metropolises become major celebrities; watch from a distance and be mindful of fencing or any other protections for the birds. Be mindful too of the other birders and the general public, often attracted in large numbers to Snowy Owl "stakeouts." Courteous and respectful birders are excellent ambassadors for nature study.

### Northern Hawk Owl

*Surnia ulula* | NHOW | L 16" (41 cm)

The bird's name suggests a blend between two kinds of predators, and that is indeed the case with this species. It hunts by day like a hawk, but its close-set yellow eyes and round facial disc are all owl.

■ **APPEARANCE** Tail, long and tapered, imparts distinctive shape; typically seen watching for prey from prominent perch. Dark overall, including blackish facial disc.

■ **VOCALIZATIONS** Song a slow trill of short whistles, the entire series 5–10+ seconds in duration. Carries well; one of the spookier sounds of open boreal forests and bogs.

■ **POPULATIONS** Mostly resident, but a few drop south every year. Sax-Zim Bog, Minn., and the urban centers of southern Canada usually host a few in winter.

### Irruptions

Owls do not leap to mind when one thinks of the great movements of birds across the continent. Instead, one's imagination conjures up the dramatic and conspicuous biannual migrations of wood-warblers, sandpipers, seabirds, and so forth. But all our owl species are on the move in one way or another. A few are migratory in the traditional sense of the word: Burrowing and Elf Owls (p. 222) fit the mold, and the Flammulated Owl of western North America is particularly notable in this regard. Others—like the Barred Owl (p. 222)—are in the process of long-term range expansions. But the most celebrated strigine movements of all are the ones termed irruptions.

A biological irruption is a rapid population surge. It can be an actual numerical increase, as in the case of many mammals. With owls and other birds, though, an irruption usually refers

to a sudden movement of birds away from the core range. Irruptions can happen with species that are migratory to begin with, like the Snowy Owl; they also occur in species that are mostly sedentary, like the Northern Hawk Owl. Other irruptive species in the East are a taxonomic mixed bag and include the Bohemian Waxwing (p. 294), Red-breasted Nuthatch (p. 298), and various northern finches (pp. 324–331).

The proximate cause of avian irruptions is usually food availability, but the ultimate causes are trickier to pin down. Wildfires, drought, and even sunspot activity have been posited as drivers of food availability and hence irruptions. And while irruptions are a natural ecological process, climate change has added an additional element of complexity to their timing, duration, and severity. The key idea is that irruptions, which can be triggered by many factors, are less than annual and frequently difficult to predict.

**OWLS GREAT AND SMALL**
The borderland Elf Owl is shorter than many sparrows, while the Great Gray Owl of the boreal forest is close to the length of an eagle. But all of these owls are essentially the same in structure: compact, no-necked, and round-headed, with close-set staring eyes.

## Ferruginous Pygmy-Owl

*Glaucidium brasilianum* | FEPO | L 6¾" (17 cm)
Widespread in Neotropics; barely reaches our area in South Texas, where scarce in open woodlands. Quite small with warm rufous tones. Long tail barred black and orange. Often seen and heard by day; song is a long, slow, steady series of whistled toots.

## Elf Owl

*Micrathene whitneyi* | ELOW | L 5¾" (15 cm)
Tiny but well-built, with short tail and no "ear tufts." Note foxy colors: red-orange and leaden. Nocturnal and easily overlooked; widespread but uncommon in summer in desert woodlands in West Tex. and in forest along Lower Rio Grande. Song a yapping series of 8–12 notes, muffled at beginning and end, shrill in the middle: *yup-yup-yee-yee-YEE-YEE-yee-yee-yup.*

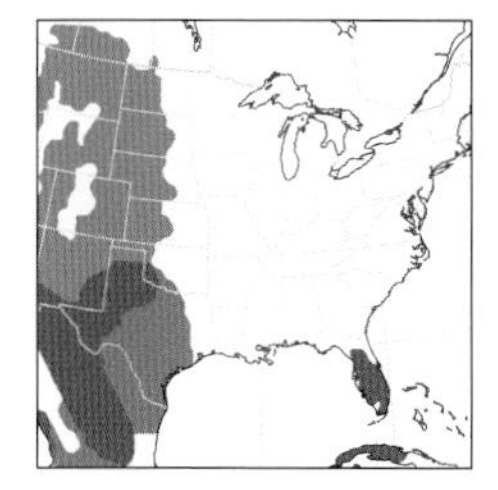

## Burrowing Owl

*Athene cunicularia* | BUOW | L 9½" (24 cm)
A long-legged owl standing alert on bare ground in midday is likely to be this species—especially if it descends into a subterranean burrow.
■ **APPEARANCE** Rotund and fairly small; by day, watches from short mounds in open country, often half below ground level. Adult heavily barred below, juvenile plain and buffy below. Family groups stand shoulder to shoulder.
■ **VOCALIZATIONS** Two-note song, first note shorter: *hu hoooo,* repeated. The effect is dovelike; also suggests Scaled Quail (p. 54).
■ **POPULATIONS** Occurs in our area in Fla. and Great Plains. The Fla. birds are darker and sharper, and mostly excavate their own burrows; browner Great Plains birds usually nest in burrows dug by prairie dogs.

## Barred Owl

*Strix varia* | BADO | L 21" (53 cm)
This assertive owl is found in hardwood bottomlands anywhere in the East, especially in the South.
■ **APPEARANCE** Medium-large with dark eyes; breast densely barred, belly broadly streaked. Mostly nocturnal, but hoots in late afternoon.
■ **VOCALIZATIONS** Exceedingly vocal. Its famous song (*Who cooks for you? Who cooks for you all?*) can be heard even in broad daylight.
■ **POPULATIONS** Aggressive and adaptable; increasingly seen in wooded districts even in cities.

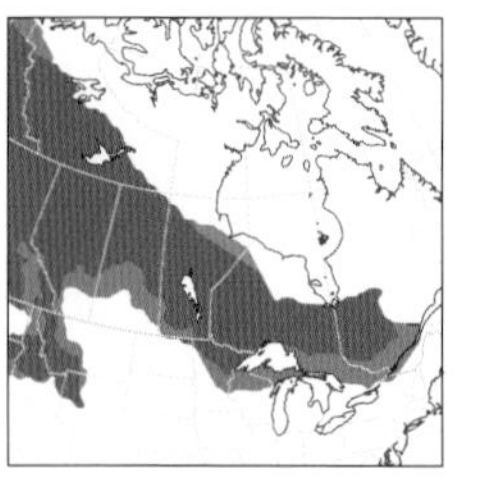

## Great Gray Owl

*Strix nebulosa* | GGOW | L 27" (69 cm)
This awesome predator hears animals *beneath* the snow, then smashes into a snowbank and grabs a meal.
■ **APPEARANCE** Appears huge, but weighs less than Great Horned (p. 218). Mostly dark gray with yellow eyes, concentric rings on face, and white "bow tie."
■ **VOCALIZATIONS** Deep, well-spaced hoots, up to 10 per series.
■ **POPULATIONS** Weakly irruptive; a few, often young birds, drift south to northern Great Lakes and northern New England most winters.

"false eye"
spots
Ferruginous
Pygmy-Owl
yellow
eyes
Elf Owl
Burrowing
Owl
juvenile
no dark
barring
below
long
legs
black and white on facial
disc gives "bow tie" effect
Barred
Owl
black
eyes
Great
Gray Owl

## OWLS OF FIELD AND FOREST

Our two owls in the genus *Asio* hunt in mostly open habitats, whereas our two *Aegolius* owls restrict their activities largely to extensive forest tracts.

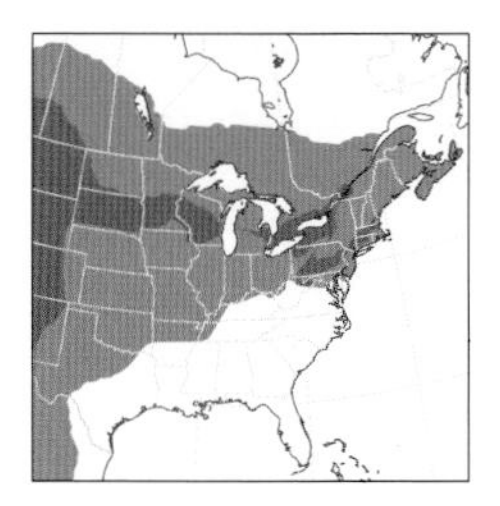

### Long-eared Owl

*Asio otus* | LEOW | L 15" (38 cm)

Encounters with this secretive species typically occur at day roosts—maybe one owl perched in a conifer, maybe 10+ in a hedgerow in farm country.

■ **APPEARANCE** Smaller than Great Horned Owl (p. 218); roosting, assumes elongate posture. Orange face crossed vertically by black and white bands. Flight erratic with jerky wingbeats; note orangey patch across outer wing.

■ **VOCALIZATIONS** Male, a rising *hooOO,* repeated slowly; claps wings in display. Coming in to roost at dawn, gives cackles, whines, and hoots.

■ **POPULATIONS** Widespread but hard to find. Listed as a Species of Special Concern in several eastern states.

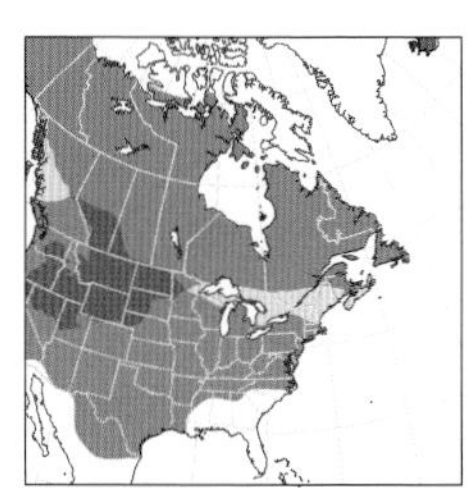

### Short-eared Owl

*Asio flammeus* | SEOW | L 15" (38 cm)

This owl of open country perches on haystacks, fences, and small outbuildings; it hunts by day, especially when skies are cloudy.

■ **APPEARANCE** A bit paler and stockier than Long-eared Owl; the "ear tufts," set close, are tiny. Beats wings slowly, like a giant butterfly; yellow-buff patch across upperwing and black bar on underwing prominent in flight.

■ **VOCALIZATIONS** Gives short barks when interacting: *chyurr* or *chyaap.* Male courtship song a series of simple hoots, faster than Long-eared.

■ **POPULATIONS** Wintering birds readily take to artificial prairie and other created habitats; reclaimed strip mines heavily used in Ohio River Valley.

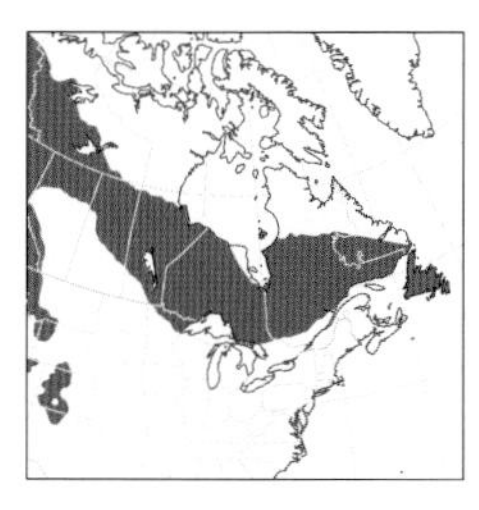

### Boreal Owl

*Aegolius funereus* | BOOW | L 10" (25 cm)

Because it breeds far from large human population centers, this owl is seldom seen and little known. Adding to its mystique, the Boreal is strictly nocturnal and partial to the densest forest tracts.

■ **APPEARANCE** Darker and larger than Northern Saw-whet, but female saw-whets approach the size of some male Boreals. Bill pale (dark on saw-whets); body of Boreal streaked and spotted dark-chocolate and white. Whitish face bordered by black; forecrown speckled black-and-white. Rarely seen fledgling mostly uniform dark brown, but with white markings on face.

■ **VOCALIZATIONS** Song a series of *hu* notes, increasing in loudness; similar to winnowing of Wilson's Snipe (p. 122), common in boreal bogs.

■ **POPULATIONS** Favors spruce-fir forests with aspen-birch admixtures. Migrates south every winter, with occasional irruptions.

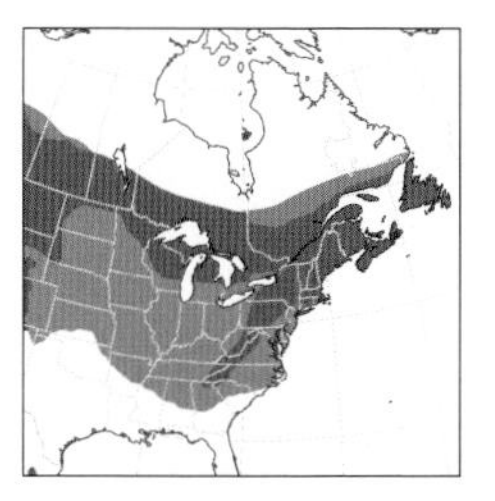

### Northern Saw-whet Owl

*Aegolius acadicus* | NSWO | L 8" (20 cm)

If there is a gnome of the Northwoods, this is it. The bird is widespread in diverse forest types, but finding one is always a special treat.

■ **APPEARANCE** Smallest owl in much of the East; similar length as an Eastern Screech-Owl (p. 218), but weighs less; lacks "ear tufts." Bill dark (pale on Boreal). Adult chestnut, coarsely streaked below. Fledgling rich buff-brown with white Y on dark face.

■ **VOCALIZATIONS** Song a very long succession of short whistles.

■ **POPULATIONS** Movements complex, with much interannual variation; more common in winter than widely appreciated.

niform barring
n upperwings;
ompare to
hort-eared
Long-eared Owl
rufous facial disc similar to Great Horned
Caribbean
*domingensis*
darker facial disc border
rare in South Florida
Short-eared Owl
*flammeus*
blotchy upperwing markings
very short "ear tufts"
dark "wrist" mark
dark border to facial disc
Boreal Owl
rk
own
dy
juvenile
Northern Saw-whet Owl
light brown body
juvenile

## KINGFISHERS

With their bright colors, blocky heads, oversize bills, and unkempt crests, these birds are easily recognized. And if any doubt remains, note their behavior: Kingfishers watch from a snag above the water, then plunge face-first after a fish beneath the surface.

### Ringed Kingfisher

*Megaceryle torquata* | RIKI | L 16" (41 cm)

The "King Kong Fisher" is one of the showier specialties of South Texas. It is the largest kingfisher in the Americas.

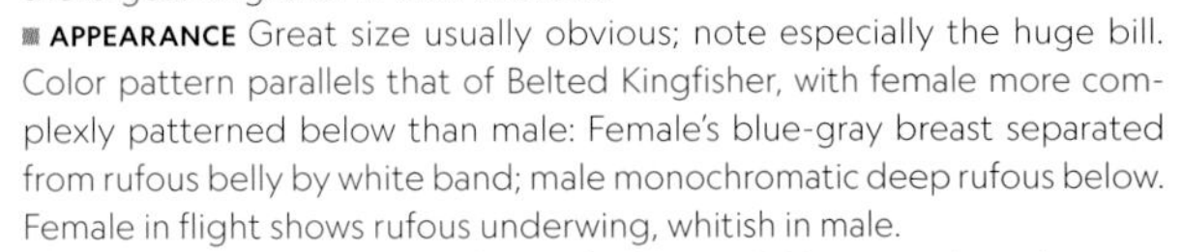

**■ APPEARANCE** Great size usually obvious; note especially the huge bill. Color pattern parallels that of Belted Kingfisher, with female more complexly patterned below than male: Female's blue-gray breast separated from rufous belly by white band; male monochromatic deep rufous below. Female in flight shows rufous underwing, whitish in male.

**■ VOCALIZATIONS** A long, loud, steady chatter, 5-10+ seconds in duration, usually given perched; sometimes shortened to just one or two seconds. In flight, utters a short, muffled *chuh* or doubled *chuh-chuh.*

**■ POPULATIONS** Recent addition to our avifauna. Began nesting in Tex. in 1970, now well established in Lower Rio Grande Valley; still expanding.

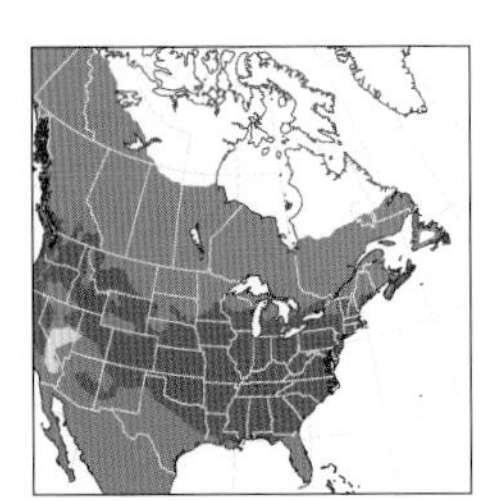

### Belted Kingfisher

*Megaceryle alcyon* | BEKI | L 13" (33 cm)

Throughout much of the U.S. and Canada, this is the only kingfisher. The species never flocks, but individuals and pairs may be seen practically anywhere there is open water.

**■ APPEARANCE** Unmistakable in most of range, although unrelated Blue Jay (p. 274) shares overall scheme of blue above and crested head, white below with breastband. Ringed Kingfisher, restricted in East to South Tex., notably larger with underparts mostly rufous. Adult male Belted has single blue-gray breastband; juvenile and adult female have blue-gray and rufous bands.

**■ VOCALIZATIONS** Loud rattle, often given in flight, less steady than call of Ringed Kingfisher. Perched, gives a quieter, faster, more muffled rattle, starting with a quick burst and then slowing: *ch'CH'CH'CH'-chuh-chuh-chuh.*

**■ POPULATIONS** Many migrate considerable distances, but others winter as far north as there is open water. Wary and difficult to approach, but accepting of built-up landscapes: parks, marinas, fish hatcheries, etc.

### Green Kingfisher

*Chloroceryle americana* | GKIN | L 8¾" (22 cm)

They're both Tex. specialties, but this species and the Ringed Kingfisher are temperamentally as different as can be. The demure Green Kingfisher sits quietly at the edge of wooded waterways, snags a small fish with little fanfare, then flies off quickly.

**■ APPEARANCE** Much smaller than Belted Kingfisher. Upperparts dark green; bill long and thin. Male has rufous breast; breast of female splotchy green. Unlike *Megaceryle* kingfishers, is not given to hovering out in the open; flies swiftly along streams or pond edges, low to the water, flashing white in tail.

**■ VOCALIZATIONS** Soft, two-note clicking, *tik ... tik ...*, repeated slowly by perched bird; easily overlooked. In flight, gives single notes, more widely spaced and erratic than those produced by the bigger kingfishers.

**■ POPULATIONS** Following near extirpation from Tex. in mid-20th century, has recovered substantially. Not confined to Lower Rio Grande Valley; range reaches San Antonio and Austin metro areas.

♀
huge bill with bone-colored base
♂
solid rufous belly
Ringed Kingfisher
primarily rufous underparts
white wing patches
single blue-gray band
♂
♀
♀
Belted Kingfisher
♂
blue-gray and rufous bands
large bill with gray base
short crest pointed at rear
♀
♂
extensive white in tail
♂
Green Kingfisher

## THREE CLASSY WOODPECKERS

Many woodpeckers in the East are modestly attired in black and white with a dash of color, but these three go all out. Adding to their flamboyance, they forage right out in the open.

### Lewis's Woodpecker

*Melanerpes lewis* | LEWO | L 10¾" (27 cm)

Striking western woodpecker, oily green above with beet-red face and lovely rosy cast to belly; breeds east into S. Dak., Nebr., Okla. Many are migratory, wandering east in Great Plains, especially fall. Quiet but industrious; favors recently burned woods, where it spends much time out in the open catching aerial insects. Appears black and crowlike on the wing.

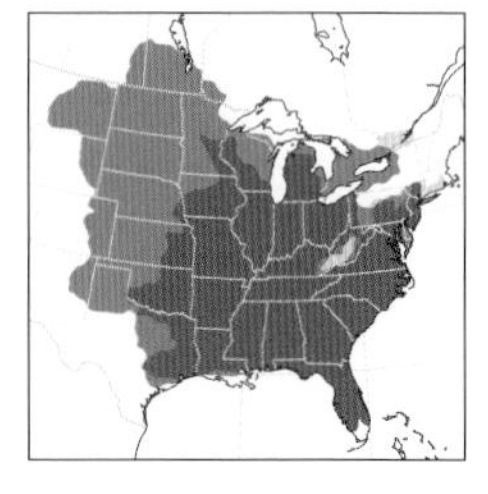

### Red-headed Woodpecker

*Melanerpes erythrocephalus* | RHWO | L 9¼" (23 cm)

There are many red-headed woodpeckers, but this is *the* Red-headed Woodpecker, its entire head and neck imbued in blazing bright red.

■ **APPEARANCE** Adults of both sexes instantly recognized by completely red head; in flight, black-and-white pattern above is obvious. Juvenile, often seen on fall migration away from breeding grounds, may cause brief confusion; note gray-brown hood, blackish upperparts with white on wings, and dusky-white underparts.

■ **VOCALIZATIONS** Primary call an urgent *quee-uh!* Shriller and sharper than that of Red-bellied Woodpecker (p. 230), suggestive of nocturnal flight call of Gray-cheeked Thrush (p. 314). Also gives a soft, muffled, scolding rap, four to six notes like a wren or frog.

■ **POPULATIONS** Like Lewis's, typically occurs in very open woods—including tracts that are burned, diseased, or storm-damaged; also in orchards. Local; often absent from seemingly appropriate habitat. Has declined substantially in northeastern U.S.

### Acorn Woodpecker

*Melanerpes formicivorus* | ACWO | L 9" (23 cm)

Widespread in oak-pine woods from western U.S. to northern S. Amer.; resident in Davis Mts. and Big Bend N.P. in Tex., sparingly east to Edwards Plateau. Note bright eye on boldly patterned face. Highly social and incessantly vocal around communal granaries, where the birds store acorns. Mostly resident, but a few wander to western Great Plains.

### Woodpeckers That Don't Peck Wood

One of the most impressive spectacles in all of eastern birding has to be the sight—and sound—of a mighty Pileated Woodpecker (p. 236) obliterating an old stump, wood chips flying everywhere. But many woodpeckers employ feeding techniques that do not require pecking at or pounding on wood. Those in the genus *Melanerpes* (pp. 228–231) catch insects on the wing, gobble down berries, and store nuts. Flickers (p. 236) feast on ants in the soil, and their specialized tongues share features with those of anteaters! Downy Woodpeckers (p. 234) in winter are peckers for sure, but they're often pecking on mullein stalks, not tree trunks. Three-toed woodpeckers (p. 232) flake bark, sapsuckers (p. 230) tend sap wells, and so forth.

Watch a woodpecker long enough, and you'll appreciate its ability to mix and match foraging strategies. Even the habitual fly-catchers and ant-eaters in the woodpecker clan make good use of their chisel-like bills when the opportunity arises. Practically any woodpecker in the East will visit a suet feeder, and some do so routinely. And even the powerful Pileated isn't so exclusive after all: It readily accepts nuts and berries, especially in winter, and is unable to resist a well-placed block of suet.

juvenile
Lewis's Woodpecker
dark overall
silvery-gray collar
slow crowlike wingbeats
adults
adult
pure white rump and secondaries
Red-headed Woodpecker
♂
white
rimary
atches
♀
brownish head
juvenile
barred secondaries
♂
adult
Acorn Woodpecker

## FOUR MIDSIZE WOODPECKERS

The genera *Melanerpes* (pp. 228–231) and *Sphyrapicus* (sapsuckers) are sister taxa, nearest relatives of one another. The Golden-fronted and Red-bellied Woodpeckers are a close pairing, the Yellow-bellied and Red-naped Sapsuckers another.

### Golden-fronted Woodpecker

*Melanerpes aurifrons* | GFWO | L 9¾" (25 cm)

This mostly Mid. Amer. woodpecker adds a boisterous element to the mesquite woodlands of central and southern Tex.

■ **APPEARANCE** Named for small tuft of yellow at base of bill, but it is the extensive golden wash on the hind nape that attracts notice. Adult male has small red cap, lacking in female and juvenile. Tail clean black, rump and uppertail coverts clean white; compare with Red-bellied Woodpecker.

■ **VOCALIZATIONS** Noisy; common call a short, raspy trill, *pr'r'r'r'r',* less musical than corresponding call of Red-bellied.

■ **POPULATIONS** Hybridizes with Red-bellied, especially from Austin to San Antonio; but note that pure Red-bellieds may also show yellow on the head.

### Red-bellied Woodpecker

*Melanerpes carolinus* | RBWO | L 9¼" (23 cm)

This bird's trademark reddish-tinged belly is difficult to see, but its brilliant red crown will catch your attention.

■ **APPEARANCE** Away from Tex., the only woodpecker in the East that combines extensive red on the head with a zebra-back pattern above; red on female limited to nape, absent on juvenile. Tail shows some black-and-white barring; whitish uppertail coverts and rump flecked with black.

■ **VOCALIZATIONS** All year, muffled *chivf,* often doubled or trebled, and muffled whinny; and bright, rolling *cheerr,* heard mostly late winter to summer.

■ **POPULATIONS** Range has been expanding north for 50+ years; is now widespread and fairly common in hardwood forests of southern New England.

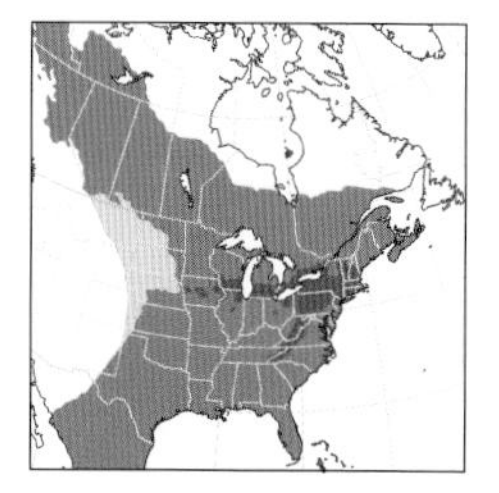

### Yellow-bellied Sapsucker

*Sphyrapicus varius* | YBSA | L 8½" (22 cm)

In summer, the rallentando drumming of this woodpecker is one of the characteristic sounds of the northern hardwood forest. In winter, sapsuckers are unobtrusive denizens of diverse forest types farther south.

■ **APPEARANCE** Broad white wing patch, appearing vertical on bird when upright, contrasts with otherwise mottled plumage. Male throat and forecrown red; female throat white. Juveniles, widely noted fall and winter, have messy gray-brown heads.

■ **VOCALIZATIONS** Call a whiny, descending *pyeeer,* like distant Red-shouldered Hawk's (p. 214). Drum distinctive: a few quick raps, then quickly slowing, *ratta-tatta-tat ... tat ... tat ....*

■ **POPULATIONS** Drills tidy rows of holes in trunks; known as sap wells, they ooze phloem, providing nourishment to sapsuckers and other birds and animals.

### Red-naped Sapsucker

*Sphyrapicus nuchalis* | RNSA | L 8½" (22 cm)

Replaces closely related Yellow-bellied in Interior West; many adults of both sexes of Red-naped have small but distinct patch of red on nape. Hard to separate juveniles in winter in western Great Plains, where both species scarce on migration. Note molt timing: First-winter Red-naped acquires adultlike plumage by mid-autumn, unlike first-winter Yellow-bellied.

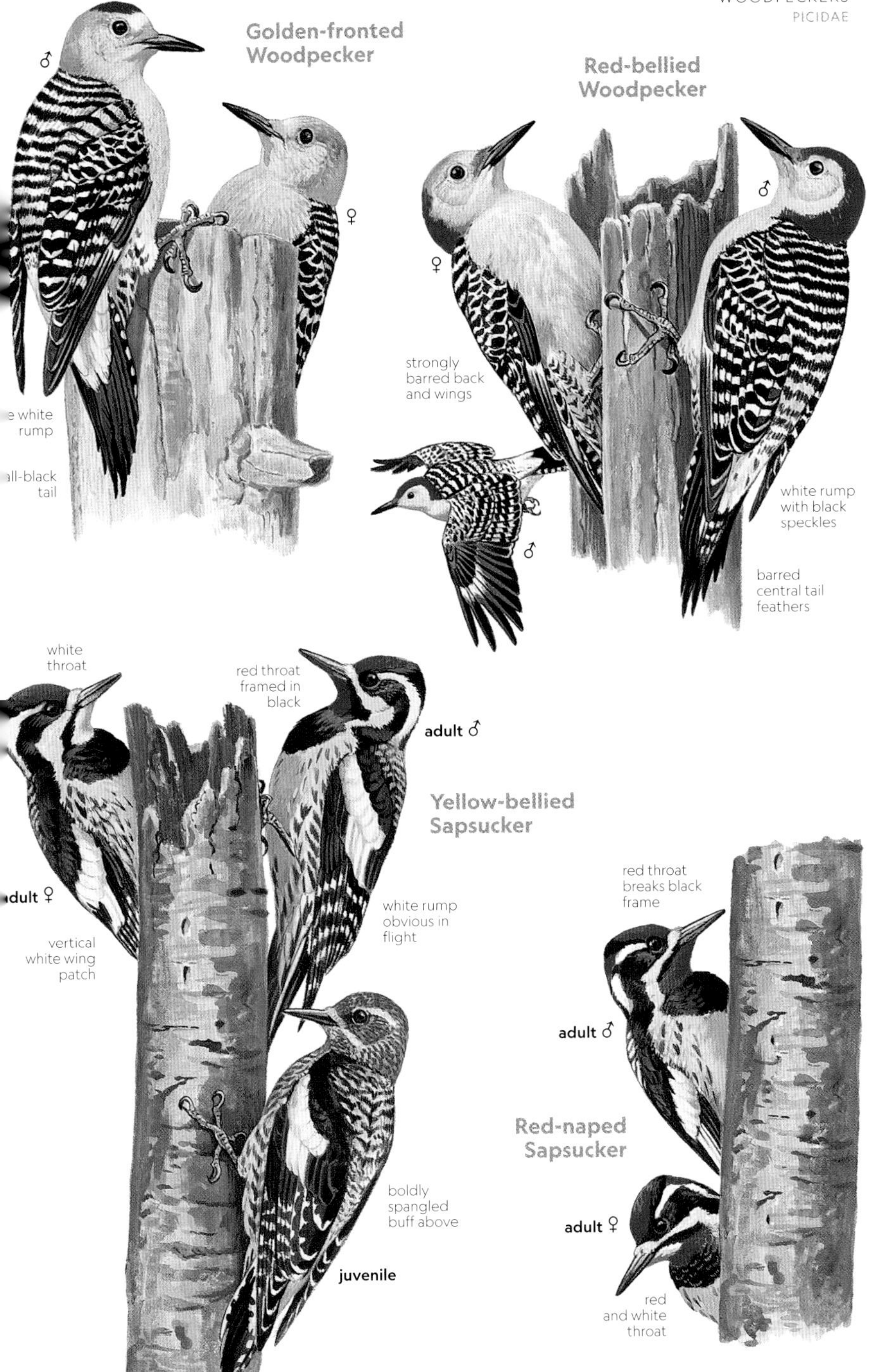
Golden-fronted Woodpecker
♂
♀
white rump
ll-black tail
Red-bellied Woodpecker
♀
♂
strongly barred back and wings
♂
white rump with black speckles
barred central tail feathers
white throat
red throat framed in black
adult ♂
Yellow-bellied Sapsucker
dult ♀
vertical white wing patch
white rump obvious in flight
boldly spangled buff above
juvenile
red throat breaks black frame
adult ♂
Red-naped Sapsucker
adult ♀
red and white throat

## THE THREE-TOED WOODPECKERS

Unlike all other woodpeckers in the East, those in the genus *Picoides* have three toes per foot, rather than the usual four. These two are the most northerly of our woodpeckers, and they are adorned with yellow, not red, on the crown.

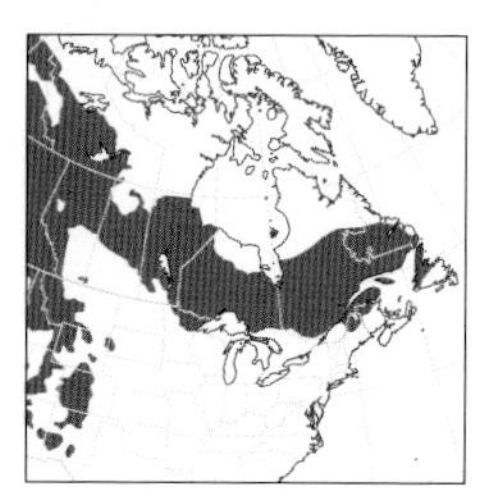

### American Three-toed Woodpecker

*Picoides dorsalis* | ATTW | L 8¾" (22 cm)

Closely associated with fir and spruce forests, this woodpecker is often espied at eye level. The foraging bird works one particular part of the tree for a long time, flaking off bark and picking off arthropods.

■ **APPEARANCE** About the same build as the familiar Hairy Woodpecker (p. 234). Has three toes and barred flanks like Black-backed; adult males of both species are yellow-capped. Back and nape of American Three-toed barred black and white (white in Black Hills breeders), solidly black on Black-backed. Compare female with female Hairy.

■ **VOCALIZATIONS** Generally quiet; *plick* call like that of Hairy. Drum trails off, suggesting Pileated but not as powerful; both sexes drum.

■ **POPULATIONS** Widespread but scarce; even in favored habitat, often outnumbered by other woodpeckers. Found in forests that are some combination of old, burned, and diseased. Resident, but some movement toward recently burned tracts. Three subspecies. Most in East are *bacatus;* nominate *dorsalis* of Rocky Mountains, resident in Black Hills, is white-backed, resembling female Hairy Woodpecker; northwestern *fasciatus,* similar to *bacatus,* extends east to Sask.-Man. border.

### Black-backed Woodpecker

*Picoides arcticus* | BBWO | L 9½" (24 cm)

Both of the three-toed woodpeckers in the genus *Picoides* are attracted to burns, this one especially so. A single burned tree in an otherwise healthy forest may well have a Black-backed Woodpecker!

■ **APPEARANCE** The solid black upperparts are thought to be cryptic, an adaptation for blending in with the charred trunks on which the bird forages. Like American Three-toed, has yellow crown (male), barred flanks, and three toes.

■ **VOCALIZATIONS** More vocal than American Three-toed. Common call an odd, resonant, cluck: *tock,* often with weak initial stutter, *t'tock.* Also gives rattles, squeals, and a grinding, ternlike snarl. Drum similar to that of American Three-toed; both sexes drum.

■ **POPULATIONS** Scarce, but generally outnumbers American Three-toed; somewhat more prone to wandering, especially in search of burns, than American Three-toed.

### Woodpecker Physiology

Collisions with structures are often fatal for birds—with tallies into the hundreds of millions annually in the U.S. alone. Yet a drumming woodpecker bashes its bill full-force against a tree trunk at a rate of 10–20 impacts per second. How does the bird not sustain a concussion, dislodge its eyeballs, or, at the very least, get a nasty headache? The physiology is complex, but a key adaptation is the woodpecker's very long tongue, which acts like a seat belt encircling the brain. Specialized spongy bone tissue further protects the brain, and the woodpecker's blunt bill actually blunts the force of the impact. Powerful muscles—unusually thick for a bird—in the neck and in the eyelids also play a role.

Clearly, these adaptations work when pecking for food, but it is woodpeckers' powerful drumming that pushes the limits of vertebrate physiology. For many woodpeckers, the primary

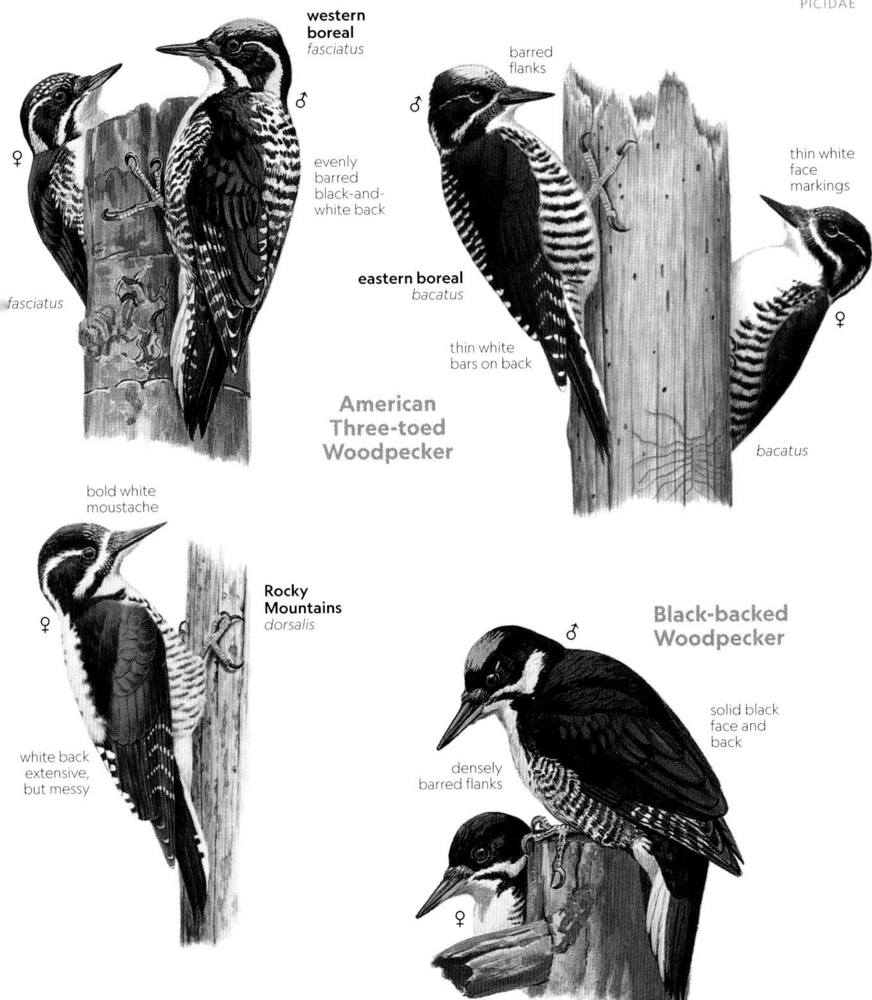


song is drumming—a nonvocal utterance that is functionally identical to the dulcet offerings of a sparrow or a thrush. Referring to a woodpecker's drumming as a song may be initially surprising, and there is a large literature on what actually constitutes birdsong; in recent years, the functional view of song—any sound that is used for attracting mates and defending territories—has enjoyed wide support.

Distinguishing among the different woodpeckers' drums is difficult, but a few species are distinctive: The resounding drum of the Pileated Woodpecker (p. 236) fades toward the end, for example, and the drum of the Yellow-bellied Sapsucker (p. 230) slows down. On top of their drumming, many woodpeckers sing "normal" songs with the syrinx, and some perform elaborate curtseying in display. Woodpeckers, typically thought of as diligent and industrious, go all out with courtship.

## GENUS *DRYOBATES*

If there is a template for the classic woodpecker, it might be this genus. The *Dryobates* woodpeckers are pied, compact, medium to small in size, and hardworking.

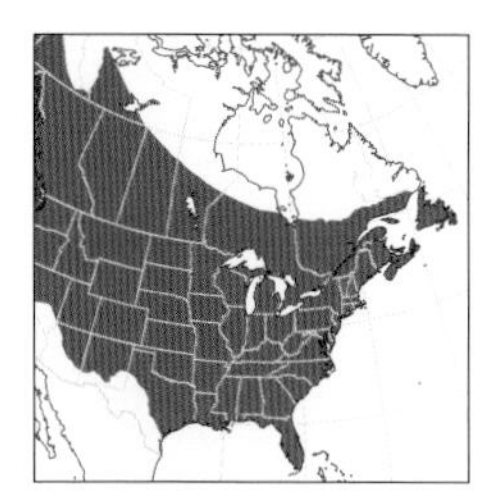

### Downy Woodpecker

*Dryobates pubescens* | DOWO | L 6¾" (17 cm)

The smallest woodpecker in the East, and perhaps the most catholic in its habitat choices. "Downies" may live in planted trees in our largest cities, in unbroken tracts of forest wilderness, and everywhere in between.

■ **APPEARANCE** Small and short-billed with an unmarked white back; adult male has red patch on hindcrown, lacking on female. Look-alike Hairy has proportionately longer bill and unmarked outer tail feathers; outer tail feathers of Downy often flecked black. Compare also with Ladder-backed.

■ **VOCALIZATIONS** When foraging, a simple *pik,* heard year-round; also a rapid whinny of such notes, often given on landing.

■ **POPULATIONS** Adaptable, accepts human-modified habitats; visits feeders.

### Ladder-backed Woodpecker

*Dryobates scalaris* | LBWO | L 7¼" (18 cm)

A woodpecker of arid woodlands and shrublands, this close relative of the Downy is at home where there is cactus, creosote, yucca, and mesquite.

■ **APPEARANCE** A bit larger than Downy and proportionately longer-billed. Zebra-backed in all plumages; adult male has extensive red on crown, absent in female. Underparts buff-white with irregular spotting.

■ **VOCALIZATIONS** *Pik* note and whinny like Downy's, but less flat.

■ **POPULATIONS** Like Downy, accepting of developed habitats in towns and cities; the two species also co-occur in riverside bosques.

### Red-cockaded Woodpecker

*Dryobates borealis* | RCWO | L 8½" (22 cm)

The red "cockade" is hard to see. Fortunately, this near-threatened bird is distinctive in other aspects of morphology and especially ecology.

■ **APPEARANCE** Smaller than Hairy, larger than Downy. White cheek patch framed in black is distinctive in all plumages. Spotted on flanks with zebra-back pattern like Ladder-backed, but ranges do not overlap. Adult male's red "cockade" normally invisible; juvenile male has red forecrown.

■ **VOCALIZATIONS** Call a bright, abrupt *preer* or *preert,* buzzy and wavering, like a sped-up Red-headed Woodpecker (p. 228).

■ **POPULATIONS** Restricted to carefully managed longleaf pine forests; occurs in clans of closely related individuals who favor living but diseased trees.

### Hairy Woodpecker

*Dryobates villosus* | HAWO | L 9¼" (23 cm)

In much of the East, this is the wilder counterpart of the familiar Downy. Although it favors larger forest tracts than its smaller congener, the Hairy is not above a quick snack at a suet feeder.

■ **APPEARANCE** Plumage closely matches Downy. Larger than Downy, but overall size hard to assess in many settings. Hairy has proportionately longer bill, and the white outer tail feathers are unmarked. Juvenile male has smudgy red forecrown, as in other *Dryobates* woodpeckers.

■ **VOCALIZATIONS** Sharp, shrill, far-carrying *pleek!* Rapid whinny of similar notes can sound more like whinny of Red-bellied (p. 230) than Downy.

■ **POPULATIONS** Often absent from smaller tracts used by Downy. Considered nonmigratory, but many wander widely in winter; attracted to burns.

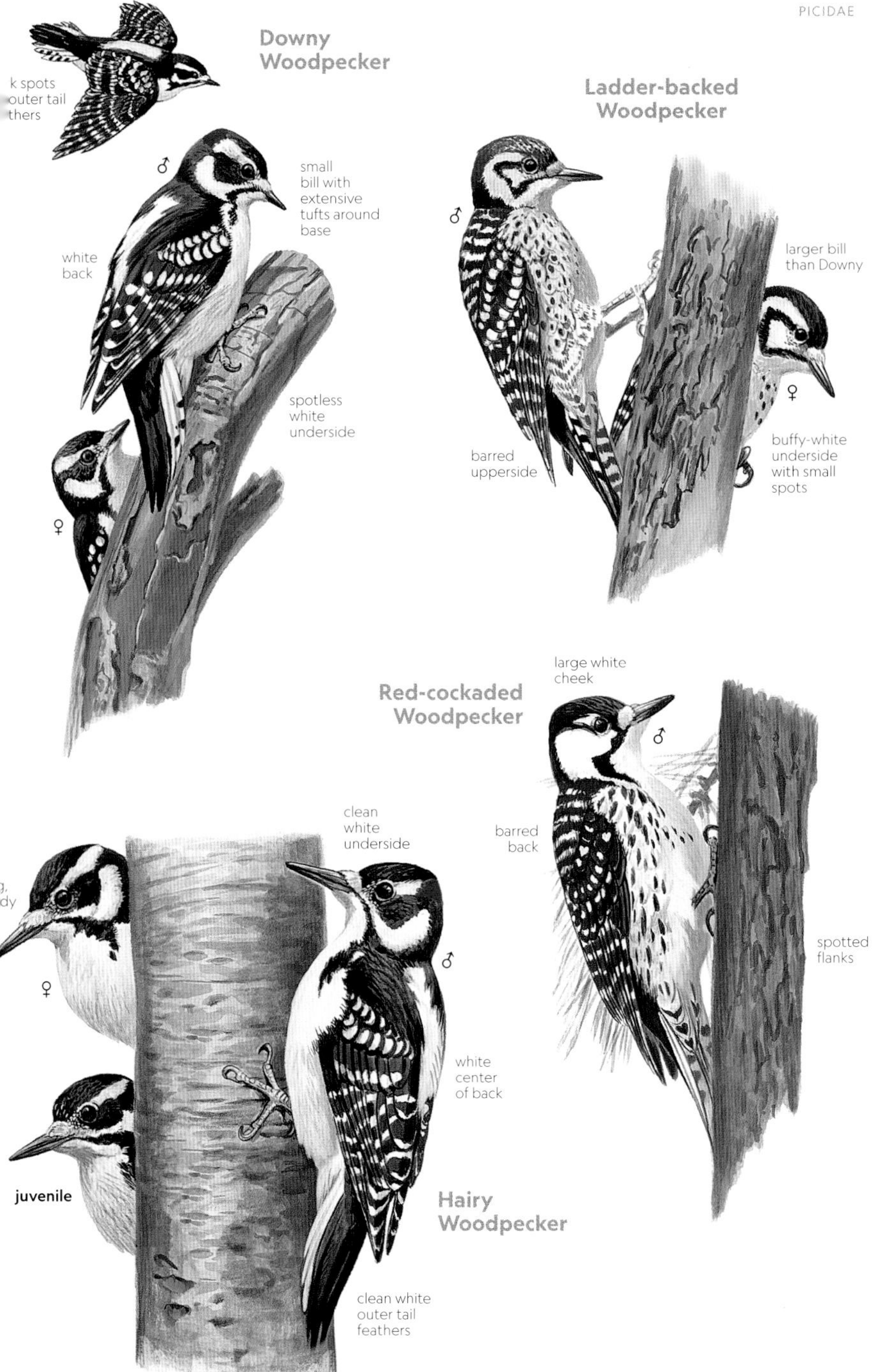
Downy Woodpecker
small bill with extensive tufts around base
white back
spotless white underside
Ladder-backed Woodpecker
larger bill than Downy
barred upperside
buffy-white underside with small spots
Red-cockaded Woodpecker
large white cheek
barred back
spotted flanks
clean white underside
white center of back
juvenile
Hairy Woodpecker
clean white outer tail feathers

## OUR TWO LARGEST WOODPECKERS

Although they are very different in appearance, the Northern Flicker and the Pileated Woodpecker can sound similar. Another point of convergence is that both species have adapted well to human-modified landscapes.

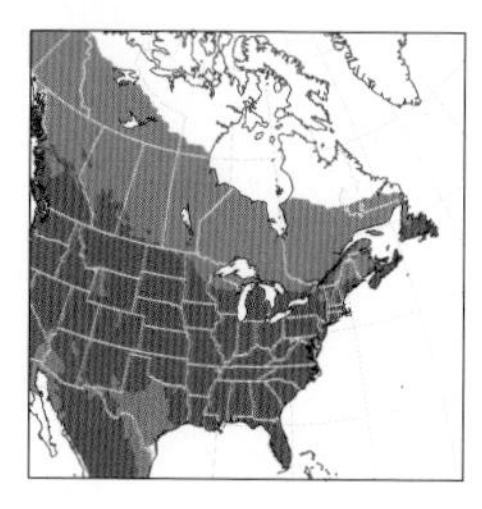

### Northern Flicker

*Colaptes auratus* | NOFL | L 12½" (32 cm)

This flicker is a woodpecker, but it could be called an anything-pecker. This species routinely drums on vents, gutters, and siding, and it spends at least as much time probing the soil for ants as pecking on wood. It is also a habitué of feeding stations.

■ **APPEARANCE** Readily identified to species, but sorting out geographic variation is a fun challenge. All are brown above with black barring, whitish below with big black spots and a thick breastband; white rump striking in flight. In most of the East, flickers are "Yellow-shafted": Both sexes have tan face, gray crown, and bright yellow flight feathers, prominent on bird in flight; male has red crescent on nape and blobby black stripe on side of face; female has red crescent on nape but is plain-faced. Western "Red-shafted," with mostly gray head and salmon-pink flight feathers, is seen regularly east to Great Plains; male has unmarked brown nape and blobby red stripe on side of face; nape and face of female plain. "Yellow-shafted" and "Red-shafted" intergrade broadly in western Great Plains, and close study reveals that many individuals there show blended traits. An emerging complication farther east, especially along coast, is "Yellow-shafted Flickers" with some red feathers, the result of a diet shift to nonindigenous invasive plants with red fruits.

■ **VOCALIZATIONS** Highly vocal. Year-round gives a bright *keeer* or *kee-yer*. Spring and summer, sings constantly both vocally and nonvocally. Vocal songs are fast series of repeated *wick!* or *wake-up!* elements; drum long and steady.

■ **POPULATIONS** Nests in tree cavities, but frequently forages in nearby clearings; common in towns and cities. "Yellow-shafted" more migratory than "Red-shafted."

### Pileated Woodpecker

*Dryocopus pileatus* | PIWO | L 16½" (42 cm)

"Stump Breaker," "Tree Cleaver," *"Le Grand Pic"*—those are among the countless colloquial names for this impressive woodpecker. "The Big P" is on everybody's short list of the grandest of birds in the East.

■ **APPEARANCE** Notably larger than a flicker; the body is mostly black, although the wings flash white in flight. Long-necked and long-billed; head and neck are boldly striped red, white, and black. Both sexes emblazoned with fiery crest; male has red stripe on side of face, black on female.

■ **VOCALIZATIONS** Drum is resonant and powerful, carries far; trails off at end. Call, heard year-round, a hurried series of *wuk* notes, speeding up in the middle; shorter overall and less steady than song of Northern Flicker. Also gives a slower series of *wuk* or *wek* notes, the overall effect rather flicker-like.

■ **POPULATIONS** A natural engineer, the Pileated is important in nutrient cycling and the provisioning of nest cavities for owls, ducks, squirrels, and even pine martens. Remarkably, has taken to woods in urban districts in recent decades; the largest cities along the I-95 corridor have Pileateds now.

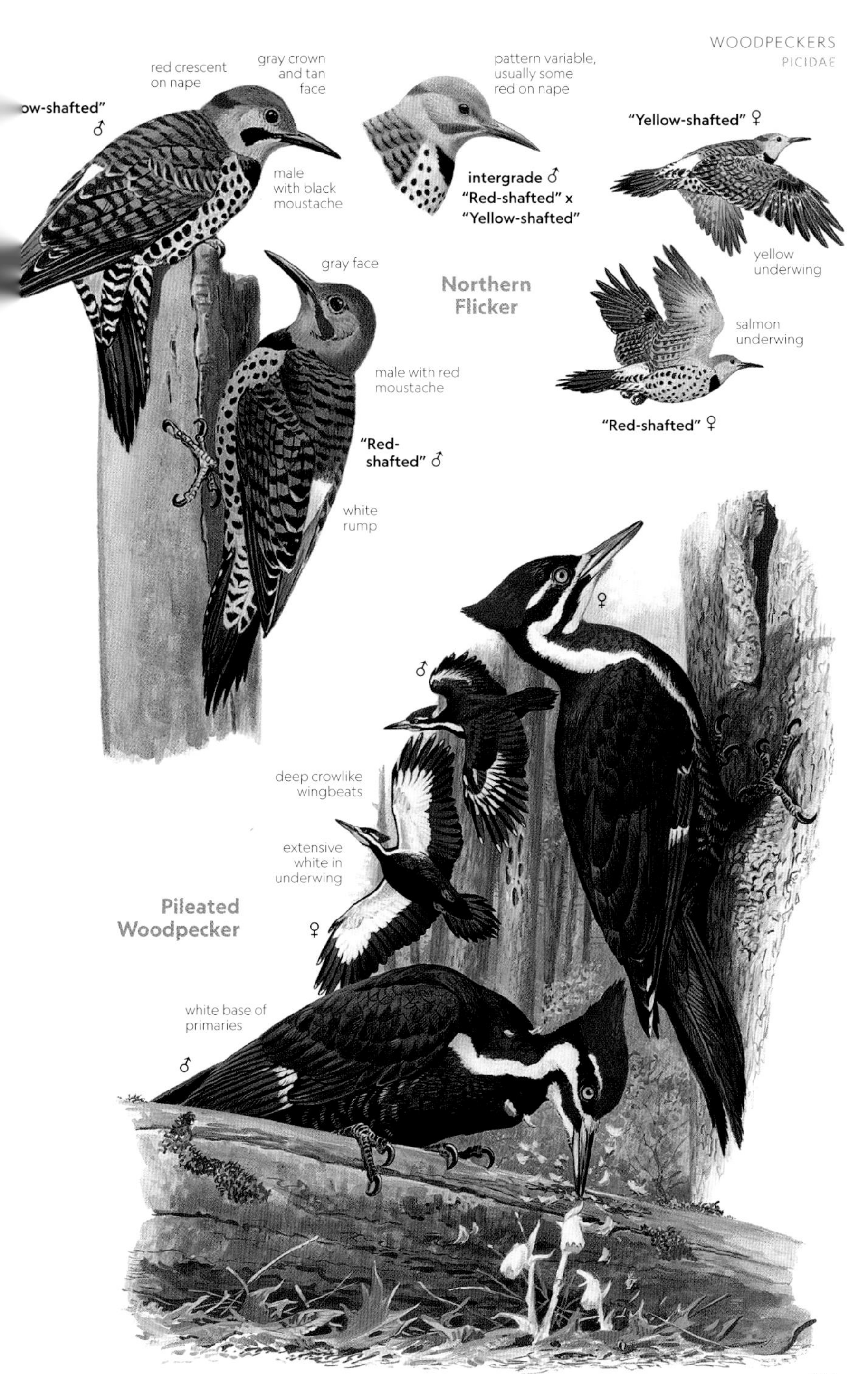
red crescent
on nape
gray crown
and tan
face
ow-shafted" ♂
male
with black
moustache
pattern variable,
usually some
red on nape
intergrade ♂
"Red-shafted" x
"Yellow-shafted"
"Yellow-shafted" ♀
yellow
underwing
gray face
Northern
Flicker
salmon
underwing
male with red
moustache
"Red-shafted" ♀
"Red-
shafted" ♂
white
rump
♀
♂
deep crowlike
wingbeats
extensive
white in
underwing
Pileated
Woodpecker
♀
white base of
primaries
♂

## COLORFUL FALCONS

Although usually grouped with hawks, falcons are more closely related to parrots (pp. 242–245) and passerines (pp. 246–409). Except for the unusual caracara, our falcons are fast-flying and pointy-winged.

### Crested Caracara

*Caracara plancus* | CRCA | L 23" (58 cm) WS 50" (127 cm)

This long-legged scavenger is often found on the ground, and the bird's wings are rounded and typically held flat. Yet the caracara has long been classified as a falcon—a result reaffirmed in recent studies.

■ **APPEARANCE** Large-headed and big-billed; long-necked and long-legged. Flies fairly low. On adult in flight, dark body contrasts with mostly pale tail and wing tips; colorful bill and dark cap distinctive. Young patterned like adult, but colored brown and beige.

■ **VOCALIZATIONS** In interactions with other scavengers, especially Black Vultures (p. 198), a dry ratcheting, rather froglike.

■ **POPULATIONS** Favors dry shrublands, including ranchlands. Range in East disjunct; in Fla., classified as threatened.

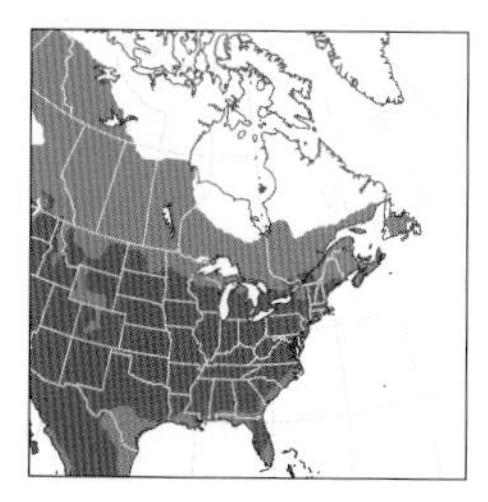

### American Kestrel

*Falco sparverius* | AMKE | L 10½" (27 cm) WS 23" (58 cm)

A little bird is hovering in the wind. Its wings flutter above its body like angel's wings. It's a kestrel, our smallest and most colorful falcon.

■ **APPEARANCE** Very colorful: Chalk-blue wings of male contrast with bright russet back; female orange-winged. Both sexes have vertical black lines on face. Perched, often jerks its tail. Pointed wings in flight quite different from rounded wings of Sharp-shinned Hawk (p. 204).

■ **VOCALIZATIONS** Shrill *klee klee klee.* Call may induce scared rodents to urinate; seeing a bright UV trace in the urine, the kestrel pounces.

■ **POPULATIONS** Usually seen singly in open country, especially on roadside wires; has declined in northeastern U.S.

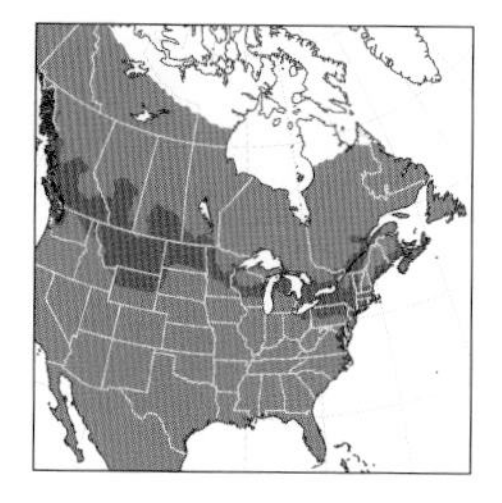

### Merlin

*Falco columbarius* | MERL | L 12" (30 cm) WS 25" (64 cm)

Despite its small size, this is a fearsome predator. Like the larger falcons (p. 240), the Merlin catches birds on the wing.

■ **APPEARANCE** Low-contrast plumage overall compared to Peregrine Falcon (p. 240) and American Kestrel. Male bluish above, female and juvenile brownish above; both sexes have faint vertical stripe on face, not nearly as prominent as kestrel's. Tail mostly dark with thin white bars. Flight fast and powerful on pointed wings, more suggestive of Peregrine than kestrel.

■ **VOCALIZATIONS** Calls shrill and rapid; heard mostly in connection with territorial defense.

■ **POPULATIONS** Two subspecies in East. Widespread nominate *columbarius* darker than *richardsonii* ("Prairie Merlin") of mid-continent. Breeding range expanding south, especially *columbarius;* pioneers often nest in towns and cities.

### Aplomado Falcon

*Falco femoralis* | APFA | L 15–16½" (38–42 cm) WS 40–48" (102–122 cm)

Formerly ranged north from Neotropics to grasslands in N. Mex. and Tex.; sightings today are chiefly of birds from a reestablishment program in coastal Tex. All are long-winged and long-tailed, with dark bulge on flanks. Adult male has boldly marked face, white breast, charcoal gray upperparts; juvenile and female more buffy.

Crested Caracara
pink facial skin
long neck and legs
adult
four white areas: head, tail, and wing tips
juvenile
adult ♂
adult ♀
bright rufous tail
uniformly rufous-brown back and tail
intricate head pattern with two vertical black stripes
frequently hovers
adult ♀
American Kestrel
chalk-blue wings
juvenile ♂
adult ♂
Merlin
indistinct face pattern
powerful flight with no hovering
♀
dark brown above
adult ♂
♀
bluish above
taiga
columbarius
dult ♂
"Prairie"
richardsonii
♀
♂
dark tail with thin, pale bars
paler overall, peach-toned below
bold supercilium
juvenile
♀
adult
♂
Aplomado Falcon
juvenile
lead-gray upperside
dark belly band
very long tail
adult ♀
orange lower body

## POWERFUL LARGE FALCONS

Observing these three is like watching supersonic aircraft: The bird doesn't even seem to be going all that fast, yet in an instant it is already on the other side of the sky.

### Gyrfalcon

*Falco rusticolus* | GYRF | L 20–25" (51–64 cm) WS 50–64" (127–163 cm)

Discovering an "ice falcon" is one of the greatest thrills of winter birding. Although generally rare in winter south of the Arctic, Gyrfalcons have the convenient habit of hanging around for a while.

■ **APPEARANCE** Larger and bulkier than Peregrine and Prairie Falcons. Wings relatively broad-based in flight, but nevertheless pointed as in other falcons; tail, with many bands, also broad-based, somewhat tapered, and long. Three color morphs: white, gray, and dark. White morph distinctive, but beware occasional leucistic Red-tailed Hawk (p. 212). Gray morph and dark morph, especially juveniles, suggest huge Merlin (p. 238) in overall plumage. Also compare gray morph with American Goshawk (p. 204).

■ **VOCALIZATIONS** Winter wanderers to Canada-U.S. borderland essentially silent. Around Arctic nest sites, gives nasal squawks, quacks, or honks; gruffer and rougher than other falcons.

■ **POPULATIONS** Most remain at high latitudes in winter, where Arctic warming is an emerging threat to the species.

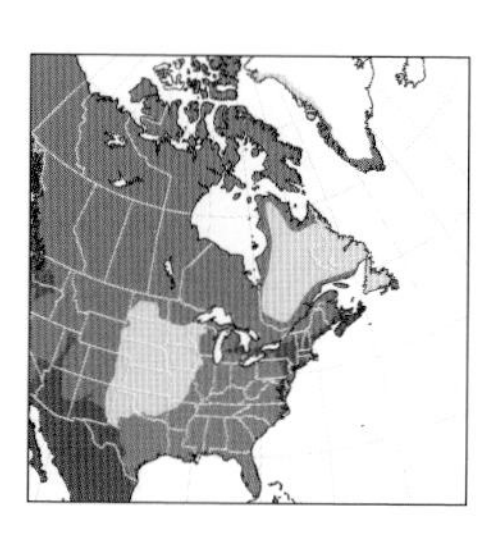

### Peregrine Falcon

*Falco peregrinus* | PEFA | L 16–20" (41–51 cm) WS 36–44" (91–112 cm)

The iconic Peregrine is famously the fastest animal on the planet. It is highly migratory and one of the most cosmopolitan of all animals.

■ **APPEARANCE** Sleek and powerful with long, pointed wings. Wing tips extend to tail tip on bird at rest; Prairie Falcon in flight appears relatively shorter-winged and longer-tailed. Adult barred below and bluish above with boldly marked "helmet." Juvenile browner, with streaking, not barring, below. Wings from below uniformly dusky, lacking dark "wingpits" of Prairie.

■ **VOCALIZATIONS** Long, nasal squeals, rising in pitch, often in slow series: *rraaaahh ... rraaaahh ... rraaaahh ....* Typically heard around eyries, especially right after fledging.

■ **POPULATIONS** Status complex; recovering in East from sharp decline in 20th century. Nests on sea cliffs and canyon walls, but also on skyscrapers; urban birds may be "Pseudogrines," mixes of buff-washed subspecies *anatum* and paler *tundrius*. Migrates across broad front, even out over the ocean.

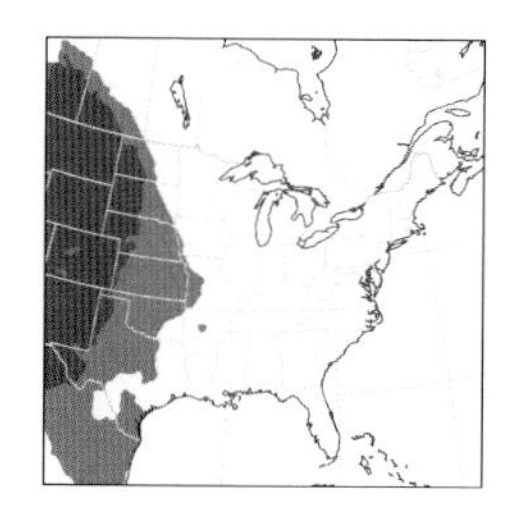

### Prairie Falcon

*Falco mexicanus* | PRFA | L 15½–19½" (39–50 cm) WS 35–43" (89–109 cm)

Though it's less familiar than *Falco peregrinus*, the Prairie is just as impressive as the Peregrine. In the East, this mighty falcon is chiefly a winter visitant to farm country, badlands, and grasslands.

■ **APPEARANCE** Pale; sandy-brown plumage matches arid landscapes favored by the species. About the size of Peregrine, but relatively longer-tailed and shorter-winged. Black "wingpits" prominent in flight. Juvenile a bit darker than adult, streaked below; adult more spotted below. Brown moustache weaker than Peregrine's, stronger than Merlin's; compare especially with female "Prairie Merlin," subspecies *richardsonii*.

■ **VOCALIZATIONS** Noisy at eyrie like Peregrine; call similar, but averages shorter and higher in Prairie. Usually silent in winter.

■ **POPULATIONS** In winter, feeds on Horned Larks (p. 286) but will also eat mammals; found in same winter landscapes as Ferruginous Hawk (p. 216).

weak moustache
gray-morph juvenile
gray-morph adult
Gyrfalcon
broad wings, breast, and tail
dark-morph juvenile
low-contrast overall
tail extends well beyond wing tips
white-morph adult
eastern
anatum
adult
Peregrine Falcon
uniform underwing
crisp black helmet
juvenile
northern
tundrius
eastern
anatum
adult
adult
juvenile
relatively short tail
more extensively bright white
juvenile
slate gray above
pale face with thin brown markings
sandy brown above
rounder wing tips than Peregrine
adults
Prairie Falcon
dark "wingpits"
long tail

## PSITTACIFORMES I

Parrots and parakeets are widely known, but field ID is tricky, taxonomy is in flux, and distributional data are elusive. All are in the order Psittaciformes, and only one, the extinct Carolina Parakeet (p. 426), is unequivocally indigenous to the East.

### Monk Parakeet

*Myiopsitta monachus* | MOPA | L 11½" (29 cm)

It's the dead of winter in Brooklyn or the South Side of Chicago, and a flock of quaker gray and kelly green birds goes screeching by. They're Monk Parakeets, indigenous to S. Amer. south of the Amazon Basin and the most extensively established Psittaciformes species in the U.S.

■ **APPEARANCE** Larger than a "Budgie" (p. 244). Gray underparts and blue remiges separate it from other parrots and parakeets likely seen in the East.

■ **VOCALIZATIONS** Soft, rolling *kreet* and *prrrt* notes; birds in chorus at nest colonies sound like small sandpipers.

■ **POPULATIONS** Widely established around major southern metropolises (Tampa, Houston, New Orleans, etc.), but also well north (especially Chicago, New York). Huge stick nests in electrical infrastructure, considered a fire hazard, are routinely torn down. Escapes may be seen anywhere; Toronto has seven records of free-flying Monks and even a nesting attempt.

### Nanday Parakeet

*Aratinga nenday* | NAPA | L 13¾" (35 cm)

Indigenous range centered on Paraguay; "Nanday" is the Guarani name for the species. Well established in Florida, with largest concentration in St. Petersburg metro area; other strongholds are St. Johns Co. and Palm Beach Co. Larger than Monk Parakeet. Note black head and black bill; breast washed blue, "trousers" red. Flight feathers mostly black below, tinged blue above. Calls high, clipped, and grating: *kraah!* and *rrraaay!*

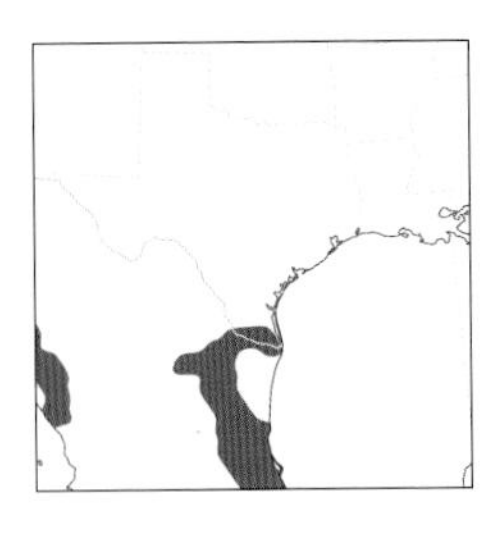

### Green Parakeet

*Psittacara holochlorus* | GREP | L 13" (33 cm)

Park your car near palm trees at a shopping center in the Lower Rio Grande Valley, and it's quickly apparent that Green Parakeets own the place. But they are relatively recent arrivals, first detected in the late 1960s.

■ **APPEARANCE** Essentially a green parakeet, just as the name implies; most have irregular red spots on face and breast. Flight feathers golden from below.

■ **VOCALIZATIONS** Calls relatively squeaky and clear, suggesting Laughing Gull (p. 142).

■ **POPULATIONS** Indigenous to Mid. Amer. north to Tamaulipas. First recorded in Tex. in 1969; many, but not necessarily all, are assumed to be derived from introduced stock. Sticks to palms; Red-crowned Parrots (p. 244) in South Tex. often in oaks. A few also in Fla., especially Miami area.

### Mitred Parakeet

*Psittacara mitratus* | MIPA | L 15" (38 cm)

Indigenous to the east slope of the Andes from Peru to Argentina; well established around Miami, with reports elsewhere in Fla. (both coasts). A large parakeet; green overall, including leading edge of wing, with extensive but splotchy red on face. Calls clipped and squeaky, like a toy horn. Compare with similar Red-masked Parakeet (also pictured), *P. erythrogenys*, indigenous to west slope of Andes, also in Miami area, but not considered established; Red-masked has solidly red face and red on leading edge of wing.

gray face
and breast
Monk
Parakeet
blue flight
feathers
black face
and bill
Nanday
Parakeet
tail and flight
feathers
black below
Green
Parakeet
uniformly
yellow-green
red on face
variable
large pale
bill
Mitred
Parakeet
Red-masked
Parakeet for
comparison

Parrot or parakeet? As with pigeons and doves (pp. 70–75), the names do not reflect a clean taxonomic break. In general, parrots are stocky and short-tailed, parakeets leaner and longer-tailed.

## White-winged Parakeet

*Brotogeris versicolurus* | WWPA | L 8¾" (22 cm)

Formerly lumped with Yellow-chevroned as a single species, Canary-winged Parakeet. In indigenous range, occurs in stupendous flocks along Amazon R. Established but declining in and around Miami. White triangle on wing prominent in flight. Small, with correspondingly light, chirpy calls.

## Yellow-chevroned Parakeet

*Brotogeris chiriri* | YCPA | L 8¾" (22 cm)

Not on the official Fla. list, but more common there than White-winged Parakeet. Calls a bit sharper, shriller than White-winged's. Yellow wing bar prominent in flight. Indigenous range does not overlap with White-winged, but both occur in Miami area; hybrids complicate field ID.

## Red-crowned Parrot EN

*Amazona viridigenalis* | RCPA | L 13" (33 cm)

Endemic to northeastern Mexico; established in South Texas, where population may be descended from both introduced birds and strays. A few also in Fla., especially Palm Beach. Red crown extensive on male, reduced on female. Note also blue wash on nape; red secondaries prominent in flight. Calls include harsh, down-slurred whistle and pounding squawks.

## Budgerigar

*Melopsittacus undulatus* | BUDG | L 7" (18 cm)

The familiar pet store "Budgie" is in a different family, Psittaculidae, than the other species on this page. Indigenous to Australia; thousands nested in central Fla. in the 20th century, but now extirpated; the decline of the Budgerigar had not been widely anticipated during the species' heyday here. Escapes are still noted practically anywhere there are humans. Colors variable (greens and yellows, even blues and whites); nearly all are barred above, plain below.

### Parrots and Parakeets in the East

On top of the challenge of field identification, parrots and parakeets present us with the thorny question of whether they're "countable." Individual birds and even whole flocks of them may be seen or heard, but are their populations considered to be biologically established? If so, they are said to "count" for official state and provincial lists.

In Florida alone, an astonishing 81 species in the order Psittaciformes have been credibly documented in the wild by field ornithologist Bill Pranty and his energetic colleagues. Pranty and others have persuasively shown that a few species are well established, a great many clearly are not, and several are in a no man's land in between. Ascertaining population status is an important matter elsewhere in the East too, especially Texas.

Population establishment is an uncertain and often unstable matter. This guide includes all species on the official lists of eastern states and provinces. A few others make the cut as well. But scores of species of Psittaciformes are found in the wild in the East, far too many for inclusion here. Study data in the eBird and iNaturalist databases, but also consider contributing new knowledge. The data are incomplete, parrot populations are dynamic, and crowdsourced range maps are only as good as the birders who create them.

White-winged Parakeet
bold white triangle
flight of both *Brotogeris* erratic and fluttery
Yellow-chevroned Parakeet
dark green flight feathers
flight direct, with shallow wingbeats
Red-crowned Parrot
red patch in secondaries
immature
adult
short tail with yellowish tip
long, pointed tail
yellow variant
escaped birds highly variable
blue variant
usually has yellow face, barred upperparts
Budgerigar

## TEXAS TYRANNI

These tropical species reach the northern limits of their range in our area in South Texas. The becard and beardless-tyrannulet are scarce and often overlooked, the kiskadee common and conspicuous.

### Rose-throated Becard

*Pachyramphus aglaiae* | RTBE | L 7¼" (18 cm)

Rare but annual visitor to our area; the only regular representative of its family, Tityridae, in the East. Recorded year-round, but most are in winter, maybe reflecting observer bias. Stocky and relatively short-tailed. Adult male dark gray above, paler gray below with rosy throat; female warm buff all over with extensive black cap. Call a weak whistle, *pip-s'WEEeeeee,* easily missed amid the louder natural sounds of Rio Grande woodlands.

### Northern Beardless-Tyrannulet

*Camptostoma imberbe* | NOBT | L 4½" (11 cm)

Our smallest flycatcher and among the smallest passerines in the East. Upright posture, slight crest, and wing bars suggest a tiny *Contopus* (p. 254). Pale lores and eyebrow aid in separation from similarly diminutive but unrelated Verdin (p. 284); compare also with genus *Empidonax* (pp. 256–261). Uncommon year-round resident; fairly active, but often in dense thickets, where it gleans arthropods with its bill—short, thick, and orange. Song a series of four to five rich whistles: *pyee pyee pyee pyee.*

### Great Kiskadee

*Pitangus sulphuratus* | GKIS | L 9¾" (25 cm)

Just on looks, the kiskadee is one of the great birds of South Texas. But its calls are arguably even more impressive: All day long, pairs and small groups scream *KIS-ka-DEE!* wherever there is a bit of cover.

■ **APPEARANCE** Our biggest-bodied flycatcher, bobtailed and blockheaded. Unlike any other bird in Tex.: tail and wings rufous; underparts bright yellow; head strikingly patterned.

■ **VOCALIZATIONS** Eponymous call accented on the first and especially third syllables. Also gives monosyllabic squeals and longer bursts of chatter.

■ **POPULATIONS** Widespread, but most common at edges of woods, particularly around waterways, where it feeds on both flying insects and small fish snagged from the water's surface. Range slowly expanding north, with vagrants in recent years to several midwestern states and even Ont.

### Suborder Tyranni

Well over half the world's bird species are in a single order: the Passeriformes, or passerines. The passerines found in the East span from this page to the end of the main text. Why has this order radiated so spectacularly? The short answer is that we don't know. The passerines' success in leafy, often wooded habitats is likely at play; their small size and even the vagaries of geologic history probably have something to do with it too.

Within this vast order are three well-delineated suborders. One, represented by only six extant species, is restricted to New Zealand. A second, the Tyranni (pp. 246–263), with over 1,000 species, is found only in the New World; if the Tyranni were their own order, they would be the second largest extant order of birds. The rest are in the huge suborder Passeri (pp. 264–409), with close to 5,000 species globally.

The Tyranni get their name from the tyrant flycatchers, one of the best-represented groups within the suborder. They're also known as suboscines, a term differentiating them from the oscines in the suborder Passeri. The oscines are technically songbirds, and suboscines are, technically, not. But many suboscines, including some eastern species, have lovely songs.

♂
short,
stout bill
gray overall,
paler below
Rose-throated
Becard
contrasting
dark cap
♀
warm buff,
brighter on
back and wings
small curved
bill with pale
base
worn
adult
fluffy crest
short, pale
eyebrow
Northern
Beardless-Tyrannulet
low-contrast
wing markings
fresh
Great Kiskadee
bright
rufous
wings
and tail
Flycatchers

**GENUS *MYIARCHUS***

These woodland flycatchers are long-tailed and weakly crested, with black bills, gray heads and breasts, washed-out yellow bellies, and rufous in the wings and tail. Bill structure and vocalizations are valuable in field ID. *Myiarchus* means "lord of the flies."

## Ash-throated Flycatcher

*Myiarchus cinerascens* | ATFL | L 7¾" (20 cm)

Widespread and common in the western U.S., this is a bird of arid woodlands, particularly those with a juniper component. Like other *Myiarchus* flycatchers, it nests in holes, especially those drilled by flickers.

■ **APPEARANCE** Smaller overall, smaller-billed, and paler than the Great Crested Flycatcher, its counterpart in the East. Belly especially pale; in close view or photos, note dark-tipped tail from below.

■ **VOCALIZATIONS** Call a short, rough, rolling *prrrp.* On breeding grounds, gives disyllabic *k'brick* or *k'breer,* not unlike Cassin's Kingbird (p. 250).

■ **POPULATIONS** Regular in small numbers to Atlantic seaboard late fall to winter, especially in recent years.

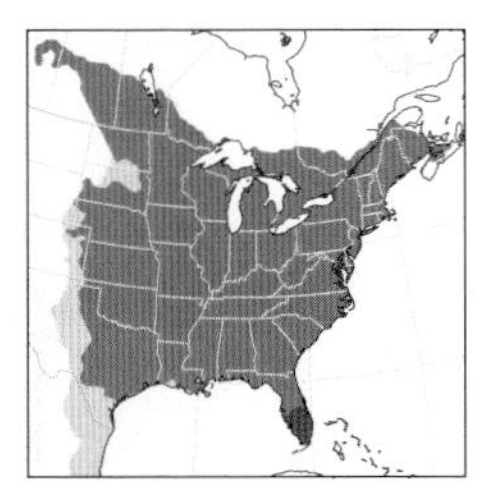

## Great Crested Flycatcher

*Myiarchus crinitus* | GCFL | L 8½" (22 cm)

In much of the East, this is the only *Myiarchus* flycatcher. The Great Crested is distinctive, but it haunts dense, leafy broadleaf forests, where its loud calls are the best clue to its presence.

■ **APPEARANCE** Darkest *Myiarchus* in the East, with bill a bit bigger than that of Ash-throated; yellow belly of Great Crested contrasts well with cold gray breast. In close view or review of photos, note white-edged tertials, mostly rufous undertail, and a bit of color at base of otherwise black bill.

■ **VOCALIZATIONS** Two main calls: a far-carrying, whistled *wheee-ip,* ending abruptly, and a buzzy, somewhat shrill *bzhrrrrrp,* often doubled or trebled.

■ **POPULATIONS** Breeding range expanded north and west in 20th century. Most leave midlatitudes in the East by mid-Sept.; an East Coast *Myiarchus* from Nov. onward is likely Ash-throated.

## Brown-crested Flycatcher

*Myiarchus tyrannulus* | BCFL | L 8¾" (22 cm)

Across a fair swath of southern Tex., three *Myiarchus* flycatchers are reasonably common. Even more than the others, the Brown-crested stays in dense woods with ash, elm, and hackberry.

■ **APPEARANCE** Color intermediate between pale Ash-throated and dark Great Crested. Bill much bigger than Ash-throated's, a bit bigger than Great Crested's. In close view, tail less broadly rufous below than Great Crested's.

■ **VOCALIZATIONS** Song a rolling *prrrr da-rrrr* (*What we'll do!*), often run out to more syllables. Also a short *whip,* like a clipped Great Crested.

■ **POPULATIONS** Breeding range apparently expanding north and west in Tex. Annual vagrant east to South Fla.

## La Sagra's Flycatcher

*Myiarchus sagrae* | LSFL | L 7½" (19 cm)

Annual vagrant from Cuba and the Bahamas, mostly fall to early spring, to Fla., especially Miami area. Our smallest and palest *Myiarchus,* with trace of yellow below, reduced rufous in wings and tail; bill thinner, slighter than even Ash-throated's. Call a simple *wheet* or *whee-it.* Vagrants found in hammocks and parks, often in well-birded, somewhat urban districts.

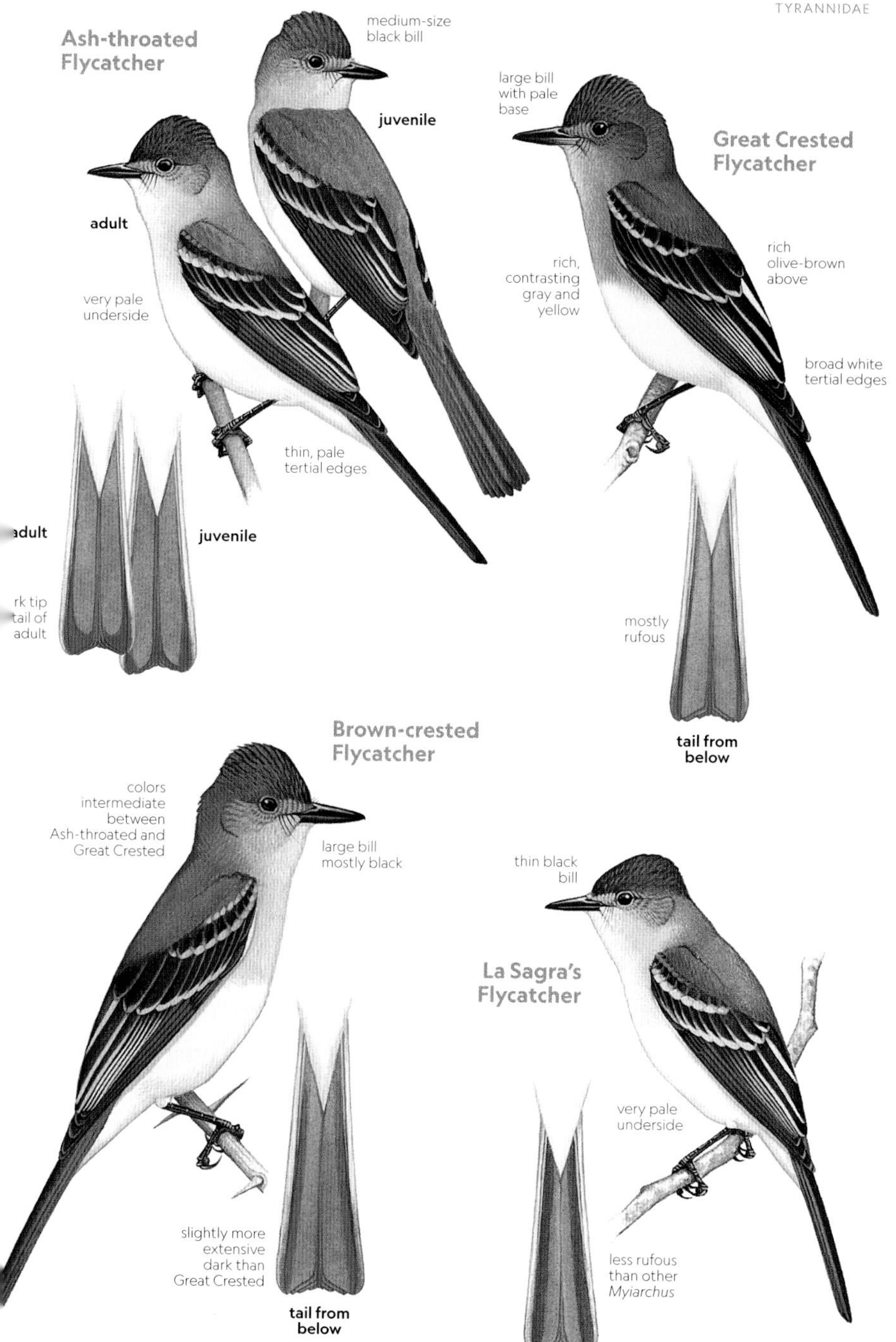
Ash-throated Flycatcher
medium-size black bill
juvenile
adult
very pale underside
thin, pale tertial edges
adult
juvenile
rk tip tail of adult
large bill with pale base
Great Crested Flycatcher
rich, contrasting gray and yellow
rich olive-brown above
broad white tertial edges
mostly rufous
tail from below
Brown-crested Flycatcher
colors intermediate between Ash-throated and Great Crested
large bill mostly black
slightly more extensive dark than Great Crested
tail from below
thin black bill
La Sagra's Flycatcher
very pale underside
less rufous than other Myiarchus

## YELLOW-BELLIED KINGBIRDS

Flycatchers in the genus *Tyrannus* (pp. 250–253) give their family a bad name. Literally. The kingbirds in this type genus for the family Tyrannidae (pp. 246–263) are aggressive, even tyrannical.

### Tropical Kingbird

*Tyrannus melancholicus* | TRKI | L 9¼" (23 cm)

This feisty flycatcher is increasing in South Texas, where its status is clouded by confusion, especially historically, with the look-alike Couch's Kingbird.

■ **APPEARANCE** Fairly large-billed, with olive green back and notched tail. Similar Western Kingbird has slighter bill and unnotched black tail with white edges. Exceedingly similar Couch's Kingbird has shorter, thicker bill; listen for the calls, notably different.

■ **VOCALIZATIONS** Calls short and sharp, with mechanical quality; *pleek* and *plick* notes uttered in series, often stuttering.

■ **POPULATIONS** Favors open areas near humans (parking lots, golf courses, etc.); Couch's prefers woodland edges. Annual vagrant well north of Tex., especially Atlantic coast and Great Lakes.

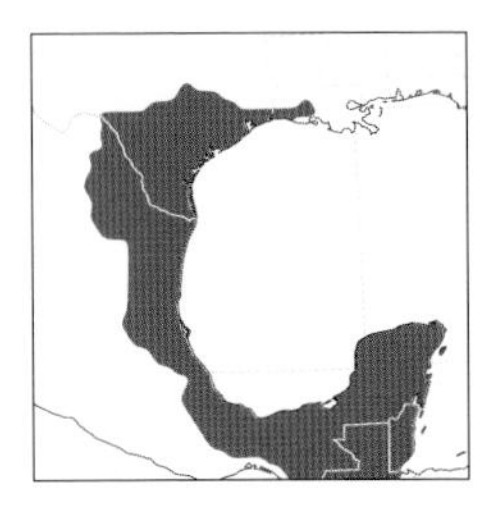

### Couch's Kingbird

*Tyrannus couchii* | COKI | L 9¼" (23 cm)

Lumped with the Tropical Kingbird for decades, this species was assumed to be the only lowercase-t tropical kingbird in southern Tex. Questions remain.

■ **APPEARANCE** Nearly identical to Tropical, but Couch's bill shorter and thicker; know the calls. Couch's larger and darker overall with larger bill than widespread Western; lacks prominent white tail edges of Western.

■ **VOCALIZATIONS** Calls distinct from Tropical: a shrill *bzheeer,* suggesting a distant pauraque (p. 82), and a series of squeaky disyllabic notes, *t'wee! t'wee t'WEE! tweeooo.* Also gives a single *kip,* not unlike Western.

■ **POPULATIONS** Present in South Tex. year-round but mostly summer; wanders annually to La. Expanding north; vagrates well out of range.

### Cassin's Kingbird

*Tyrannus vociferans* | CAKI | L 9" (23 cm)

A migratory kingbird of arid western woodlands. Eastern range limit complicated by interannual variation and apparent expansion; breeds east at least to Black Hills, Nebr. panhandle, and Trans-Pecos of Tex. Same size and build as Western, but darker overall. Cassin's tail lacks white edges of Western's; small white throat patch contrasts with darker hood. Call an emphatic *ch'BEW,* unlike Western's but similar to Ash-throated Flycatcher's (p. 248).

### Western Kingbird

*Tyrannus verticalis* | WEKI | L 8¾" (22 cm)

Across much of its range, this is the common roadside flycatcher, perched on signs, fences, and utility wires.

■ **APPEARANCE** Yellow belly and white-edged (not white-tipped) tail instantly separate it from Eastern Kingbird (p. 252); note also contrast between pale gray back and nearly black tail. Compare carefully with Cassin's and especially Tropical and Couch's, all prone to vagrancy. Compare also with Say's Phoebe (p. 262) and, oddly, American Robin (p. 318).

■ **VOCALIZATIONS** Inordinately vocal, starts at the first hint of dawn. Listen for short *rik!* and *reek!* notes, often in longer series: *reek! a-rik! arikaree!*

■ **POPULATIONS** Not the only "western" kingbird; outnumbered by Eastern in wet valleys as far west as western Mont., but outnumbers Eastern in drier habitats. Rare in fall and winter to Fla., very rare farther north along coast.

Cassin's
Western
Tropical
Couch's
Tropical Kingbird
tail all-dark and notched
adults
rge bill,
irly thin
base
greenish breast and back
spring adult
adult primary tips
Couch's Kingbird
bill slightly shorter and thicker than Tropical
best identified by call
spring adult
adult primary tips
Cassin's Kingbird
contrasting white throat
juvenile
adults
dark gray
spring adult
worn fall adult
dark tail with obscure pale tip
dark wings and tail contrast with gray back
small bill
juvenile
adults
spring adult
worn fall adult
square black tail with narrow white edges
Western Kingbird

## GRAY-BREASTED KINGBIRDS

These mostly gray kingbirds of open country lack the bright yellow bellies of those on the preceding page. The two long-tailed species are effectively kingbirds, despite not going by that name.

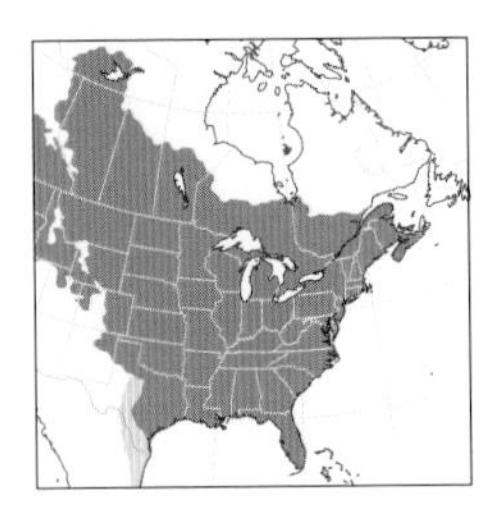

### Eastern Kingbird

*Tyrannus tyrannus* | EAKI | L 8½" (22 cm)

With a name like *Tyrannus tyrannus,* it comes as little surprise that this species is pugilistic. Watch as it drives off larger species in midair assaults.

■ **APPEARANCE** Dark above, white below; tail tip white. Head and tail of adult blackish, rest of upperparts gray; a red patch on crown is usually concealed. Juvenile browner above, a bit grayer below. Eastern Phoebe (p. 262) smaller and drabber, lacks white tail tip.

■ **VOCALIZATIONS** *Dzzt* or *bzzeert*, shrill and buzzy; often run together in sputtering series, like a downed electrical wire.

■ **POPULATIONS** The only regularly occurring kingbird in much of the East. Where range overlaps with the Western's, the Eastern favors wetter habitats.

### Gray Kingbird

*Tyrannus dominicensis* | GRAK | L 9" (23 cm)

Human "snowbirds" may not ever see this large gray flycatcher. It occurs in the East only in the warmer months, mostly in southern peninsular Florida.

■ **APPEARANCE** Similar in habits, habitats, and plumage to Eastern Kingbird; Gray is paler and a bit larger overall, with a bigger bill. Tail of Gray slightly notched, lacking white terminal band of Eastern.

■ **VOCALIZATIONS** Common call a rapid series of rising *p'* or *pip!* notes intermixed with fast trills: *p'pip! peeee p'p'peeee.*

■ **POPULATIONS** Although a Fla. specialty, breeds north to St. Simons I., Ga., and expanding to north-central Gulf Coast. Regular vagrant north along Atlantic coast to mid-Atlantic, mostly fall.

### Scissor-tailed Flycatcher

*Tyrannus forficatus* | STFL | L 13" (33 cm)

Some U.S. states have questionable state birds, but not Oklahoma: The Scissor-tailed Flycatcher is one of the superb birds of N. Amer. and a splendid emblem for the Sooner State.

■ **APPEARANCE** Adult male, with long outer tail feathers, unmistakable; splays tail out in flight, especially when chasing insects. Adult female's tail shorter, juvenile's shorter still. Adult, with salmon and orangey highlights, distinctive; juvenile paler, some resemble pale juvenile Western Kingbird (p. 250).

■ **VOCALIZATIONS** Calls similar to those of Western Kingbird, overlapping in timbre; individual notes of Scissor-tailed a bit sharper, *pleep!* and *pleek!*

■ **POPULATIONS** Relatively range-restricted, breeding mostly in the southern Great Plains; wanders widely however, wintering regularly in small numbers in Fla. and sometimes breeding well out of range.

### Fork-tailed Flycatcher

*Tyrannus savana* | FTFL | L 14½" (37 cm)

Widespread in Neotropics, this austral migrant often overshoots, vagrating as far north as Canada. Annual to East, especially coasts and Great Lakes, with preponderance of records in fall. All are dark above and pale below, suggesting the bulkier Eastern Kingbird; relatively short-tailed immature could be passed off as that species. Spectacularly long-tailed adult sometimes confused with Pin-tailed Whydah, *Vidua macroura,* a popular cage bird from Africa occasionally seen in the wild in the East.

Eastern Kingbird
broad white tip on tail
Gray Kingbird
dark mask
long, heavy bill
pale gray back and crown
notched, unmarked tail
immature
Fork-tailed Flycatcher
sharp black cap
adult
pure white below
ribbon-like tail with thin white edges
juvenile
adult ♂
salmon below, brightest in "wingpit"
gid forked tail
ith extensive
hite
juvenile
Scissor-tailed Flycatcher
very pale
adult ♂
female shorter-tailed

## GENUS *CONTOPUS*

Flycatchers in this genus typically perch out in the open on a dead snag, where they wait motionless for aerial prey. They are dusky overall, weakly crested, and a bit larger and longer-winged than *Empidonax* flycatchers (pp. 256–261).

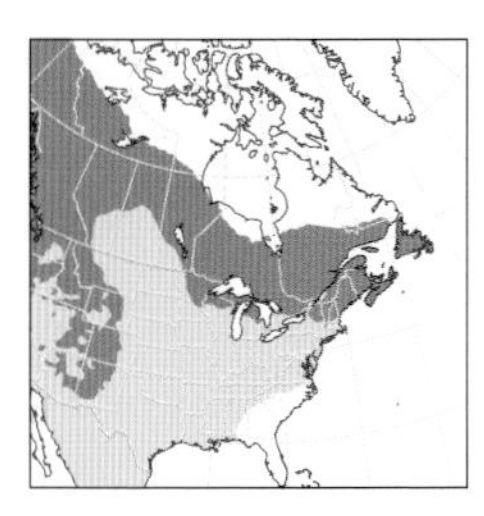

### Olive-sided Flycatcher

*Contopus cooperi* | OSFL | L 7½" (19 cm)

A bird sitting straight up atop the very highest dead snag in a row of trees often proves to be this species, especially if it sallies out a considerable distance, snags a flying insect, and returns to the same perch.

■ **APPEARANCE** Our largest pewee, quite long-winged and rather short-tailed; large-headed, with weak wing bars and no eye ring. Most wear a "vest," with dark breast broken by vertical white stripe. White tufts poking out of back can be revealed or concealed in a matter of seconds.

■ **VOCALIZATIONS** Song a strange, memorable *pip PREEeee preeer,* notoriously mnemonicized as *Quick! Three beers!* Calls include a single, sharp *peep* and a longer series, *pip pip pip.*

■ **POPULATIONS** Late spring migrant, into June, through midlatitudes of U.S.; first "fall" migrants show up again by late July. Almost never occurs in winter in East.

### Western Wood-Pewee

*Contopus sordidulus* | WEWP | L 6¼" (16 cm)

In most of the U.S. and Canada, wood-pewee ID is a straightforward matter of geography. But the species overlap at the western edge of the East, and silent wood-pewees on migration are vexing there.

■ **APPEARANCE** The name *sordidulus* means "dirty little thing," and this species indeed averages muddier than Eastern Wood-Pewee. Bill of Western more extensively dark beneath than that of Eastern; Western also has fainter wing bars, dirtier undertail coverts, and a browner breast.

■ **VOCALIZATIONS** All day long on breeding grounds, gives descending buzz ending in weak whistle, *pZZZzzeee;* dawn song, heard only on breeding grounds, mixes short whistles with buzzier phrases. Migrants and vagrants fairly quiet, but confusingly give purer calls, like Eastern.

■ **POPULATIONS** Common migrant a bit east of breeding range into western and maybe even central Great Plains. Breeding ranges of the two wood-pewees largely nonoverlapping; some contact in Missouri River Valley in N. Dak., though.

### Eastern Wood-Pewee

*Contopus virens* | EAWP | L 6¼" (16 cm)

All day long in hardwood forests, even around midday on hot summer days, this drab flycatcher sings its trademark song. Both wood-pewees sit patiently in micro-clearings, watching intently for aerial insects.

■ **APPEARANCE** Told from smaller empids (pp. 256–261) by lankier shape, longer wings, and lack of eye ring. Similar Western Wood-Pewee has darker lower mandible, smudgier breast, and splotchier undertail coverts; *virens* denotes green, and Eastern averages brighter and greener than Western.

■ **VOCALIZATIONS** Slowly alternates between two main song elements: *PEEee-uh-weee,* exaggeratedly slow and with explosive first syllable, and wimpier *pyeeeeuh,* falling in pitch.

■ **POPULATIONS** Wanders annually late spring–summer almost to Rockies. Like Western Wood-Pewee and Olive-sided Flycatcher, returns fairly late in spring, departs before mid-autumn. Neither wood-pewee winters in U.S.

large bill
dark hood and "vest"
Olive-sided Flycatcher
adults
usually vaguely streaked
contrasting white stripe from throat to belly
white flank tufts often visible
adult
very long wings
short tail
mpids (pp. 256–261) have more compact bodies, longer legs, and shorter primary projection than wood-pewees
Western Wood-Pewee
weak wing bars
st Flycatcher comparison
darker gray-brown than Eastern, especially on breast
variations
mostly dark bill, occasionally with pale orange
smudgy undertail coverts
paler adult
darker adult
brighter wing bars than Western
Eastern Wood-Pewee
mostly pale below
variations
mostly orange lower mandible
fresh spring adult
worn late summer adult
juvenile

## TWO FOREST EMPIDS

In the breeding season, these relatively colorful *Empidonax* flycatchers, or empids, occur within forests, often with a mesic (moist) element. Both are fairly long-winged for their genus.

### Yellow-bellied Flycatcher

*Empidonax flaviventris* | YBFL | L 5½" (14 cm)

For most birders in the East, this relatively bright empid is seen only on migration. Yellow-bellied Flycatchers nest in the Northwoods, typically amid ferns and peat mosses at ground level in the dark understory. Migrants are usually found in dense vegetation.

■ **APPEARANCE** Most are obviously yellow-bellied, but an even better mark is the yellow throat. (But always keep in mind the effect of lighting on our perception of color, especially in forests.) Eye ring also yellow, and wing bars tinged yellowish. Bill small and entirely orange-yellow below.

■ **VOCALIZATIONS** Two main sounds: a rough *rrr-bunk*, similar to Least Flycatcher's song (p. 258), but burrier; and a whistled *prree-rreee*, like a weak call of Eastern Wood-Pewee (p. 254). Sometimes vocalizes on spring migration.

■ **POPULATIONS** Migrates quite late in spring (until mid-June in Ohio River Valley) and early in fall. A circum-Gulf migrant; thus, rare in much of southeastern coastal plain north to Carolinas. In recent decades has experienced some retraction of breeding range in northern U.S.

### Acadian Flycatcher

*Empidonax virescens* | ACFL | L 5¾" (15 cm)

The name Acadian, referring to parts of Maine and eastern Canada, is a misnomer. This is the most southerly of eastern empids, with a preference for Carolinian hardwood forests.

■ **APPEARANCE** Like Yellow-bellied, a fairly bright empid; large-billed and long-winged. The most greenish eastern empid, with the usual caveats about lighting and feather wear. Eye ring thin and complete; throat off-white. Juveniles and freshly molted adults on fall migration resemble Yellow-bellied, but note white or grayish throat and large bill.

■ **VOCALIZATIONS** Song an explosive *pEET-seeick* (*pizza!*), uttered 5–10 times per minute; on breeding grounds, also a rapid series of *pee* notes. Call a snappy *psEET*, sometimes given by migrants.

■ **POPULATIONS** Nests in leafy broadleaf woods, especially along streams and at swamp edges, even in major cities like Philadelphia. The nest, notably messy, is placed high in a tree, often on a thin branch precariously overhanging a ravine or waterway.

### Field ID of *Empidonax* Flycatchers

The five empids widespread in the East (pp. 256–259) are perhaps the most notorious identification challenge in our region. All are about the same size (small) and the same color (drab), and all have the same pattern (not much). What is a birder to do?

Start by asking, Is the bird even an empid? The similar wood-pewees are a bit longer and especially longer-winged; they sit motionless on dead snags, whereas empids typically forage from leafier microhabitats. The drab, tail-wagging Eastern Phoebe (p. 262) lacks the empids' wing bars and eye rings and usually occurs in more open habitats.

Next, pay attention to what you're hearing: The songs, although short, are usually diagnostic; field ID on the breeding grounds can typically be established by song alone. Many empids call on migration, and separating Alder and Willow Flycatchers (both on p. 258) when

migrating often requires hearing them call.

Also be sure to take a close look. Differences in body shape and proportion are reasonably consistent among species on all but the youngest empids. The throat and eye ring are often helpful for field ID. Look around too: Acadian and Willow Flycatchers, say, may nest within a few hundred feet of each other, but they are found in entirely different microhabitats.

Finally, try getting experience with empids on the breeding grounds. There you will find them conveniently singing and usually in the right habitat. Time spent with empids of known ID builds confidence for encounters with empids on migration, even silent birds in undifferentiated stopover habitat. But some empids, even on the breeding grounds, are going to be impossible to identify. You'll find that eBird provides categories like "Alder/Willow Flycatcher," "*Empidonax* sp.," and even "flycatcher sp." Experienced birders routinely make good use of those options.

**THREE SUCCESSIONAL EMPIDS**
Look for these three in less heavily wooded habitats than those where Yellow-bellied and Acadian Flycatchers (both on p. 256) are found. Alder and Willow breed in thickets with small trees, Least in second-growth (successional) woodlots.

## Alder Flycatcher

*Empidonax alnorum* | ALFL | L 5¾" (15 cm)

Formerly lumped with the Willow Flycatcher as Traill's Flycatcher, the Alder is the more northerly of the two, favoring wetter microhabitats. Abundant across its huge breeding range, it is oddly underdetected on migration.

■ **APPEARANCE** Nearly identical to Willow, and silent migrants of either species are often reported as "Traill's." Both are fairly large and broad-billed, with eye rings reduced or absent. Alder is greener overall, especially on back; crown more rounded, bill a bit smaller. Alder has very thin eye ring; Willow usually lacks eye ring.

■ **VOCALIZATIONS** Song a harsh, low *rrree-beeeuh,* given 8–20 times per minute; lower-pitched with less variation in pitch than song of Willow. Call a high, bright *peep.*

■ **POPULATIONS** Because it is a circum-Gulf migrant, Alder is rare in the Southeast; western limits of normal migration route poorly known. Migrates late in spring, with many still traveling in June.

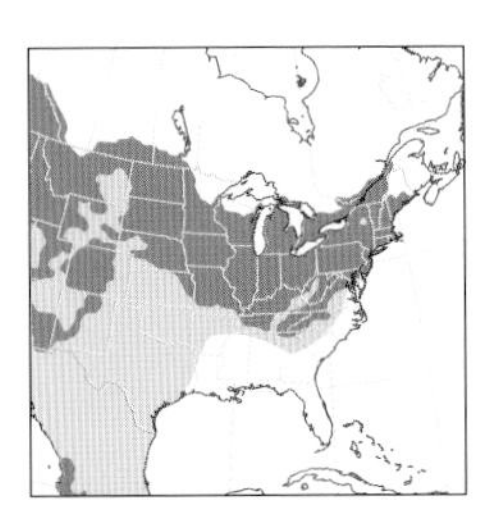

## Willow Flycatcher

*Empidonax traillii* | WIFL | L 5¾" (15 cm)

Among all the empids, the Willow is the most pewee-like. The species is relatively large, its eye ring is basically nonexistent, and it even forages from dead snags in the manner of pewees.

■ **APPEARANCE** Nearly identical to Alder Flycatcher—but first make sure you're not looking at a pewee! Both Alder and Willow are largish and broad-billed. Willow is browner overall than Alder, especially above; Alder often shows a very thin but complete eye ring, typically lacking on Willow.

■ **VOCALIZATIONS** Song an abrupt *peep-bew* or *fitz-bew,* given 5–15 times per minute; also a simpler *brrrew,* burry and rising. Call, often given by migrants, a liquid *whit,* distinct from *peep* of Alder.

■ **POPULATIONS** Like Alder, a circum-Gulf migrant that arrives late in spring and departs early in fall. Breeds in old fields with thickets and small trees.

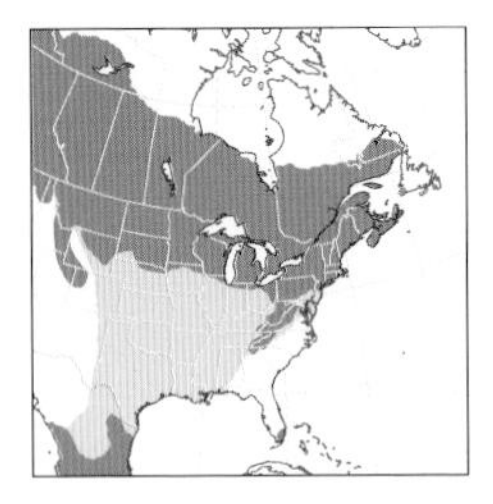

## Least Flycatcher

*Empidonax minimus* | LEFL | L 5¼" (13 cm)

As its name suggests, this is the smallest of the *Empidonax* flycatchers. But it is rather active and sometimes quite vocal, making it easy to spot and (relatively) easy to identify.

■ **APPEARANCE** Boldly marked for an empid, with prominent eye ring and contrasting dark and light on wings; large head, short wings, and short but broad bill impart a compact look overall. In close view, note pale throat and orange lower mandible.

■ **VOCALIZATIONS** Song a simple, two-note *t'tek (Quebec),* rough and grating, often repeated rapidly, one per second. Call a dry *wit.* Migrants frequently call and even sing.

■ **POPULATIONS** Earliest empid to return in spring in the East, last to leave in autumn; adults molt after fall migration, but young before. Breeds in diverse northern hardwood and mixed broadleaf-needleleaf woods, often around edges and clearings; locally common, but also absent from seemingly suitable swaths of habitat.

greenish gray above
1st fall
Alder Flycatcher
worn fall adult
moderate primary projection
head
y thin
e ring
spring
Willow Flycatcher
sloping forehead
worn fall adult
greenish brown above
1st fall
minimal eye ring
spring
broad, square tail
moderate primary projection
broad bill, entirely pale below
nall, broad
bill, usually
oped dark
1st fall
round head
worn fall adult
short
primary
jection
strong eye ring
spring
Least Flycatcher
grayish overall, slightly greener on back
short, narrow tail

## FOUR WESTERN EMPIDS

All four breed regularly right up to the eastern flank of the Rockies. They are understandably regular—but also difficult to identify—on migration at the western limits of the East.

### Hammond's Flycatcher

*Empidonax hammondii* | HAFL | L 5½" (14 cm)

Smallest of the western empids, the same heft as a Least Flycatcher (p. 258). Blocky head and short tail impart front-heavy look; note also small bill, long wings, and gray head contrasting with olive breast and back. Regular on migration on plains just east of Rockies; annual vagrant, late fall to winter, in recent years to East Coast. Call a high *peep,* like Alder's, but most fall migrants silent.

### Gray Flycatcher

*Empidonax wrightii* | GRFL | L 6" (15 cm)

Plumage low-contrast and grayish overall; head small and smoothly rounded. Bill long; wings relatively short, tail long. Key behavioral mark is gentle tail dipping, like that of Eastern Phoebe (p. 262); other empids flick tail in jerky movements. Regular on migration on western Great Plains; very rare but increasing farther east. Migrants stay close to ground at edges of clearings and waterways, sometimes giving toneless *wit* notes.

### Dusky Flycatcher

*Empidonax oberholseri* | DUFL | L 5¾" (15 cm)

Similar to Hammond's; Dusky is a bit larger, with relatively longer tail, longer bill, and shorter wings. Timing of fall molt differs from that of Hammond's: Dusky molts after fall migration, so fall migrants appear dingy; Hammond's molts before fall migration, so fall migrants appear bright. Breeds east to Black Hills. Regular migrant to western Great Plains, casual fall and winter to East Coast. Migrants favor brushy habitats, sometimes give dull *wit* call.

### Western Flycatcher

*Empidonax difficilis* | WEFL | L 5½"–5¾" (14–15 cm)

With its yellowish tones, suggests Yellow-bellied Flycatcher (p. 256). Eye ring of Western fairly thick, pinching behind eye in "teardrop"; head more crested than Yellow-bellied's. Breeds east to Black Hills, migrates broadly but sparingly across western Great Plains. Vagrants east to Atlantic seaboard, fall to winter, likely belong mostly to Pacific Slope population, formerly treated as a separate species (Pacific-slope Flycatcher).

### Identifying Empids Out of Range

Most *Empidonax* flycatchers in the U.S. and Canada are impressive migrants, and vagrancy has been well established in the genus. So it is not out of the question that you will observe an empid far out of range. In such situations, the best course of action is to get lots of photos. Videos, even cellphone videos of birds small in the frame, can be extremely helpful. In some cases, an audio recording might seal the deal.

The bulk of records of rare empids in the East occur in fall and early winter. These birds, *after* their fall molt, may look appreciably different from "normal" (more worn) migrants earlier in the year. Freshly molted adult Least Flycatchers in late fall are impressively bright and might suggest a rarity from the West. Here's a final consideration: Recent climate change models, notably one by the National Audubon Society, forecast significant range shifts in several empids—especially in winter. Careful observations of vagrants, especially from the West, are potentially valuable for conservation science.

small, dark bill
large, blocky head
Hammond's Flycatcher
eye ring wider behind eye
gray and throat
long primary projection
fall
spring
short, notched tail
long bicolored bill
Gray Flycatcher
winter
short primary projection
spring
long tail, often dipped like Eastern Phoebe
bill narrower than Least, longer than Hammond's
Dusky Flycatcher
1st fall
long tail
short primary projection
worn fall adult
winter
spring
distinct peak at back of head
eye ring pointed behind eye
spring
Western Flycatcher
large bill orange below
lower-contrast wing pattern than Yellow-bellied
yellow-toned overall
long tail

## PHOEBES

These occur singly in open areas with scattered trees, especially around standing water and outbuildings. They migrate early in spring, and they winter farther north than most flycatchers.

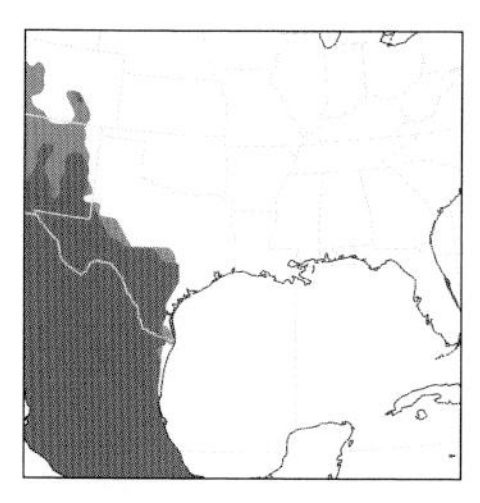

### Black Phoebe

*Sayornis nigricans* | BLPH | L 6¾" (17 cm)

Stunningly pied, this phoebe is a prominent denizen of southwestern waterways, extending east in winter to the Gulf Coast of southern Tex. Watch as it catches insects from the water's surface.

■ **APPEARANCE** Adult's snow-white belly contrasts with black plumage overall; young show variable warm buff above, especially on wings.

■ **VOCALIZATIONS** Squeaky song, *ts'wee! t'wee;* call, a down-slurred *swip.*

■ **POPULATIONS** Seasonal movements complex, including some winter dispersal northward. Range expanding north overall.

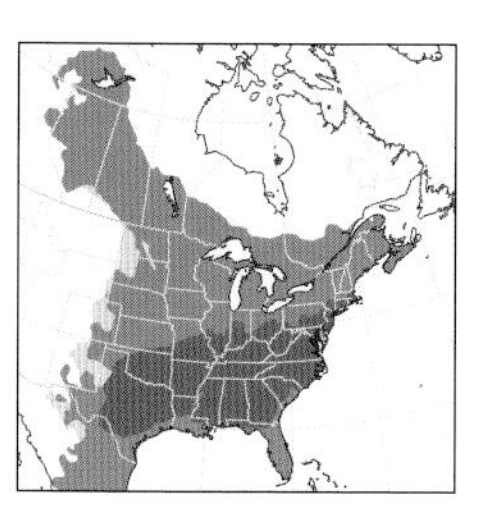

### Eastern Phoebe

*Sayornis phoebe* | EAPH | L 7" (18 cm)

Find this earliest of spring-migrant flycatchers in the East by its eponymous song—*phoebe!*—often around bridges, cabins, and outbuildings.

■ **APPEARANCE** Dark overall; muddy brown above, off-white below. Stereotyped tail-dipping distinctive. Similarly drab wood-pewees (p. 254) have prominent wing bars; smaller empids (pp. 256–261) sport wing bars, eye rings. Freshly molted adults in early fall can be extensively yellow below, suggesting Say's Phoebe.

■ **VOCALIZATIONS** Song alternates between emphatic *fee-bee!* and rolling *frree-brree* elements; bright *chip!* note recalls Swamp Sparrow (p. 354) and Palm Warbler (p. 394).

■ **POPULATIONS** Well adapted to human-modified landscapes. Winter range expanding north; breeding range expanding west toward Rockies.

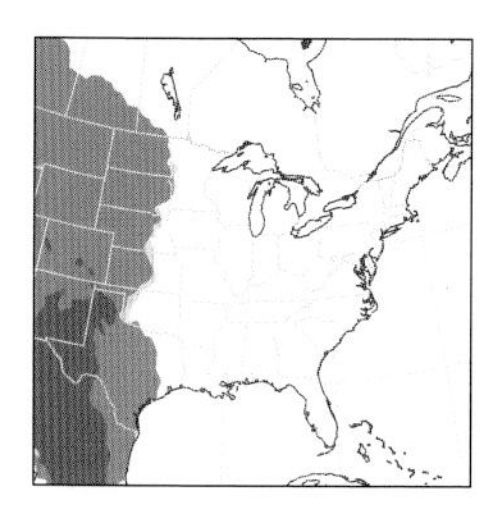

### Say's Phoebe

*Saynornis saya* | SAPH | L 7½" (19 cm)

The Eastern's western counterpart, this species favors drier habitats and is even more hopelessly drawn to barns, sheds, and homes.

■ **APPEARANCE** Apricot underparts contrast with dark plumage overall; tail is blackish. Like Eastern, dips tail, but not as much. Distinctive, but can be confused with Western Kingbird (p. 250) in brief view.

■ **VOCALIZATIONS** Sweet and plaintive, evocative of the lonely country favored by the species. Down-slurred *see-ur* given year-round, more complex *ch'pee-dar* on breeding grounds.

■ **POPULATIONS** The most migratory phoebe; returns early in spring, departs late in fall. Winter range expanding north; annual vagrant to East Coast.

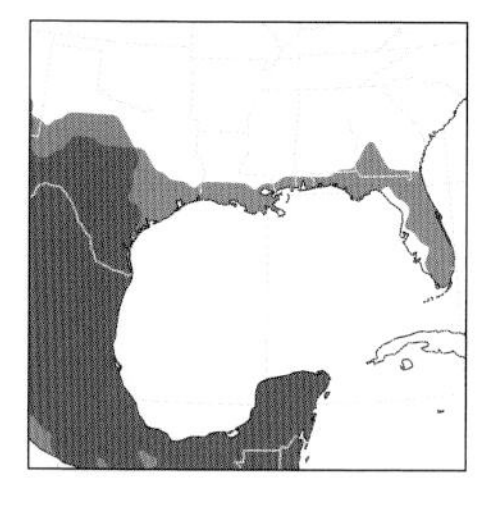

### Vermilion Flycatcher

*Pyrocephalus rubinus* | VEFL | L 6" (15 cm)

The adult male missed the memo that flycatchers tend to be drab! Although not called a phoebe, this species is thought to be a close relative.

■ **APPEARANCE** Adult male intense vermilion and dark brown; extent of vermilion reduced on immature male. Drabber females, especially immatures, told by salmon underparts, streaky breasts, well-patterned face. Dips tail, phoebe-like.

■ **VOCALIZATIONS** Song, given in flight, rapid, stuttering; call, a rising *psee!*

■ **POPULATIONS** Declining in U.S., but ranges widely in Latin America. Mostly sedentary, but some wander far north and east, mostly fall to early winter.

adult
Black Phoebe
cinnamon scaling above, especially on wings
juvenile
plain dark head
Eastern Phoebe
worn summer adult
fresh fall
long tail
indistinct wing pattern
warm yellow belly in fall
Say's Phoebe
warm brown contrasts with black tail
cinnamon belly
whitish eyebrow
Vermilion Flycatcher
immature ♀
lightly streaked breast
immature ♂
ale yellow or pinkish belly
adult ♀
juvenile
adult ♂
smaller and shorter-tailed than phoebes

## SMALLISH VIREOS

The vireos, small birds of mostly wooded habitats, are related to corvids (pp. 274–281) and shrikes (p. 272). Like corvids and shrikes, vireos have hooked bills.

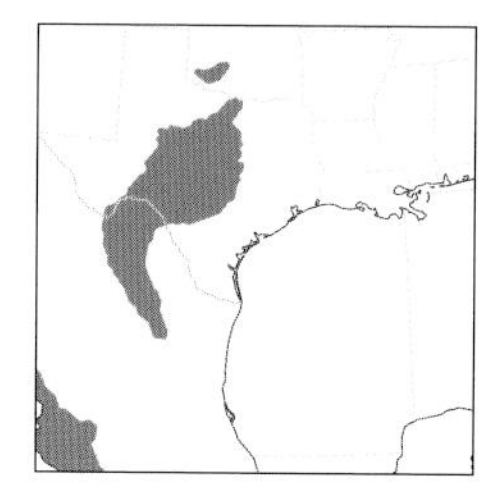

### Black-capped Vireo

*Vireo atricapilla* | BCVI | L 4½" (11 cm)

Look and listen for this range-restricted vireo in hot, shrubby habitats with an oak component. It's in an "eye-ringed" clade with White-eyed and Bell's.

■ **APPEARANCE** Small. Mostly black head of adult male contrasts with white on face and around eye; eyes red. Female slightly grayer, young more so.

■ **VOCALIZATIONS** Complex song bright and cheery, a mix of short trills, short buzzes, and abrupt chirps: *rip cheep t't't't' bzz slee! t't't't',* etc.

■ **POPULATIONS** Now restricted as a breeder largely to central Tex., although an outpost is holding out in Wichita Mountains, Okla. Development, fire suppression, and cowbird parasitism (see p. 370) are factors in its decline.

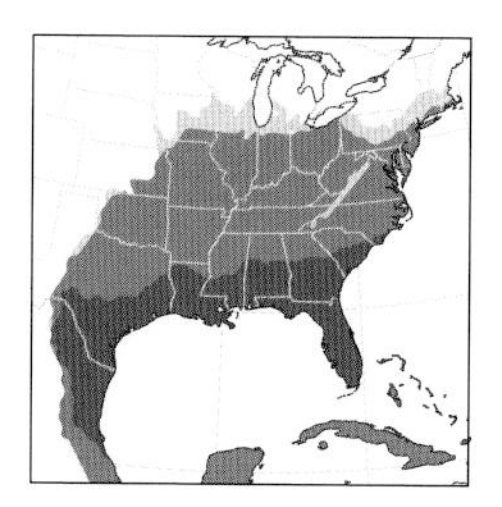

### White-eyed Vireo

*Vireo griseus* | WEVI | L 5" (13 cm)

Residents of the Southeast have undoubtedly heard the White-eyed Vireo. But seeing one is harder: The species loves brambles, tangles, and swamps.

■ **APPEARANCE** Bright and small; more fidgety than most vireos. Head gray; yellow around bill and eye. Adult's white eye is distinctive; juvenile's is dark.

■ **VOCALIZATIONS** Variable song quick and plucky, often bookended with abrupt *chick!* notes: *chick! uh-duh-WEEE-oh chick! (Pick up the beer check!).*

■ **POPULATIONS** Breeding range expanded north in Pa. and Ohio in 20th century, but has since stabilized; overall numbers increasing range-wide.

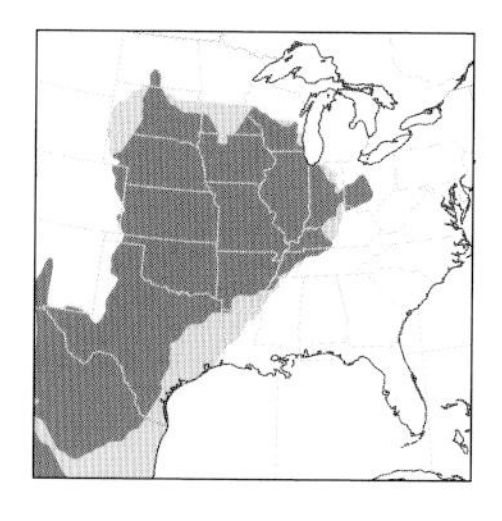

### Bell's Vireo

*Vireo bellii* | BEVI | L 4¾" (12 cm)

Like the closely related Black-capped and White-eyed Vireos, the tiny Bell's inhabits thickets and tangles in brushy habitats. Listen for its odd song.

■ **APPEARANCE** It's as if a brighter vireo had faded into this species; Bell's has a faint eyeline, indistinct eye ring, and reduced wing bars.

■ **VOCALIZATIONS** Song fast and grating; final note often audibly different from phrase to phrase: *ch'reedu- j'riddy-jurry-zheer? ... ch'reedu-j'riddy-jurry-chroo!*

■ **POPULATIONS** Declining, most sharply in the Southwest; midwestern breeders affected by habitat loss and cowbird parasitism (see p. 370) but also opportunistically establish in early successional habitats.

### Gray Vireo

*Vireo vicinior* | GRVI | L 5½" (14 cm)

Barely reaches our area as a breeder in arid shrublands of Edwards Plateau. Low-contrast plumage suggests Bell's, but mostly gray like Plumbeous (p. 266). More active than Plumbeous, often flipping its long tail about as it forages. Slurred song elements like Plumbeous, but sweeter and faster.

### Hutton's Vireo

*Vireo huttoni* | HUVI | L 5" (13 cm)

Like Gray Vireo, reaches Edwards Plateau; closely related to Gray, in a "spectacled" clade of vireos. Hutton's more frequently confused with Ruby-crowned Kinglet (p. 296) than other vireos; distinguished from kinglet by thicker bill, blue-gray legs, different wing pattern, and completely different song: simple falling phrases, *z'weeooo* and *zooweech,* endlessly repeated.

Black-capped Vireo
adult ♀
adult ♂
bold white "spectacles" broken over eye
immature ♀
White-eyed Vireo
griseus
yellow "spectacles"
dark eyes
immature
yellow flanks
thin, clean wing bars
Florida Keys
maynardi
duller than mainland population, with larger bill
ort white
ebrow and
distinct arc
der eye
pale
bill
small and long-tailed
Bell's Vireo
lower wing bar crisp, upper one faint
indistinct eye ring and wing bars
Gray Vireo
long tail flipped actively
wings lack black primary bar of Ruby-crowned Kinglet
worn summer
Hutton's Vireo
stout gray bill and legs
fresh fall

## LARGE VIREOS WITH GOGGLES

Despite different color schemes, these four are closely related, part of a "spectacled" clade with the Gray and Hutton's Vireos (p. 264). They forage methodically and sing exaggeratedly slow songs.

### Yellow-throated Vireo

*Vireo flavifrons* | YTVI | L 5½" (14 cm)

In leafy hardwood forests, this brightly colored vireo sings its comically slow song. The species is so yellow that it invites comparison with warblers more than with other vireos.

■ **APPEARANCE** Large and stocky, with eye-catching yellow throat and breast; note conspicuous yellow "spectacles," white wing bars, and white belly. Superficially similar to distantly related Pine Warbler (p. 394), which has thinner bill and streaked flanks.

■ **VOCALIZATIONS** Rising and falling phrases, widely spaced, most with burry quality: *churro ... Detroit ... surreal ....*

■ **POPULATIONS** A broadleaf generalist, but partial to lowlands with oaks. Appears to be in slow recovery from losses in the 20th century stemming especially from spraying pesticides in forests.

### Cassin's Vireo

*Vireo cassinii* | CAVI | L 5" (13 cm)

Breeds in forests of far western U.S. and southwestern Canada, but wanders farther east. An elliptical migrant, going south in fall along a more easterly trajectory than in spring; regular in late summer to western Great Plains. Like a washed-out Blue-headed Vireo, with weaker contrast overall, especially on face and throat. Worn individuals quite dull, approaching Plumbeous. Like Plumbeous, sings on fall migration; song of Cassin's a bit sweeter and faster.

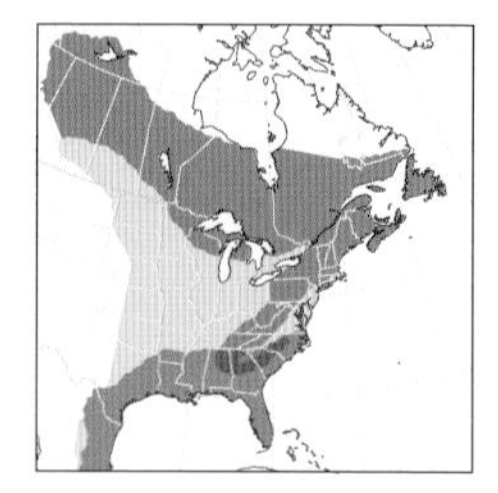

### Blue-headed Vireo

*Vireo solitarius* | BHVI | L 5" (13 cm)

Along with Cassin's and Plumbeous Vireos, the Blue-headed was known as the Solitary Vireo until 1997. They are now treated as three different, albeit very closely related, species; this one is the snazziest of the lot.

■ **APPEARANCE** Unmistakable in most of range, with its blue-gray head, greenish back, yellow flanks, and bright white throat; note also prominent wing bars and "spectacles." Overlaps on migration in western Great Plains with Cassin's and Plumbeous Vireos; all three are uncommon to rare there.

■ **VOCALIZATIONS** Song phrases rich and sweet, slowly delivered: *swee-uh-wee* and *siri?,* plus burrier notes. Slower and lower than Red-eyed's (p. 270), faster and sweeter than Yellow-throated's.

■ **POPULATIONS** Earliest eastern vireo in spring, latest in fall. Breeding habitat overlaps with Yellow-throated Vireo, but Blue-headed favors northern hardwoods: maple, beech, birch. Numbers increasing overall, probably in response to gradual expansion of eastern hardwood forests.

### Plumbeous Vireo

*Vireo plumbeus* | PLVI | L 5¼" (13 cm)

The western interior representative of the "Solitary Vireo" complex, breeding east to the Pine Ridge of Nebr. and Black Hills of S. Dak. On migration through western Great Plains, the most common of the three. Freshly molted Plumbeous in fall, tinged yellow and green, very similar to dull Cassin's in worn plumage. Plumbeous migrates later in fall than Cassin's. Song of Plumbeous especially hoarse, approaching Yellow-throated.

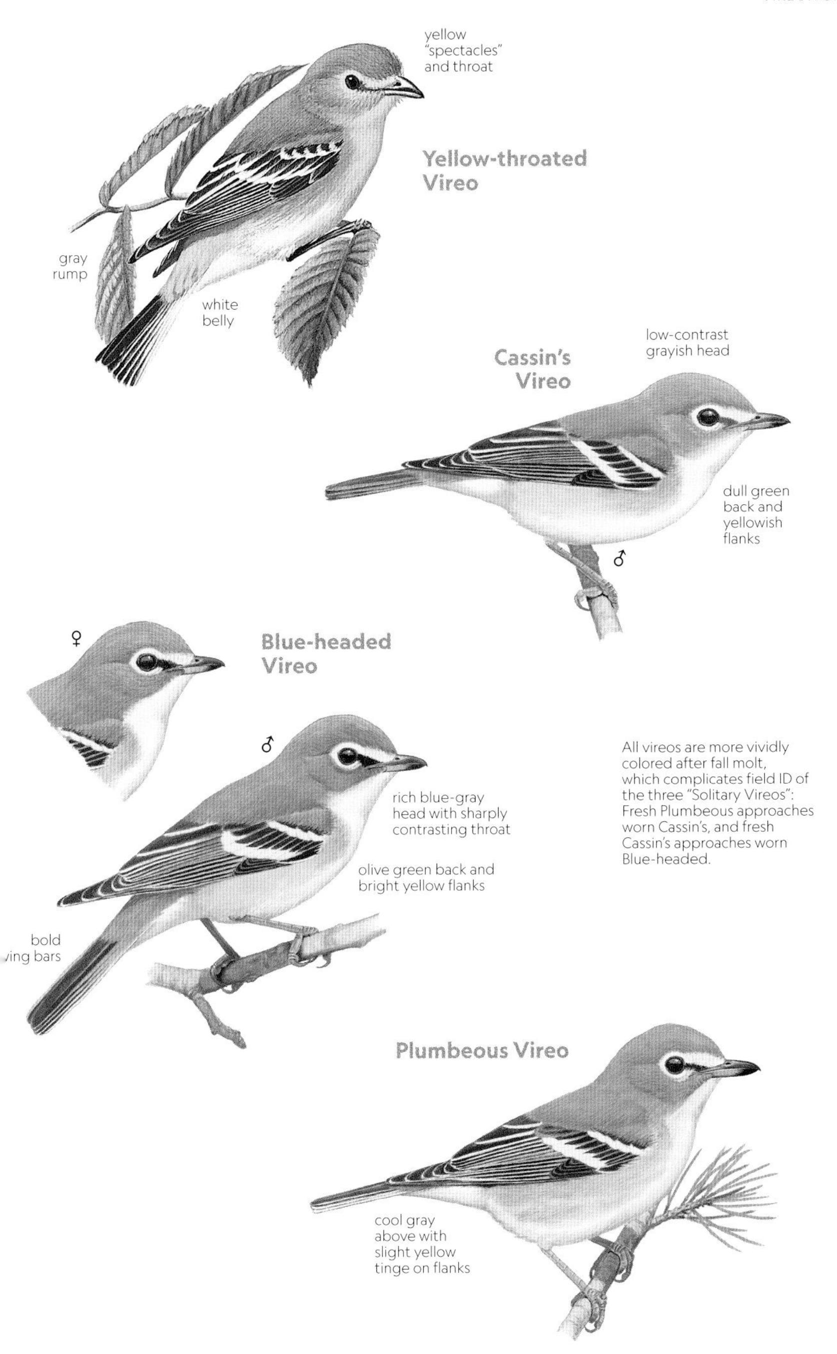


All vireos are more vividly colored after fall molt, which complicates field ID of the three "Solitary Vireos": Fresh Plumbeous approaches worn Cassin's, and fresh Cassin's approaches worn Blue-headed.

## VIREOS WITH EYELINES I

Within an "eyelined" clade (pp. 268–271) of related vireos, the Philadelphia and Warbling are particularly close. All the vireos in this clade are forest-dwellers with plain wings and striped faces.

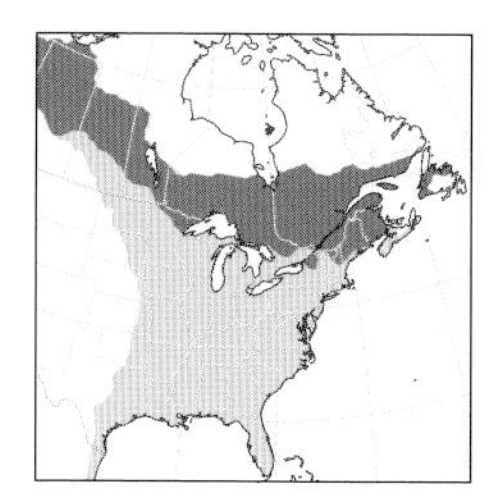

### Philadelphia Vireo

*Vireo philadelphicus* | PHVI | L 5¼" (13 cm)

Among the vireos of the eastern hardwood forest, this one is generally the least common. It looks like a Warbling Vireo but sounds like a Red-eyed.

■ **APPEARANCE** Yellowish below with olive topside and grayish crown. Note dark line through eye (transocular); Warbling may show dark behind eye, but not in front of eye. Yellow of Philadelphia variable below, but most intense at center of breast; on Warbling, yellow usually most intense on flanks. Tennessee Warbler (p. 382), especially breeding female, similar but has thin bill and white undertail coverts.

■ **VOCALIZATIONS** Song completely different from look-alike Warbling's, but very similar to Red-eyed's (p. 270). Philadelphia's phrases a bit slower and higher than Red-eyed's, but much overlap. Note variation in phrasing: Philadelphia has limited repertoire, repeats song phrases regularly; Red-eyed has larger repertoire, delivers phrases seemingly randomly.

■ **POPULATIONS** Fall migration more easterly than spring migration. Breeders defend territories against both conspecifics and Red-eyeds.

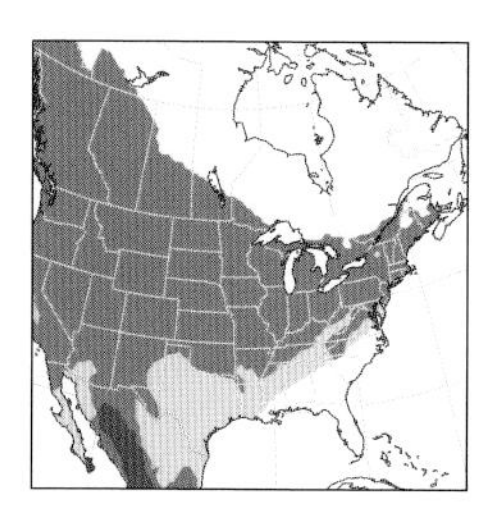

### Warbling Vireo

*Vireo gilvus* | WAVI | L 5½" (14 cm)

Few birds have a higher ratio of heard-to-seen than this species. It is drab and slight of build, seemingly always in dense treetops, and constantly vocal on the breeding grounds.

■ **APPEARANCE** Pale olive overall, darker above than below. Most are weakly capped with a broad but diffuse white eyebrow. Smaller-billed Philadelphia usually brighter and bolder, but fresh Warbling in fall can be rather yellow; even a dull Philadelphia has yellow wash on center of breast and dark line in front of eye, lacking on Warbling.

■ **VOCALIZATIONS** Most in the East sing a frenetic, herky-jerky series, around two seconds long, with a sharp and rising terminal note: *I can see you, I will seize you, I will squeeze you, 'til you SQUIRT!* Call a peevish *waaah* or *gwaaay,* harsh and unmusical.

■ **POPULATIONS** Circum-Gulf migrant; thus, rare at best in Fla. and the Southeast's coastal plain. Eastern *gilvus* partial to woods near water.

### Cryptic Species

Distinguishing between Philadelphia and Warbling Vireos is tricky, especially in fall when migrants are mostly silent. But there's more: Two Warbling subspecies, eastern *gilvus* and western *swainsoni,* are nearly identical in appearance. Yet they exhibit consistent differences in song and habitat, and some ornithologists suspect they are valid species. In particular, they may be cryptic species, so named because we can barely discern their visual differences.

Both occur in the East, with *swainsoni* breeding east to the Black Hills and migrating across the western Great Plains. The song is critical for ID: That of *swainsoni* is lazy and burry, shorter than that of *gilvus* and typically lacking the sharp terminal note. Western birds are somewhat smaller and shorter-billed, slightly less colorful, and a bit more prominently capped; their habitat preferences are also more catholic, whereas *gilvus* is partial to the edges of waterways.

Determining whether these are separate species will likely depend on the contributions of birders. Make audio recordings or, even better, videos of singing birds, and upload them to eBird's Macaulay Library.

dark gray
crown
dark line through
eye to bill
fall
all vireos
more richly
colored in fall
Philadelphia
Vireo
yellow variable,
but brightest
in center of
breast
spring
shorter bill
and tail than
Warbling
male often
sings from nest,
a rare behavior
among birds
pale grayish
crown
spring
eastern
gilvus
pale
lores
Warbling Vireo
fall
whitish below,
buffy yellow
on flanks
western
swainsoni
western populations
slightly smaller-billed
and darker-capped

## VIREOS WITH EYELINES II

These three are members of a superspecies complex, closely related yet distinct enough to be considered separate species. All have red eyes, gray crowns, and fairly prominent stripes on the face.

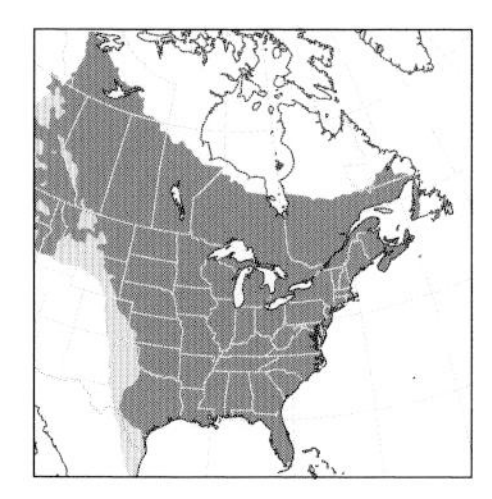

### Red-eyed Vireo

*Vireo olivaceus* | REVI | L 6" (15 cm)

It's the middle of the afternoon on a hot, hazy, muggy day in July, and the woods seem practically devoid of birdlife—except for the Red-eyed Vireos! The species is also known as the "Preacher Bird," for its habit of droning on endlessly.

■ **APPEARANCE** Across extensive breeding range, red eye and black stripes on face separate it from all other vireos, but those marks usually hard to see. The bird is often a sluggish form in the treetops, dark olive above and pale olive-white below. Juvenile, seen in our area in fall, has brownish eye, richer underparts, and weaker facial pattern, inviting comparison with Philadelphia and Warbling Vireos (p. 268).

■ **VOCALIZATIONS** Individual songs comprise pure-tone phrases of two to four syllables, separated by pauses lasting one-half to two seconds: *cheerio ... cheery ... j'reery ... j'cheer-cheery ....* Compare with similar-sounding Blue-headed (p. 266) and especially Philadelphia Vireos. Call, an unmusical *gwaaah,* almost identical to Warbling Vireo's.

■ **POPULATIONS** Common in broadleaf woodlands, even in city parks. Migrates at night; many are killed in collisions with buildings and towers.

### Yellow-green Vireo

*Vireo flavoviridis* | YGVI | L 6" (15 cm)

The Mid. Amer. version of the Red-eyed Vireo. Breeds in the Lower Rio Grande Valley, where infrequently detected but perhaps more common than reports suggest; annual vagrant elsewhere on Gulf Coast. More extensively yellow than even the yellowest Red-eyed Vireos; bill even longer than Red-eyed's, and stripes on head less bold than on Red-eyed. Song like Red-eyed's, but phrases even faster; phrases multisyllabic like Red-eyed's, but so fast that we hear them as monosyllabic chirps.

### Black-whiskered Vireo

*Vireo altiloquus* | BWVI | L 6¼" (16 cm)

The Caribbean counterpart of the widespread Red-eyed Vireo, this species is fairly common within its U.S. range. But finding one requires some stamina, as the bird seems to prefer the muggiest and buggiest of mangrove swamps.

■ **APPEARANCE** The "whisker," if pronounced and seen well, is diagnostic; but the mark is variable, being diffuse and grayer on some birds. Gray crown only weakly bordered by black; compare with Red-eyed Vireo's. Like Yellow-green Vireo, Black-whiskered is a bit longer-billed than Red-eyed.

■ **VOCALIZATIONS** Song comprises slurred elements like those of Red-eyed, but pattern of delivery quite different: bundled in quick bursts of two to four phrases, separated by pauses of one to three or more seconds. Call a nasal, down-slurred *raauuh;* shorter than that of Red-eyed Vireo, a bit reminiscent of Gray Catbird's (p. 306).

■ **POPULATIONS** Regular along coast halfway up the Fla. peninsula; annual vagrant on Gulf Coast to Tex., nearly annual to Carolinas on Atlantic coast. May not have occurred in Fla. until 19th century, but historical record sparse.

## Red-eyed Vireo

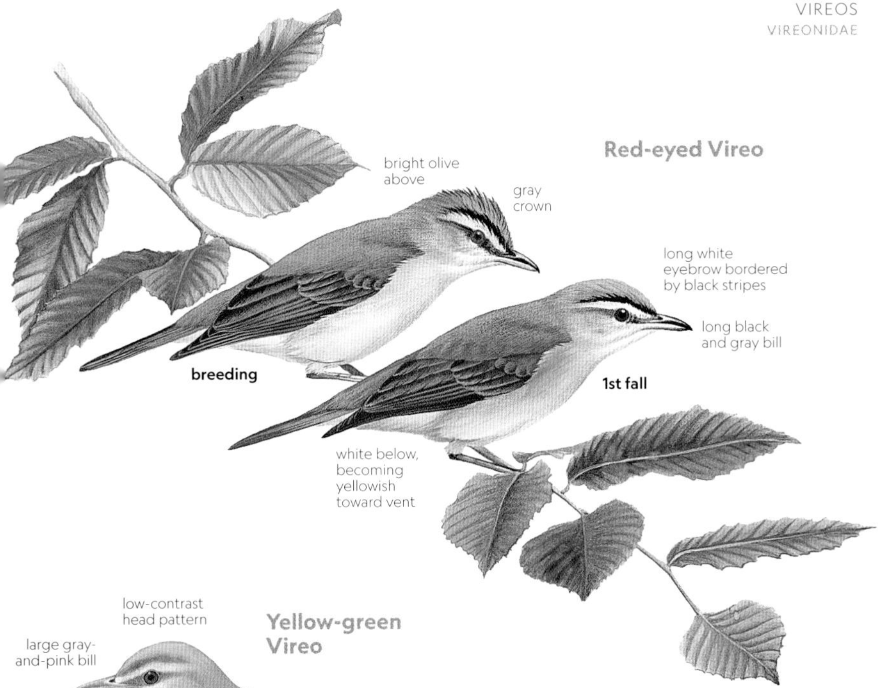

bright olive above
gray crown
long white eyebrow bordered by black stripes
long black and gray bill
**breeding**
**1st fall**
white below, becoming yellowish toward vent

## Yellow-green Vireo

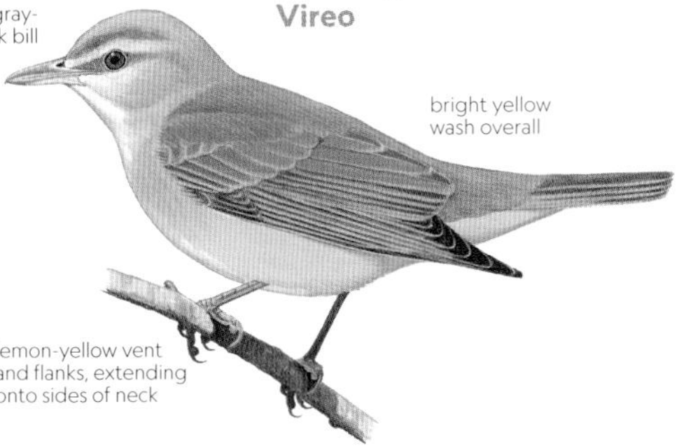

low-contrast head pattern
large gray-and-pink bill
bright yellow wash overall
lemon-yellow vent and flanks, extending onto sides of neck

## Black-whiskered Vireo

head less contrasting than Red-eyed, with larger gray bill
brownish olive above
"whisker" can be indistinct
**Florida**
*barbatulus*
buffy face
bill even longer than Florida birds
vagrant to Florida and Louisiana
**Greater Antillean**
*altiloquus*

## SHRIKES

Their mild appearance belies alarming feeding behavior. Shrikes have the macabre habit of impaling their prey—grasshoppers, beetles, even small vertebrates—on thorns and barbed wire fences.

### Loggerhead Shrike

*Lanius ludovicianus* | LOSH | L 9" (23 cm)

Both shrikes are birds of open country with perches for hunting. They go after prey, usually on the ground, in sudden sorties just above the surface.

■ **APPEARANCE** A bit smaller and darker than Northern; black mask of Loggerhead is thicker than Northern's, especially around bill. Loggerhead also has shorter bill and tail. Adult clean gray, black, and white; juvenile weakly barred below and scaly above. Northern Mockingbird (p. 308), similar in overall pattern, has lower-contrast plumage and yellow eyes.

■ **VOCALIZATIONS** Varied calls, heard year-round, include harsh *raaah* or *rraack*, but also sweeter *ch'peee*, given slowly or in quick succession. Songs especially varied, typically with short trills and whistles; often sound subdued or understated.

■ **POPULATIONS** Has declined steadily at northern limit of range in East; extirpated from New England and much of mid-Atlantic.

### Northern Shrike

*Lanius borealis* | NSHR | L 10" (25 cm)

When this species is in sight, you can almost always be certain that it's wintertime. South of its remote taiga breeding grounds, this powerful and predatory songbird occurs only during the colder months.

■ **APPEARANCE** Larger, paler, and longer-billed than Loggerhead Shrike. Black face mask is thinner than Loggerhead's, especially in front of eye. Many sightings in winter are of juveniles, weakly barred below. Compare also with Northern Mockingbird (p. 308).

■ **VOCALIZATIONS** Like Loggerhead, but generally silent in winter, when most are found. One common call is a startling, retching *RRRAAAA*. Extraordinary is the species' practice of "acoustical luring": singing softly and ventriloquially, but beautifully, to catch the attention of small birds—which it then eats.

■ **POPULATIONS** Widespread but scarce in winter, usually seen singly. An irruptive species (p. 220) whose numbers vary annually, probably in connection with cyclical food resources on the breeding grounds.

### Infraorder Corvides

All eastern birds in the huge order Passeriformes can be placed in one of two suborders, the Tyranni (pp. 246-263), and the Passeri (pp. 264-409). In the East, the Passeri species are grouped into two infraorders: the Corvides (pp. 264-281) and the Passerides (pp. 282-409).

The Corvides are represented in the East by three families: Vireonidae (vireos, pp. 264-271), Laniidae (shrikes, p. 272), and Corvidae (jays and crows, pp. 274-281). Recognition of the Corvides is relatively recent, a result of the avian genomics revolution starting in the late 20th century. And small wonder: At first glance, the Corvides are a grab bag indeed, encompassing the arboreal vireos, the predatory shrikes of open terrain, and the wily jays and crows. In the Old World, drongos and even birds-of-paradise contribute to the diversity of the Corvides.

A useful analogy is with the marsupials. Koalas, kangaroos, and opossums are not obviously kin. But morphology and molecular biology confirm their relationship. As the Corvides expanded globally from their ancestral home, which was probably in Australia, they encountered novel niches in new environments. Multiple convergent evolution (p. 136) events ensued and were major drivers in the striking radiation of these New World birds with Old World origins.

Loggerhead Shrike
stubby black bill
juvenile
quick, snappy wingbeats
small white wing patch
broad mask connects over bill
adult
slow, lazy wingbeats
extensive white
dark gray back
Northern Mockingbird for comparison

thin mask, especially in front of bill
Northern Shrike
pale gray back
longer bill than Loggerhead, with distinct hook
mostly pink lower mandible
brown-tinged overall and barred below
immature
juvenile

**JAYS**
In the East, birds called jays tend to be some combination of the following: smaller, slimmer, and more brightly colored and patterned than other species in the family Corvidae (corvids).

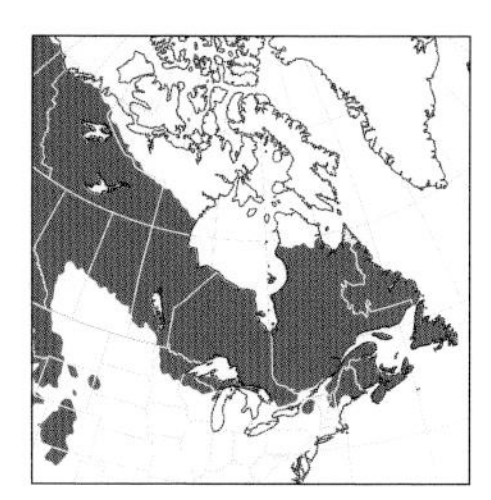

## Canada Jay

*Perisoreus canadensis* | CAJA | L 11½" (29 cm)

This is the infamous "Camp Robber" of the boreal forest. Singles or small groups sneak into the day-use area and snag a bit of tofu or s'mores.

▪ **APPEARANCE** Slim and relatively small-bodied; has small bill, a cold-weather adaptation. Was until recently known as Gray Jay, an efficient if unexciting name. Adult has black crown, large white cheek, and mostly gray plumage otherwise.

▪ **VOCALIZATIONS** A variety of chuckles, whistles, and sighs; they carry well, but sound soft. A "strange sound" in the Northwoods is often this bird.

▪ **POPULATIONS** Nearly sedentary; irruptions infrequent, involving small numbers of birds wandering short distances.

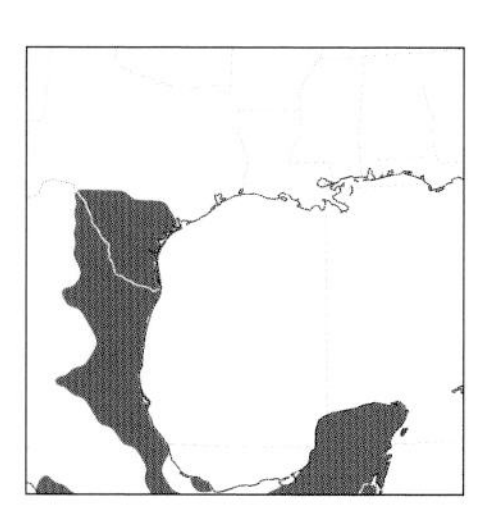

## Green Jay

*Cyanocorax yncas* | GRJA | L 10½" (27 cm)

The Green Jay might be the most stunning of all the South Texas specialties. But it favors well-shaded woodlots, where it is surprisingly easy to overlook.

▪ **APPEARANCE** A mostly green jay, with yellow in tail and hood patterned in black and blue. In size, tail length, and bill shape, similar to other jays.

▪ **VOCALIZATIONS** Calls varied, but one stands out: a series of four to six nasal clanks, uttered in quick succession, *claa! claa! claa! claa! claa!*

▪ **POPULATIONS** A Lower Rio Grande Valley specialty until late 20th century, but expanding northwest recently. Reaches north and west of San Antonio now; still expanding.

## Pinyon Jay

*Gymnorhinus cyanocephalus* | PIJA | L 10½" (27 cm)

Unsurprisingly, given the name, a piñon pine specialist, regularly ranging east in sizable numbers to the Black Hills (S. Dak.) and Pine Ridge (Nebr.); very rarely vagrates farther east. Lovely sky blue all over, deepest on the head. Bill long and straight, with "nostrils" exposed. *Gymnorhinus* means "naked nose," indicating an adaptation for feeding on sticky piñon rosin. Call an arresting *hya!* or *howay*, suggesting Laughing Gull (p. 142).

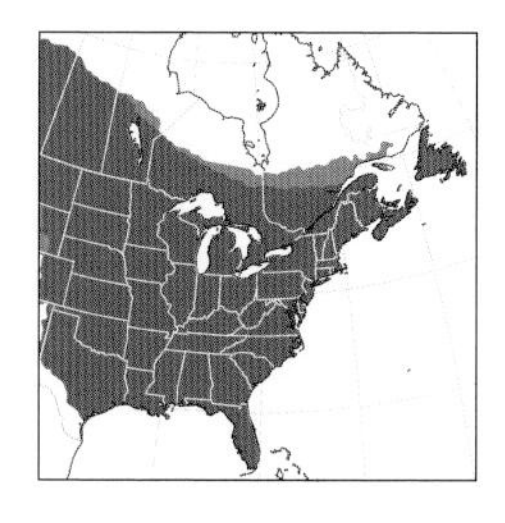

## Blue Jay

*Cyanocitta cristata* | BLJA | L 11" (28 cm)

Around most human population centers in the East, this is the only jay. It flourishes in towns and cities, but also occurs deep in the forest interior.

▪ **APPEARANCE** Bold, brilliant; extensively blue, with prominent crest and high-contrast black "necklace" on paler breast. In flight, white in tail and trailing edge of wing obvious. Scrub-jays (p. 276) longer-tailed with uncrested, smoothly rounded heads. Juvenile, often seen midsummer, can be quite gray, but still shows some blue, along with necklace and crest.

▪ **VOCALIZATIONS** Classic call is a nasal shriek, descending slightly, *aaaayyy*, often doubled or trebled. Also a variety of whistles and squeaks—and superb imitations of hawks. Infrequently heard is a "whisper song," given in early spring, a run-on jumble of slurred whistles.

▪ **POPULATIONS** Although present year-round in much of range, performs spectacular daytime migrations. The northbound spring migration across the southern Great Lakes is stirring.

sooty overall, with paler moustache
small bill
whitish head with dark nape
Canada Jay
juvenile
Green Jay
long, spike-like bill
Pinyon Jay
short tail
yellow outer tail feathers conspicuous in flight
Blue Jay
pale gray below with black "necklace"
intricately patterned with white spots and black bars
paddle-shaped wings with white trailing edge

## LONG-TAILED CORVIDS

The magpie gets the corvids' award for longest tail, but the two eastern scrub-jays are also notably long-tailed. The scrub-jays are closely related to the Blue Jay (p. 274), but magpies are probably closer to crows and ravens (pp. 278–281).

### Florida Scrub-Jay

*Aphelocoma coerulescens* | FLSJ | L 11" (28 cm)

"It takes a village" is an apt adage for this Florida endemic. The Florida Scrub-Jay is an obligate cooperative breeder, occurring in extended families whose members help raise young other than their own.

■ **APPEARANCE** Restricted to Fla., where the only other jay is Blue Jay. Scrub-jay lacks crest of Blue Jay; breastband of scrub-jay blue and diffuse, but black and contrasting on Blue Jay. Scrub-jay relatively longer-tailed and shorter-winged than Blue Jay, with sand-colored back.

■ **VOCALIZATIONS** Calls harsh and simple, *raaa* and *aaahr*, not nearly as diverse as the Blue Jay's repertoire.

■ **POPULATIONS** In long-term decline; requires frequently burned oak scrub on sandy soils, a habitat imperiled by development. At some birding hotspots, is ridiculously tame.

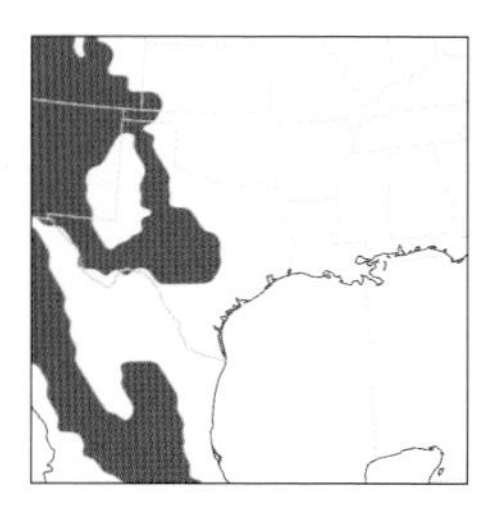

### Woodhouse's Scrub-Jay

*Aphelocoma woodhouseii* | WOSJ | L 11" (28 cm)

Part of a four-species complex that includes the Florida Scrub-Jay, this one is widespread in arid woodlands and scrublands of interior Mexico and the U.S. It does not normally practice cooperative breeding.

■ **APPEARANCE** Similar to Florida Scrub-Jay, but ranges do not overlap. Differs from Blue Jay by proportionately longer tail, uncrested crown, indistinct breastband, and uniformly blue wings and tail; the overall effect is that Woodhouse's is less dashing than Blue Jay.

■ **VOCALIZATIONS** Most common call a rising, screeching, somewhat *nasal scraay-aay!* Also a pounding series, *yeet! yeet! yeet! yeet! yeet!*

■ **POPULATIONS** Range in East mostly Edwards Plateau; largely sedentary, but some wander in fall. Where Blue Jay and Woodhouse's Scrub-Jay co-occur, the latter tends to occur in "wilder" habitats: rocky hillsides, lonely juniper woodlands, etc.

### Black-billed Magpie

*Pica hudsonia* | BBMA | L 19" (48 cm)

Although versatile and adaptable, the Black-billed Magpie is fundamentally restricted to cold and dry climes, from shrublands to forests to farms. It is bold and sometimes aggressive, but also wary and flighty.

■ **APPEARANCE** Size (large), shape (that tail!), and plumage contrast (extreme) alone are sufficient for field ID. Appears black and white in many views, but in good light shows metallic blue on wings and glistening green on tail.

■ **VOCALIZATIONS** Call a rising *maaag?* or shorter *mag*, often run together in series: *mag mag mag mag*. Song, heard year-round, a jumble of whistles, rattles, and clacks, interspersed with *mag* notes.

■ **POPULATIONS** Ranges east almost to Lake Superior. Records farther east, all the way to Atlantic seaboard, often dismissed as escapes from captivity; but the species is a strong flier and prone to considerable dispersal, and many occurrences out of range may be of natural vagrants.

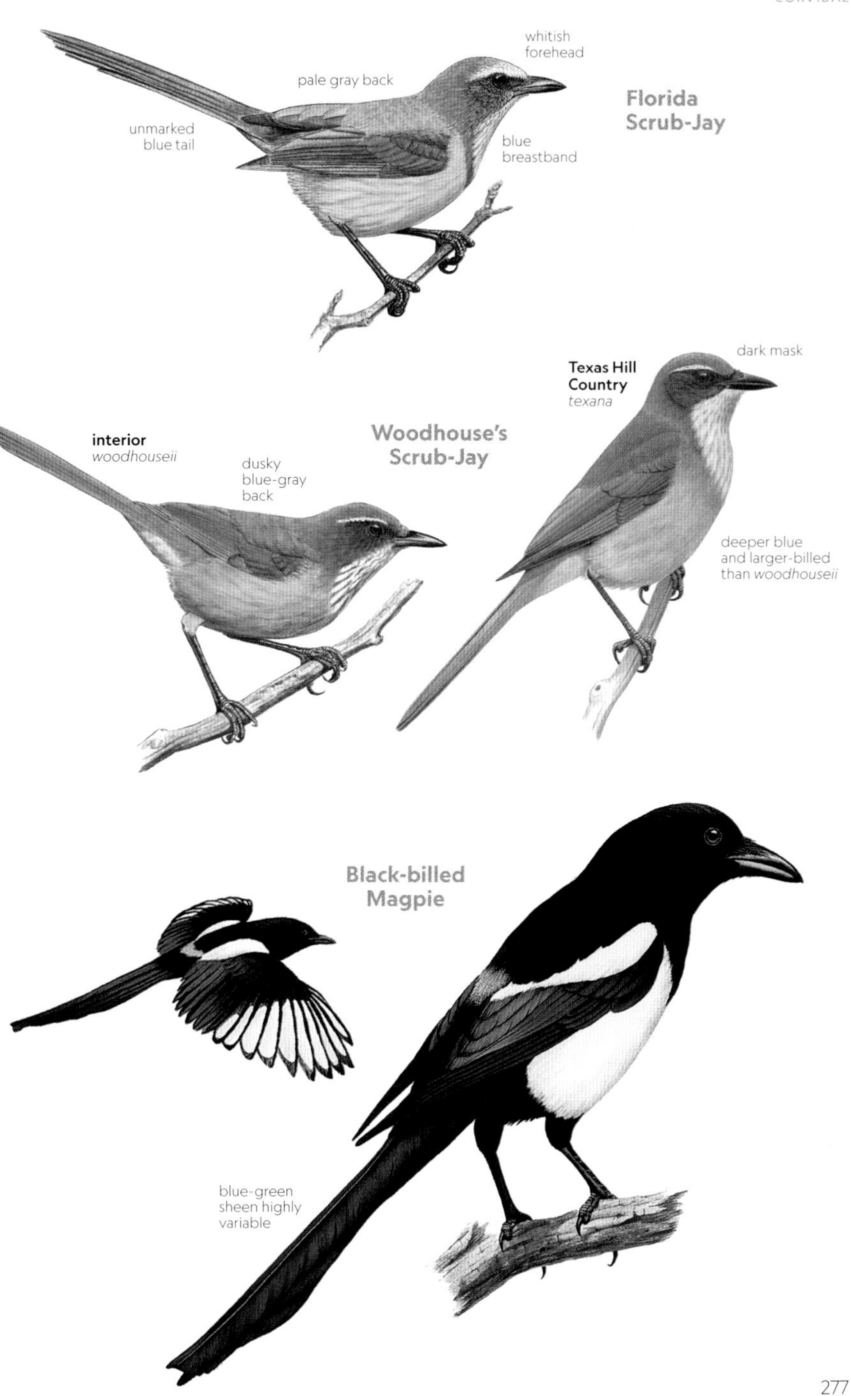
whitish
forehead
pale gray back
Florida
Scrub-Jay
unmarked
blue tail
blue
breastband
dark mask
Texas Hill
Country
texana
interior
woodhouseii
Woodhouse's
Scrub-Jay
dusky
blue-gray
back
deeper blue
and larger-billed
than woodhouseii
Black-billed
Magpie
blue-green
sheen highly
variable

## CROWS

The generic crow is known to all. Species-level ID, however, is difficult. The names crow and raven are biologically imprecise, not necessarily implying placement in one group or another.

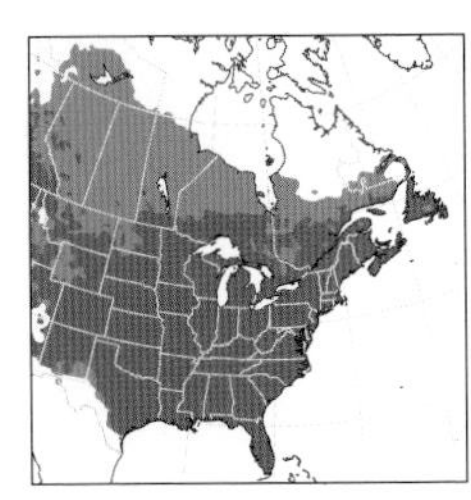

### American Crow

*Corvus brachyrhynchos* | AMCR | L 17½" (44 cm)

This ID used to be so easy: Away from the coasts and major riverways, this was the only expected all-black corvid across much of the East. But recent range expansions by the Common Raven (p. 280) and Fish Crow have made things complicated.

■ **APPEARANCE** Well-built and stocky; entirely black. Bill strong and sturdy, but not as massive as that of ravens. In flight, wings and tail appear rounded; ravens have longer wings and ample tails, often fanned. Crows never soar, ravens often do. Compare with Fish Crow.

■ **VOCALIZATIONS** Well-known call a straight-up *caw*, often in series. Also bill clicks and spooky clattering.

■ **POPULATIONS** Until mid-20th century, found mostly in farm country; has dramatically urbanized in recent decades. Enormous roosts in cities are an urban gothic spectacle.

### Tamaulipas Crow

*Corvus imparatus* | TACR | L 14½" (37 cm)

Scarce and erratic in extreme southern Tex., especially Brownsville, mostly fall and winter. Smaller, slighter-billed, and more iridescent than American Crow, inviting confusion with Great-tailed Grackle (p. 374), abundant in the region. Does not overlap with American Crow, but compare with sympatric Chihuahuan Raven (p. 280). Call is a wimpy croak or quack.

### Fish Crow

*Corvus ossifragus* | FICR | L 15½" (39 cm)

In the East, a playful approach to separating this from American Crow is to ask, "Are you an American Crow?" If the answer is "Uh-uh" or "Uh-oh," it is likely a Fish Crow.

■ **APPEARANCE** Identical in plumage and very similar in structure to American Crow; Fish Crow smaller overall, with proportionately shorter bill and feet. Wings of Fish Crow slightly longer; wingbeats a bit faster. Male Boat-tailed Grackle (p. 374), nearly as large, occurs in same habitat; body shape and vocalizations very different.

■ **VOCALIZATIONS** All calls nasal; listen for a short *ca* or *car,* and two-note *ca-ha (uh-oh),* first note rising, second falling. Any American Crow, especially young, can give nasal notes like Fish's, but the latter's *ca-ha* is distinctive.

■ **POPULATIONS** Flocky, often around dumps, beaches, and salt marshes; also suburbs. Range rapidly expanding inland.

### Field ID of Entirely Black Birds

The crows and ravens, genus *Corvus,* are the most monochromatic of all birds. They are completely black. So how do we tell them apart? Vocalizations are useful, but with these caveats: (1) Crows and ravens often mix it up, giving atypical calls and sometimes mimicking other species, and (2) they are sometimes frustratingly silent. Differences in body structure are real and consistent—but generally subtle.

Crows and ravens are always *doing* something, and watching one—or, ideally, a whole flock—is a superb discipline. In a mixed-species crow roost in the mid-Atlantic, try to discern the faster wingbeats and floppier flight style of the Fish Crow. Where crows and ravens gather

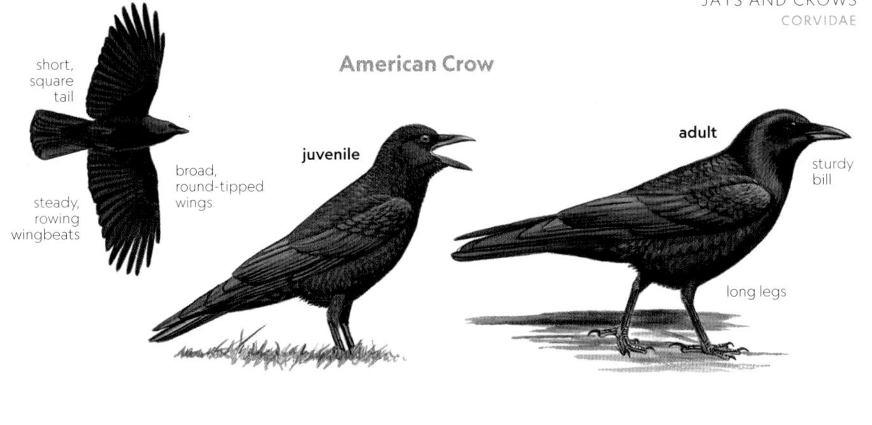


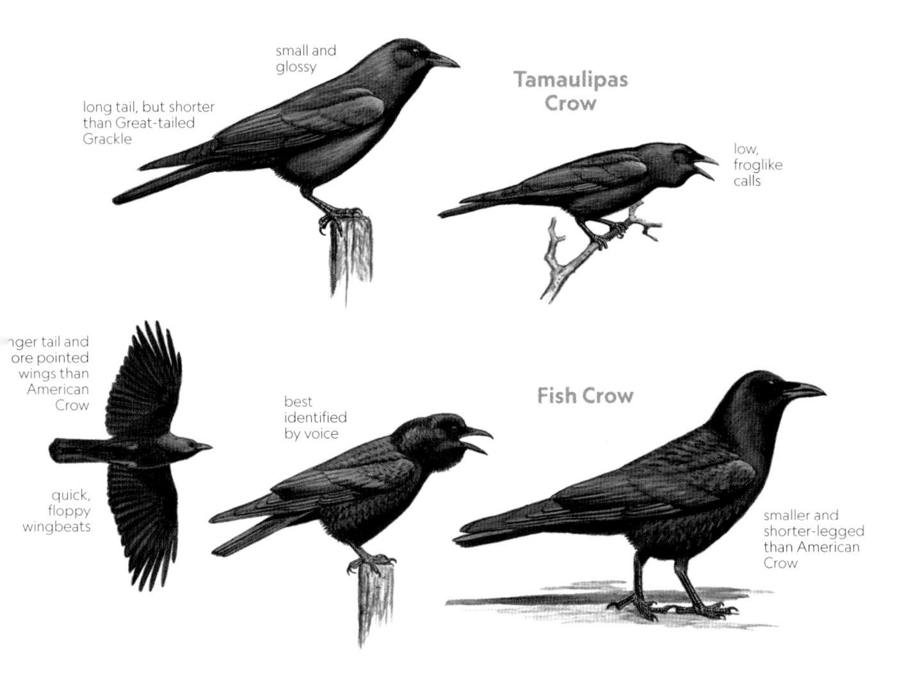


amid the knobs and ridges of the Appalachians, compare the rowing wingbeats of the former with the soaring flight of the latter. And even the notoriously difficult Common and Chihuahuan Ravens may be provisionally distinguished by differences in microhabitat and social behavior.

Finally, note this exception to the rule: It is not quite correct to say that crows and ravens are entirely black. When they blink, their eyes sparkle luminously blue! This happens when the bird's nictitating membrane, which wipes across the eye, reflects ambient light. The effect is hard to see in real life, for it happens literally in the blink of an eye. But it is striking in photographs—and affirms that, with these "all-black" birds, there is more than meets the eye.

## RAVENS

Everybody knows ravens from scary movies and Edgar Allan Poe's famous poem, but it can be hard to recognize them in real life. Separating them from crows is not straightforward, and distinguishing between the two raven species is particularly fraught.

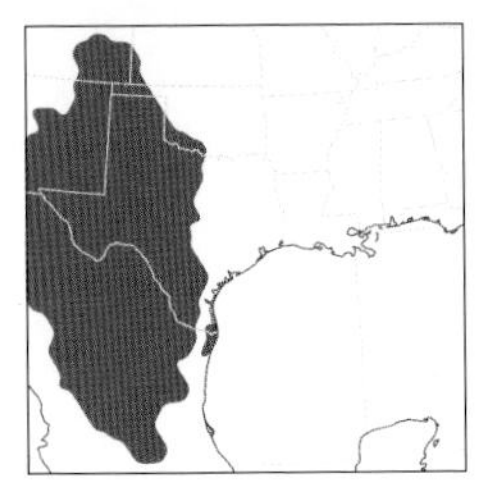

### Chihuahuan Raven

*Corvus cryptoleucus* | CHRA | L 19½" (50 cm)

Both of our ravens can be found in arid shrublands and grasslands, but this smaller of the two is restricted to such habitats. Bulky stick nests on roadside utility poles are often the handiwork of a pair of Chihuahuan Ravens.

■ **APPEARANCE** Formerly known as White-necked Raven; *cryptoleucus* means "hidden white," and the white bases of the Chihuahuan's neck feathers are a nearly useless field mark. Chihuahuan is smaller overall than Common Raven, with proportionately shorter wings and tail and a less massive bill; in close view or photographic review, note that the nasal bristles ("nose hairs") of Chihuahuan extend farther out on the bill.

■ **VOCALIZATIONS** Call harsh and nasal, like quack of Mallard (p. 36): *rraack, rruuhh,* etc.

■ **POPULATIONS** Restricted mostly to creosote bush–dominated desert in Tex. and shrub-steppe of southern High Plains. Wanders north, sometimes in flocks, in winter.

### Common Raven

*Corvus corax* | CORA | L 24" (61 cm)

The largest passerine on the planet, the Common Raven surpasses the Red-tailed Hawk (p. 212), a good-size raptor, in body length, body mass, and wingspan. The species occurs year-round in hot canyons, in the high Arctic, and, increasingly, in big cities.

■ **APPEARANCE** Enormous yet slim. In direct flight, wings and tail notably pointed; long bill obvious even at some distance. When getting a close look, note long and unkempt throat feathers; in bright glare, the all-black throat feathers of Common Raven glisten brilliant white, suggesting Chihuahuan Raven. Nasal bristles don't extend as far out on long bill as those of shorter-billed Chihuahuan. Habitually soars with tail splayed out; Chihuahuan also soars, but not as often. Crows (p. 278) sometimes stall out when banking, but never soar.

■ **VOCALIZATIONS** Classic call is a powerful, far-carrying, somewhat wavering croak: *kr'r'r'rck.* Also short whistles and rattles, including notes not unlike those of American Crow (p. 278) and Chihuahuan Raven.

■ **POPULATIONS** Range expanding, slowly but steadily, in northeastern U.S.; has become established in big cities, where it nests on bridges, tall buildings, and big box stores.

### Evolution of Crows and Ravens

A key idea in the evolution of biodiversity is that of adaptive radiation, whereby an ancestral species gives rise to a multitude of descendant species, typically specialized in behavior and distinctive in morphology. Famous examples *are* "Darwin's finches" (which aren't finches) on the Galápagos Islands and the Hawaiian honeycreepers (which actually *are* finches). Enter the crows and ravens, genus *Corvus.*

Unlike Darwin's finches and the Hawaiian honeycreepers, crows and ravens are similar-looking, similar-sounding habitat generalists. In many areas, two or more species co-occur without interbreeding. And their molecular biology generally affirms full-species rank for the several

species in the East and many more worldwide. It's a paradox: an adaptive radiation of sameness. Or is that just a narrow human conception? Seemingly small differences in morphology and behavior may be much more significant for crows and ravens, among the most cognitively and culturally advanced of all animals.

An oddity in the story of avian adaptive radiation involves the familiar Common Raven. Surprisingly, it is in the process of reverse speciation, whereby two ancestral species have merged into one. A range-restricted California clade and a widespread Holarctic clade, formerly separate species, have conjoined in recent geologic time, reversing their earlier separation. Avian biology is rich and dynamic, even across evolutionary landscapes that may seem almost featureless to the human eye.

## CHICKADEES

These proverbial cute little birds are also hardy, adaptable, and sometimes aggressive. While all tend feeders, they are eminently capable of fending for themselves far from human habitation.

### Carolina Chickadee

*Poecile carolinensis* | CACH | L 4¾" (12 cm)

In most of its range, this U.S. endemic is the only chickadee—and a cinch to identify. But it comes into contact with the Black-capped Chickadee at the northern limit of the range and at higher elevations in the southern Appalachians, where field ID is fraught and fascinating.

■ **APPEARANCE** Classic chickadee, with black cap, white cheek, and black throat. Similar Black-capped Chickadee larger overall and longer-tailed, with white frosting to secondaries and secondary coverts; wings of Carolina more uniformly gray. Flanks of Black-capped usually tinged warm pinkish brown, more grayish on Carolina.

■ **VOCALIZATIONS** Song a series of four or more short whistles alternating in pitch: *see-wee see-way*, etc. Onomatopoetic "chickadee" call faster and thinner than that of Black-capped: *tsick-tseck ch'ch'ch'ch'ch'ch'ch'ch*.

■ **POPULATIONS** Across much of range, shifting slowly and steadily northward and upslope. In narrow contact zone with Black-capped, hybrids are legion. Vagrants north of contact zone exceedingly rare.

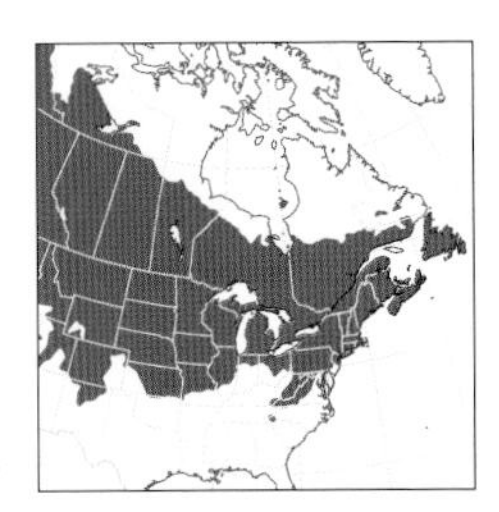

### Black-capped Chickadee

*Poecile atricapillus* | BCCH | L 5¼" (13 cm)

This northern counterpart of the Carolina Chickadee was Ralph Waldo Emerson's favorite bird—for its supposed virtues of industry and ingenuity. Like other chickadees, this species often forms the core of mixed-species foraging flocks.

■ **APPEARANCE** Larger and longer-tailed than Carolina. Secondaries and secondary coverts frosted white; wings of Carolina plainer gray. Note also buffy flanks of Black-capped, grayer on Carolina.

■ **VOCALIZATIONS** Song simple: two short whistles, the second lower. One of the first birds to sing each year. *Chick-a-dee-dee-dee* call averages lower and slower than that of Carolina; the number of *dee* notes has been shown to communicate levels of danger to flockmates.

■ **POPULATIONS** Southern limit of range withdrawing in perfect lockstep with northward push of Carolina Chickadee; one-way competitive exclusion by Carolina has been posited but not proven. A few wander south in winter, well into the range of Carolina.

### Boreal Chickadee

*Poecile hudsonicus* | BOCH | L 5½" (14 cm)

Both the Black-capped and Boreal Chickadees occur throughout the boreal forest biome, but the latter species is more specialized, tending to stick to conifer woods. Spruce and fir forests are especially favored.

■ **APPEARANCE** Basic shape and pattern like Black-capped Chickadee's, but browner overall; many look "messy." Cap chocolate (not black), flanks orange-brown, wings plain gray. White in cheek reduced compared to Black-capped's.

■ **VOCALIZATIONS** Lacks the whistled song of Black-capped and Carolina Chickadees; instead, gives a short gurgle, often ending in a loose trill. "Chickadee" call of Boreal especially nasal and tinny.

■ **POPULATIONS** Most winters, just a small push south from southern edge of core range; vagrants farther south, to mid-Atlantic, not even annual.

westernmost
*atricapilloides*
grayer overall than *extimus*
Carolina Chickadee
gray tinge on rear cheeks
low-contrast wings
eastern
*extimus*
fresh fall
dull flanks
worn summer
eastern
*extimus*
shorter-tailed than Black-capped

Black-capped Chickadee
uniform white cheeks
fresh fall
worn summer
white-edged secondaries
bright buff flanks
brown cap
juvenile
Boreal Chickadee
extensively gray cheeks
worn summer
fresh fall
bright rusty flanks

## TITMICE AND PENDULINE TITS

These species have similar names and a shared evolutionary heritage. The titmice are in the same family, Paridae, as the chickadees (p. 282). The Verdin is in a different family, Remizidae, thought to be the closest relative of the Paridae.

### Tufted Titmouse

*Baeolophus bicolor* | TUTI | L 6¼" (16 cm)

A bit larger than chickadees, the titmice are distinctively crested. Along with chickadees, they often form the core, or nucleus, of mixed-species foraging flocks, especially in winter.

■ **APPEARANCE** Slim and gray overall, with crest usually prominent; like chickadees, has buff-orange flanks. Note black forehead and gray crest, the opposite of Black-crested Titmouse. Juveniles, frequently seen all summer, lack black forehead.

■ **VOCALIZATIONS** Despite small stature, one of the loudmouths of eastern hardwood forests. Whistled song, strong and ringing, typically in triplets: *beer! beer! beer!* and *beeda! beeda! beeda!* Calls rather chickadee-like, but varied, comprising *tsick* and *dee* elements.

■ **POPULATIONS** Formerly restricted to the Carolinian life zone, but has expanded steadily northward since mid-20th century, reaching the U.S.-Canada border. Frequently tends feeders.

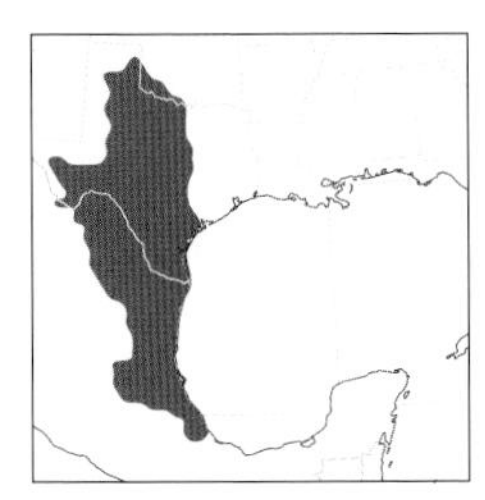

### Black-crested Titmouse

*Baeolophus atricristatus* | BCTI | L 5¾" (15 cm)

This close cousin of the widespread Tufted Titmouse is restricted in the U.S. essentially to Tex. The two species' songs and ecologies are similar, and they hybridize where their ranges come into contact.

■ **APPEARANCE** Resembles Tufted Titmouse, but head pattern reversed: Black-crested has black crest and pale forehead. Head pattern of juvenile muted, with weaker contrast between pale forehead and darker crown. Crest of adult hybrids gray with blackish suffusions; forehead buffy brown. Hybrid juveniles may not be separable in the field.

■ **VOCALIZATIONS** Song elements like those of Tufted, but five or more per song bout; Tufted usually sings only three or four.

■ **POPULATIONS** Found in oak-mesquite woodlands, shrublands, and suburbs; as mesquite has spread northward, so has this titmouse. Hybrids with Tufted occur from around Ft. Worth, Tex., to near Corpus Christi.

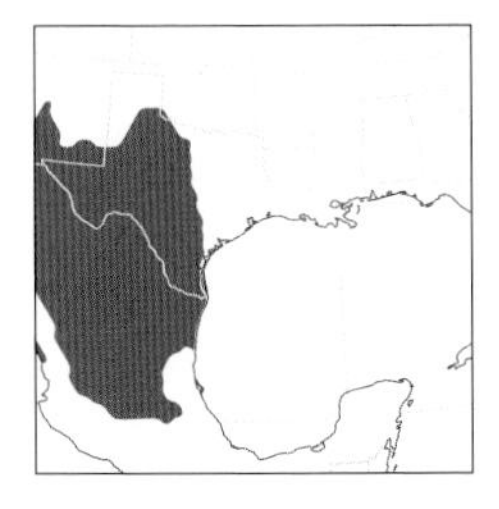

### Verdin

*Auriparus flaviceps* | VERD | L 4½" (11 cm)

Little bird, big nest. This tiny desert insectivore builds impressive spheroid nests, plural, of thorns, cactus spines, and other plant matter. Some nests are used for breeding, others for roosting.

■ **APPEARANCE** One of the smallest songbirds in the East; all are gray with sharply pointed bill. Adult male has bright yellow head and dark red "shoulders"; yellow and red reduced on adult female. Juvenile, with featureless gray plumage, recognized by bill structure and tiny size.

■ **VOCALIZATIONS** Calls constantly while foraging, which behavior it also performs constantly: a descending *beef* or *byeev.* Also a ringing *byeeo,* doubled or trebled.

■ **POPULATIONS** Like Black-crested Titmouse, strongly affiliated with northward-spreading mesquite; has likewise expanded northward a bit in recent decades, barely reaching Okla.

juvenile
Tufted Titmouse
Tufted x Black-crested hybrid adult
gray crest with black forehead
adult
variable, with intermediate features
Black-crested Titmouse
juvenile
black crest with pale forehead
adult
adult
sharply pointed bill
Verdin
tiny, with featureless plumage
juvenile

## LARKS

A number of New World birds have lark in their name, but this species is the only "true" lark in the East. Superficially similar to pipits (p. 322) and longspurs (pp. 332-335), the "true" larks are probably more closely related to swallows (pp. 288-293), bulbuls (p. 294) and the Bushtit (p. 296).

### Horned Lark

*Eremophila alpestris* | HOLA | L 6¾-7¾" (17-20 cm)

Often regarded as the "most common bird you never heard of." In open country in the Midwest, especially in winter, flocks of larks work the stubble fields and other waste places along roadsides. But they're flighty and unforthcoming, just "little brown jobs," and easily ignored.

■ **APPEARANCE** Up close, adult is exquisite, but views are often at some distance. Especially in the nonbreeding season, occurs in compact flocks, sometimes quite large; the sandy brown birds meander randomly by walking, bodies low to the ground, not hopping. Flocks flush constantly, putting up just above ground level, then putting back down and resuming their random walk. Key mark in flight is the mostly black tail, distinct from the pipits and longspurs with which larks often occur. Juvenile, plump and pale and scaly, resembles Sprague's Pipit (p. 322); the secretive pipit occurs in denser grass, sounds different, and wears streaks, not spots, above.

■ **VOCALIZATIONS** Flight call, heard year-round, a descending *ts'tseee,* sweet and jingling, not unlike American Tree Sparrow (p. 344). Flight call of American Pipit, with which it often associates, more strongly descending, less musical. Song, given perched or in flight, a stutter and then a short jumble: *pik plek p'TEE'd'DEEDle'eedle'ek;* sings on moonlit nights, spring to early summer.

■ **POPULATIONS** Breeds across much of Northern Hemisphere, from the high Arctic and highest mountaintops to below sea level in hot deserts, with considerable geographic variation. Dark subspecies *alpestris* breeds across southeastern Canada, winters widely in East. Paler subspecies *praticola* breeds across midlatitudes of eastern U.S., wintering south to Fla.; *praticola* declining, probably in response to retreat of traditional agriculture across East.

### Infraorder Passerides

In the East, the order Passeriformes comprises two suborders: the Tyranni (pp. 246-263) and the Passeri (pp. 264-409). Within the latter are two further subdivisions, the infraorder Corvides (pp. 264-281) and the infraorder Passerides (pp. 282-409).

To put that in perspective, the infraorder Passerides—a subdivision within a division within a single order—takes up fully one-third of this field guide. In contrast, the entire order Gaviiformes (loons) contains five extant species (four in the East), and the order Phaethontiformes (tropicbirds), three (two in the East).

The infraorder Passerides constitutes one of the most spectacular adaptive radiations (p. 280) in the history of life on Earth. In this book, the Passerides begin with the cold-adapted chickadees (p. 282) and then progress to the desert-loving Verdin (p. 284) and the Horned Lark, equally at home in deserts and on tundra; next up, the swallows (pp. 288-293), specialized aerial insectivores; and so forth and so on, all the way to the nectivorous Bananaquit and granivorous Morelet's Seedeater (p. 408).

The secret of their success has been their ability to quickly establish in diverse terrestrial habitats, to exploit all available niches, and to construct (or, in some cases, to borrow) elaborate nests of every sort imaginable. The recognition of the Passerides is recent, dating to the late 20th century, but it is now essentially unquestioned: Independent analyses based on everything from molecular biology to paleontology to biogeography affirm the monophyly (p. 54) of this immense assemblage.

Arctic
hoyti
large and pale
♂
rich brown, with yellow eyebrows
winter
♂
♂
♀
"Northern Horned Lark"
alpestris
Lapland Longspur for comparison
Horned Lark
"Northern"
alpestris
juvenile
heavily spotted above
Great Plains
enthymia
juvenile
"Northern"
alpestris
mostly black tail with brown center, thin white edge
"Northern"
alpestris
Great Plains
enthymia
breeding ♂
pale, but warmly colored
winter
♂
Great Plains
enthymia
winter
♀
"Prairie Horned Lark"
praticola
medium-pale, with whitish eyebrow
very yellow below
Gulf Coast
giraudi

## SWALLOWS WITH WHITE THROATS

Because they feed heavily to exclusively on flying insects, the swallows (pp. 288-293) occur in most of the East only in the warmer months. Although frequently on the wing, swallows also have the convenient habit of landing on wires and tending easily observed nests.

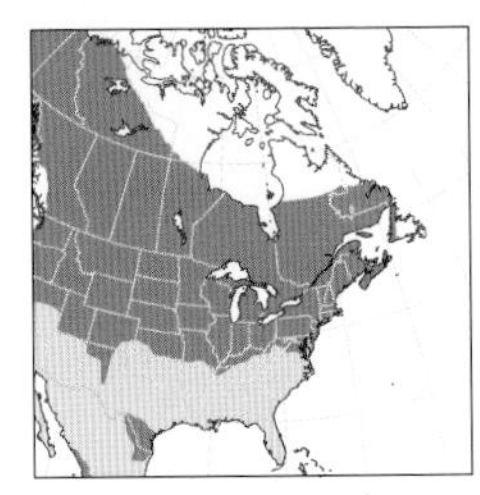

### Bank Swallow

*Riparia riparia* | BANS | L 4¾" (12 cm)

The bird's name is a dead giveaway: Bank Swallows excavate nests in earthen embankments. *Riparia* is likewise informative, indicating the species' preference for riverside (riparian) microhabitats for nesting.

■ **APPEARANCE** Our smallest swallow, its body length as short as that of our slightest sparrows; perched shoulder to shoulder with other swallows, the size difference is obvious. Adult brown above and white below, with broad brown breastband sharply set off from gleaming white throat. Northern Rough-winged Swallow (p. 290) is chunkier, brown above and paler brown-gray below, has dusky throat, and lacks breastband. Juvenile Bank not as crisply marked below as adult, inviting comparison with Northern Rough-winged and juvenile Tree Swallows.

■ **VOCALIZATIONS** Call a harsh buzz, typically given in flight, in rapid series, accelerating slightly: *bzhh bzhh bz bz'bz'bz'bz'bz*. Sometimes just a single *bzhh*.

■ **POPULATIONS** Nests in small to large colonies, especially along rivers; also accepts human-made habitats like quarries and roadcuts. Summer distribution quite local, reflecting habitat availability.

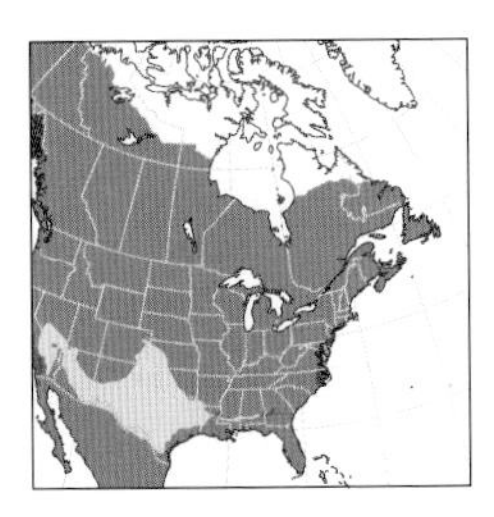

### Tree Swallow

*Tachycineta bicolor* | TRES | L 5¾" (15 cm)

As a descriptive name, Tree Swallow works, but "Nest Box Swallow" would be even better. Favored nest sites are around water, but individuals on foraging bouts may be found almost anywhere.

■ **APPEARANCE** Adult steely green or blue above, underparts snow-white; gloss of upperparts dependent on lighting. Juveniles, fully capable of flight and widely noted by midsummer, are solid brown above; many have diffuse breastband, suggesting Bank Swallow. Juvenile Bank, smaller and sleeker, usually shows thin wing bars, lacking on Tree.

■ **VOCALIZATIONS** Song a short jumble of *slweet* and *tlit* notes, with liquid or gargling quality. Frequently calls in flight, giving one to a few of the song elements as it goes.

■ **POPULATIONS** Hardy; arrives earlier in spring than other swallows, departs later in fall. Migratory flocks, compact and immense, stage over bays, dunes, salt marshes, etc. Like Yellow-rumped Warbler (p. 396), has specialized biochemistry for digesting waxy coating of bayberries, which it devours in cold weather when bugs aren't flying.

### Violet-green Swallow

*Tachycineta thalassina* | VGSW | L 5¼" (13 cm)

Colorful close cousin of the widespread Tree Swallow, breeding as far east as Badlands, S. Dak. Rare but regular migrant, spring and fall, western Great Plains; vagrates annually much farther east, reaching the Atlantic seaboard most years. In flight, even at a distance, white on cheek and sides of rump prominent; Tree Swallow has dark cheeks and all-dark rump. Calls like Tree's, but drier, mnemonically matching Violet-green's preference for more arid habitats than Tree's.

thin pale
wing bars
juvenile
Bank
Swallow
adults
rapid, shallow
wingbeats
slender,
notched
tail
white throat
wraps around
auriculars
narrow
pointed
wings
adult
1st spring ♀
broad
wings
Tree Swallow
juveniles
often with
diffuse
breastband
adult
strong
contrast
between
cheek and
throat
pale tertial
tips
juvenile
white
sides of
rump
fall adult
adults
spring adult
juvenile
low-contrast
face
extensively
white face
Violet-green
Swallow
small and
short-tailed
adult ♂

## UNPATTERNED SWALLOWS

Despite their different English names, swallows in the genus *Stelgidopteryx* and martins in the genus *Progne* are closely related. The name martin simply refers to largish swallows.

### Northern Rough-winged Swallow

*Stelgidopteryx serripennis* | NRWS | L 5" (13 cm)

Like the Bank Swallow (p. 288), the Northern Rough-winged nests in embankments, typically near water. But it doesn't excavate its own nest sites, instead appropriating the old or abandoned burrows of other animals—often Bank Swallows.

■ **APPEARANCE** Our drabbest swallow. Adult dull brown above; sandy gray-brown below, becoming whiter toward belly and undertail coverts. Juvenile like adult, but with rufous suffusion on wings. The name corresponds to serrations on the outer primaries, visible in good digital photographs, their function unknown.

■ **VOCALIZATIONS** Simple calls by a plain swallow: flatulent *rrrpp* notes, often in fast series. Not as harsh as Bank's; suggests flight call of Dickcissel (p. 408).

■ **POPULATIONS** Winter range expanding north, with a remarkable flock overwintering in Philadelphia since 2004.

### Purple Martin

*Progne subis* | PUMA | L 8" (20 cm)

Centuries ago, Choctaw and Chickasaw communities began attracting martins by hanging hollowed-out gourds. Possibly more than any other species in the East, martins are now completely won over by nest boxes, including lavish "apartments" that house multiple families.

■ **APPEARANCE** Our largest swallow; a bit smaller than a starling (p. 310), which it resembles in flight. Adult male dark blue-purple, appearing black in poor light; male takes two years to acquire full coloration. Adult female has scaly gray underparts; juvenile like female, but paler.

■ **VOCALIZATIONS** Calls, low and rich, include a down-slurred whistle and a descending chatter. Song, given incessantly around martin apartments, a pleasing gargle.

■ **POPULATIONS** Migrants return to southern U.S. very early, in midwinter, then slowly fan northward; "fall" migration in some places begins in meteorological spring. Numbers crash regionally following severe weather events.

### Identifying Swallows in Flight

When perched on wires or ledges or while tending nest sites, swallows are easy to identify. But they spend much of their time on the wing, zipping by quickly, making field ID more challenging. Fortunately, swallows in flight often come in close enough for observers to see field marks, like the Cliff Swallow's "headlights" (p. 292) or the Bank Swallow's broad breastband. Pay particular attention to body shape: The Purple Martin's forked tail separates it from the European Starling; the Barn Swallow's deeply forked tail (p. 292) differentiates it from other orange-and-blue swallows.

Flight style can be very useful, especially on birds at a distance. Being large swallows, martins often glide and may even soar. Compare the choppy wingbeats of Bank Swallows with the more languid wingbeats of Northern Rough-wings. A disorganized cluster of swallows rising up from a culvert is often Cliff Swallows, whereas a dense array of thousands of swallows over a *Phragmites* marsh is invariably Tree Swallows.

As always, listen! Swallows on the wing are typically vocal, and their flight calls can be learned with practice. Several species—notably Purple Martin, Barn Swallow, and Violet-green Swallow (p. 288)—habitually call before dawn. It is bewitching to hear and be able to recognize their pleasant utterances while it is still night.

dull brown, blending to white belly
Northern Rough-winged Swallow
adult
broad wings
slow, floppy wingbeats
broad, square tail
rufous wing bars
low contrast overall
juvenile
Purple Martin
1st-summer ♂
♀
deeply notched tail
dark throat and pale belly
♀
adult ♂
adult ♂
soars more than other swallows

## ORANGE-AND-BLUE SWALLOWS

They're elegant, with their bright and warm color schemes, but they also like to play in the mud: These three are part of a "mud-nester" clade of closely related swallow species.

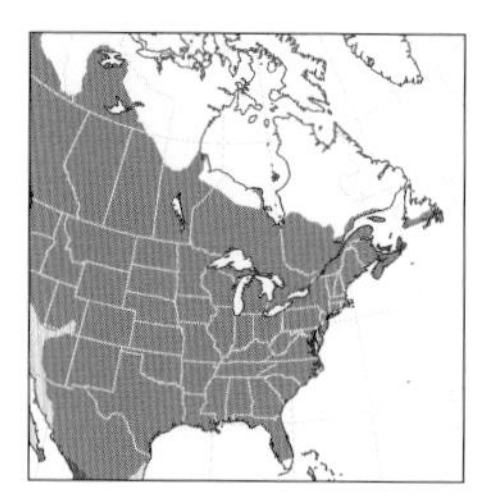

### Barn Swallow

*Hirundo rustica* | BARS | L 6¾" (17 cm)

Few things say "summer" more than Barn Swallows tending their earthen nests under bridges, verandas, and lampposts. They're classically associated with farm country but may be found almost anywhere.

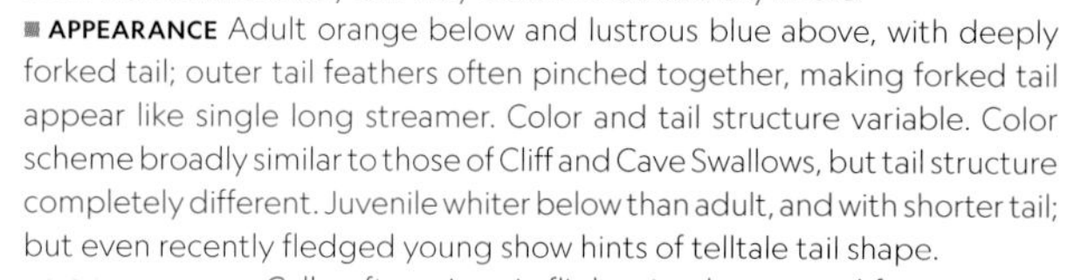

■ **APPEARANCE** Adult orange below and lustrous blue above, with deeply forked tail; outer tail feathers often pinched together, making forked tail appear like single long streamer. Color and tail structure variable. Color scheme broadly similar to those of Cliff and Cave Swallows, but tail structure completely different. Juvenile whiter below than adult, and with shorter tail; but even recently fledged young show hints of telltale tail shape.

■ **VOCALIZATIONS** Calls, often given in flight, simple *vree* and *frit* notes, rising sharply, often in short, fast series. Song, given by both sexes, a lengthy outpouring of chirpy phrases ending in a twangy, rising note and dry rattle.

■ **POPULATIONS** Tolerant of humans; range has been slowly expanding for several centuries. Emerging threat is rapid loss of the swallows' prey base of aerial insects due to pesticides.

### Cliff Swallow

*Petrochelidon pyrrhonota* | CLSW | L 5½" (14 cm)

Finding Cliff Swallow nests on cliffs is possible, but you are vastly more likely to discover them under bridges and culverts. The species breeds in densely packed colonies often exceeding 100+ earthen nests.

■ **APPEARANCE** Bluish above and orange below like Barn Swallow, but tail short and squared off. Bright buff-orange rump prominent in flight; forehead pale cream, known as "headlights" for its conspicuousness on birds flying toward observer. Juvenile has dark forehead, but nevertheless presents pale rump. See Cave Swallow.

■ **VOCALIZATIONS** Flight calls breathy and nasal, typically down-slurred: *reer, vreen,* etc. Complex song comprises grinding and cranking notes, interspersed with call notes.

■ **POPULATIONS** Mostly a western species historically; has expanded greatly in East in past 150 years.

### Cave Swallow

*Petrochelidon fulva* | CASW | L 5½" (14 cm)

Until the late 20th century, Cave Swallows were essentially unheard of along the mid-Atlantic and eastern Great Lakes. Today they're regular, sometimes in large flocks, in late fall.

■ **APPEARANCE** Similar to Cliff Swallow, but head pattern reversed: Adult Cave has dark forehead and pale throat; adult Cliff has pale forehead and dark throat. Juvenile duskier on head; juvenile Cliff can be quite similar, but note richer rump and paler throat of juvenile Cave.

■ **VOCALIZATIONS** Calls varied, including rising notes (like Barn Swallow) and falling notes (like Cliff Swallow). Notes tend to be purer and less nasal than those of Cliff, but much variation.

■ **POPULATIONS** Fla. breeders are subspecies *cavicola,* with relatively dark rump. Tex. breeders, subspecies *pallida,* have paler rump. Flights in late fall to mid-Atlantic and eastern Great Lakes, including Canada, occasionally 100+, are *pallida,* not *cavicola.*

Barn Swallow
dark
males more orange below, females and juveniles more white
juvenile
long forked tail with white band
juvenile
pale forehead
ale buff
ımp
dark throat
Cliff Swallow
pyrrhonota
adult
Mexican
melanogaster
rare in East; has dark forehead like Cave
chunky and square-tailed
buff-orange rump
Cave Swallow
pale cheeks and throat
cinnamon forehead
juvenile
cinnamon face and throat
rich orange rump
darker cinnamon tones, especially on rump
Cuban
cavicola
adult
southwestern
pallida
southwestern
pallida

## MIDSIZE PASSERINES WITH CRESTS

These arboreal frugivores are among our most stylish songbirds—sleek and crested, with bright splashes of color. Even their nests, woven with fine moss and cobwebs, are elegant.

### Red-vented Bulbul

*Pycnonotus cafer* | RVBU | L 8½" (22 cm)

Indigenous to the Indian subcontinent, this bulbul has recently become established in and around Houston, Tex. The bird is dark overall, nearly black on the crested head, with medium-long tail; undertail coverts bright red. Vocalizations low and rich: calls include a sharp *aaawrp,* twangy and ascending; song a short gargle, *rawrp reer-a-a-rip.* Found mostly in wooded urban districts: parks, greenways, neighborhoods, etc.

### Red-whiskered Bulbul

*Pycnonotus jocosus* | RWBU | L 8" (20 cm)

Indigenous across large swath of southern Asia; established in Miami, Fla., area since mid-20th century. Dark brown above, whitish below; crest long and pointed. Red "whisker" hard to see; red undertail coverts often prominent. Smaller and brighter than Red-vented; similar habitats, but they do not co-occur in the East. Vocalizations higher and sweeter than Red-vented's: call a descending *ch'whee-ur;* song a short, whistled jumble.

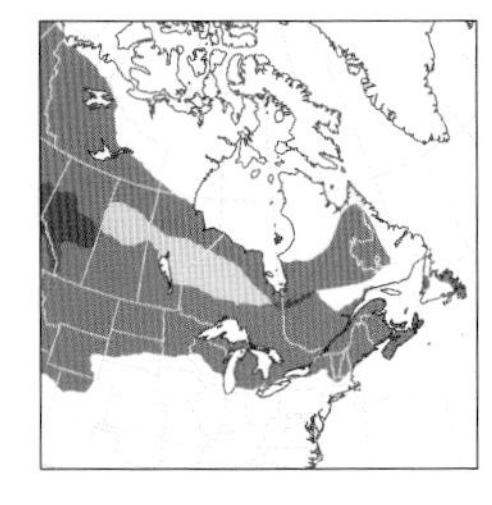

### Bohemian Waxwing

*Bombycilla garrulus* | BOWA | L 8¼" (21 cm)

Named for the old duchy in Europe, they're also bohemians of a sort, unpredictable and impulsive. And they're noisy! *Garrulus* is Latin for "talkative."

■ **APPEARANCE** Identifiable as a waxwing by soft plumage, black mask, crest, and waxy tips of flight feathers. Larger and grayer than Cedar, with chestnut undertail coverts. Both have yellow-tipped tail, but wing of adult Bohemian more complexly marked with red, yellow, and white. Young grayer than adult, variably splotched below, but shows telltale chestnut under tail.

■ **VOCALIZATIONS** Flocks call constantly: a loose trill, dropping in pitch—lower, louder, and more jangling than Cedar's. Ironically, given their talkativeness, waxwings apparently lack a song.

■ **POPULATIONS** Mostly western breeder, but nests irregularly east to Nfld. Irruptive in winter, sometimes occurring south to New England and upper Midwest in large flocks.

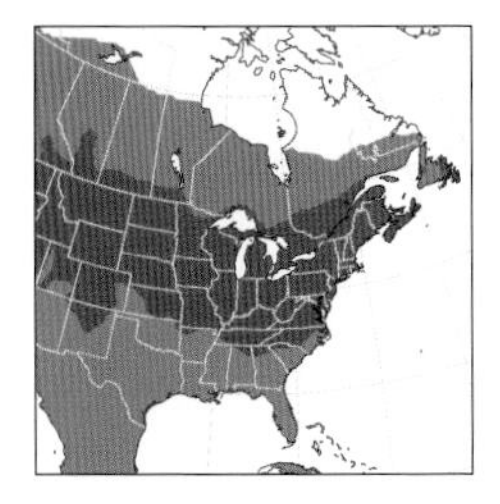

### Cedar Waxwing

*Bombycilla cedrorum* | CEDW | L 7¼" (18 cm)

This exquisite songbird is commonly portrayed on Christmas tree ornaments. Its color palette is variable, depending on age, sex, and diet.

■ **APPEARANCE** Smaller and warmer than Bohemian; undertail coverts whitish (chestnut on Bohemian). All ages, including juvenile, usually have yellow tail tip. Waxy red tips to secondaries acquired with age; male usually has more than female. Flocks bunch tight in treetops and fly in compact "balls." Where Bohemians and Cedars co-occur, they usually do not mix. Groups of either species, perched or flying, suggest starling (p. 310) flocks.

■ **VOCALIZATIONS** Like Bohemian, evidently lacks song. Flocks trill and twitter, especially in flight; calls higher and thinner than Bohemian's.

■ **POPULATIONS** Complex movements include broad-front migration in late spring, frequent irruptions in fall, and unpredictable midwinter wanderings. Variation in plumage color, especially tail tip, influenced by diet; birds feeding on exotic berries show orange or even red tail tips.

Red-vented Bulbul
very dark overall
enile
adult
white rump and tail tips
Red-whiskered Bulbul
upswept crest
dark spur at side of breast
adult
juvenile
white-tipped tail
Bohemian Waxwing
intricate wing pattern
grayish overall
bulky and short-tailed
adult
juvenile
white-tipped tail
chestnut undertail coverts
juvenile
messy gray streaks
Cedar Waxwing
both waxwings with triangular wings and yellow-tipped tails
white undertail coverts
yellowish belly
adult
smooth brown with gray rump

**OUR SMALLEST PASSERINES**

They're tiny and always on the move. The Bushtit weighs, on average, less than any other songbird in the East; the kinglets, although slightly bulkier, are the shortest songbirds in the East.

## Bushtit

*Psaltriparus minimus* | BUSH | L 4½" (11 cm)

The sociable Bushtit is our only representative of the otherwise Eurasian family Aegithalidae. Energetic flocks roam shrubby habitats, whether in arid wildlands or suburban neighborhoods with suitable plantings.

■ **APPEARANCE** Tiny; like a dirty cotton ball with a toothpick tail. Variable brown ear patch contrasts with paler crown. Age and sex classes nearly identical in plumage. Eye yellow on adult female, black on adult male.

■ **VOCALIZATIONS** Notes short and sharp: *spik, speek,* etc. Flocks sputter constantly as they forage. A ringing *lililililili* may function as song, and a descending *WHEE-whee-whee-whi-whi-whi* signals a predator.

■ **POPULATIONS** Restricted in East mostly to Tex. Hill Country. Core range farther west expanding, with recent wandering to western Great Plains.

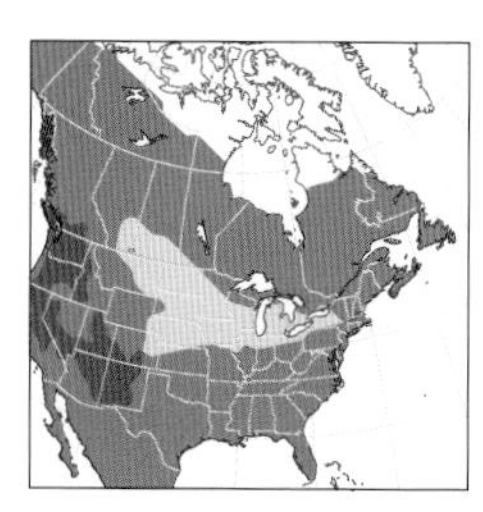

## Ruby-crowned Kinglet

*Corthylio calendula* | RCKI | L 4¼" (11 cm)

The kinglets, family Regulidae, restricted to the Northern Hemisphere, are a small family of very small birds. Their taxonomic status, relative to other passerines, is unclear; they fall within the huge infraorder Passerides (pp. 282–409), but beyond that, things are murky.

■ **APPEARANCE** A tiny ball of a bird, dull olive green, always fidgety. Male's ruby crown may be dazzlingly displayed, barely visible, or, more often than not, entirely hidden; female lacks ruby on crown. Wing bars thick and white, with wide black border below. Bold eye ring broken above and below; face lacks black stripes of Golden-crowned. Similar Hutton's Vireo (p. 264) has thicker bill, blue-gray legs, different wing markings, and different song.

■ **VOCALIZATIONS** Song loud, starts with high notes, inaudible to some, followed by lower elements: *see see see see cheer cheer cheer cheer ch'whooda ch'whooda ch'whooda*. Call, given year-round, a rough *digit* or *fidget*.

■ **POPULATIONS** Breeds in boreal forests. Winters widely; occasionally tends feeders. Migrates fairly early in spring, fairly late in fall.

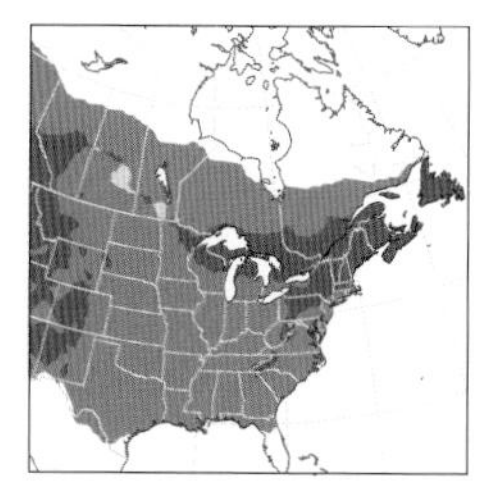

## Golden-crowned Kinglet

*Regulus satrapa* | GCKI | L 4" (10 cm)

Even in the dead of winter, this metabolic marvel subsists mostly on animal matter. During cold nights in the Maine woods, Golden-crowned Kinglets huddle together to preserve warmth; individuals lose body weight overnight, but regain it while feeding continually the next day.

■ **APPEARANCE** A bit smaller than Ruby-crowned Kinglet, with different face pattern: Golden-crowned has white eyebrow bordered black above and below; lacks eye ring of Ruby-crowned. Both male and female show yellow crown, variably suffused with orange on male.

■ **VOCALIZATIONS** Very high call a series of three quick hisses: *sssss sssss sssss.* Song like Ruby-crowned's, but even higher, with less elaborate ending: *see see see si si si si si sick-sick-sick-suck-suck.*

■ **POPULATIONS** Breeding range formerly restricted mostly to spruce-fir boreal forests, but has expanded south in recent decades, nesting in Christmas tree farms. Winters farther north and farther inland than Ruby-crowned, roaming broadleaf woods with parids (pp. 282–285).

Bushtit
interior ♂
interior ♀
female with
pale eye,
dark in male
"Black-eared Bushtit"
juvenile ♂
tiny, with long
thin tail
white wing
bar with black
rectangle below
♀
Ruby-crowned
Kinglet
crown bordered
by black
whitish
belly
♂
broken
eye ring
♂
dull olive; more
buffy below
both kinglets
hover and
flick their
wings while
foraging
both kinglets
have yellow feet,
thin legs, delicate
bills, and short
notched tails
Golden-crowned
Kinglet
gray-and-white
striped face
♀

## TREEHUGGERS

Three species of nuthatches (family Sittidae) and one treecreeper (Certhiidae) occur in the East. Treecreepers spiral up large branches and boughs; nuthatches habitually walk *down* tree trunks.

### Red-breasted Nuthatch

*Sitta canadensis* | RBNU | L 5" (13 cm)

For many eastern birders, this is the winter nuthatch. It joins mixed-species foraging flocks and reliably tends feeders.

■ **APPEARANCE** A bit smaller than White-breasted Nuthatch. Red-breasted is bluish above with dark cap, white eyebrow, black eyeline; all are washed orangey below, males a darker hue.

■ **VOCALIZATIONS** Call a petulant *aaank* or *ehhnk*, often repeated slowly. Agitated birds, mobbing an owl or responding to pishing, call constantly.

■ **POPULATIONS** Annual movements complex; common in lowlands some winters, scarce others. Breeding range expanding.

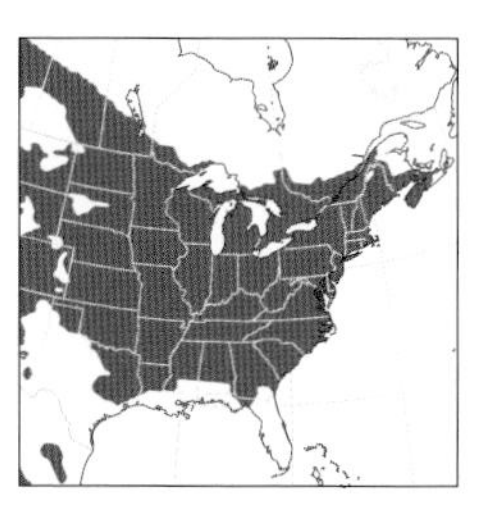

### White-breasted Nuthatch

*Sitta carolinensis* | WBNU | L 5¾" (15 cm)

The largest of our nuthatches, this species occurs year-round in diverse forest types, favoring older broadleaf tracts, especially those with standing deadwood.

■ **APPEARANCE** Blue-gray above; mostly white below, with orangey undertail coverts. Black eye stares from white face; crown and nape dark.

■ **VOCALIZATIONS** Far-carrying call a slightly nasal *yarnk* or *yank*, often doubled; lower, less whiny than Red-breasted's. Song a sonorous series of reedy whistles, curiously suggestive of a distant flicker (p. 236).

■ **POPULATIONS** Present year-round in much of range, but many disperse in fall. Favors larger woodlands; regional declines linked to forest fragmentation.

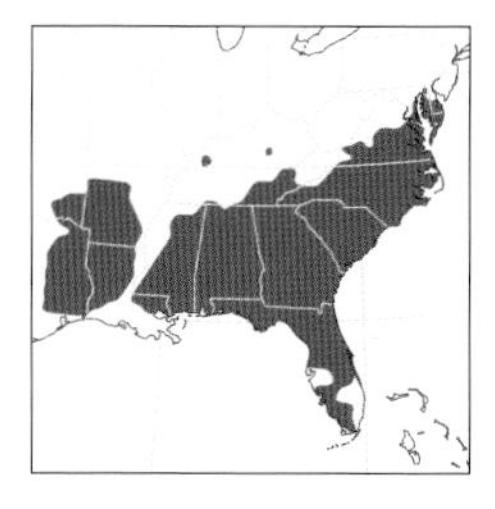

### Brown-headed Nuthatch

*Sitta pusilla* | BHNU | L 4½" (11 cm)

A southeastern specialty, this nuthatch occurs only in open pine woods. The species is highly social, invariably found in small flocks.

■ **APPEARANCE** Smallest eastern nuthatch. Differs in plumage from Red-breasted and White-breasted by its warm brown cap. Bluish above like the larger species, but cream-buff below.

■ **VOCALIZATIONS** Call high and excited (*squeak it!*), oddly and convincingly similar to the squeak of a rubber ducky.

■ **POPULATIONS** Depends on healthy southeastern pine forests; responds well to management practices like reforestation and snag preservation.

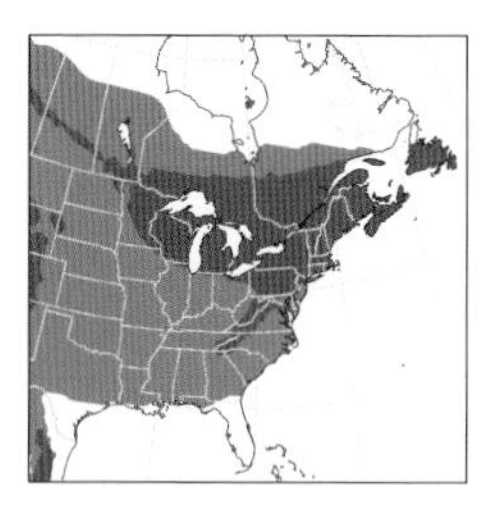

### Brown Creeper

*Certhia americana* | BRCR | L 5¼" (13 cm)

Our only representative of the Old World treecreeper family, this master of camouflage occurs singly wherever there are old trees with rugose trunks.

■ **APPEARANCE** Flat and long-tailed; upperparts patterned like tree bark. Gleaming white below reflects light into crevices, aiding the search for food; bill thin, long, decurved. Jerks its way up tree trunk, then flutters weakly to base of tree and ascends again. Sexes and ages alike.

■ **VOCALIZATIONS** Call a wavering trill, *ssseee*, loud and high-pitched. Song a short jumble of high, thin whistles.

■ **POPULATIONS** Like Red-breasted Nuthatch, a winter visitant for most eastern birders. Breeding range expanding south.

♀
female
paler
below
Red-breasted
Nuthatch
white
eyebrow
♂
♂
extensive
white face
checkered
rufous undertail
coverts
White-breasted
Nuthatch
Brown-headed
Nuthatch
tiny and
large-headed
♀
grayer crown
than male
solid brown
cap
thin,
curved bill
camouflaged
brown
upperparts
buff wing
markings
Brown
Creeper
long,
spiky tail

## GNATCATCHERS AND WRENS

Gnatcatchers, family Polioptilidae, and wrens, family Troglodytidae (pp. 300–305), may be each other's closest relatives. Their tails attract attention, cocked up or flipped about expressively.

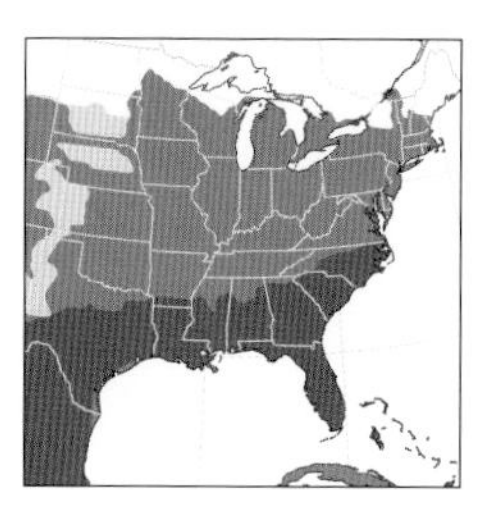

### Blue-gray Gnatcatcher

*Polioptila caerulea* | BGGN | L 4¼" (11 cm)

Slender, long-tailed, and grayish overall, this species suggests a tiny mockingbird (p. 308)—although it is slighter than the smallest wrens.

■ **APPEARANCE** Narrow tail often cocked; bill small. Pale blue overall, brighter above than below. Tail is black above like Black-tailed's, but mostly white below. Bluish head of breeding male Blue-gray marked with thin black eyeline and complete white eye ring; female has eye ring but lacks eyeline.

■ **VOCALIZATIONS** Call a tinny *bzzzeeee,* nasal and falling in pitch. Song a giddy jumble of thin notes.

■ **POPULATIONS** In most of East, the only gnatcatcher; in zone of overlap in Tex., Blue-gray often gets into desert habitat favored by Black-tailed.

### Black-tailed Gnatcatcher

*Polioptila melanura* | BTGN | L 4" (10 cm)

This is the desert gnatcatcher, frequenting washes and pond edges, occasionally well out into sparse creosote bush monocultures; range in Tex. doesn't reach Gulf of Mexico. Cap of breeding male black, unlike Blue-gray's, but females more similar; all plumages of Black-tailed have mostly black tail above and below. Calls and songs harsher than corresponding vocalizations of Blue-gray; compare with sounds of Bewick's Wren (p. 304).

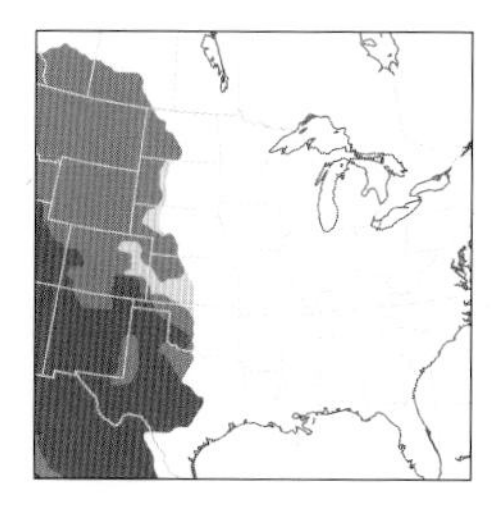

### Rock Wren

*Salpinctes obsoletus* | ROWR | L 6" (15 cm)

In arid country extending east into Great Plains, this fairly large wren occurs widely but usually singly on buttes and mesas with loose rock. Perched on a ledge or outcropping, it bops up and down while surveying its domain.

■ **APPEARANCE** Rotund, with low-contrast gray-brown plumage. Broad, buff-tipped tail often splayed out in short flights among boulders.

■ **VOCALIZATIONS** Song incredibly varied, but always involving repeated elements, like a slowed-down mockingbird: *bee ... bee ... bee ... b'dee ... b'dee ... b'dee ... bzheer ... bzheer ... bhzeer ....* Far-carrying call note an explosive *ch'PEE!*

■ **POPULATIONS** Strays annually to East Coast, fall to early winter; vagrants somehow find their way to quarries, rock piles, even stone masonry.

### Canyon Wren

*Catherpes mexicanus* | CANW | L 5¾" (15 cm)

Although they broadly overlap, Canyon and Rock Wrens sort out by microhabitat: The former favors sheer cliffs, often with a sandstone component to match the bird's orangey plumage. Canyon Wrens scurry up rock faces.

■ **APPEARANCE** Similar in build to Rock Wren, but slimmer with longer bill; throat gleaming white, crown blotchy gray. Feeds on arthropod prey, mostly spiders and centipedes, extracted with long bill, proportionately longer than that of any other songbird in the East; like Brown Creeper (p. 298), reflects light into crevices with white throat.

■ **VOCALIZATIONS** Song as invariant as the Rock Wren's is varied: a cascade of falling whistles, rated by many as one of the most haunting birdsongs on the continent. Call a shrill *beet!*

■ **POPULATIONS** Nonmigratory; in harsh winters, subsists on arthropods.

Blue-gray Gnatcatcher
western
obscura
duller gray than eastern
breeding ♂
pure white outer tail feathers
♀
distinct eye ring
eastern
caerulea
bright blue-gray above
♂
Black-tailed Gnatcatcher
restricted white in tail
black cap
dull gray
♀
♂
Rock Wren
sandy gray with fine speckles
stands tall and bobs body
buff tail corners
very long bill
rich rufous
contrasting white throat
Canyon Wren

## SHORT AND DRAB WRENS

Species in the genera *Troglodytes* and *Cistothorus* are our smallest wrens. They favor some of the densest habitats but have the convenient habit of often coming in quite close.

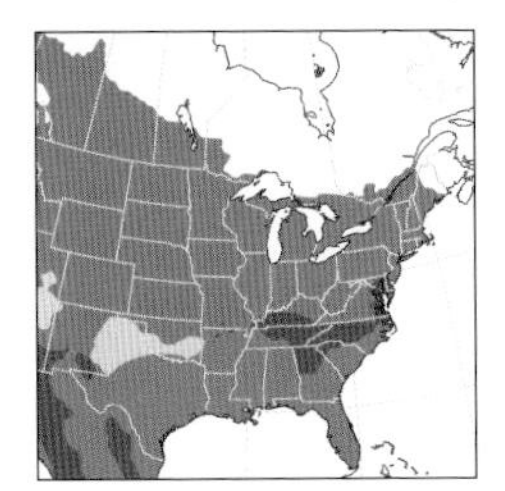

### House Wren

*Troglodytes aedon* | HOWR | L 4¾" (12 cm)

Perhaps the most generalist wren in the East, this species can be found anywhere there is a bit of cover: from clearings deep in the forest to old pastures to urban districts with adequate plantings.

■ **APPEARANCE** Plain and plump. Similar Winter Wren smaller, with shorter tail and more obvious barring on flanks. Many House Wrens in fall and winter make their way to cattail marshes, where confusion with *Cistothorus* wrens is possible.

■ **VOCALIZATIONS** Vigorous song a bubbly, exuberant chatter, fading or falling off. Calls variable, but always harsh, often run together in short series: *chet, chich,* etc., suggesting Marsh Wren.

■ **POPULATIONS** Readily takes to birdhouses. House Wrens from Great Plains westward average paler, but much variation within regions.

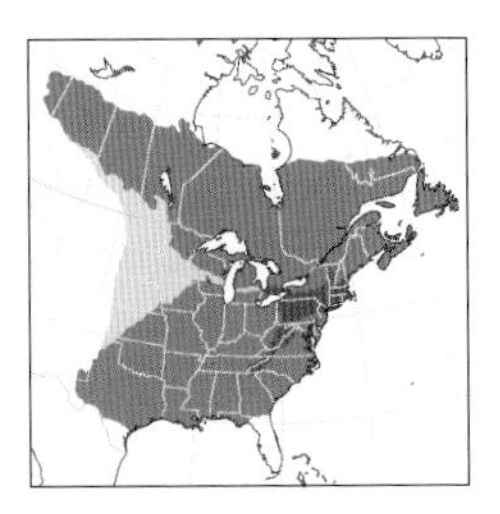

### Winter Wren

*Troglodytes hiemalis* | WIWR | L 3½" (9 cm)

This tiny bird has an incredible song. Winter Wrens are practically mouselike as they dart about streamside tangles at ground level.

■ **APPEARANCE** A little brown bird with tiny tail often cocked straight up. Darker overall with bolder barring on flanks than on larger House Wren.

■ **VOCALIZATIONS** Song a long series of tinkling trills and twitters, high-pitched and musical. Call a nasal *chimp,* similar to Song Sparrow's (p. 354), but Winter Wren's usually delivered in incessant doublets.

■ **POPULATIONS** Hardy; winters north to Great Lakes and New England.

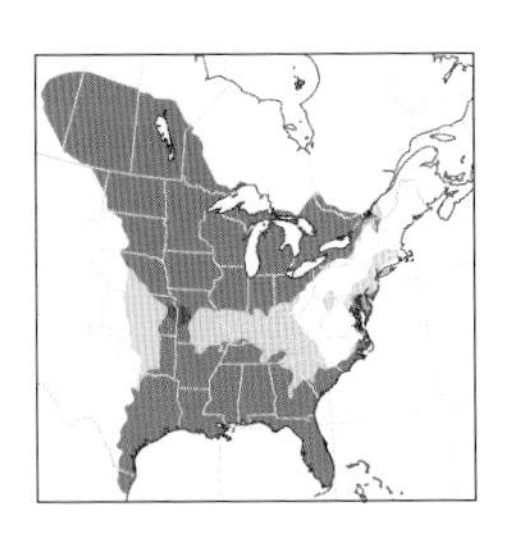

### Sedge Wren

*Cistothorus stellaris* | SEWR | L 4½" (11 cm)

Even more reclusive than the closely related Marsh Wren. Well named, the Sedge Wren haunts wet meadows with sedges and graminoid vegetation.

■ **APPEARANCE** Smaller and slighter than Marsh Wren, with striped (not solid) crown; wings boldly barred and checkered. In fleeting glimpse, note bright tan tones overall; flushing, appears ruddy-tailed.

■ **VOCALIZATIONS** Song comprises bright *chap* notes, accelerating. Call, like a single element from song, cleaner and clearer than Marsh Wren's.

■ **POPULATIONS** Often migrates within breeding season to start second brood; despite much annual variation, numbers stable in the long term.

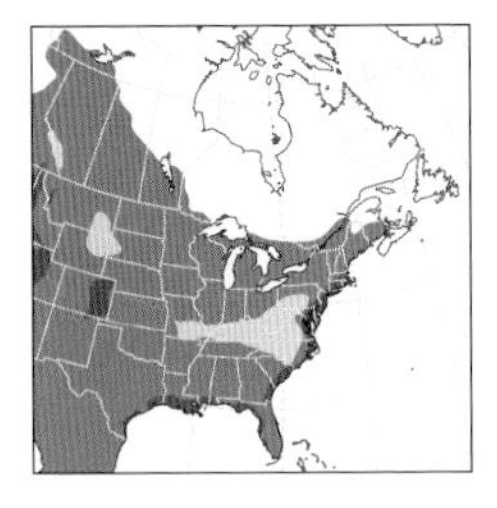

### Marsh Wren

*Cistothorus palustris* | MAWR | L 5" (13 cm)

This chatterbox of the cattails sings round the clock and throughout much of the year. Marsh Wrens stick close to the mud, venturing briefly atop cattail stalks, then retreating to ground level.

■ **APPEARANCE** A wren of average proportions; dark cap contrasts with white eyebrow. Back dark with white streaks; wings dark rufous. Juvenile duskier and lower-contrast, inviting comparison with House Wren.

■ **VOCALIZATIONS** Gurgling song has run-on quality; starts with a few stutters, then rambles on; sings at night. Call harsh and scratchy, *chit.*

■ **POPULATIONS** Regional differences in song; dividing line in Great Plains. Southern coastal plain subspecies ("Worthington's Wren") dull and pale.

juvenile
averages paler, grayer to the west
western
parkmanii
House Wren
aedon
indistinct head pattern
medium length tail
Winter Wren
short tail held vertically
pale eyebrow
brown-barred belly
boldly barred wings
soft eyebrow
bright tan overall
Sedge Wren
western
plesius
western birds slightly paler
Marsh Wren
bold black and white streaks on back
bright rufous wings and tail
white eyebrow
eastern
palustris
grayish and pale
southern Atlantic coast
griseus

**BOLDLY MARKED WRENS**

They're in three different genera, but this trio shares several points in common: They have well-marked supercilia (eyebrows), their tails are notably long, and they are mostly southern. But their wondrous songs are different as can be.

## Carolina Wren

*Thryothorus ludovicianus* | CARW | L 5½" (14 cm)

Among the loudmouths of southern hardwood forests, this is one of the loudest. At first light, as well as in the heat of the afternoon, in summer and even in the dead of winter, Carolina Wrens are always vocalizing.

■ **APPEARANCE** In most of East, the largest wren; patterned like Bewick's, but warmer and proportionately shorter-tailed. Long white eyebrow prominent; rich earth tones also obvious.

■ **VOCALIZATIONS** Repetitious song bright and whistled: *cheerily cheerily cheerily*... and *mediator mediator mediator*.... Sometimes sing antiphonally, the female initiating a slow buzz while the male whistles away. Varied calls include a descending trill and a froglike stutter.

■ **POPULATIONS** Gradually expanding north; harsh winters cause diebacks at northern edge of range, followed by recovery within a few years.

## Bewick's Wren

*Thryomanes bewickii* | BEWR | L 5¼" (13 cm)

Birders joke that a "mystery" sound in the shrublands and open woodlands of the southern Great Plains (and farther west) is this bird, the great deceiver. Bewick's Wrens occur in backyards and brushy habitats, especially those with a trashy, decrepit character.

■ **APPEARANCE** Slender build and bold eyebrow suggest Carolina Wren, but gray-brown overall, never as warmly colored as Carolina. Long tail, edged white, expressively flipped sideways, constantly in motion; even among wrens, tail of Bewick's notably long.

■ **VOCALIZATIONS** Many songs include an inhaled-sounding buzz followed by a jangling trill, but variations are endless. Common call a growling, nasal, slightly rising *grree*.

■ **POPULATIONS** Birds at eastern edge of range redder than on southern Great Plains, but not as brightly rufous as Carolina Wren. Occurred north to Pa. and N.Y. well into 20th century, but now absent east of Appalachians and largely absent from Ohio River Valley. Reasons for decline essentially unknown; nest destruction by House Wrens (p. 302) a possibility.

## Cactus Wren

*Campylorhynchus brunneicapillus* | CACW | L 8½" (22 cm)

Nearly twice the mass of the good-size Carolina Wren, the Cactus Wren calls to mind a thrasher (pp. 306–309). The nests are strikingly conspicuous—but almost impossible to get close to, ingeniously protected by thorns and cactus spines.

■ **APPEARANCE** Large and spotted, with sturdy legs and bill, and a broad tail. Eyebrow prominent; only wren in East with dense spots on breast. Adult mostly brown, black, and white; juvenile buff below. Flying away, with white-cornered tail splayed out, suggests Sage Thrasher (p. 308).

■ **VOCALIZATIONS** Song a rapid series of harsh *chuh* or *chugga* notes, increasing in loudness; sounds like an old car starting up in the distance. Calls varied, mostly rough and monosyllabic.

■ **POPULATIONS** At home in subdivisions and remote desert, but thorny plants required. Nonmigratory; rarely wanders from core range.

bold
eyebrow

rich rufous
and buff

**Carolina Wren**

long tail with
white barred
corners

bold eyebrow

drab gray-brown
above

**Bewick's Wren**

**western**
*eremophilus*

**eastern**
*bewickii*

pale gray
below

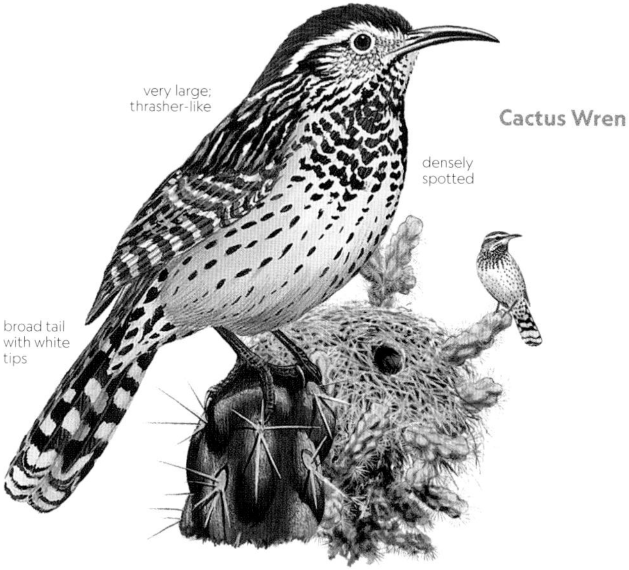

very large;
thrasher-like

**Cactus Wren**

densely
spotted

broad tail
with white
tips

## MIMIDS I

These midsize to large songbirds are slim with long tails, and several have notably curved bills. They are fantastic vocalists, and some are accomplished mimics.

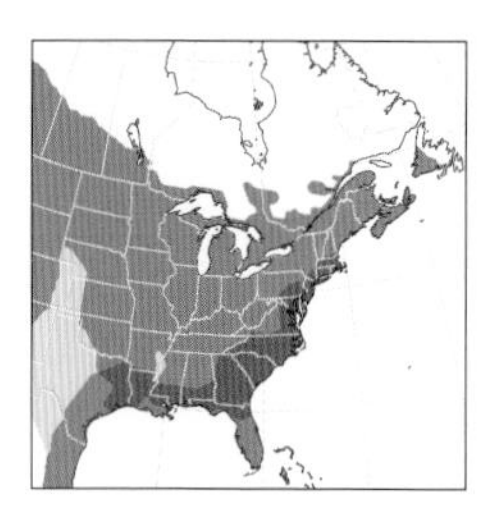

### Gray Catbird

*Dumetella carolinensis* | GRCA | L 8½" (22 cm)

Among wooded tangles, the catbird is one of the most common birds in the East. It is at home in southern swamps, in coastal thickets on Long Island, even in wooded canyon country at the western periphery of our region.

■ **APPEARANCE** Slim, long-tailed, and gray, with black cap and blackish tail; undertail coverts deep rufous. Skulks in thickets, where it appears all-dark.

■ **VOCALIZATIONS** Famous call a descending, nasal *rrwaaaa,* like an irritated cat; also a muffled *chutch* and a harsh cackle. Song a jumble of squeaks, squawks, and rich whistles.

■ **POPULATIONS** Many migrate to tropics, but many others stay behind, especially coastally; winterers in the north subsist where berries abound.

### Curve-billed Thrasher

*Toxostoma curvirostre* | CBTH | L 11" (28 cm)

This is the easternmost representative of a "desert thrasher" complex concentrated in the southwestern U.S. and northwestern Mexico. Its range and habitats in the East broadly match those of the Cactus Wren (p. 304).

■ **APPEARANCE** Bill of adult long and decurved; bill of juvenile can be much shorter. Eyes yellow-orange. Dirty olive-brown; breast diffusely spotted.

■ **VOCALIZATIONS** Song a disorganized, halting warble, a bit like Gray Catbird's, but less wild and squeaky. Call an explosive *WEET! WEET! WEET!*

■ **POPULATIONS** Considered nonmigratory, but some dispersal following nesting. Range has slowly spread northward since mid-20th century.

### Brown Thrasher

*Toxostoma rufum* | BRTH | L 11½" (29 cm)

This species combines the looks of the Wood Thrush (p. 316) with the body structure and melodies of a mockingbird (p. 308).

■ **APPEARANCE** Only thrasher in much of East; superficially similar Wood Thrush smaller and plumper with shorter tail, dark eye, and black spots (not brown streaks) below. Compare with Long-billed Thrasher.

■ **VOCALIZATIONS** Run-on song suggests Northern Mockingbird, but Brown Thrasher sings elements in pairs, unlike mockingbird. Call a smacking *tchack,* like a loud Fox Sparrow's (p. 344).

■ **POPULATIONS** Range expanded in 20th century but has since stabilized. Migrates fairly early in spring.

### Long-billed Thrasher

*Toxostoma longirostre* | LBTH | L 11½" (29 cm)

This Tamaulipan counterpart of the Brown Thrasher is a lover of thickets and tangles, especially around streams and ponds. Its habits and behaviors are similar to those of its northern relative.

■ **APPEARANCE** Size and body shape like Brown Thrasher's, but has longer bill and deeper orange eyes. Long-billed less colorful than Brown; note especially cold gray face and black-and-white underparts.

■ **VOCALIZATIONS** Song squeakier than Brown's, with fewer doubled elements. Smacking call like Brown's; also gives crowlike *rraaww,* puppy-like yelp, and descending whistle *rreeer.*

■ **POPULATIONS** Range expanding; now occurs to north of San Antonio, Tex.

Gray Catbird
black cap
overall steel gray
gray-brown above
diffuse spots
Curve-billed Thrasher
small white tail corners
bright rufous above
Brown Thrasher
long tail
off-white below with dark brown streaks
duller brown above than Brown Thrasher
gray face
white below with black streaks
Long-billed Thrasher

## MIMIDS II

Two go by the name mockingbird, and the third might as well. The Sage Thrasher, formerly known as the Mountain Mockingbird, is more closely related to mockingbirds than to *Toxostoma* thrashers (p. 306).

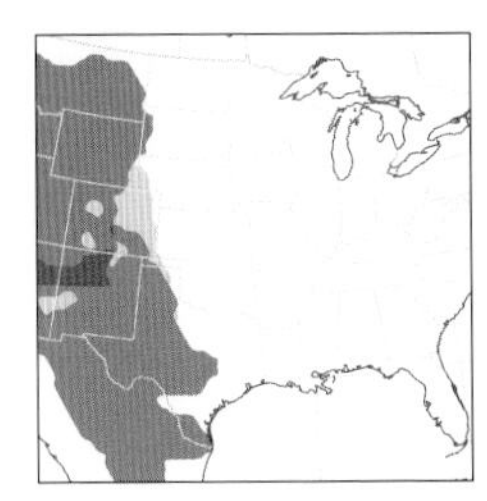

### Sage Thrasher

*Oreoscoptes montanus* | SATH | L 8½" (22 cm)

Although restricted on the breeding grounds mostly to sagebrush-dominated habitats, this species is encountered in the East chiefly in fall and winter in a diversity of open habitats.

■ **APPEARANCE** Small and stout for a thrasher, with slight build and unimpressive bill. On adult in fresh fall plumage, sharp arrow marks on underparts are striking, but plumage becomes quite worn by the following summer, at which time it resembles a small, short-billed Curve-billed Thrasher (p. 306). Juvenile like adult but grayer above with more diffuse markings below.

■ **VOCALIZATIONS** Song a long, run-on warble; delivery is random, but not as herky-jerky as other thrashers', with little variation in pitch throughout. Calls soft; include muffled *chuck* and whistled *churr.*

■ **POPULATIONS** Migrates early in spring. Wanders well east, reaching Atlantic seaboard most years; vagrants recorded primarily fall to winter.

### Bahama Mockingbird

*Mimus gundlachii* | BAMO | L 11" (28 cm)

Resident in the Bahamas and nearby islands in West Indies; vagrant to Fla., annual in recent years to Miami region, with multiple occurrences farther north on both Fla. coasts. Larger, duskier, and streakier than Northern Mockingbird; wings lack white flash of Northern, and white on tail only at tip. Brings to mind an oversize Sage Thrasher, but ranges do not overlap. Song rich and rambling like other thrashers' but also distinctive, with most elements burry and down-slurred; unlike its famous mainland counterpart, does not habitually imitate other birds.

### Northern Mockingbird

*Mimus polyglottos* | NOMO | L 10" (25 cm)

Other birds in the East may sing subjectively more beautiful songs, but the Northern Mockingbird has got to be the most talented vocalist of them all.

■ **APPEARANCE** Long and slender; gray overall. Adult has white in wings and tail, striking in flight, and thin black eyeline; compare with shrikes (p. 272). Adult plain below with golden eye; juvenile, with darker eye, spotted below, suggesting Sage Thrasher. Generally animated; prances around, flashing wings. Even without the astonishing song, a terrific bird.

■ **VOCALIZATIONS** Song impressive, sometimes going on for a quarter hour, right through the middle of the night (and all day long too). Individual song elements repeated three to six times, with amazing imitations: Single birds skillfully impersonate 20+ other species, exceptionally up to 30, in high-fidelity imitations of their songs, also repeated three or more times. The song is simply astounding. Calls include a powerful *tsuck,* a descending *tssseeep,* and a harsh *rrrwwuuh;* all calls sound annoyed.

■ **POPULATIONS** Defends winter feeding territories, unusual among eastern songbirds; particularly proprietary about stands of invasive *Rosa multiflora.* Many withdraw from northern portion of range, especially Midwest and Great Plains.

short, straight bill

## Sage Thrasher

crisply streaked, becoming indistinct in late summer

**worn**

dark moustache

## Bahama Mockingbird

mostly dark tail and wings

brown-tinged and streaky overall

small white tips

## Northern Mockingbird

**juvenile**

large white wing patches

long tail with white edges

## STARLINGS AND MYNAS

Old World starlings and mynas (family Sturnidae) have a long history of contact with human cultures. Introduced populations in the East are particularly conspicuous in built-up urban areas.

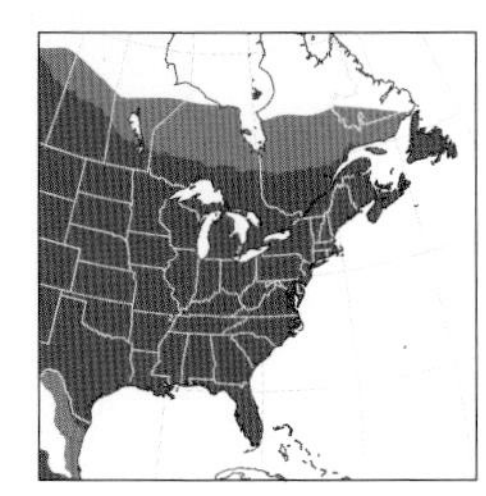

### European Starling

*Sturnus vulgaris* | EUST | L 8½" (22 cm)

Introduced to New York City in the late 1880s, the starling quickly swept across the lower 48 and southern Canada. Starlings are notorious for usurping the nests of indigenous cavity-nesting birds, but they are undeniably beautiful and industrious.

■ **APPEARANCE** Dark and compact, often appearing all-black; tail short, legs sturdy, bill fairly long. Distinctive in flight, with outspread wings triangular. Adult in fresh fall plumage black with bright white spots; bill black fall and winter. White spots wear off during the colder months, and adult's feathers become glossy purple and green; bill yellow by spring. Sexes differ in color of base of bill, spring and summer: blue in male, pink in female. Fresh juvenile dull dusky gray-brown, but proportioned like adult. In transition to adult plumage, a mosaic of juvenile and adult feathers.

■ **VOCALIZATIONS** Fantastically varied; almost as expert as Northern Mockingbird (p. 308). Most songs and calls squeaky and high-pitched, many with squealing, tinny quality. Remarkable full-on song, with rising whistles interspersed with clacks and rattles. Readily imitates other birds, especially those with whistled calls: Killdeer (p. 100), Sora (p. 90), Eastern Wood-Pewee (p. 254), etc. Call a squeaky *cheek,* often in quick series, and rougher *chuuh;* flight call a purring *zzzzzzrrr.* Juvenile begs with downslurred *jeeeuurrr.*

■ **POPULATIONS** Especially in nonbreeding season, gathers in immense roosts at bridges, warehouses, and such. Starlings migrate spring and fall, even within portions of range in which the species is present year-round.

### Common Hill Myna

*Gracula religiosa* | CHMY | L 10½" (27 cm)

Indigenous to Southeast Asia; introduced and established around Miami, where declining. Current holdout in Coral Gables, where it breeds in well-wooded districts with woodpecker cavities for nesting. Bulkier, blacker overall than Common Myna. Bright orange bill quite large; yellow wattles on face and nape variable but usually obvious. Remarkable song has artificial, electronic quality, with loud, fairly low-pitched, clanging and piercing notes: *zweep! leoooong laaaang whee-urr,* and so on, like a synthesizer. Captive birds habitual mimics; birds in the wild not as much.

### Common Myna

*Acridotheres tristis* | COMY | L 10" (25 cm)

Handsome but worrisome—that's the gist of the Common Myna situation in the East. The species, indigenous to South Asia and recently established in Fla., is notably aggressive and invasive.

■ **APPEARANCE** Smaller and browner than Common Hill Myna; white on wings and tail inconspicuous on perched bird, prominent in flight. Exposed skin around eye yellow; bill also yellow. More of a city bird than Common Hill, in flocks around strip malls, boulevards, etc.

■ **VOCALIZATIONS** Varied calls mostly sharp and whistled, often run together in short series, occasionally with imitations of other birds.

■ **POPULATIONS** Established in Fla. late 20th century; quickly increased, but range expansion may have stalled out.

long
straight bill
fall
European
Starling
triangular
wings
winter
short,
square
tail
drab,
but with
distinctive
structure
breeding ♂
juvenile

heavy
orange bill
Common Hill
Myna
striking
white in wings
and tail
short
yellow
bill
Common
Myna

## BLUEBIRDS AND A SOLITAIRE

Thrushes, family Turdidae (pp. 312–319), are a diverse assemblage of frugivorous and insectivorous birds with notable songs. Bluebirds and solitaires are closely related within the thrush family.

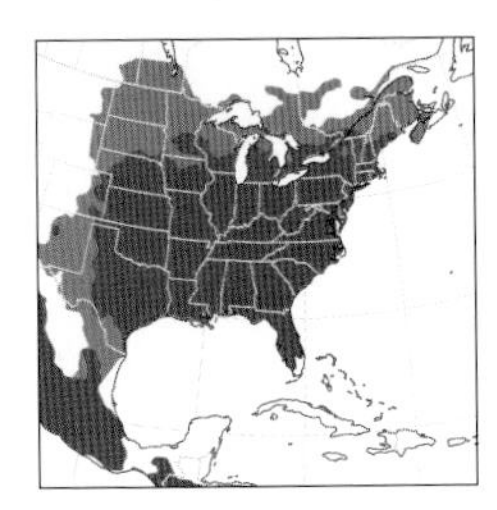

### Eastern Bluebird

*Sialia sialia* | EABL | L 7" (18 cm)

A fabled icon of happiness, harbinger of spring, and metonym for clear skies, this species is at home around farms, forest edges, and old pastures.

■ **APPEARANCE** Small and plump. Adult male bright blue above, orange-red below; female grayer above, dusky orange below. Heavily spotted juvenile mostly brown, with some blue already showing in tail and wings.

■ **VOCALIZATIONS** Song, a simple but joyous gurgle. Most often detected by loud, twangy, musical flight call, *cheer-lee* or *chur-lee*.

■ **POPULATIONS** The only bluebird normally expected in much of East. Declined in 20th century; rebounded due to nest-box provisioning.

### Western Bluebird

*Sialia mexicana* | WEBL | L 7" (18 cm)

Darker counterpart in the West of Eastern Bluebird; belly blue-gray (white on Eastern). Male has darker, violet-tinged blue hood, without rufous intrusions to throat and neck as on Eastern; unlike Eastern, dark rufous extends to scapulars and back. Female has grayish throat (whitish on Eastern). Annual fall and winter to western Kans., Okla. Panhandle, and central Tex., especially near fruiting trees; mostly quiet in winter, but a few give muffled *pew* and *pyoof* notes. Unlike Mountain, almost never vagrates well east.

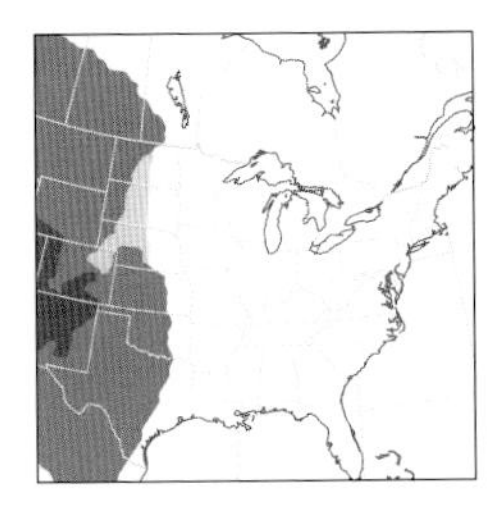

### Mountain Bluebird

*Sialia currucoides* | MOBL | L 7¼" (18 cm)

The male of this species is as brilliant as a shard of sky. Despite the name, occurrences in our area, mostly of singles and small flocks, are on the plains.

■ **APPEARANCE** Long-winged; thin-billed. Adult male intense sky blue; perched on wires, may seem camouflaged against the sky. Female much grayer, with blue highlights in wings and tail. Many in winter suffused rusty below, suggesting Western; Western has stouter bill and shorter wings.

■ **VOCALIZATIONS** Calls, weak and muffled, intermediate in timbre between ringing Eastern's and gruff Western's.

■ **POPULATIONS** Early spring migrant; wanders widely, fall through winter, in southern Great Plains. Winter range, delimited in part by availability of tiger moth caterpillars, shifting north as climate warms.

### Townsend's Solitaire

*Myadestes townsendi* | TOSO | L 8½" (22 cm)

It's fall or winter in the western Great Plains, and a gray bird is perched upright on a tree in a shelterbelt. Then it puts up in twisting flight and flutters in front of a fruiting juniper. The bird is almost assuredly a solitaire.

■ **APPEARANCE** Slender overall; long-tailed and small-headed. Buff-orange in wings inconspicuous on perched bird, prominent in flight. Gray tones, eye ring, and wing markings may invite confusion with wood-pewees (p. 254) and *Empidonax* flycatchers (pp. 256–261).

■ **VOCALIZATIONS** Call a whistled *eee*, repeated in long, slow sequence; flight call a thin, drawn-out buzz. Song finchlike, a halting, singsong warble.

■ **POPULATIONS** Breeds widely in western mountains, disperses to western Great Plains in fall; annual vagrant much farther east. Aggressive toward conspecifics year-round; both sexes defend feeding territories in winter.

Eastern Bluebird
all juvenile bluebirds heavily spotted
juvenile
rufous wraps around sides of neck
♀
stout bill
♂
Western Bluebird
gray neck
♀
blue throat
♂
usually grayish, occasionally more rusty below
Mountain Bluebird
thin bill
♀
♂
juvenile
white eye ring
Townsend's Solitaire
long tail with white edges
buff wing markings

## *CATHARUS* THRUSHES

Thrushes in the genus *Catharus* (pp. 314–317), along with the closely related Wood Thrush (p. 316) in the genus *Hylocichla*, are among the most ethereal songsters in the East.

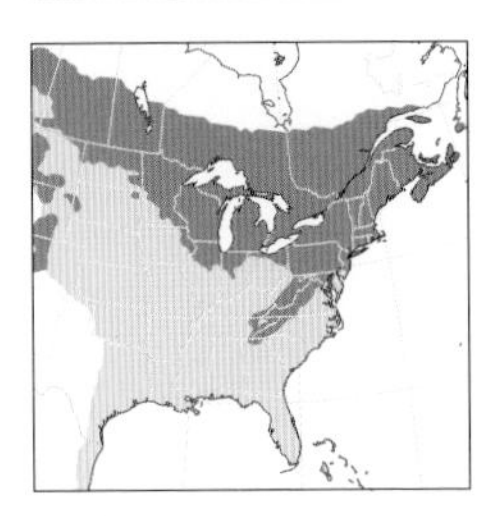

### Veery

*Catharus fuscescens* | VEER | L 7" (18 cm)

An old name for this bird is Willow Thrush, indicating the species' preference for wet thickets. But Veery, from the bird's call, is at least as good.

■ **APPEARANCE** Plainest and warmest *Catharus* thrush in East. Uniform warm brown above; plain face has faint broken eye ring like Gray-cheeked's. Below, warm butterscotch wash across lightly spotted breast; belly white.

■ **VOCALIZATIONS** Song starts with a gruff rising note (*rrwaa?*), followed by slurred *ree-ur* notes that cascade downward: *rrwaa? ree-ur ree-ur ree-ur ree-ur.* Low-pitched flight call rough and burry: *vrruurrr.*

■ **POPULATIONS** Powerful flier; winters mostly southern Brazil, migrates in broad front over Caribbean Sea and Gulf of Mexico.

### Gray-cheeked Thrush

*Catharus minimus* | GCTH | L 7¼" (18 cm)

Lucky birders will see one of these long-distance migrants in the forest understory, on or close to the ground, during stopovers in spring and fall.

■ **APPEARANCE** Coldest and grayest of eastern *Catharus* thrushes; breast densely spotted, eye ring broken. Swainson's brighter and buffier on breast and face.

■ **VOCALIZATIONS** Song, descending weakly, comprises one or two rough notes followed by slurred phrases, thin and nasal: *uh uh ssweew sweew sweeah sweeah sweeeurp.* Buzzy flight call an urgent *queeerrr!*

■ **POPULATIONS** Incredible migrant, with some crossing Bering Strait to nest in Russia; night flights, especially in fall, impressive.

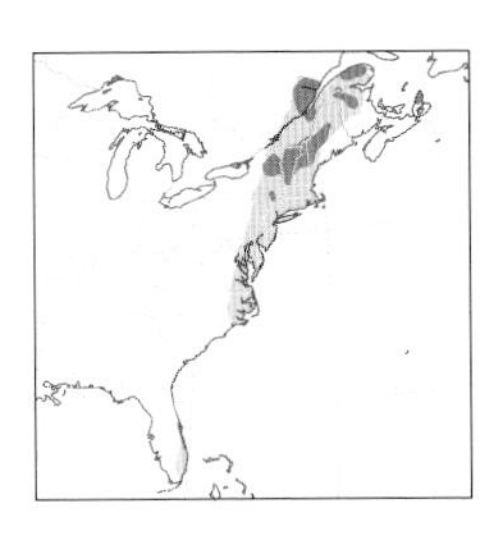

### Bicknell's Thrush

*Catharus bicknelli* | BITH | L 6¼" (16 cm)

This sought-after thrush, closely related to the Gray-cheeked, nests in windswept conifer forests near timberline.

■ **APPEARANCE** Very similar to Gray-cheeked, but a bit smaller, with warmer red-brown tail. When separating a presumptive Bicknell's from Gray-cheeked, make sure the bird isn't "just" a Hermit Thrush (p. 316)!

■ **VOCALIZATIONS** Song like Gray-cheeked's, but final element in Bicknell's typically rises. Flight call higher and longer than Gray-cheeked's.

■ **POPULATIONS** Winters in Caribbean; poorly documented migration is along Atlantic seaboard.

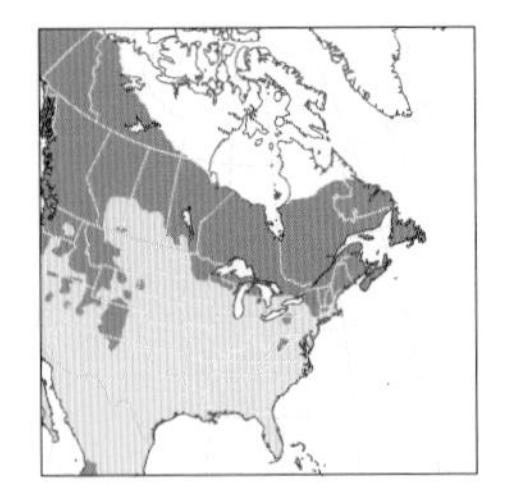

### Swainson's Thrush

*Catharus ustulatus* | SWTH | L 7" (18 cm)

This species emblematizes the bewitching phenomenon of nocturnal migration—especially on chilly nights in early autumn.

■ **APPEARANCE** Olive-brown upperparts contrast with warm buff wash on breast; buffy eye ring and lores create "spectacles." Compare with Gray-cheeked's plain gray face, vague eye ring, cold gray breast.

■ **VOCALIZATIONS** Song spirals upward, intensifying: *eewee eweee eweeah eeweewee eeeweewee! aurelia!* Daytime call note a resonant *pilp,* like a water droplet; nocturnal migrants give a mellow, far-carrying, rising *pweee.*

■ **POPULATIONS** Common breeder in wet northern woods; apt former name, Mosquito Thrush. Widespread on migration, sometimes away from woods.

Veery
duller brown than eastern
western
salicicola
bright reddish brown above
lightly marked face and breast
eastern
fuscescens
gray flanks
Gray-cheeked Thrush
cool gray-brown upperside
1st fall
gray face with indistinct eye ring
Newfoundland
minimus
boreal
aliciae
gray-brown flanks
slightly warmer brown
slightly smaller and warmer brown than Gray-cheeked
Bicknell's Thrush
buffy eye ring and lores
olive-brown above
buff wash to breast and face
Swainson's Thrush

**TWO SPOT-BREASTED THRUSHES**

All our thrushes (pp. 312-319) are spotted or scaled as juveniles, but only our *Catharus* (pp. 314–317) and *Hylocichla* thrushes have spots as adults. For this reason, birders often group them together as the spot-breasted thrushes.

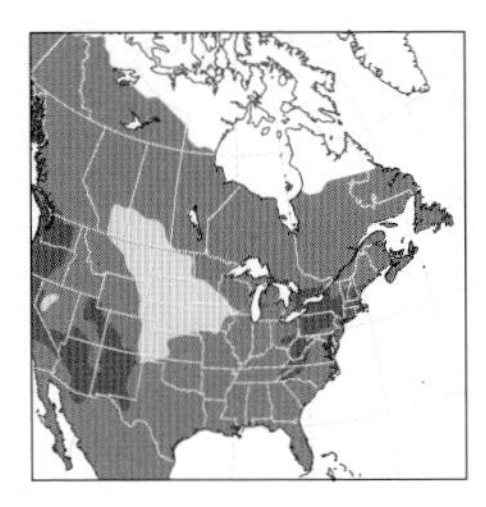

## Hermit Thrush

*Catharus guttatus* | HETH | L 6¾" (17 cm)

A small brown bird in the shadows slowly raises, then quickly lowers, its tail. That behavior alone strongly suggests a Hermit Thrush. And if there's enough light to make out the tail's rufous hue, that clinches it.

■ **APPEARANCE** Built like other *Catharus* thrushes (pp. 314–317), but habit of raising and lowering tail distinctive. Eye ring thin but prominent; breast and throat have small black spots. Most in East warm reddish brown, broadly converging on color scheme and pattern of Fox Sparrow (p. 344). Interior West subspecies, regular on migration far western Great Plains, grayer above and on flanks. An underappreciated field ID problem is Gray-cheeked (p. 314) vs. Interior West Hermit; check for the rufous tail, raised then lowered, of Hermit.

■ **VOCALIZATIONS** Even among *Catharus* thrushes, Hermit is lauded for its musicality. Haunting song opens with achingly pure whistle, then goes into an airy, tinkling jumble like wind chimes—each song different from the preceding one. Flight call, heard at night, a simple, monotone whistle; by day, gives hollow *tuck* and rising, whining *rrweeeen*.

■ **POPULATIONS** Earliest *Catharus* thrush in spring, latest in fall. The only spot-breasted thrush expected in winter in East.

## Wood Thrush

*Hylocichla mustelina* | WOTH | L 7¾" (20 cm)

In lowland hardwood forests in the East, this is often the only spot-breasted thrush in summer. On migration, especially in fall, it joins with the more northerly *Catharus* thrushes (pp. 314–317) in stirring night flights.

■ **APPEARANCE** Larger and plumper than *Catharus* thrushes. Rich rufous-brown above; white underparts boldly spotted black. Compare with Ovenbird (p. 376), which inhabits forest understory, and Brown Thrasher (p. 306), of thickets and tangles.

■ **VOCALIZATIONS** Song, rich and powerful, typically starts with two short twangy notes, then a complexly slurred whistle, then a wavering trill: *ur-ur EEEooooAAAY bzzzzzzz*. Like Hermit Thrush, never sings the same song in succession. Diurnal calls include an explosive *yip! yip! yip!* and low, murmuring *prr prr prr prr;* migrants at night give a harsh, abrupt *breeeh!*

■ **POPULATIONS** Declining; sensitive to cowbird parasitism (p. 370), forest fragmentation, and, more recently, ruin of woodlands by white-tailed deer.

### Nocturnal Flight Calls

Few phenomena are more mesmerizing than the unseen, but eminently audible, nocturnal migration of land birds over eastern North America. On cool nights around the autumn equinox, one can hear hundreds of migrants per hour—a mighty river of birds carrying energy and biomass from the boreal forest to the Neotropics. Biologist Bill Evans, who pioneered the field ID of birds by their nocturnal flight calls, has famously remarked that a person might hear more Gray-cheeked Thrushes in one hour on nocturnal migration than one might see in a lifetime of daytime bird-watching. But how does one identify unseen migrants in the dark?

Well, Evans and other experts have made it easy: Go to *oldbird.org* (created by Evans and ID expert Michael O'Brien), and navigate to the

"Spot-breasted Thrushes" (genera *Catharus* and *Hylocichla*) are heard far more often than seen; their songs and even their nocturnal flight calls carry well and are easily learned.

species index, which has curated recordings and spectrograms of the flight calls of all eastern land birds known to give them. The rapidly burgeoning libraries of sound recordings at Xeno-Canto and eBird are also excellent: Search for "flight call" to bypass songs and other vocalizations.

In the field, the spot-breasted thrushes are a superb starting point for learning flight calls. Their calls are loud, low-pitched, and, with some practice, readily separated. They give flight calls by day, although infrequently, so that a bird of known identity may be compared with its unseen counterparts at night. Always keep in mind that identifying monosyllabic utterances in the dark is perforce somewhat conjectural, but don't let that interfere with the enjoyment of one of the most enchanting and awe-inspiring experiences in all of nature study.

## LARGE THRUSHES

These three are bruisers, more than twice the mass of bluebirds (p. 312) and *Catharus* thrushes (pp. 314–317). The closely related Clay-colored Thrush and American Robin are in the worldwide genus *Turdus*, while the Varied Thrush gets its own genus, *Ixoreus*.

### Clay-colored Thrush

*Turdus grayi* | CCTH | L 9½" (24 cm)

Why is the modestly attired Clay-colored Thrush the national bird of Costa Rica, home to visually stunning quetzals and motmots and toucans? It's all about the song.

■ **APPEARANCE** Proportioned like familiar American Robin, but dull and drab. Fairly pale in good light, but appears dark in favored woodland haunts. Throat streaked brown and white, but otherwise plain-faced. Eye dull red, bill dull yellow.

■ **VOCALIZATIONS** Song slow and languid, with rich, halting whistles; more restrained than American Robin.

■ **POPULATIONS** Neotropical species that reaches north to South Tex.; apparently increasing there for close to a century.

### American Robin

*Turdus migratorius* | AMRO | L 10" (25 cm)

The robin is so familiar that it is easy to take the species for granted. But it is one of our greatest birds, with extraordinary migrations, an incredible song, and endlessly fascinating feeding and breeding behaviors.

■ **APPEARANCE** Large thrush with rufous below (thus "robin redbreast"), dark on males, orangey on females; juveniles, common and conspicuous mid to late summer, heavily spotted below. All have white marks on darker head; red or orange color below extends to underwing coverts, obvious in flight. Plumage varies considerably with age, sex, and feather wear.

■ **VOCALIZATIONS** Song loud, complex, and exuberant: a series of four to eight slurred whistles, followed by a pause, then another series, the whole performance running on for 5–10 minutes. Has many calls, including a descending *WHEE WHEE whew whew whew whew,* a scolding *tut tut tut,* an explosive *PLEEP,* and, in flight, a buzzy *dzeeeet.*

■ **POPULATIONS** Common to abundant in New York City and Chicago, in wilderness in the Ozarks to Labrador, and elsewhere. Most subspecies migratory, some impressively so; movements by night in fall along Atlantic coast can be spectacular. Devours worms in summer, berries in winter; readily nests on houses and other structures, affording close-up study of breeding biology.

### Varied Thrush

*Ixoreus naevius* | VATH | L 9½" (24 cm)

A beautiful denizen of coniferous forests to our west; although most stay in the West, a few wander well east every winter, especially from upper Midwest to New England. Stockier and shorter-tailed than American Robin, with high-contrast orange-and-dark plumage overall. Adult male has striking black breastband and face mask; breastband and face mask duskier on adult female. Both sexes have extensive orange in wings, prominent in flight. Weird, "electronic" song not given by wanderers eastward, but birds in winter sometimes give a hollow, down-slurred *tsooh* or *chook.*

unmarked
brown head and
upperparts
greenish bill
Clay-colored
Thrush
heavily
spotted
juvenile
small white
tail corners
broken white
eye ring
♀
American
Robin
variable;
some much
paler or darker
♂
juvenile
orange
eyebrow
♀
♂
Varied
Thrush
dark
breastband
striking wing
pattern

## WHEATEARS AND OLD WORLD SPARROWS

The families Muscicapidae (including wheatears) and Passeridae (including House and Eurasian Tree Sparrows) occur mostly in the Old World. The former are related to thrushes (pp. 312–319), while the latter are allied with pipits (p. 322) and nine-primaried oscines (pp. 324–409).

### Northern Wheatear

*Oenanthe oenanthe* | NOWH | L 5¾" (15 cm)

A restless denizen of dry tundra, this species breeds in our area widely but sparingly across the eastern Canadian Arctic. The bird's odd name derives from the Old English words "white" and "arse," denoting the white rump.

■ **APPEARANCE** Small and long-legged, sticks close to ground; portly build and active demeanor impart distinctive gestalt. All plumages show white rump and contrastingly patterned tail; adult male boldly marked, nonbreeders of both sexes less so.

■ **VOCALIZATIONS** Song a short, stuttering jumble. Calls short, sharp whistles and clucks. Vagrants in fall usually silent.

■ **POPULATIONS** Breeding grounds in East rarely visited by humans; very rare but annual in fall in northeastern U.S. and Atlantic Canada. Eastern subspecies an astonishing migrant, with most overflying Atlantic Ocean to Africa.

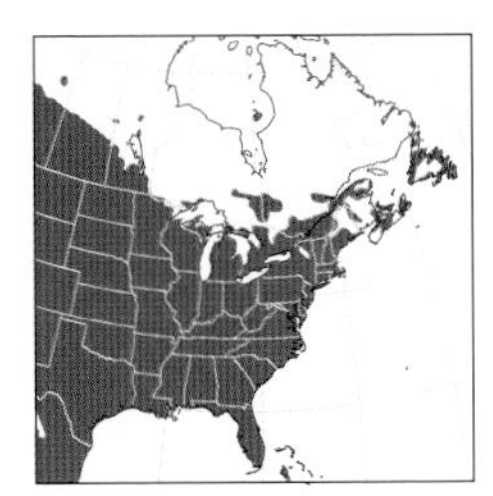

### House Sparrow

*Passer domesticus* | HOSP | L 6¼" (16 cm)

A textbook example of a commensal organism, the House Sparrow is always found near humans. It flourishes in huge metropolises, around barns and grain elevators, and at roadside rest stops everywhere.

■ **APPEARANCE** Cheery and sociable; compact, with relatively short wings and tail. Breeding male has whitish cheeks, gray crown, chestnut nape, and black bill and bib; as with European Starling (p. 310), spiffy breeding garb paradoxically acquired through feather wear. Sandy brown female has broad buffy eyebrow, white wing bar, and plain breast.

■ **VOCALIZATIONS** Song sounds like monotonous chirps to humans, but each utterance caroled like a robin; however, our human brains can't distinguish these fast, complicated sounds and process the variation and musicality in the song. Calls include a muffled *chiff,* nasal *laah* or *laah-laah,* and dry rattle.

■ **POPULATIONS** Introduced from Europe in 19th century; quickly spread, but numbers declining of late. House Sparrows in the Americas have diverged morphologically from those in Eurasia, an important result for evolutionary biology.

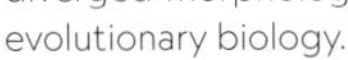

### Eurasian Tree Sparrow

*Passer montanus* | ETSP | L 6" (15 cm)

Like the widely established House Sparrow, this species is a European import. Less commensal than the House Sparrow, the Eurasian Tree Sparrow is more inclined to suburban and agricultural districts than the urbanite House Sparrow is.

■ **APPEARANCE** Slightly smaller but longer-tailed than House Sparrow, with sexes similar. All have solid brown cap and white cheeks with black splotch.

■ **VOCALIZATIONS** Song like that of House Sparrow, but clearer and higher; calls include a rattling *tss'VVIT* and a dry *tik,* often doubled or trebled.

■ **POPULATIONS** Introduced in 19th century to St. Louis, Mo., area, but range hasn't expanded much; some wander far, though, with records to Canada and East Coast. Even in core range, often outnumbered by House Sparrows.

## Northern Wheatear

## House Sparrow

## Eurasian Tree Sparrow

## PIPITS

These ground-dwelling species are brown and streaky, suggesting New World sparrows (pp. 336–359). But they differ from sparrows in wing structure and bill shape; they also sound different and walk differently.

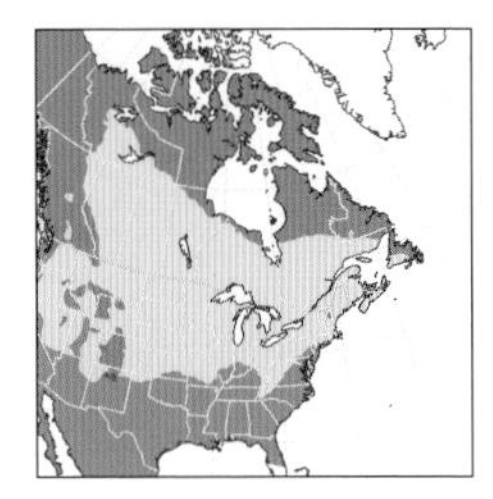

### American Pipit

*Anthus rubescens* | AMPI | L 6½" (17 cm)

The most widespread pipit in the Americas, this species is known in the East mostly as a migrant and winterer. Away from the breeding grounds, it favors beaches, muddy pastures, and lake edges.

■ **APPEARANCE** Slim and long-tailed; pumps tail as it walks. Freshly molted in fall, shows bold streaking below; by following spring and summer, plainer below with pumpkin-orange wash. Both pipits have long tertials, covering primary tips on bird at rest. In flight, flashes white edges of dark tail.

■ **VOCALIZATIONS** Flight call, often the first cue to the bird's presence, a sharp down-slurred *psleet* or *psleeit,* often doubled, *pipit.* Flight call of Horned Lark (p. 286) sweeter and thinner. Song a pulsing series given in flight low above the ground.

■ **POPULATIONS** Most in East breed on low-elevation tundra far north, but a few range south to highest mountaintops in New England—where conditions match those in the Arctic.

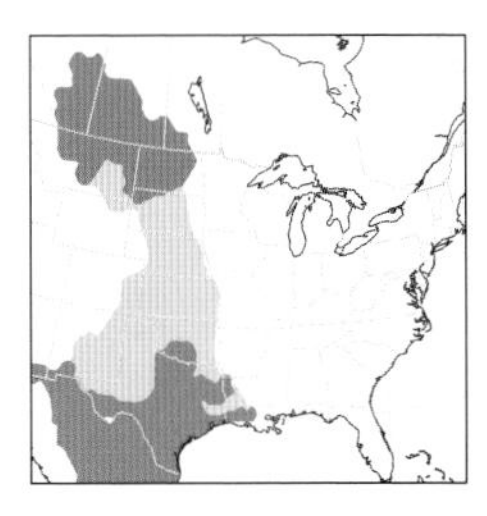

### Sprague's Pipit

*Anthus spragueii* | SPPI | L 6½" (17 cm)

England has its skylark, and N. Amer. has the Sprague's Pipit. The towering flight display and intense song of this plump passerine are among the greatest allures of prairie birding.

■ **APPEARANCE** Stockier and shorter-tailed than American Pipit. Beady black eye stares from plain face. Upperparts streaked, legs pink; American has plain back, dark legs. Flashes white in tail when flushing; does not pump tail when walking. Juvenile Horned Lark (p. 286), common where Sprague's occurs, similarly streaked and scaly; lark has dark legs, longer tail, and different calls. Compare also with juvenile Vesper Sparrow (p. 348).

■ **VOCALIZATIONS** Song, heard only on breeding grounds, gushes downward, a bit like Veery's (p. 314), but Veeries don't sing hundreds of feet high in the sky; skylarking Sprague's Pipits get so high that they are often hard to see. Call a sharp *pleet,* often doubled, typically given as bird flushes.

■ **POPULATIONS** Secretive away from breeding grounds; declining. Migrants and winterers, often in dense grass, are best detected by their flight calls.

### Flight Songs

Birdsong, all by itself, is one of the most wondrous things in the natural world. And the bird in flight is, for many nature lovers, the ultimate emblem of wildness. Put the two together—singing and flying—and you have something truly special.

Flight songs are performed across a broad range of taxa. The most famous performer of all is no doubt the Eurasian Skylark, so much so that it has an actual verb named after it. "Skylarking" is most common in birds of open country, with celebrated examples in the East including snipes and woodcocks (p. 122), longspurs (pp. 332–335), certain New World sparrows (pp. 336–359), and others.

Along with singing and being airborne, there is a third element to most flight displays: choreography. Not only are our two pipits' flight songs dissimilar, but their choreographies are also totally different. Sprague's Pipits rise to tremendous heights and just hang there for minutes at a time, so high up that they are difficult to see; in contrast, American Pipits sing while flapping furiously in short-duration, low-altitude flights above the tundra.

long
tertials
breeding
bobs tail
frequently
winter
"necklace"
of streaks
American
Pipit
long tail
with narrow
white edges
plain pale
face
Sprague's
Pipit
scaly back
long tertials
pink legs
short tail with
broad white
edges

spotted above with
"normal" tertials
short bill
juvenile
Horned Lark
for comparison
conical bill
strong face
pattern
dark legs
juvenile
Vesper Sparrow
for comparison

**A MEDLEY OF FINCHES**
The finches (pp. 324–331) trend toward northern latitudes, high elevations, and wintertime visitations. They eat plant matter, and many are flocky and nomadic.

## Evening Grosbeak

*Coccothraustes vespertinus* | EVGR | L 8" (20 cm)

They're endearingly known as "Flying Pigs" or "the Motorcycle Gang." A flock of voracious Evening Grosbeaks at a winter feeding station is a riot of sound and color.

■ **APPEARANCE** Large, stocky, big-billed finch. All have golden hues and dark wings with white patches. Adult male relatively dark, with yellow brow. Female paler overall, with gray on head and back, and apple-green bill.

■ **VOCALIZATIONS** Far-carrying, descending *CLEEP,* often given in flight. Milling around at feeders, gives burry *zrrrrt*.

■ **POPULATIONS** Expanded east in 1800s with planting of box elder trees, *Acer negundo,* but has withdrawn somewhat recently. Geographic variation complex, with multiple "types" possibly indicating different species.

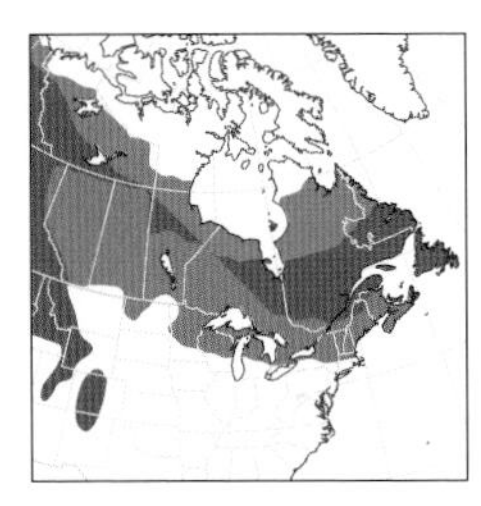

## Pine Grosbeak

*Pinicola enucleator* | PIGR | L 9" (23 cm)

This longest of our finches is probably also the tamest. Along trails in coniferous forests and at feeders, singles and small flocks are easily approached.

■ **APPEARANCE** Large and long-tailed, with stubby, rounded, black bill. Older males extensively red-pink with gray highlights. Females and first-year males variably clad in yellows and olives; some greenish females, with their slow movements and rounded bills, almost parrotlike.

■ **VOCALIZATIONS** A short, desultory warble, befitting this slow-moving bird: *t'leedle ... t'lweeooli ....*

■ **POPULATIONS** Like most finches, wanders after breeding, but irruptions generally weak; scarce, even in winter, south of U.S.-Canada border.

## Gray-crowned Rosy-Finch

*Leucosticte tephrocotis* | GCRF | L 5½–8¼" (14–21 cm)

The only rosy-finch to reach the East, Gray-crowned is annual but unpredictable in the western Great Plains and accidental farther east. Tends feeders. Like all "rosies," is dark with pink highlights. Head has variable gray, ranging from extensive on adult male "Hepburn's" subspecies to just a sliver on female and young of Interior West breeders. In flight, flashes pink and gray in wings; listen for muffled *chfff* and brighter *cheer* from birds on the wing.

### Birds at the Back of the Book

The ongoing "checklist shuffle," as bird-book aficionados have come to know it, started decades ago—and it is advancing at a particularly frenetic pace in the 21st century. And as if to save the best for last, taxonomic change has been perhaps the most extreme among the familiar groups that come at the end of North American field guides: finches, warblers, sparrows, tanagers, blackbirds, and so on.

As has long been established, all these groups are closely related. They make up an evolutionarily cohesive assemblage of nine-primaried oscines, so designated because they share the trait of presenting nine easily seen primary flight feathers. But it's within the nine-primaried oscines clade where so much of the taxonomic change has been happening. This has involved not only reassignments within families, but also many assignments to different families—and even, excitingly, the recognition of entirely new families.

Evening Grosbeak
♂
♀
striking white wing patch
massive ivory bill
adult ♂
juvenile ♂
white patches in wing
short tail
breeding ♀
♀
stubby blackish bill
white wing bars
adult ♂
variable amounts of pink
Pine Grosbeak
russet variant ♀
long, notched tail

Gray-crowned Rosy-Finch
"Hepburn's" ♂
*littoralis*
bill yellow in winter, becoming black in spring
interior ♀
*tephrocotis*
interior ♂
*tephrocotis*

## GENUS *HAEMORHOUS*

They're known as "American rosefinches," not to be confused with rosy-finches in the genus *Leucosticte* (p. 324). Older males are extensively red, and both sexes are gifted songsters.

### House Finch

*Haemorhous mexicanus* | HOFI | L 6" (15 cm)

Although it nests around houses and even much larger structures, this adaptable finch can be found far from human habitation in forests and deserts. The species is among the most common in the East.

■ **APPEARANCE** A bit sleeker, longer-tailed, and longer-winged than House Sparrow (p. 320). Plumage variation extreme. Typical adult male combines broad reddish eyebrow with concolorous "breastplate"; extent and hue of red variable, with some individuals orange and, more rarely, yellow. Younger males and most females dirty brown all over with blurry streaking below. Compare with Purple Finch.

■ **VOCALIZATIONS** Song, given year-round, a bright, twangy warble, typically comprising 12–20 elements; song often ends with a rising, nasal *chweeEER*. Up-slurred call, frequently given in flight, a rough *wriitt* or brighter *wreeet*.

■ **POPULATIONS** Indigenous to dry, open habitats of western N. Amer., but introduced to New York City around 1940; quickly spread across eastern U.S. More than most birds, House Finches are susceptible to a debilitating conjunctivitis, causing local mortality events.

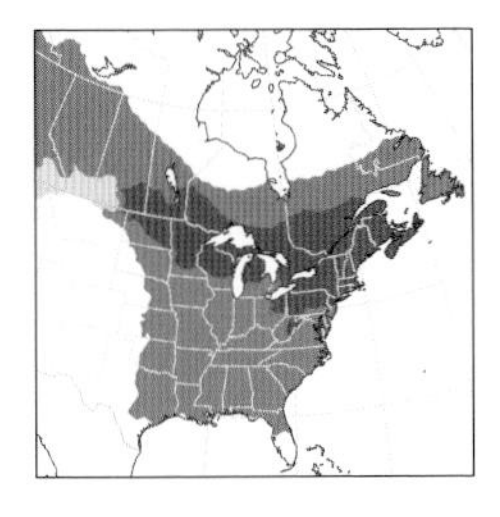

### Purple Finch

*Haemorhous purpureus* | PUFI | L 6" (15 cm)

This is the wilder and "finchier" counterpart in the East of the familiar House Finch. On the nesting grounds, Purple Finches gravitate toward woods with a coniferous component; outside the breeding season, their preferences are less specific.

■ **APPEARANCE** Similar in build to House Finch, but has longer wings, pointier bill, and strongly notched, shorter tail. Older males suffused in raspberry; belly and undertail coverts lack fine, dark streaking of House Finch. Females and first-year males have strongly marked face pattern, suggesting that of much larger female Rose-breasted Grosbeak (p. 402). Compare also with Cassin's Finch, occurring at the periphery of our area.

■ **VOCALIZATIONS** Song a bright warble, faster and richer than House Finch's; rises a bit in pitch as it goes, but lacks up-slurred terminal note. Calls quite unlike those of House Finch: a hard *plick* given in flight and a slurred *slee-weeoo* like a Blue-headed Vireo's (p. 266) song element.

■ **POPULATIONS** Winters widely in eastern U.S., but numbers vary from year to year.

### Cassin's Finch

*Haemorhous cassinii* | CAFI | L 6¼" (16 cm)

This is the "American rosefinch" of the Interior West, regular in fall and winter to the western Great Plains; breeds in small numbers in Black Hills. Similar to Purple Finch, but more crisply marked; relatively straight bill and oft-crested crown contribute to the distinction. Sharp streaking below extends to undertail coverts. Red crown of adult male well set off from rest of head; raspberry on head of male Purple more diffuse. Female and younger male have blurrier face pattern than Purple's. Calls, sometimes heard fall and winter, include a short chuckle and a wheezy whistle.

variant ♂
typical ♂
red eyebrow and throat/breast
brown streaks on belly

## House Finch

rather square-ended tail
plain gray-brown head and back
curved culmen
messy gray streaks
♀

rose red, brightest on crown

## Purple Finch

short, thick bill with lightly curved culmen
dark cheek and pale facial stripes
dark streaks above
distinct, broad streaks
♂
♀
notched tail

bright red cap
large bill with straight culmen
♂
less distinct face pattern than Purple
long primary projection
thin, distinct streaks
♀

## Cassin's Finch

streaked undertail coverts

## REDPOLLS AND CROSSBILLS

Early each fall, birders eagerly await the annual Winter Finch Forecast (*finchnetwork.org*), with species in these two genera often the best bet for memorable irruptions southward.

### Common Redpoll

*Acanthis flammea* | CORE | L 5¼" (13 cm)

On the taiga breeding grounds, this small finch is widespread and reliably found. But it is highly unpredictable farther south: scarce many winters, altogether absent in others, and downright common in "invasion years."

■ **APPEARANCE** Smaller than House Finch (p. 326). Pale overall and streaky with well-defined red cap (or poll) and small black patch on throat; bill yellow. Adult male has extensive pink on breast, reduced on female; juvenile, plain and streaky, differs from juvenile Pine Siskin (p. 330) in bill structure.

■ **VOCALIZATIONS** Dry, descending *chit* notes as birds flush or fly over. Murmurs while feeding, with twangy, siskin-like *eeeawee* and *eeyay*.

■ **POPULATIONS** Feeds heavily on birch catkins; also visits feeders.

### Hoary Redpoll

*Acanthis hornemanni* | HORE | L 5½" (14 cm)

This sought-after finch of the far north is treated by some experts as a subspecies, albeit a distinctive one, of the closely related Common Redpoll.

■ **APPEARANCE** Plumage and morphology reflect its high-latitude biology: larger-bodied than Common (to retain body heat) but smaller-billed (to prevent heat loss), with frostier plumage overall. Streaking on rump and undertail coverts reduced compared to Common's.

■ **VOCALIZATIONS** Calls average lower than Common, but much overlap.

■ **POPULATIONS** Breeds far north, with most wintering within the Arctic; rare to southern Canada and northern U.S.

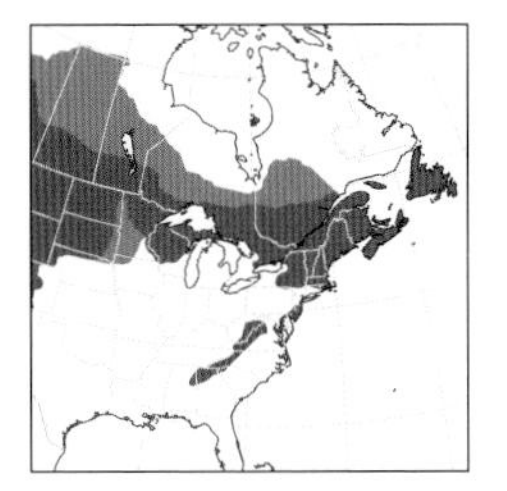

### Red Crossbill

*Loxia curvirostra* | RECR | L 5½–7¾" (14–20 cm)

Even among the peripatetic finches, the Red Crossbill is notably nomadic. Recent research indicates it may comprise around 10 cryptic species!

■ **APPEARANCE** Block-headed and short-tailed. Crossed bill tips are an adaptation for extracting conifer seeds. All plumages of Red are normally plain-winged, lacking bold white wing bars. Adult male brick red all over, adult female olive-yellow; juvenile streaked brown, separated from juvenile siskins (p. 330) and redpolls by larger size and bill shape.

■ **VOCALIZATIONS** Flight calls monosyllabic, often doubled: *kyew!* or *tyip*, etc., depending on population.

■ **POPULATIONS** Around 10 "types" occur in N. Amer.; eastern specialties include Type 1 ("Appalachian") and Type 8 ("Newfoundland"). Types differ in flight calls, feeding ecology, and bill morphology.

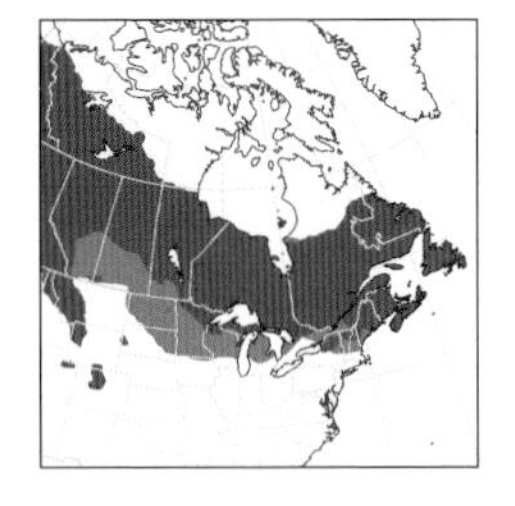

### White-winged Crossbill

*Loxia leucoptera* | WWCR | L 6½" (17 cm)

Like the Red Crossbill, the White-winged is a conifer specialist—with spruces particularly important in the biology of this species.

■ **APPEARANCE** All have black wings with thick white wing bars. Adult male extensively pink, adult female yellow-gray, juvenile streaky.

■ **VOCALIZATIONS** Flight call a muffled *chut*, down-slurred; not as bright and ringing as most Red Crossbills' flight calls.

■ **POPULATIONS** Availability of spruce seeds, intrinsically variable, plays a key role in driving irruptions.

juvenile
breeding ♀
breeding ♂
variable pink breast
streaked undertail coverts
Common Redpoll
flammea
very rare
"Greenland"
rostrata
winter ♀
winter ♀
larger and heavily streaked
winter ♂
lightly streaked rump
frosty pale overall
Arctic
exilipes
Hoary Redpoll
winter ♂
Arctic Islands
hornemanni
stubbier bill than Common
larger and paler
vinter ♀
Arctic
exilipes
winter ♂
mostly white rump and undertail coverts
large head and bill
typical adult ♀
immature ♂
younger males yellowish; may show thin wing bars
juvenile
Red Crossbill
brick red
typical adult ♂
short, notched tail
juvenile
immature ♂
obvious wing bars
♀
smaller-billed than Red Crossbill
adult ♂
White-winged Crossbill

**GENUS *SPINUS***

With their cheery songs and bright yellows, siskins and especially goldfinches are often called "wild canaries." They eagerly tend feeders, and their flight calls are distinctive.

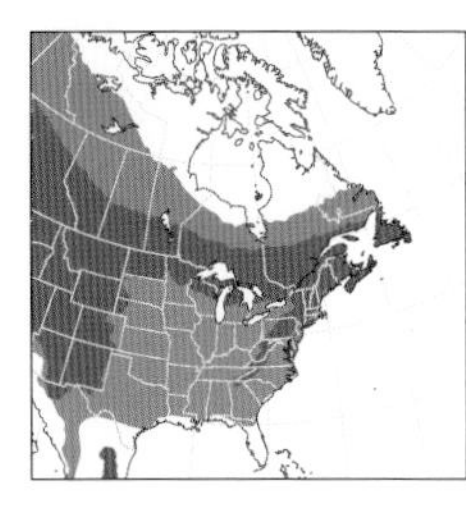

## Pine Siskin

*Spinus pinus* | PISI | L 5" (13 cm)

Pound for pound, these little finches may be the most expensive of all our birds to feed. Pine Siskins flock to seed socks filled with costly thistle seed.

■ **APPEARANCE** Thin bill, specialized for eating small seeds, separates it from closely related goldfinches and all other finches in the East. Plumage, brown and streaked, suggests female House Finch (p. 326), but Pine Siskin smaller with slighter bill. All have at least some yellow in wings and tail, prominent in flight; older adult males extensively yellow, other plumages less so.

■ **VOCALIZATIONS** Flight call a rising, nasal *zraaay;* especially around feeders gives wavering, very buzzy *zzzhhrrRRR*, rising in pitch and amplitude, then suddenly ending. Song twittery and twangy, sometimes including short imitations of other birds' calls.

■ **POPULATIONS** Breeds mostly in coniferous forests. Winters widely but erratically; frequently tends feeders. Irruptive like most finches.

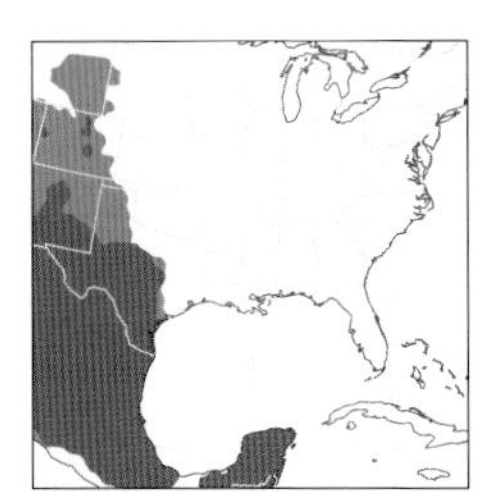

## Lesser Goldfinch

*Spinus psaltria* | LEGO | L 4½" (11 cm)

The winter finch or northern finch moniker doesn't apply to our most southerly finch. It occurs in the East chiefly in Tex. and ranges well into S. Amer.

■ **APPEARANCE** Smaller than American Goldfinch; bill usually dark. Breeding male, with extensive black above, distinctive, but other plumages, especially females and young in winter, much less so; on such birds, note dusky yellow above and below, extending to undertail coverts. White flash at base of primaries prominent in flight.

■ **VOCALIZATIONS** In flight, alternates pure-tone whistles, rising or falling, with rough chatter: *chleee ... ch'ch'ch'ch ... plweee ....* Song rapid and tinkling, almost as fast as Winter Wren (p. 302), with imitations of other species' calls frequently mixed in.

■ **POPULATIONS** Erratic but annual summer and fall to western Great Plains. Winter range expanding north; vagrants noted far east of core range, often at feeders. Tex. breeders mostly of ill-defined black-backed morph, formerly called Arkansas Goldfinch.

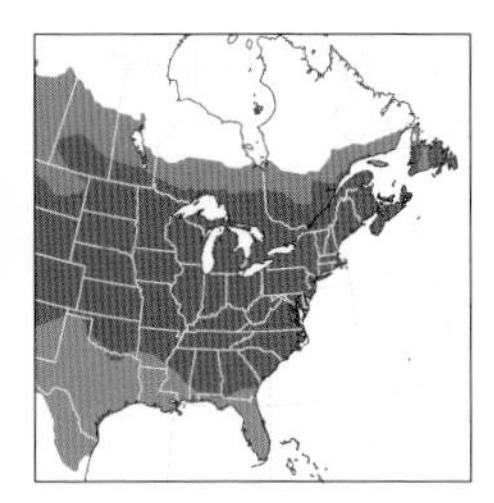

## American Goldfinch

*Spinus tristis* | AGOL | L 5" (13 cm)

Among the best known birds in the East, the American Goldfinch is hopelessly addicted to gardens with sunflowers and other composites. The species nests late in the summer.

■ **APPEARANCE** Breeding male, searing yellow with black cap and wings, unmistakable, but other plumages trickier. Even the dullest birds in winter usually show broad yellowish wing bars on brownish wings; undertail coverts clean white (dirty yellow on Lesser). Bill yellow-pink on most, darker on Lessers.

■ **VOCALIZATIONS** Flight call a rapid stutter of three to five, often four, falling notes, *chih-chih-chih-chit (potato chip)*; rising *tsuWEEE* given perched or in flight. Run-on song, bright and fast, comprises sharply rising and falling notes: *s'wee tweet swee tyeep syeet tsweep tswee-ch-ch,* etc.

■ **POPULATIONS** Year-round presence in East masks strong annual movements, especially coastally in fall.

Pine Siskin
yellow wing stripe
green morph
thin, sharply pointed bill
juvenile
yellowish below, extending to undertail coverts
♀
pale ♀
olive above
Black-backed
psaltria
adult ♂
Green-backed
hesperophilus
adult ♂
Lesser Goldfinch
short, dark bill
white at base of primaries
immature ♂
breeding ♀
brownish above
pale bill
breeding ♂
winter ♂
American Goldfinch
winter ♀
white undertail coverts
bold wing bars
juvenile

## *CALCARIUS* LONGSPURS

"True" longspurs, genus *Calcarius,* have a long claw on the hind toe, an adaptation for ambling over uneven terrain. Breeding males sport colorful napes, and the three species are sometimes referred to as the "collared" longspurs.

### Lapland Longspur

*Calcarius lapponicus* | LALO | L 6¼" (16 cm)

Like all longspurs, the Lapland is known to most birders in the East as a winter visitant to open country, often keeping company with Horned Larks (p. 286) and Snow Buntings (p. 334).

■ **APPEARANCE** Sparrowlike in size and general appearance, but gait and other behaviors more like Horned Lark. Most encounters are with drab birds in winter, told by their long primaries, reddish panel on wings, and face with black markings; flushing, reveals white outermost tail feather. Males begin to acquire breeding garb in early spring, just as the last of them are decamping the wintering grounds.

■ **VOCALIZATIONS** Calls include a descending *pyoo* and short rattle, alternated in flight. Jangling song heard mostly on breeding grounds.

■ **POPULATIONS** By far the most widely distributed longspur; only species expected in winter from Appalachians eastward.

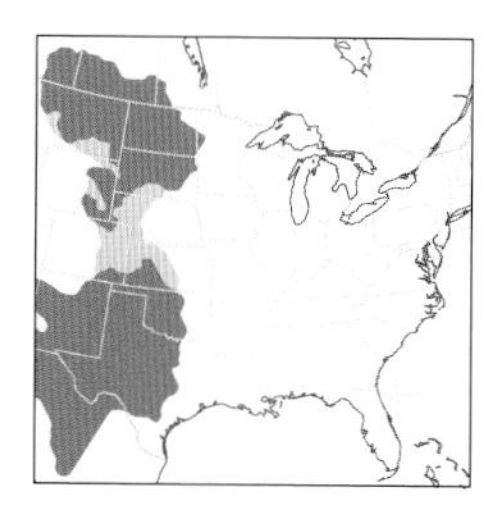

### Chestnut-collared Longspur

*Calcarius ornatus* | CCLO | L 6" (15 cm)

The two prairie-nesting longspurs—Thick-billed (p. 334) and Chestnut-collared—sort out on the breeding grounds by microhabitat preferences. This species prefers taller grasses, often in lower and wetter settings.

■ **APPEARANCE** Relatively petite; slimmer-bodied and smaller-billed than Thick-billed. Outside breeding season, told from Smith's and Lapland by pale gray plumage, short primaries, and, in flight, mostly white tail with broad black tips on inner rectrices. Winter male retains hint of black breast; all plumages relatively dark-billed in winter.

■ **VOCALIZATIONS** Rattling flight call averages shorter (two to three syllables) than other longspurs'; but call often runs long, like other longspurs'. Song rich and gurgling; shorter than song of Thick-billed.

■ **POPULATIONS** In long-term and worrisome decline, but still fairly common in lush grasslands in upper Midwest. Short-distance migrant; only rarely strays from normal migration routes.

### Smith's Longspur

*Calcarius pictus* | SMLO | L 6¼" (16 cm)

Ecologically similar to the widespread Lapland, Smith's is a tundra breeder with long, dark primaries—an adaptation for long-distance migration.

■ **APPEARANCE** The most warmly colored longspur; even in its drab winter plumage, Smith's is washed buff below. Tail mostly dark, but two outermost rectrices edged white; Lapland has only the outermost rectrix edged white. Broad white upper wing bar forms crescent on bird at rest; often prominent in flight. Like Lapland, begins to acquire breeding attire—male has harlequin face pattern and bright butternut breast—by end of winter.

■ **VOCALIZATIONS** Rattling flight call not as rapid as other longspurs'. Song, heard mostly on breeding grounds, tinkling and rising, recalls Horned Lark's (p. 286).

■ **POPULATIONS** In winter and on migration favors disturbed open habitats: corn stubble, grazed pastures, airfields, and so on. Unusual among passerines, is polyandrous, like phalaropes (p. 126).

Lapland Longspur
tail from above
dark tail
th narrow
ite edges
long primary
projection
breeding ♂
breeding ♀
reddish
secondary coverts
winter ♀
dark
border
to cheek
buffy fall ♀
Chestnut-collared Longspur
winter ♂
small
darkish bill
sandy gray with
blurry streaks
winter ♀
tail from above
mostly white
tail with dark
triangle
breeding ♂
short
primary
projection
Smith's Longspur
breeding ♂
thin
bill
tail from above
broad
white sides
breeding ♀
broad
white
upper
wing bar
long primary
projection
buffy overall
winter ♂

## THICK-BILLED LONGSPUR, SNOW BUNTING

Despite its name, the Thick-billed Longspur is more closely related to the Snow Bunting than to *Calcarius* longspurs (p. 332). Formerly classified as New World sparrows (pp. 336–359), longspurs and Snow Buntings are now in the family Calcariidae.

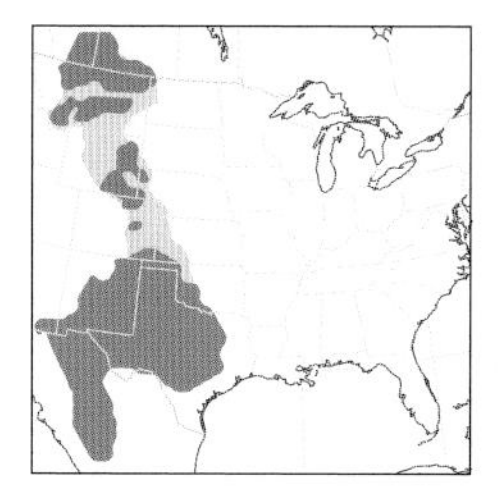

### Thick-billed Longspur

*Rhynchophanes mccownii* | TBLO | L 6" (15 cm)

Along with the Chestnut-collared Longspur (p. 332), this is one of the two prairie longspurs: short-distance migrants with relatively short wings. The Thick-billed favors dry, often disturbed, shortgrass prairie.

■ **APPEARANCE** Pale and gray overall, with large bill. Tail short and mostly white, but tips black on all but outermost rectrix; has shorter "spur" than other longspurs. Breeding adult, especially male, distinctive. Differs in winter from Chestnut-collared by plain gray (male) or gray-buff (female) breast and large pale bill; even drab birds in winter usually show chestnut patch on wings; larger, stockier body a good mark year-round.

■ **VOCALIZATIONS** Flight call a short rattle like other longspurs', but also gives a mellow *plip*. Song a loose, run-on twittering, given in flight.

■ **POPULATIONS** Tolerates cattle grazing, but range has withdrawn considerably due in part to fire suppression.

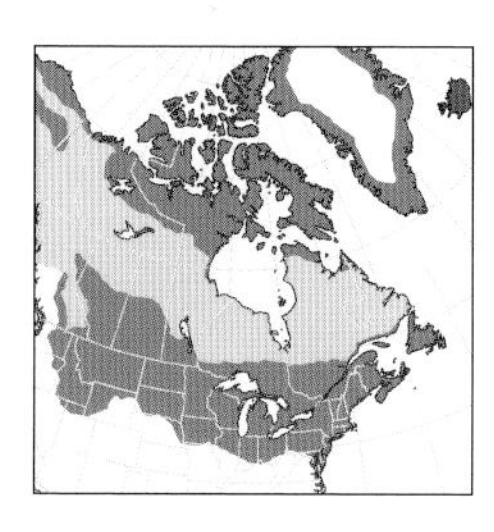

### Snow Bunting

*Plectrophenax nivalis* | SNBU | L 6¾" (17 cm)

The bird's name is doubly appropriate: It is whiter than any other songbird in the East, and it is known to most birders as a winter visitant to the northern U.S. and southern Canada.

■ **APPEARANCE** Slightly larger than longspurs, but similarly flocky and inclined toward open habitats; small-billed. In winter, when most are seen, buffy white above and mostly white below; in breeding garb, starting to show by late winter, whiter overall, with males black above, females grayer. All plumages flash huge white patches on wings in flight. Aberrantly whitish (leucistic) juncos (p. 348) and other songbirds can suggest Snow Bunting.

■ **VOCALIZATIONS** Like Lapland Longspur (p. 332), alternates short rattle with down-slurred whistle when flying over. Song a run-on twitter, tinkling and high-pitched.

■ **POPULATIONS** Nests in rock crevices on tundra, where well protected from predators; recent declines may result from high-latitude climate change.

### What Is a Bunting?

Only four birds on Earth are called longspurs; three are in the genus *Calcarius,* and the other, although in a different genus, is nevertheless in the longspur family Calcariidae. The scientific names derive from Latin *calcar,* meaning "spur."

But things get messy when we consider the two other species in the six-species family Calcariidae: the widespread Snow Bunting and the range-restricted McKay's Bunting, breeding only on Bering Sea islands. The word "bunting" proved irresistible to English-language namers of birds, resulting in such designations as Indigo Bunting (p. 406), Lark Bunting (p. 340), Rustic Bunting (annual in Alaska), and Snow Bunting—in four different families!

Such muddles seem most pronounced within the nine-primaried oscines (pp. 324–409), a speciose assemblage whose relationships are still being worked out. Coining new group names—for example, "Piranga" has been proposed for certain eastern tanagers (p. 400) that aren't, in fact, tanagers—seems ill-advised, as names like Scarlet Tanager and Snow Bunting are familiar to so many nature lovers. But staying on top of the fascinating and fast-developing taxonomic literature is enthusiastically recommended.

Thick-billed Longspur
breeding ♂
medium primary projection
white tail with black T
breeding ♀
unstreaked breast
winter ♀
large pinkish bill
plain face pattern
winter ♂
mostly white wings with black tip
breeding ♂
Snow Bunting
winter ♂
breeding ♂
1st winter ♂
breeding ♀
1st winter ♀
warm cinnamon tones
uvenile ♂
white wing patch
winter ♂
1st winter ♂

**GENUS *PEUCAEA***

Three species in this mostly Middle American genus occur in the East. Our three are large and plain. They share the trait of singing bewitching songs—although their songs are very different.

## Botteri's Sparrow

*Peucaea botterii* | BOSP | L 6" (15 cm)

In the shrubby grasslands of South Texas, Botteri's and Cassin's Sparrows occur side by side. They differ subtly in appearance but strikingly in songs.

■ **APPEARANCE** Like all *Peucaea* sparrows in the East, plain-breasted and large-billed with flat forehead. Tail of Botteri's lacks white edges and transverse barring of Cassin's; wings plainer than Cassin's. Streaky juvenile, with plain plumage overall, very similar to Cassin's; all Botteri's have slightly larger bills than Cassin's. Compare also with Grasshopper Sparrow (p. 338), warmer and somewhat smaller, and Brewer's Sparrow (p.342), colder and notably smaller.

■ **VOCALIZATIONS** Long song begins with tentative sputtering notes, erratically uttered, ending in accelerating trill. Some birds omit the trill, others the sputtering, and yet others change the order.

■ **POPULATIONS** Numbers vary naturally from year to year, but overgrazing on coastal plain not helping at northern edge of species' range in Tex.

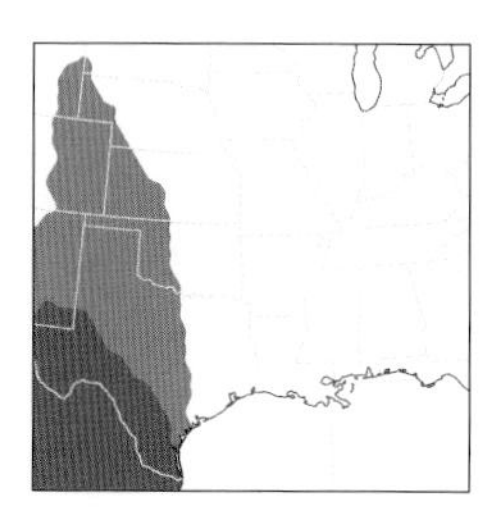

## Cassin's Sparrow

*Peucaea cassinii* | CASP | L 6" (15 cm)

One of the plainest birds of the prairie has one of the sweetest songs. These birds are hard to see when they're perched or foraging, but they have the helpful habit of singing on the wing.

■ **APPEARANCE** Built like Botteri's Sparrow, which occurs with Cassin's in the East only in South Tex. Fresh Cassin's are scaly-backed with white edgings on wing; tail barred and white-edged, and most show thin white eye ring.

■ **VOCALIZATIONS** Song high and sweet. Typically begins with one or two pure whistles, then a high trill, then two lower whistles; the latter actually comprise four notes, not two, obvious on a spectrogram, hard for humans to discern. When not singing, gives high, weak *phit* notes in slow, stuttering series.

■ **POPULATIONS** Highly variable from year to year, even within a single breeding season; birds move around in response to rainfall or lack thereof.

## Bachman's Sparrow

*Peucaea aestivalis* | BACS | L 6" (15 cm)

Formerly known as the Pinewoods Sparrow, this southeastern specialty is restricted to open, grassy pine woods and sandhills. Solitary and secretive, the Bachman's prefers to run, rather than fly, from danger.

■ **APPEARANCE** Only *Peucaea* sparrow found in its range. Adult warm overall; reddish gray above with thin white eye ring, buffy below. Some bear superficial resemblance to reddish sparrows like Field (p. 344) and Swamp (p. 354), but note structural differences: Bachman's long-billed and long-tailed, with flat forehead. Streaky juvenile shows eye ring.

■ **VOCALIZATIONS** Song a breathy buzz followed by a musical trill; similar to Bewick's Wren's (p. 304), but that species now largely gone from range of Bachman's Sparrow.

■ **POPULATIONS** Shares a strange history with Bewick's Wren: Both species expanded north around turn of 20th century, only to withdraw just as quickly. Bachman's Sparrows in western part of range brighter and redder than eastern counterparts.

Botteri's Sparrow
dull wing pattern
unstreaked flanks
juvenile

Cassin's Sparrow
hite tail corners
white eye ring
scaled back pattern
tail barred
juvenile
western
bachmani/illinoensis
rufous upperparts
eastern
aestivalis
Bachman's Sparrow
juvenile

## TWO PLAIN-BREASTED SPARROWS

The widespread Grasshopper Sparrow was until recently thought to be related to the marsh sparrows (p. 350) but now is believed to be closer to the *Peucaea* sparrows (p. 336) and Olive Sparrow.

### Grasshopper Sparrow

*Ammodramus savannarum* | GRSP | L 5" (13 cm)

Although Grasshopper Sparrows do eat grasshoppers, they get their name from their strange insectlike song. They are habitual night singers, especially vocal around the full moon in early summer.

■ **APPEARANCE** Rotund; short-tailed, flat-headed, and large-billed. Adult buffy overall, brightest on unstreaked breast. Note also white crown stripe and thin white eye ring of adult; most have ochre patch in front of eye. Juvenile has streaks on buff-brown breast; compare with Cassin's Sparrow (p. 336) and marsh sparrows (p. 350).

■ **VOCALIZATIONS** Primary song comprises two to four high chips followed by a thin buzz like a grasshopper: *p't'tk'ZZZZZZZZZZ*. Sounds feeble, but carries far; is notably ventriloquial. Secondary song an ill-formed jumble of short, high buzzes; does not carry well, hard to hear. Flight call a high, shrill, thin whistle.

■ **POPULATIONS** Geographically variable; all nest in tallgrass, often rather dry. Most in East are subspecies *pratensis,* often a "pioneer species" in reclaimed strip mines; paler western subspecies *perpallidus* ranges to Midwest. Endangered subspecies *floridanus,* relatively dark above, requires burned prairie on central Fla. peninsula.

### Olive Sparrow

*Arremonops rufivirgatus* | OLSP | L 6¼" (16 cm)

The bouncing-ball sound in thickets and woodlands in southern Tex. is invariably the song of this bird. It is the only species in the tropical genus *Arremonops* to reach the U.S.

■ **APPEARANCE** Stocky and short-winged. Adult plain olive above and plain gray below; rusty crown has thin gray central stripe, hard to see. Green-tailed Towhee (p. 356), similarly colored and with bright rusty crown, is longer-tailed and longer overall, with white throat. Juvenile Olive fairly bright yellow-olive with blurry streaks above and below.

■ **VOCALIZATIONS** Song a series of about 20 sharply down-slurred notes, accelerating: *chip ... chip ... chip-chip-chip-chip-chipchipchipchipchip.*

■ **POPULATIONS** Common in woody vegetation within range, but secretive; heard far more often than seen. Range in Tex. dynamic, contracting and expanding in response to clearing and recovery of wooded habitat.

### Learning Sparrow Songs

The New World sparrows (pp. 336–359), family Passerellidae, are perhaps the ultimate "little brown jobs," or LBJs. Forty-four species are on the U.S. list, with records of almost all from the East. Identification based on plumage is tricky to begin with (New World sparrows traffic heavily in shades of brown) and is complicated by seasonal feather wear and, in some species, strong geographic variation. Body structure is important to assess, especially with look-alike juveniles once fully fledged. Habitat, especially on the breeding grounds, often brings the number of possibilities down to a small handful. To be sure, New World sparrows can be identified by what they look like and where they occur. But in so many instances, they don't have to be seen at all. Almost all sing distinctively, with the songs of many among birders' favorites.

The complex songs of Song (p. 354) and Vesper Sparrows (p. 348), say, are a daunting starting point; and the sound-alike songs of the

Dark-eyed Junco (p. 348) and Chipping Sparrow (p. 342) can trip up even the experts. Fortunately, though, many sparrows sing rather stereotyped songs that are easily learned: White-throated Sparrows (p. 346) sing several thin whistles, Clay-colored Sparrows (p. 342) sing a few raspy buzzes, and the Eastern Towhee (p. 358) sings *"Drink your tea!"*

Even the tricksters can be identified with a bit of discipline. The opening notes of a Vesper Sparrow's songs are typically slurred whistles, whereas the opening notes of a Song Sparrow's songs are typically not; the notes in a Chipping Sparrow's song are generally more strongly down-slurred than those of a Dark-eyed Junco; and so forth. In many cases, sparrows' call notes, distinct from their songs, are diagnostic: The junco's call is a toneless smacking sound, the Chipping Sparrow's a high, thin whisper. Listen to recordings online, study spectrograms, and, most important of all, spend time in the field watching and listening to these gifted songsters.

## THREE STRIKING SPARROWS

These three are exceptions to the rule that sparrows are supposed to be brown and nondescript. Adult males are unmistakable, and other plumages are usually easy to ID with a halfway decent look.

### Black-throated Sparrow

*Amphispiza bilineata* | BTSP | L 5½" (14 cm)

Even among birds adapted to arid climates, this sparrow stands out. Its kidneys have adaptations for extracting metabolic water from dry seeds, and the species flourishes in the sunbaked lowlands of our driest deserts.

■ **APPEARANCE** Slim and trim. Head and breast of adult boldly marked black and white; rest of body plain grayish. White corners of dark tail prominent as bird flushes. Juvenile not nearly as striking, but nevertheless shows basic pattern of adult: bold white eyebrow, bold white malar, and smudgy wash on breast, with white corners on tail.

■ **VOCALIZATIONS** Tinkling song comprises a few short chirps followed by a loose trill: *chit chit ch' twe-e-e-e-e-e-e*. Calls clipped and high-pitched: *tee, tsee, seet,* etc.

■ **POPULATIONS** At home in dry deserts, especially creosote flats. Reaches East primarily in Edwards Plateau of Tex., more sparingly toward Gulf Coast; casual vagrant to Atlantic seaboard.

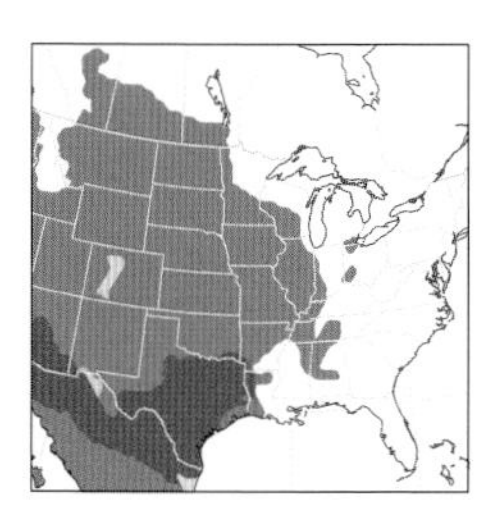

### Lark Sparrow

*Chondestes grammacus* | LASP | L 6½" (17 cm)

The face pattern alone would make the Lark Sparrow one of the most distinctive members of its family. But it's also got a huge "stickpin" on the breast, a flashy tail, and a complex song.

■ **APPEARANCE** Large and boldly marked. Harlequin face pattern of adult has extensive chestnut; more monochrome on younger birds and some winter adults. Plain breast has single dark splotch; ample tail, splayed out in flight, shows white tip and corners. Briefly held juvenile plumage streaky below, with distinctive tail of adult and hint of adult head pattern.

■ **VOCALIZATIONS** Song a hesitant jumble of short buzzes, whistles, and trills, with elements often doubled: *chit-chit zzzz syee-syee whit e-e-e-e-e seet-seet*. Flight call an abrupt *tsip*.

■ **POPULATIONS** Favors open country with sandy soils; found around hedgerows, outbuildings, and woodland edges. Erratic breeder east of mapped range; rare but regular, mostly fall to early winter, to East Coast.

### Lark Bunting

*Calamospiza melanocorys* | LARB | L 7" (18 cm)

Despite its name, the Lark Bunting is an exemplary sparrow, nested well within the family Passerellidae according to most analyses. Within its limited range, the species is common.

■ **APPEARANCE** One of the largest New World sparrows; bill large, tail fairly short. White wing panel prominent in flight in all plumages. Snazzy black-and-white breeding male unmistakable. Other plumages, especially young in winter, told by large size, chunky bill, and, in flight, white on wing.

■ **VOCALIZATIONS** Pulsing song, given in flight, mixes series of buzzes with loose trills: *rzz rzz rzz jree-jree-jree-jree-jree dzzzzzzz chre-chre-chre-chre*. Flight call a rich, rising *whooee?*

■ **POPULATIONS** Like Cassin's Sparrow (p. 336), numbers vary with precipitation. Breeds shortgrass prairie, winters grasslands and deserts. A short-distance migrant, yet manages to wander annually to East Coast.

Black-throated Sparrow
white throat
juvenile
Lark Sparrow
harlequin head pattern
white corners on tail
juvenile
long tail
early spring ♂
winter ♂
Lark Bunting
heavily streaked below
breeding ♂
♀
whitish wing patch on all plumages

## GENUS *SPIZELLA*

The name *Spiza*, from Greek, is a catchall for little brown bird, and *Spizella* is the diminutive form. Even among the little brown sparrows, these three are notably slight in build.

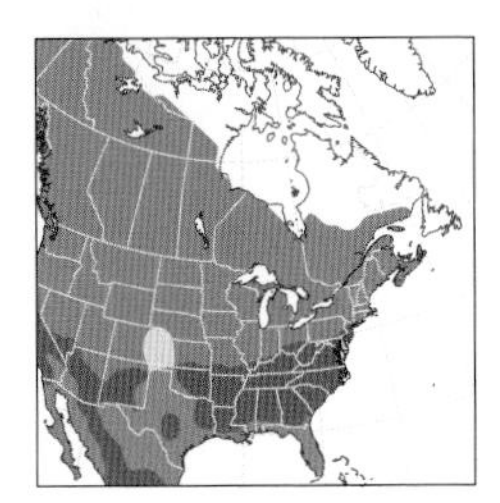

### Chipping Sparrow

*Spizella passerina* | CHSP | L 5½" (14 cm)

The widespread "Chippy" is one of the most familiar birds in the East, often seen in cities and towns, but also at home in remote pinewoods.

■ **APPEARANCE** Like all *Spizella* sparrows, potbellied, long-tailed, and quite small. Breeding adult has rusty cap, white eyebrow, and black line through eye. Brown above, but with plain gray rump; underparts plain gray. Winter adult and immature can closely resemble Clay-colored. Black line extends through eye on Chipping, but not in front of eye on Clay-colored; Clay-colored buffier overall, with brown (not gray) rump and gray nape. Juvenile Chipping, streaked below, already shows dark line through eye.

■ **VOCALIZATIONS** Song a simple trill of monotone *chip* notes, strongly down-slurred, with dry, unmusical quality. Flight call a high, rising *seen?*

■ **POPULATIONS** Almost all withdraw in colder months to main wintering grounds; reports in winter north of normal range often erroneous.

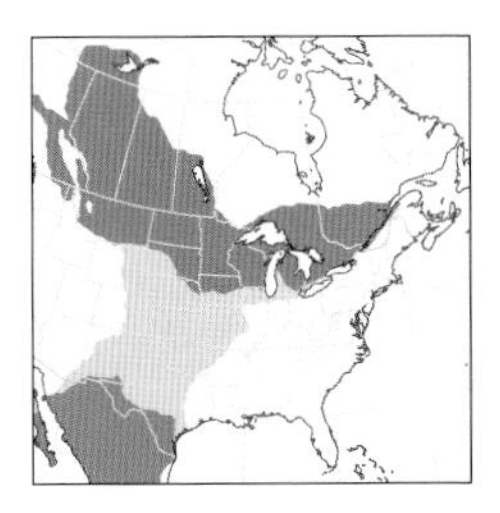

### Clay-colored Sparrow

*Spizella pallida* | CCSP | L 5½" (14 cm)

Where there is tallgrass prairie with scattered trees in the upper Midwest and southern Prairie Provinces, Clay-colored Sparrows abound as breeders.

■ **APPEARANCE** Same size and shape as Chipping Sparrow. Breeding adult has intricate head pattern; note white central crown stripe, prominent auriculars ("ear" patches), and bold "whisker." Immature and nonbreeding adult buffier overall; told from nonbreeding Chipping by brown rump (gray on Chipping), gray nape (also gray on Chipping, but not as sharply defined), and dark line behind but not in front of eye (goes through eye on Chipping). See Brewer's Sparrow.

■ **VOCALIZATIONS** Song a few pulses of white noise, *ZZZZ ZZZZ ZZZZ*, more like an insect than a bird. Flight call a weak, rising *see?*

■ **POPULATIONS** Slowly expanding to eastern Great Lakes and upper St. Lawrence Valley; settles in reclaimed strip mines, Christmas tree farms, etc.

### Brewer's Sparrow

*Spizella breweri* | BRSP | L 5½" (14 cm)

Common in sagebrush country in the Interior West, the subtly marked Brewer's Sparrow enters our area as a breeder in the western Great Plains. It is more widespread in the East in the nonbreeding season, when it mixes with look-alike Chipping and Clay-colored Sparrows.

■ **APPEARANCE** Sandy gray-brown and finely marked, suggesting washed-out Clay-colored Sparrow. Brown crown has thin white streaks; face has narrow white eye ring and thin dark "whisker." Similar to Clay-colored fall and winter; Clay-colored brighter and buffier, more boldly marked on face, and usually with pale central crown stripe and prominent gray nape. Striped juvenile plumage, held briefly, begins to show eye ring and streaky crown.

■ **VOCALIZATIONS** Long-duration song has buzzes like Clay-colored's and trills like Chipping's; overall effect is thin, dry, and gasping. Flight call like Clay-colored's.

■ **POPULATIONS** Away from breeding grounds, flocks with other *Spizella* sparrows in grasslands and shrublands. Uncommon in winter to Tex. coast, annual but very rare in recent years elsewhere in East.

dark eyeline runs through eye to bill in all plumages
breeding adult
juvenile
Chipping Sparrow
streaked plumage held into Oct.
winter adult
gray rump compared to Clay-colored
1st winter

white median crown stripe
pale gray nape
all plumages have pale lores
Clay-colored Sparrow
brown rump compared to Chipping
breeding
streaked plumage mostly gone by Oct.
juvenile
strong face pattern in all plumages
fall

brown crown, streaked black, lacks white median crown stripe
Brewer's Sparrow
white eye ring
juvenile
breeding

## THREE REDDISH SPARROWS

Their evolutionary histories were recently reassessed: The American Tree Sparrow had long been placed in the genus *Spizella,* which includes the Field Sparrow; the former is now believed to be closer to the Fox Sparrow.

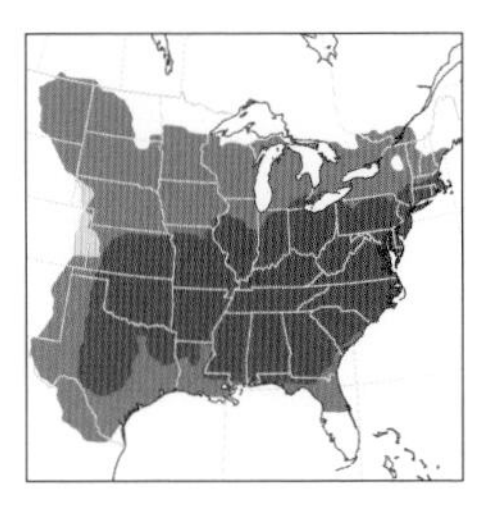

### Field Sparrow

*Spizella pusilla* | FISP | L 5¾" (15 cm)

"Field" is a suitable moniker, but "Old Field" would have been especially apt. The species abounds in weedy meadows and waste places where thickets and saplings are starting to take over.

■ **APPEARANCE** Slightly larger than our other *Spizella* sparrows (p. 342) and more reddish overall. Adult has gray face with reddish crown and reddish flare behind eye, pink bill, and white eye ring; gray breast suffused with red-orange highlights. Streaky juvenile plumage, held briefly, already shows reddish highlights and hint of eye ring.

■ **VOCALIZATIONS** Song an accelerating series of whistles; pattern much like that of Olive Sparrow (p. 338), but song of Field Sparrow more musical because notes not as strongly down-slurred. Call note a smacking *tsik;* flight call high, clear, and descending.

■ **POPULATIONS** Widespread nominate subspecies *pusilla* relatively dark and rufescent; western *arenacea* paler and grayer.

### Fox Sparrow

*Passerella iliaca* | FOSP | L 7" (18 cm)

In much of the East, this is the largest bird named sparrow. Away from the remote breeding grounds, it is seen singly or in very small flocks in the leafy understory of woodland edges.

■ **APPEARANCE** Slightly larger than Hermit Thrush (p. 316), which it resembles. Kicks around in foliage on forest floor, where Hermit Thrush also occurs. Named for foxy colors: reddish overall with gray highlights; note especially coarse streaking below and rufous tail. Thick bill has yellow lower mandible.

■ **VOCALIZATIONS** Call a solid *tsook,* like Brown Thrasher's (p. 306). Song comprises 7–15 slurred whistles, often interspersed with one or two short buzzes; starts singing in early spring while still on wintering grounds.

■ **POPULATIONS** Exhibits great geographic variation, perhaps comprising four species. Almost all in the East are in the "Red" group (*iliaca*), with the sweetest songs and reddest plumage.

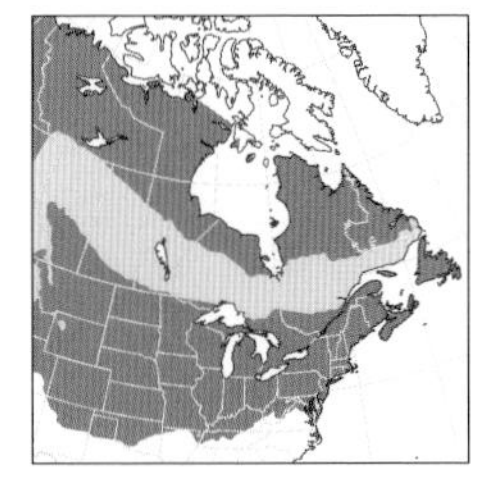

### American Tree Sparrow

*Spizelloides arborea* | ATSP | L 6¼" (16 cm)

In winter, when most birders see the American Tree Sparrow, the species occurs in marshes, meadows, and thickets—not trees. Small flocks are flighty, flushing along hedgerows with sweet tinkling notes.

■ **APPEARANCE** Known as the "winter Chippy" for its superficial similarity to the Chipping Sparrow (p. 342), which it largely replaces at midlatitudes in the colder months. Red and gray overall like Field Sparrow, but has bicolored bill and black spot on plain gray breast; outer tail feathers of American Tree frosted white. Juvenile plumage, rarely seen away from breeding grounds, streaky; note bicolored bill of juvenile.

■ **VOCALIZATIONS** All winter, gives whistled *tleet* and *tlee-eetle* calls. Rich song, rarely heard away from breeding grounds, suggests "Red" Fox Sparrow. Flight call a high, clipped *sseee.*

■ **POPULATIONS** Arrives late in fall, departs early in spring. Midwinter movements extensive, related to food availability.

Field Sparrow
plain-faced
bright
pink bill
juvenile
winter
pusilla
pink legs
breeding
arenacea

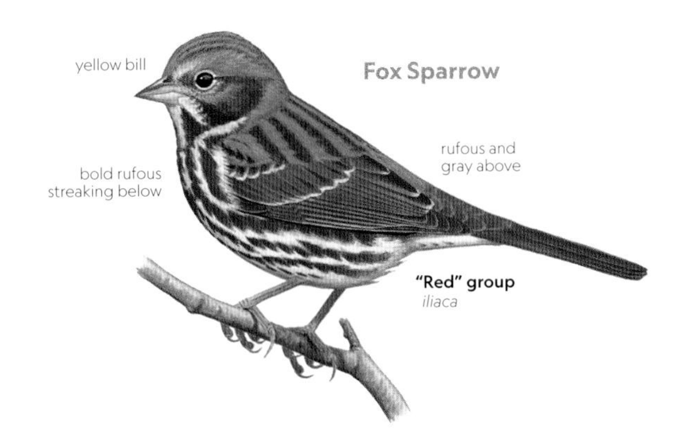
yellow bill
Fox Sparrow
rufous and
gray above
bold rufous
streaking below
"Red" group
iliaca

bicolored
bill
American Tree
Sparrow
breeding
winter
juvenile
long tail

## GENUS *ZONOTRICHIA*

Sparrows in this genus have sweet, whistled songs and high-contrast head patterns. In the East, they are known to most birders as winter visitants; they sing throughout the colder months.

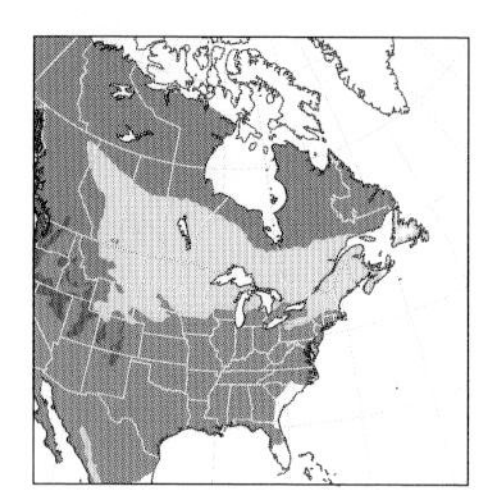

### White-crowned Sparrow

*Zonotrichia leucophrys* | WCSP | L 7" (18 cm)

With its crisp head stripes, the White-crowned ranks among our handsomest sparrows. It favors more open habitat in winter than White-throated Sparrow does.

■ **APPEARANCE** One of the largest sparrows in the East. Adult, plain gray below, has high-contrast head stripes and colorful bill. Immature plumage, held until early spring of second calendar year, told by brown-and-tan, not black-and-white, head stripes. Heavily streaked juvenile plumage rarely seen away from breeding grounds.

■ **VOCALIZATIONS** Song begins with one to three thin whistles, then goes into a short jumble of buzzy notes. Call a sharp *chink,* similar to White-throated's; flight call a high, up-slurred *tsweee.*

■ **POPULATIONS** Most in East are dark-lored, pink-billed nominate subspecies *leucophrys,* but white-lored, orange-billed *gambelii,* breeding western Canada and Alas., regular in winter east to Mississippi R.

### Harris's Sparrow

*Zonotrichia querula* | HASP | L 7½" (19 cm)

This ravishing sparrow, as large as a small towhee, is the only bird species whose breeding range is restricted entirely to Canada.

■ **APPEARANCE** Larger than White-crowned; all have pink bill and at least some black on head and breast. Adult, especially in breeding plumage, has deep black on face and breast. Black on first-winter not as extensive; note white belly, rich brown flanks, and telltale pink bill and large size.

■ **VOCALIZATIONS** Song a series of two or three long whistles, usually on same pitch and less wavering than song of White-throated. Call note a hard *chink* like other *Zonotrichia* sparrows'; flight call a steady, breathy *sseeeh.*

■ **POPULATIONS** Core wintering range a relatively small swatch of southern Great Plains, but vagrants annual to East Coast; flocks with other sparrow species, especially White-crowned, away from breeding grounds.

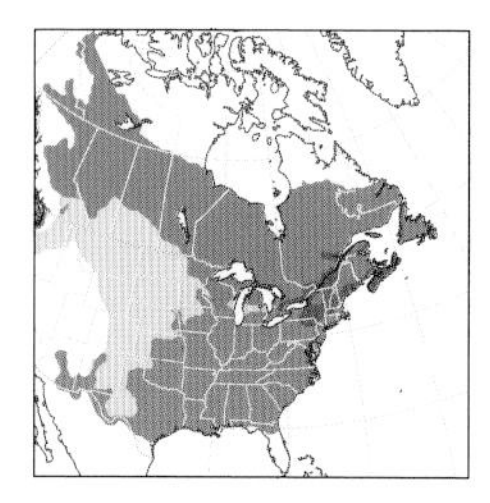

### White-throated Sparrow

*Zonotrichia albicollis* | WTSP | L 6¾" (17 cm)

This smallest *Zonotrichia* is one of the most abundant breeders in the East. Its song fills the boreal forest all day long in summer, and the species is common in wooded and brushy habitats in winter in much of the eastern U.S.

■ **APPEARANCE** Variable; adult white-striped morph unmistakable, but other plumages can be much duller. Even the drabbest birds in winter show trapezoidal white patch on throat, yellow in front of eye, and head stripes; compare with Swamp Sparrow (p. 354). Streaked juvenile plumage, seen only around breeding grounds, already shows white throat.

■ **VOCALIZATIONS** Until recently, sang two main songs, but in a striking example of cultural evolution, added a third in 2010s; all comprise wavering whistles on different pitches. Call note *chink* like Harris's and White-throated; flight call ends with a stutter, *seeet't.*

■ **POPULATIONS** Two color morphs (not subspecies) are a balanced polymorphism; white morphs mate only with tan morphs, and vice versa. White morphs are behaviorally dominant to tan morphs.

"Gambel's"
gambelii
juvenile
orange bill
White-crowned Sparrow
white lores
eastern
leucophrys
pinkish bill
immature
immature
dark lores
adult
eastern
leucophrys
adult

large pink bill
postocular spot in all plumages
Harris's Sparrow
breeding
winter adults
black on breast variable
immature

White-throated Sparrow
yellow lores
tan-striped morph
juvenile
white-striped morph

## DARK-EYED JUNCO, VESPER SPARROW

The junco and Vesper Sparrow are the only passerellids with an entirely white outermost tail feather (rectrix), quite prominent as the bird flies away. Juncos breed in coniferous forests, Vesper Sparrows in open or semi-open habitats with some woody vegetation.

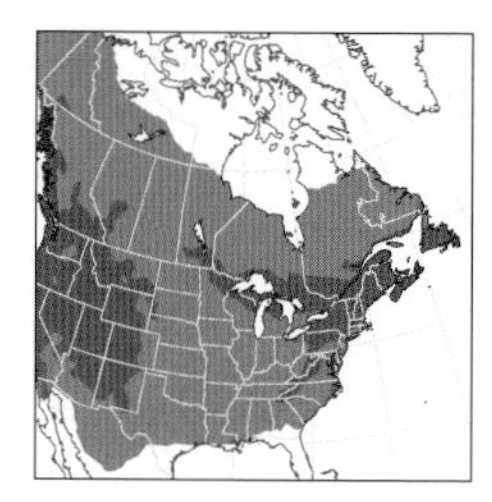

### Dark-eyed Junco

*Junco hyemalis* | DEJU | L 6¼" (16 cm)

No other bird in the U.S. and Canada exhibits as much geographic variation as the Dark-eyed Junco. It has been treated as five or more species, but current thinking is that it is just one.

■ **APPEARANCE** Size and build perfectly typical for a sparrow; bill pale pinkish white. "Slate-colored Junco" uniform gray-black (male) or gray-brown (female) above, with contrasting white belly; breeds in northern hardwood forests (to southern Appalachians), winters widely in East. "Oregon Junco," more westerly, told by "executioner's hood" (blackish on male, gray on female) contrasting with brownish upperparts and white belly; winters regularly east to Great Plains. "Cassiar Junco," comprising "Slate-colored" x "Oregon" intergrades, intermediate; widespread in Midwest in winter. "Pink-sided Junco" has pale blue-gray head with dark lores and buff-orange bulges on sides that nearly meet at center of breast; breeds central Rockies and Cypress Hills of Sask., winters to western Great Plains. Range-restricted "White-winged Junco" like a large "Slate-colored," but larger and paler, with dark lores, trademark wing bars, and more white in tail; breeds mostly Black Hills, scarce migrant through western Great Plains. Juveniles of all populations dark overall and streaked below; all show white outer rectrix and pale bill.

■ **VOCALIZATIONS** Primary song a loose trill; similar to Chipping Sparrow's (p. 342), but more jangling. "Soft song," given at end of winter, a complex jumble like a quiet or distant thrasher (pp. 306–309). Varied calls include a smacking *tsik* and a rapid series of *zing* notes.

■ **POPULATIONS** Despite striking differences in adult (especially male) plumages, introgression is extensive where ranges meet; vocalizations similar in all subspecies.

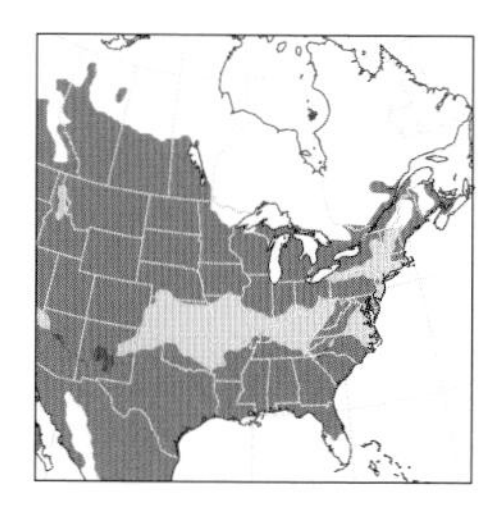

### Vesper Sparrow

*Pooecetes gramineus* | VESP | L 6¼" (16 cm)

Vespers are hymns or chants sung in the evening, and this grassland sparrow is indeed vocal at dusk. But it also sings in the morning and, often, right through the middle of the day.

■ **APPEARANCE** A hefty sparrow, bulkier than Savannah Sparrow (p. 352) and even Song Sparrow (p. 354). Breast streaking of Vesper variable, but generally sparse and fine. Brownish face has white frame below; eye ring white. Chestnut patch on wing coverts distinctive but variable. White outermost rectrix usually clinches the ID on bird in flight.

■ **VOCALIZATIONS** Song typically opens with two or three rich monotone whistles, followed by a jumble of buzzes and short trills: *tyeeee tyeeee ch'ch'ch'ch zzz brreeee ch'ch',* etc. Call note a high, light *psip;* flight call a rough, rising *sssee.*

■ **POPULATIONS** Averages paler in Midwest (subspecies *confinis*) than farther east (nominate *gramineus*); all are subject to extensive wear and bleaching by end of summer. Has declined and disappeared locally in some places east of Appalachians.

"Cassiar"
cismontanus
♂
"Slate-colored"
hyemalis
♀
♀
contrasting
hood between
head and back
♂
"Oregon"
shufeldti
bottom
of hood
convex
♂
juvenile
"White-winged"
aikeni
pale
gray bill
blue-gray
overall
♂
Dark-eyed
Junco
"Slate-colored"
hyemalis
dark
lores
♂
blue-gray
hood with
no brown on
crown
"Pink-sided"
mearnsi
pink sides
almost meet in
center of breast
Vesper
Sparrow
white
eye ring
chestnut
wing patch
white
frame on
face
western
confinis
eastern
gramineus

## MARSH SPARROWS

Many sparrows occur in marshes and wet meadows, but the four in genus *Ammospiza* are almost exclusive to such habitats. All have spiky tails and splashes of orange and yellow.

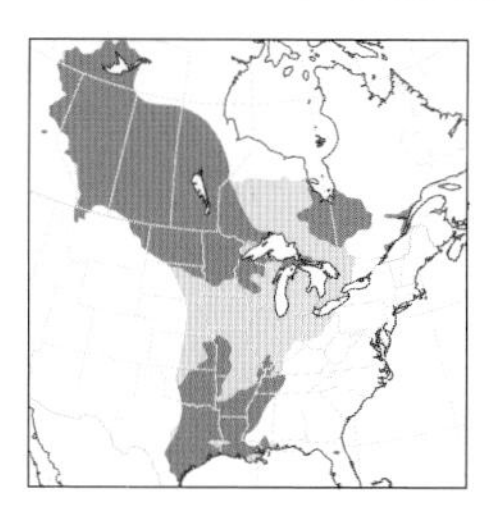

### LeConte's Sparrow

*Ammospiza leconteii* | LCSP | L 5" (13 cm)

Famously secretive, this gorgeous sparrow broadly overlaps throughout its annual cycle with the also notoriously reclusive Yellow Rail (p. 94).

■ **APPEARANCE** Slim and small-billed; brightly colored. In good view, adult shows white median crown stripe, fine dark streaks on golden breast, and unique lavender streaking on nape. Flushing, shows orange rump and black-streaked back. Juvenile, quite streaked below, has buffy crown stripe.

■ **VOCALIZATIONS** Song a few weak clicks followed by shrill buzz, like Grasshopper Sparrow's (p. 338). Flight call a fine, descending buzz.

■ **POPULATIONS** Least aquatic of *Ammospiza* sparrows, always in dense cover.

### Seaside Sparrow

*Ammospiza maritima* | SESP | L 6" (15 cm)

This habitat specialist is utterly dependent on coastal salt marshes—especially those dominated by cordgrass and haunted by Clapper Rails (p. 90).

■ **APPEARANCE** Largest and darkest *Ammospiza,* with big bill. Adult has yellow in front of eye and dark "whisker" on pale throat. Streaky juvenile told from others in genus by larger size, larger bill, duskier plumage.

■ **VOCALIZATIONS** Song starts with a flew clacks, then goes into a trailing buzz: *claa-claa cl'ZZZzzzz.* Flight call a steady, husky whistle.

■ **POPULATIONS** Widespread *maritima* dark, dusky; Gulf Coast birds, including *fisheri,* brighter; endangered *mirabilis,* only in Everglades, greenish.

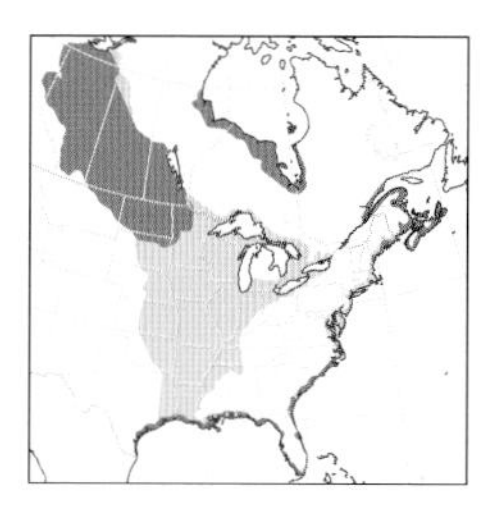

### Nelson's Sparrow

*Ammospiza nelsoni* | NESP | L 4¾" (12 cm)

Among the *Ammospiza* sparrows, Nelson's is the most catholic in its habitat preferences. Three discrete populations breed coastally and well inland.

■ **APPEARANCE** A bit larger and longer-billed than LeConte's. Adult has gray crown stripe (white on LeConte's), dull orange face, and gray nape; orange wash across breast contrasts with white belly. Juvenile bright pumpkin orange, relatively unstreaked below.

■ **VOCALIZATIONS** Song a short, sputtering buzz: *k'jzzzz-k'.* Flight call like Seaside's, but more piercing.

■ **POPULATIONS** Nominate interior subspecies brightest orange with relatively short bill; coastal *subvirgatus* duller, with diffuse streaking below; *alterus* of Hudson Bay Lowlands intermediate.

### Saltmarsh Sparrow EN

*Ammospiza caudacuta* | SALS | L 5" (13 cm)

The Saltmarsh Sparrow occurs only in tidal marshes, where it is imperiled by rising sea levels. Until 1995, Saltmarsh and Nelson's Sparrows were lumped as a single species.

■ **APPEARANCE** Averages longer-billed than Nelson's; long bills in sparrows are often adaptations for feeding in mud. Adult Saltmarsh has paler breast with heavier streaking than Nelson's; in good photos, note black streaking on broad orange eyebrow. Juvenile darker, streakier than juvenile Nelson's.

■ **VOCALIZATIONS** Half-hearted song soft and gasping; males do not defend territory. Flight call tinnier than Nelson's.

■ **POPULATIONS** Hybridizes with *subvirgatus* Nelson's from Mass. to Me.

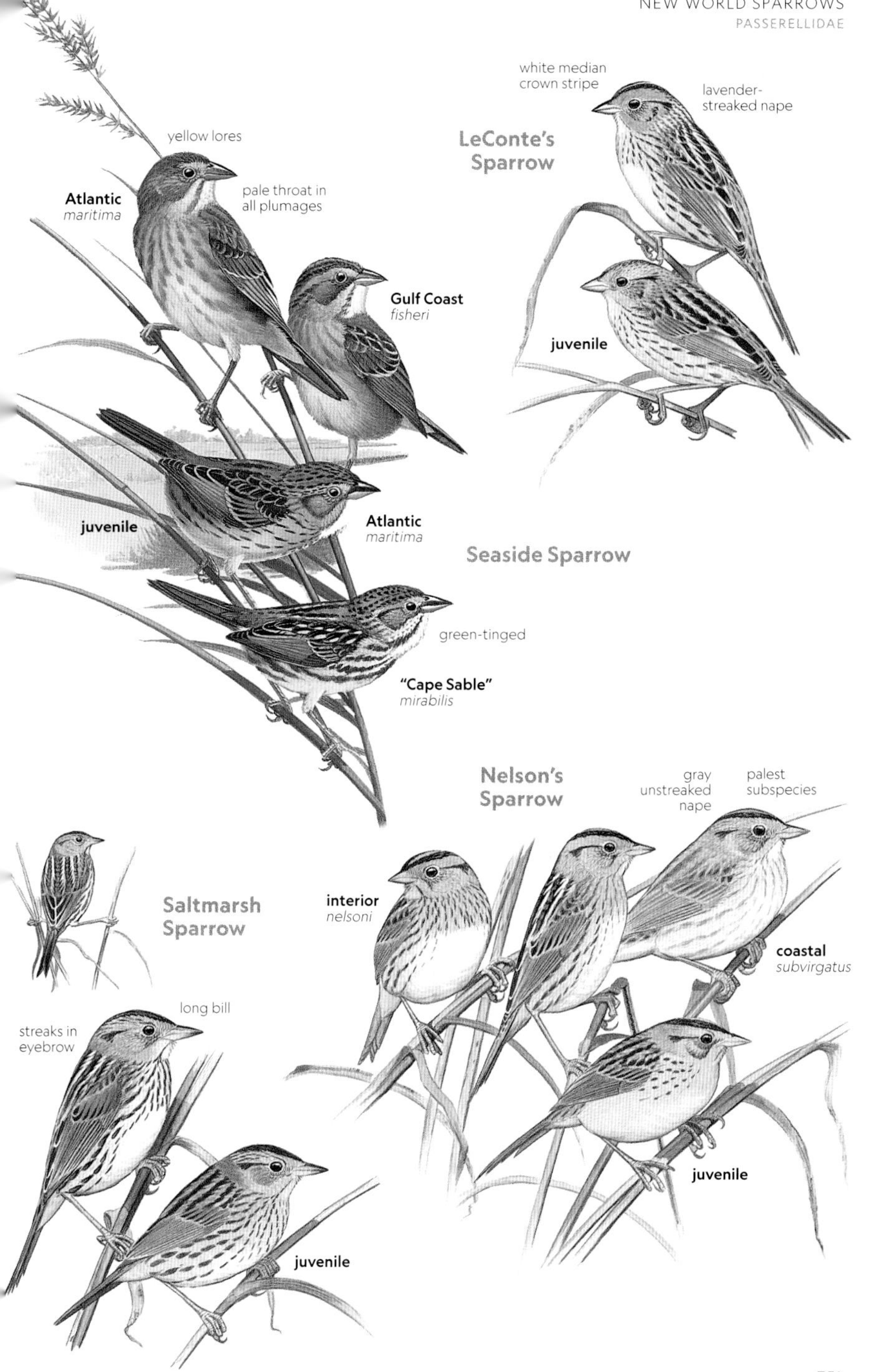
white median crown stripe
lavender-streaked nape
LeConte's Sparrow
juvenile
yellow lores
pale throat in all plumages
Atlantic
maritima
Gulf Coast
fisheri
juvenile
Atlantic
maritima
Seaside Sparrow
green-tinged
"Cape Sable"
mirabilis
Nelson's Sparrow
gray unstreaked nape
palest subspecies
interior
nelsoni
coastal
subvirgatus
juvenile
Saltmarsh Sparrow
long bill
streaks in eyebrow
juvenile

**THREE TALLGRASS SPARROWS**
The widespread Savannah Sparrow, in its own genus (*Passerculus*), is closely related to the range-restricted and secretive Baird's and Henslow's Sparrows. All have earth-toned highlights and fine streaking below.

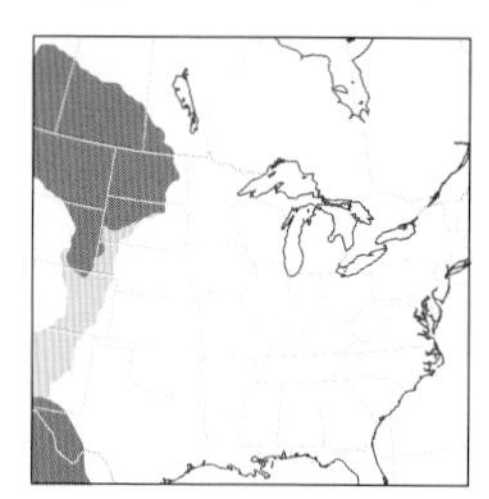

## Baird's Sparrow

*Centronyx bairdii* | BAIS | L 5½" (14 cm)

Along with the Sprague's Pipit (p. 322), the Baird's Sparrow is a major reason for many birders' summertime pilgrimages to the northern Great Plains. The two breed on wide-open prairie, and they winter in grasslands near the U.S.-Mexico border.

■ **APPEARANCE** Of medium build with pale ochre highlights. Head of adult has two dark spots on face, thin black "whisker," and ochre crown stripe. Breast has fine black streaks; wing coverts and flanks flecked chestnut. Juvenile darker, with heavier streaking below and scalloped upperparts.

■ **VOCALIZATIONS** Song, airy and tinkling, starts with one to three short whistles, then a loose trill, then a faster trill. Flight call a clipped, high *si*.

■ **POPULATIONS** In steady, long-term decline. Requires extensive tracts of undisturbed fescue and sedges.

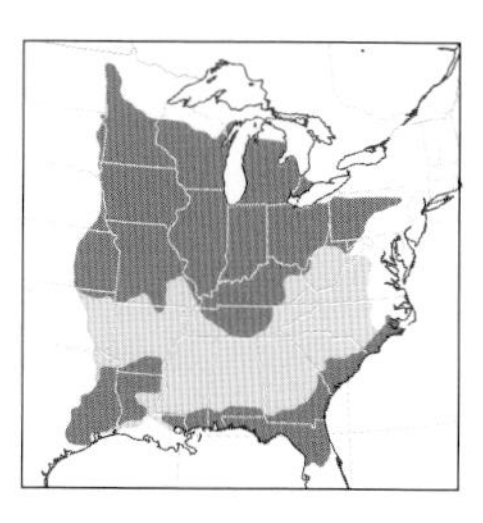

## Henslow's Sparrow

*Centronyx henslowii* | HESP | L 5" (13 cm)

Bird books from the 20th century invariably described the species' song as feeble or simple. We now know that the song is rich and beautiful—but beyond the ability of human brains to discern.

■ **APPEARANCE** Slightly smaller, relatively larger-billed, and darker than Baird's, with olive-chartreuse highlights. Dark chestnut of wings contrasts with greener head and buffier breast, extensively streaked. Finely streaked juvenile paler overall; compare with Grasshopper Sparrow (p. 338).

■ **VOCALIZATIONS** Human brains process the song as a very short *fshLCK*, but spectrograms reveal remarkable complexity; the bird hears slurred whistles and tinkling trills. Complex flight call wavering and descending, *swyeeew*.

■ **POPULATIONS** Extirpated from much of former range east of Appalachians; takes well to reclaimed strip mines in Ohio River Valley. Many in winter occur in understory of pinewoods; favors wetter microhabitats than Bachman's (p. 336).

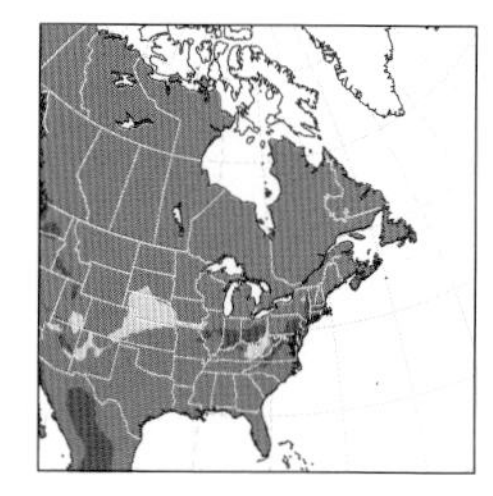

## Savannah Sparrow

*Passerculus sandwichensis* | SAVS | L 5½" (14 cm)

Is it "just" a Savannah Sparrow? In so many instances in prairie birding in the East, this is the first question to ask when encountering a streak-breasted sparrow.

■ **APPEARANCE** Small, plump, and short-tailed, with a slight bill; the default sparrow in extensive open habitats, both dry and wet. All are streaked below, lacking any obvious patterning; most show broad yellow eyebrow and thin white central crown stripe. Compare with streak-breasted sparrows of grasslands and marshes (pp. 348–353) as well as longer-tailed Song Sparrow (p. 354) of gardens and streamsides.

■ **VOCALIZATIONS** Song begins with a few clipped notes, very high, followed by long buzz then short buzz; song of Grasshopper Sparrow (p. 338) has just one buzz following introductory notes. Call note a light *tsip*, given frequently; flight call a descending *tsyee*.

■ **POPULATIONS** Ghostly "Ipswich Sparrow," subspecies *princeps*, breeds only on Sable I., N.S., but winters on outer dunes along a 1,000-plus-mile (1,600-plus km) stretch of Atlantic coast.

no postocular stripe
golden-buff head
Baird's Sparrow
juvenile
greenish head
richly colored back
Henslow's Sparrow
juvenile
yellow lores
widespread
savanna
Savannah Sparrow
northeastern
labradorius
short tail
"Ipswich"
princeps
pale secondary panel

**GENUS *MELOSPIZA***

The three sparrows in this genus are of medium stature and are dark overall. They sing three very different songs, and their chip notes are distinctive too.

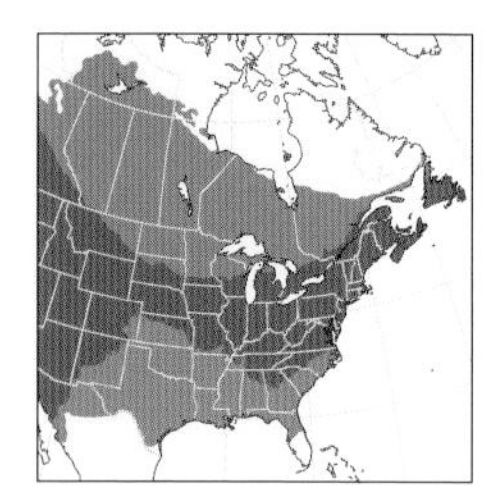

## Song Sparrow

*Melospiza melodia* | SOSP | L 5½" (14 cm)

Around towns and cities in the East, this is often the most conspicuous passerellid. It is also common away from human habitation, particularly along creeks and pond edges.

■ **APPEARANCE** Dark and streaky, with a long tail, flipped about in flight. Gray-brown overall, colder than larger Fox Sparrow (p. 344); darker than shorter-tailed Savannah Sparrow (p. 352). Breast streaking coalesces in central splotch. Juvenile like adult, but buffier with looser plumage.

■ **VOCALIZATIONS** It is indeed *the* Song Sparrow, in many urban areas the only passerellid with an obviously tuneful song. Songs highly varied, but all follow basic pattern of a few pulsing chirps, followed by a trill and a medley of short whistles. Call note a nasal *chimp* like Winter Wren's (p. 302). Flight call a high, thin *tseee.*

■ **POPULATIONS** Varies in color and brightness, with at least two dozen named subspecies across entire range; most in East are nominate *melodia.*

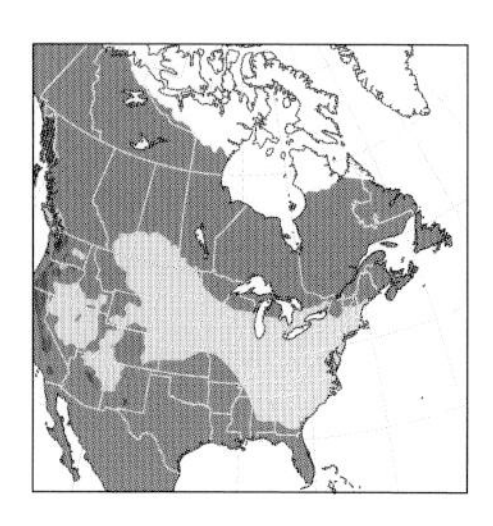

## Lincoln's Sparrow

*Melospiza lincolnii* | LISP | L 5¾" (15 cm)

Away from the breeding grounds, Lincoln's Sparrows are found singly in brushy, often swampy habitats. They nest at the edges of boreal bogs.

■ **APPEARANCE** Smaller-bodied, shorter-tailed, and smaller-billed than Song Sparrow. Adult finely streaked above and below; background color of breast yellow-buff. Thin white eye ring and crisply marked crown enhance sharp look overall—not as pixelated in appearance as Song Sparrow. Juvenile Lincoln's browner overall; similar to juvenile Song, but shorter-tailed and slighter-billed. Compare with juvenile Swamp Sparrow.

■ **VOCALIZATIONS** Song comprises three fast trills or stutters, each on a different pitch: *layda-layda-layda jurr-jurr-jurr-jurr-jurr plee-plee-plee-plee.* Call note a hard *tsik;* flight call a fine buzz.

■ **POPULATIONS** Enjoys vast breeding range in boreal N. Amer., especially West; breeders in Que. may be in decline, though.

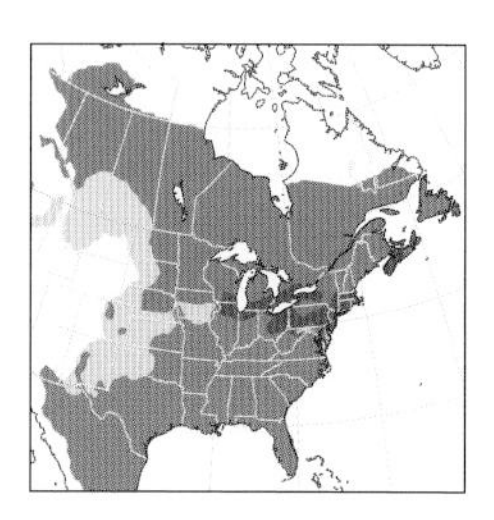

## Swamp Sparrow

*Melospiza georgiana* | SWSP | L 5¾" (15 cm)

Although a few occur in swamps, mostly in winter, "Marsh Sparrow" would have been a better name. The species abounds in densely vegetated marshes, both coastally and inland, and in winter in weedy meadows.

■ **APPEARANCE** Built like Lincoln's Sparrow, with which it sometimes occurs. All have white throat like larger White-throated Sparrow (p. 346). Breeding adult has rusty cap, browner in winter; wings have broad reddish-brown panel, flanks warm buff. First-winter dark and muddy, but all show rusty wings, buff flanks, and white throat; rusty wings already prominent on briefly held juvenile plumage.

■ **VOCALIZATIONS** Song a loud, loose trill; slower and more jangling than Dark-eyed Junco's (p. 348) and Chipping Sparrow's (p. 342). Call note a bright *chip,* sharply descending, like Palm Warbler's (p. 394) or Eastern Phoebe's (p. 262). Flight call like Lincoln's.

■ **POPULATIONS** Prone to midwinter fluctuations, likely the result of birds moving around as marshes freeze and thaw.

Song
Sparrow
dark brown
facial stripes
blotchy
streaks
long,
rounded
tail
juvenile
Lincoln's
Sparrow
buff extends
into face
often appears
crested
yellow-buff
background
fine white
eye ring
juvenile
breeding
mostly
unstreaked
gray breast
Swamp Sparrow
winter
adult
whitish
throat in all
plumages
juvenile
immature
rusty
wings

## ECLECTIC TOWHEES

Birds called towhee are basically large, long-tailed sparrows—and some of them are rather colorful. But the name is taxonomically imprecise, with the Rufous-crowned Sparrow, for example, falling within the towhee assemblage.

### Canyon Towhee

*Melozone fusca* | CANT | L 8" (20 cm)

This is a towhee of arid landscapes: dry washes, rocky slopes, and sometimes open desert dotted with cholla and yucca.

■ **APPEARANCE** Our largest sparrow, with the heft of a Red-winged Blackbird (p. 370). Adult mostly unstreaked dirty gray, but with rufous cap and undertail coverts. Throat pale buff. Upper breast has variable streaking; central breast has diffuse black splotch, also variable. Juvenile muddy brown, diffusely streaked below, with rust on crown and under tail starting to show.

■ **VOCALIZATIONS** Call note an abrupt, nasal *squeent!* or *squint!* Song a variable trill often preceded by the call note: *squint! le-le-le-le-le-le-le-le-le-le.* A fine high buzz may function as flight call.

■ **POPULATIONS** Sedentary permanent resident east to the western outskirts of Austin, Tex.; rarely wanders even short distances from core range.

### Rufous-crowned Sparrow

*Aimophila ruficeps* | RCSP | L 6" (15 cm)

Often occurring alongside the Canyon Towhee in rocky drylands, the Rufous-crowned Sparrow was recently assigned to an adjacent position in bird guides too: The genera *Melozone* and *Aimophila* are now classified as closest relatives.

■ **APPEARANCE** Of medium build; long-tailed and plain-breasted. General color scheme, including reddish crown and white eye ring, suggests smaller Field Sparrow (p. 344), but note black "whisker" and darker bill of Rufous-crowned. Juvenile, streaked below, shows dark "whisker" and hint of eye ring.

■ **VOCALIZATIONS** Bubbly song a rapid outpouring of chirps, calling to mind House Wren's (p. 302). Call a twangy, descending *gwaay* or *gweer,* heard mostly during courtship and breeding season.

■ **POPULATIONS** Particularly attracted to slopes with loose gravel; hanging on by a thread in Ark. Like Canyon Towhee, rarely strays from core range, although there is a remarkable record from Wis.

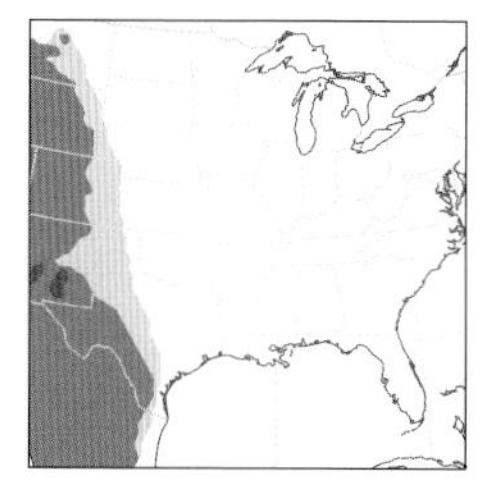

### Green-tailed Towhee

*Pipilo chlorurus* | GTTO | L 7¼" (18 cm)

Green, rufous, black, gray, and white—the Green-tailed Towhee is perhaps our most colorful sparrow. Breeding mostly in the Interior West, it winters regularly to Tex. and casually farther east.

■ **APPEARANCE** Our smallest towhee, not much bigger than Olive Sparrow (p. 338). Adult unmistakable in good view, with bright rufous crown, gleaming white throat, extensive bright olive above, and yellow undertail coverts; lurking in shadows, could be confused with Olive Sparrow. Heavily streaked juvenile suffused in dark olive-yellow.

■ **VOCALIZATIONS** Song a striking outburst of buzzes and slurred whistles, but not likely to be heard in East. Distinctive call a mewing, nasal *mreeeay.* Like other towhees, gives a fine buzz, probably a flight call.

■ **POPULATIONS** Uncommon migrant western Great Plains; regular in winter to southern Tex. Annual vagrant in recent years elsewhere in East.

rufous crown
Canyon Towhee
large overall, long-tailed
Rufous-crowned Sparrow
thick black malar stripe
juvenile

Green-tailed Towhee
spring
fall
white throat
olive wings and tail
juvenile

## RUFOUS-SIDED TOWHEES

Formerly treated as a single species, called the Rufous-sided Towhee, these two differ appreciably in voice and plumage. They sometimes hybridize where they come into contact in the Great Plains.

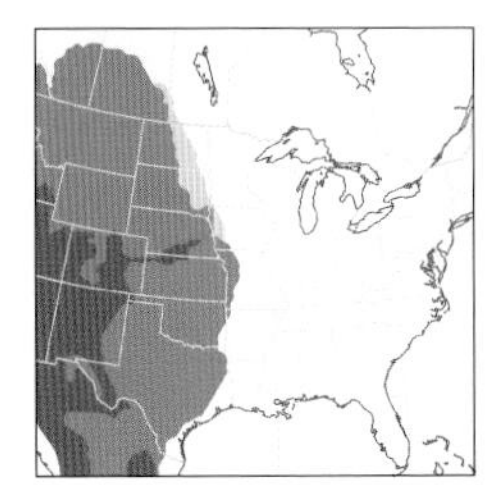

### Spotted Towhee

*Pipilo maculatus* | SPTO | L 7½" (19 cm)

In riparian habitats from Nebraska north to southern Saskatchewan, the boldly marked Spotted Towhee adds a western element to birding the Great Plains. Both rufous-sided towhees feed on the ground with jerky double kickbacks that send leaf litter flying.

■ **APPEARANCE** Large, long-tailed sparrow with red eyes. Both species in this complex are dark above with bright rufous sides and white tail corners. Chief plumage difference from Eastern is the upperparts, obviously spotted on Spotted; Eastern has only a white "pocket handkerchief" on otherwise dark wings. Males of both rufous-sided towhees have black hood. Female Spotted gray-black; female Eastern lighter, browner. Juvenile Spotted, streaked and smudgy, shows white tail corners of adult and, usually, some spotting on back and wings.

■ **VOCALIZATIONS** Classic song of subspecies *arcticus,* the breeder in our area, a few dry chips followed by an unmusical trill: *plep-plep-plep tlee-e-e-e-e-e-e-e-e.* Call note a weakly rising, very nasal *wraaah.* Like other towhees, gives a fine, high buzz, apparently a flight call.

■ **POPULATIONS** Widespread migrant on western Great Plains; singles wander regularly, fall to winter, east to Lake Michigan, often finding their way to feeders. Very rare but annual, also fall to winter, to eastern Great Lakes eastward.

### Eastern Towhee

*Pipilo erythrophthalmus* | EATO | L 7½" (19 cm)

*"Drink your tea!"* The song of the Eastern Towhee has one of the most famous mnemonics of any bird in the East. In fact, the name towhee is derived from a disyllabic variant of this species' song.

■ **APPEARANCE** Size and build identical to Spotted Towhee. Most are easily told from Spotted by unspotted upperparts; the "pocket handkerchief" is created by white at the bases of the primaries, a mark lacking on Spotted. Eye color variable: Florida breeders, subspecies *alleni,* have staring white eyes; subspecies *rileyi,* farther north in southeastern coastal plain, has rose-peach eyes; the rest are red-eyed like Spotted. Adult male blacker than the paler, browner adult female; juveniles dark and muddy, already showing pale tail corners.

■ **VOCALIZATIONS** Primary song, strongly tripartite, comprises a sharp chirp, followed by a lower whistle, followed by a higher trill: *pleenk! eeeeer tee-ee-ee-ee-ee-ee-ee* (*Drink your tea!*). Often shortened to just one introductory note, either the chirp or the whistle, followed by the trill: *tow-hee!* Call rises more, is less nasal, than Spotted's: *chreee!* (and the source of yet another name for this species, the "Chewink"). Gives a thin buzz, perhaps a flight call, like Spotted's.

■ **POPULATIONS** Hybrids in the Great Plains are relatively frequent, and full-species rank for the two rufous-sided towhees has been questioned. A typical hybrid has spotting on the wing coverts (like Spotted) and the white "handkerchief" (like Eastern); calls and songs may sound like either parental species, or a mix of the two.

## Spotted Towhee

## Eastern Towhee

*erythrophthalmus*

Florida birds with less white

*alleni*

*erythrophthalmus*

## TWO TAXONOMIC WAIFS

The spindalis and the chat, formerly in the tanager and warbler families, respectively, are now placed in their own families. The chat is closely related to the blackbirds (pp. 362–375); the spindalis is provisionally in a grouping that includes the warblers (pp. 376–399).

### Western Spindalis

*Spindalis zena* | WESP | L 6¾" (17 cm)

Compact overall; stout-billed, short-tailed. Male unmistakable with bold black and white on head and wings, plus splashes of orange-yellow on breast and back. Olive-brown female has white panel on wing. Songs and calls high, squeaky, and grating. Strays annually, mostly from the Bahamas, to southeastern Fla. Very rare most years, but more than a dozen in 2017, chiefly spring. Several subspecies have reached our area; nominate *zena* ("Black-backed") the most frequently reported.

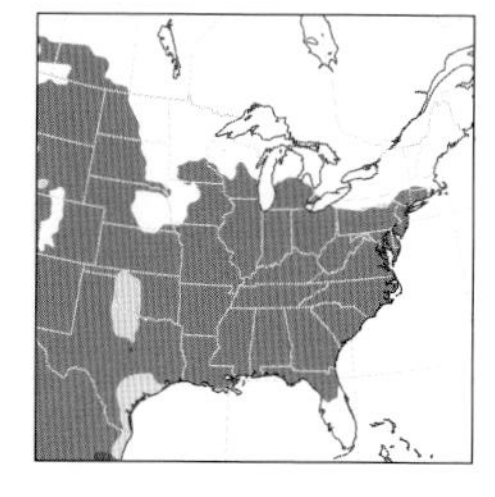

### Yellow-breasted Chat

*Icteria virens* | YBCH | L 7½" (19 cm)

Thrushes (pp. 312–319) and mimids (pp. 306–309) get all the glory, but the chat is among our most gifted vocalists. It rivals the Northern Mockingbird (p. 308) in its diversity of utterances and propensity for night singing.

■ **APPEARANCE** Larger than any warbler; long-tailed, thick-billed. Olive above with white "spectacles" and brilliant yellow below.

■ **VOCALIZATIONS** Incredible run-on song a halting mix of outbursts like the sounds laser guns make in the movies, endearingly sad sighs, and soft chuckles: *bleep! bleep! bleep!... saaad GLORP? syew! syew!... aah... true? eeee bleep! bleep! bleep!* Sings in dense vegetation, but also "skylarks" (see sidebar, p. 322); mimics birds and other animals. Calls include a harsh *raaa* and knocking *plock*. Apparently lacks a flight call.

■ **POPULATIONS** Nests in shrubs in fallow fields and at woodland edges; secretive and hard to see on migration. Regular in very small numbers in winter in coastal thickets; annual to Atlantic Provinces, where some Christmas Bird Counts record low double digits.

### New Families

The nine-primaried oscines (pp. 324–409) include some of our largest families: warblers, New World sparrows, tanagers, and, in most arrangements, finches. In the influential and comprehensive fifth edition (1957) of the *Check-list* of the American Ornithological Society (known at the time as the American Ornithologists' Union), *all* of the indigenous U.S. and Canadian nine-primaried oscines were placed in only four large to huge families. But there was always some discomfort with that scheme. For example, the Yellow-breasted Chat was classified as a warbler yet was so un-warblerlike in various respects.

The problem, it turned out, was the assumption that there were only a few large families. We learned that not only was the chat not a warbler, but, based on DNA, *it was also not anything else*. It's a chat, in its own, single-species family, Icteriidae. The situation is almost the same with the bird now called the Western Spindalis. Formerly placed in the tanager family, it was called the Stripe-headed Tanager; however, that bird was found to be not a tanager but, rather, a spindalis—another new family. (A difference from the Yellow-breasted Chat's reclassification is that the erstwhile Stripe-headed Tanager was found to be a complex of four closely related species, all christened spindalises, whereas the chat is just one species.)

The ultimate irony in all this involves the formerly huge, but now comparatively modest, tanager family Thraupidae. Some of our most beloved tanagers, including the searing Scarlet Tanager (p. 400), are now classified as cardinals, while bananaquits and seedeaters (p. 408), formerly in other families, are now tanagers!

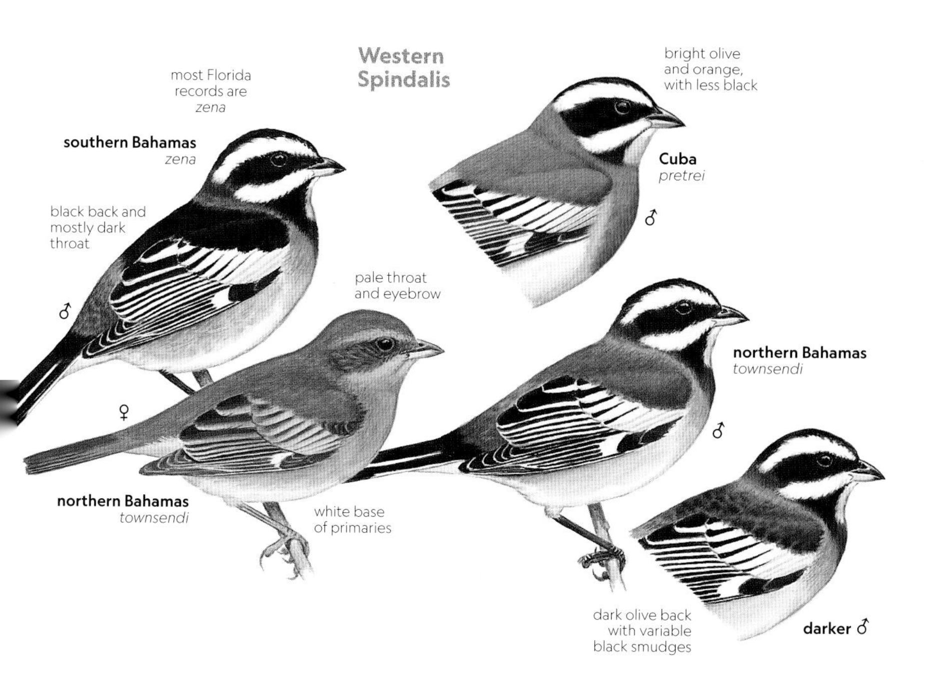
Western Spindalis
most Florida records are *zena*
southern Bahamas *zena*
black back and mostly dark throat
♂
bright olive and orange, with less black
Cuba *pretrei*
♂
pale throat and eyebrow
♀
northern Bahamas *townsendi*
white base of primaries
northern Bahamas *townsendi*
♂
dark olive back with variable black smudges
darker ♂

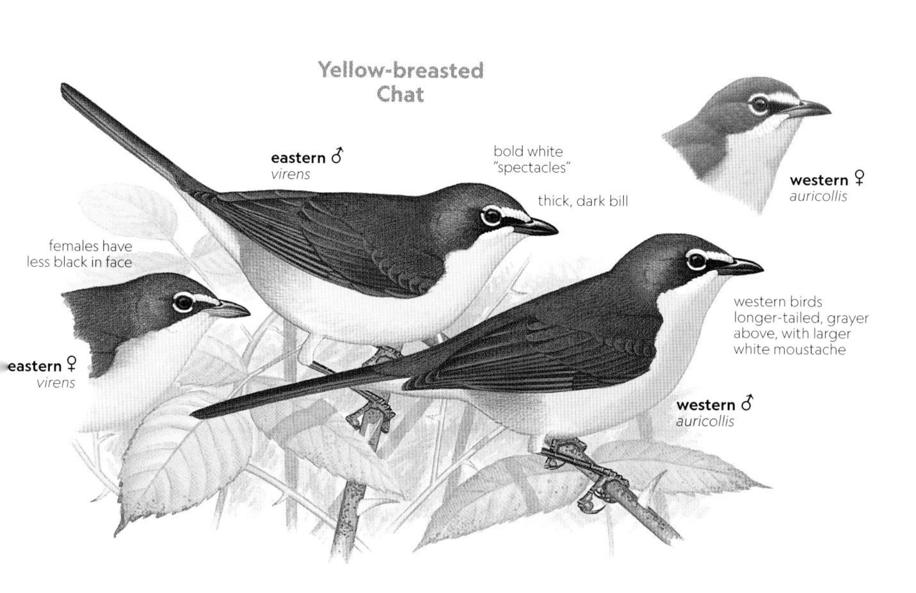
Yellow-breasted Chat
eastern ♂ *virens*
bold white "spectacles"
thick, dark bill
western ♀ *auricollis*
females have less black in face
eastern ♀ *virens*
western birds longer-tailed, grayer above, with larger white moustache
western ♂ *auricollis*

## TWO MEADOWLARK RELATIVES

The New World blackbirds (pp. 362–375), family Icteridae, are a diverse lot, with strong bills and loud vocalizations. The distinctive Yellow-headed Blackbird and Bobolink, each with their own genus, are in a grouping, or clade, with the well-known meadowlarks (p. 364).

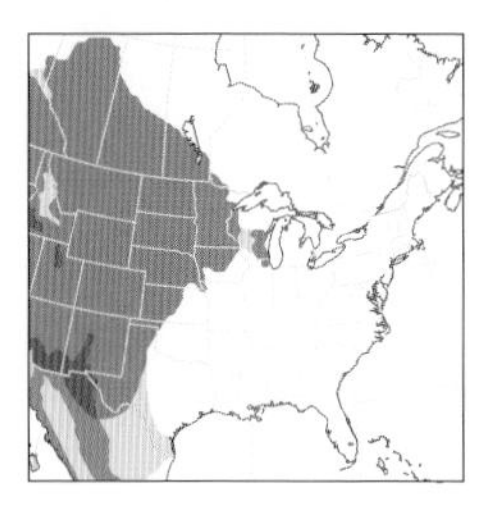

### Yellow-headed Blackbird

*Xanthocephalus xanthocephalus* | YHBL | L 9½" (24 cm)

The "Bananahead" is a loud and conspicuous denizen of prairie wetlands. It is a polygynous species, locally abundant in extensive cattail marshes.

■ **APPEARANCE** Larger and stouter-billed than Red-winged Blackbird (p. 370), with which it often co-occurs; has long, curved toenails for gripping vertical stems. In addition to trademark yellow head of adult male, note white wing patch and odd circle of yellow feathers around cloaca. Smaller adult female is browner above, with variable, usually extensive, yellow-brown below. Variable subadults told by extensive, if muted, yellow on head and breast, combined with large size and some white in wing.

■ **VOCALIZATIONS** Song of male a couple of mechanical clanks followed by a tormented scream, like a very loud and distorted marsh sparrow (p. 350). When they sing, males always turn their head to the left. Diverse female vocalizations include a rolling chatter. Both sexes give hollow clucks.

■ **POPULATIONS** Center of abundance Dakotas and Prairie Provinces, but breeds east sparingly to western Lake Erie. Can be abundant in one marsh, absent in another nearby; sensitive to both habitat quality and tract size. Marsh Wrens peck open and destroy Yellow-headed Blackbird eggs, causing significant mortality. Regular in winter, sometimes in very small flocks, to East Coast, especially Fla., usually with other blackbird species.

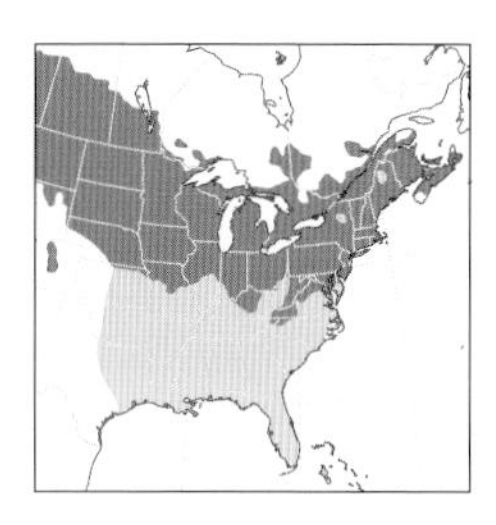

### Bobolink

*Dolichonyx oryzivorus* | BOBO | L 7" (18 cm)

One of the most astounding songbird migrants in the East, the Bobolink winters mostly in Bolivia and northern Argentina. Its unusual molt strategy reflects the species' extreme migration biology.

■ **APPEARANCE** Fairly small blackbird, about the size of Brown-headed Cowbird (p. 370); tail short and spiky. Breeding and nonbreeding males utterly different; they were originally described as separate species. Breeding male black below, with creamy nape and white on wings and rump; nonbreeding male buffy overall with streaking on back and flanks. Adult female plumage year-round like nonbreeding adult male; note pink bill of all females and nonbreeding males. Flushing from dense grass, female and nonbreeding male Bobolinks can be mistaken for buffy, spike-tailed marsh sparrows (p. 350).

■ **VOCALIZATIONS** Song of adult male squeaky and metallic, with abrupt changes in pitch (the "R2-D2 Bird"). Distinctive flight call, given day or night, a nasal, rising *weenk?*

■ **POPULATIONS** Breeds in wet meadows and hayfields, often stops in coastal marshes on migration; returns fairly late in spring, on the move south by Aug. Among U.S. and Canadian breeding birds, only the Franklin's Gull (p. 142) and Bobolink have two complete annual molts, presumably reflecting the two species' grueling migrations and constant exposure of feathers to the sun.

Yellow-headed Blackbird
variable white in wing
♂
♀
yellow breast and eyebrow
buffy head
venile ♂
adult ♂
Bobolink
♀
breeding ♂
pink bill
long wings
heavily streaked
fall
breeding ♀
breeding ♂
early spring ♂
buffy edges when fresh

## MEADOWLARKS

With brilliant yellow underparts crossed by a sharp black breastband, meadowlarks are instantly recognizable. But distinguishing between the two species is often challenging.

### Eastern Meadowlark

*Sturnella magna* | EAME | L 9½" (24 cm)

In farm country across the eastern U.S. and extreme southeastern Canada, this plump songster heralds spring with its whistled song, slurred and sweet.

■ **APPEARANCE** Both meadowlarks are rotund, short-tailed, and short-winged, with sturdy legs and long bills. They fly low to the ground in level flight, suggesting profile of European Starling (p. 310), but white outer rectrices (tail feathers) of meadowlark conspicuous as bird flushes. Visual differences between Eastern and Western are slight and somewhat variable: malar of Eastern, extending from base of lower mandible, mostly white in spring (mostly yellow on Western); outer three or four rectrices of Eastern white (on Western, outer two or three rectrices white); upperparts of Eastern darker and richer overall than upperparts of paler, grayer Western; bars on tail thicker on Eastern. Stripes on head of Eastern sharper than on Western, but both meadowlarks are darker and less sharply patterned in fall and winter than in spring and summer.

■ **VOCALIZATIONS** Song comprises four to five clear, slightly descending whistles: *see syoo see-uh-syeeer.* Primary call a harsh, flatulent *dzzrnt,* unlike corresponding primary call of Western. Both species give a buzzy chatter, typically on flushing or landing, higher in Eastern. Flight call of both a rising, nasal *wheen,* higher in Eastern.

■ **POPULATIONS** Widespread, extending to S. Amer., and geographically variable. Most in East are nominate *magna; argutula* of southeastern U.S. darker, and *hoopesi* of South Tex. lighter. **Note:** Chihuahuan Meadowlark, *Sturnella lilianae,* very recently split from Eastern, is regular in our area along far western edge of Tex. Panhandle; see p. 416.

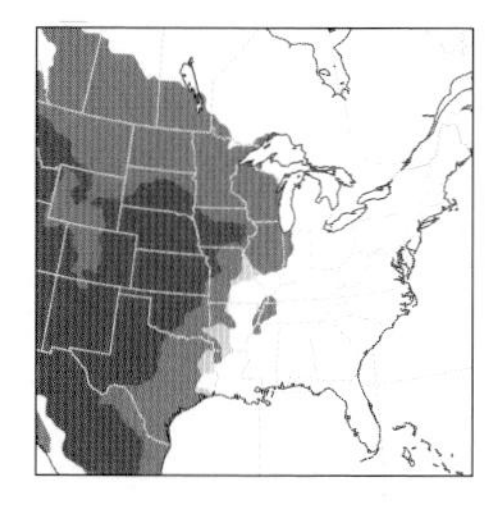

### Western Meadowlark

*Sturnella neglecta* | WEME | L 9½" (24 cm)

The scientific name *neglecta* commemorates the fact that European naturalists initially neglected to notice that this widespread and abundant meadowlark is a different species from its eastern congener.

■ **APPEARANCE** Identical in size and shape to Eastern, and very similar in plumage. Malar region of Western mostly yellow in spring (mostly white on Eastern); Western has two or three white outer rectrices (three or four on Eastern). Western paler overall, especially above, with less-striking head stripes than most Easterns.

■ **VOCALIZATIONS** Primary song a series of descending notes, like Eastern's but more complex: Western's song lower, richer, more gurgling, with six to eight or more elements (four to five in Eastern's). Primary call a rich, rather musical *clurk,* very different from corresponding call of Eastern; also gives a low, whistled *lurrr,* especially when alert or alarmed. Rattles on flushing or landing like Eastern, but rattle lower-pitched than Eastern's. Flight call rising and nasal like Eastern's, but also lower.

■ **POPULATIONS** Overlaps broadly with Eastern, but proven hybrids relatively few. Where both occur, sometimes sing the "wrong" song or, more commonly, weird and indeterminate songs. Field ID in winter especially fraught; limits of winter range of Western not fully worked out. Listen for (and make recordings of) the primary call notes year-round, especially winter.

spring
darker
upperparts
than *magna*
southeastern
*argutula*
paler upperparts
than *magna*
South Texas
*hoopesi*
spring
northern
*magna*
Eastern
Meadowlark
northern
*magna*
short
triangular
wings
white
malar in all
subspecies
spring
spring
northern
*magna*
fall
juvenile
spring
yellow
extends into
malar
less white in tail
than Eastern
flanks with
finer spots, less
connected into
streaks than Eastern
Western
Meadowlark
overall paler
above than
Eastern
fall
juvenile

## ORIOLES I

The orioles (pp. 366–369) are mostly arboreal blackbirds with bright colors and loud songs. All are in the genus *Icterus,* well represented in the Neotropics.

### Orchard Oriole

*Icterus spurius* | OROR | L 7¼" (18 cm)

Two orioles are widespread in the East: the well-known Baltimore Oriole (p. 368) and the less familiar Orchard Oriole. The latter flourishes in tall trees along rivers and lakeshores, as well as in orchards and even marsh edges.

■ **APPEARANCE** Small, active, and long-tailed; bill slender, slightly decurved. Adult male chestnut and black; mostly yellow female as likely to be mistaken for a warbler as for another oriole. Male in first spring (second calendar year) like female but with narrow black bib.

■ **VOCALIZATIONS** Song rapid and bubbly; Baltimore has slower, lower song. Calls include snappy *chack* and Baltimore-like chatter.

■ **POPULATIONS** Early fall migrant; most gone from midlatitudes by Sept.

### Hooded Oriole

*Icterus cucullatus* | HOOR | L 8" (20 cm)

This oriole of the U.S.-Mexico borderlands and points south is especially fond of palms, both in the wild and in residential districts in towns and cities.

■ **APPEARANCE** Fairly small oriole with decurved bill. Black face of male encircled by orange cowl; Altamira Oriole (p. 368) larger with straighter, stouter bill and orange "shoulder" (white on Hooded). Female Hooded yellow-orange below, including belly; female Bullock's more yellowish, with mostly white belly and straighter bill.

■ **VOCALIZATIONS** Rambling song suggests Orchard Oriole or even a thrasher (pp. 306–309). Distinctive call a rising, nasal *veent?*

■ **POPULATIONS** Subspecies *sennetti* occurs in Lower Rio Grande Valley, nominate farther west; wanders far north, with scattered records to Great Lakes.

### Bullock's Oriole

*Icterus bullockii* | BUOR | L 8¼" (21 cm)

Bullock's and Baltimore (p. 368) Orioles hybridize to some extent where their ranges overlap, but they are not each other's closest relatives. Bullock's are lovers of shade trees, especially cottonwoods.

■ **APPEARANCE** Medium-size oriole with nearly straight bill. Adult male has mostly orange face with huge white wing panels. Female similar to female Baltimore; Bullock's has more color on face, whiter belly, and thin dark line through eye. First-spring male patterned like Orchard, but larger, straighter-billed, and more orangey. Hybrid male has "messy" black-and-orange face; many hybrid females not possible to ID in field.

■ **VOCALIZATIONS** Choppy song a few squeaks followed by short whistles; not as rich and melodic as song of Baltimore. Call a dry chatter; flight call a rising, nasal *wreent?*

■ **POPULATIONS** Rare but annual in winter to East Coast, especially at feeders.

### Spot-breasted Oriole

*Icterus pectoralis* | SBOR | L 9½" (24 cm)

Southeastern Fla. only. Indigenous to dry forests of Pacific slope of Mid. Amer.; was introduced to Miami area in 1940s. Adult has black spots on orange breast, orange "shoulder," and mostly orange head; younger birds told from Baltimore (p. 368) by larger size, brighter face, and a bit of orange on shoulder. Song low and rich, runs on longer than Baltimore's.

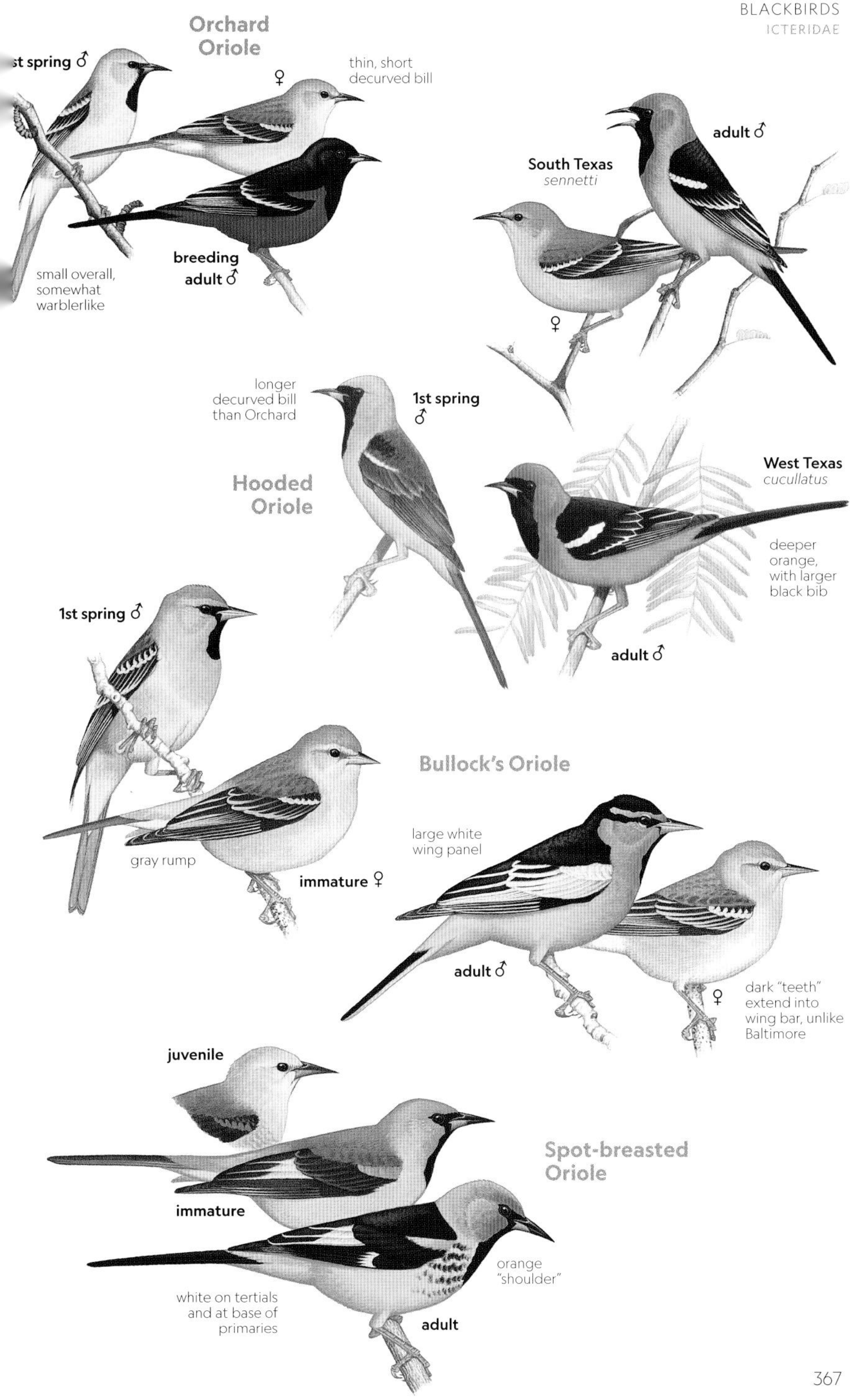

## Orchard Oriole

st spring ♂

♀

thin, short decurved bill

breeding adult ♂

small overall, somewhat warblerlike

South Texas *sennetti*

adult ♂

♀

## Hooded Oriole

longer decurved bill than Orchard

1st spring ♂

West Texas *cucullatus*

deeper orange, with larger black bib

adult ♂

## Bullock's Oriole

1st spring ♂

gray rump

immature ♀

large white wing panel

adult ♂

♀

dark "teeth" extend into wing bar, unlike Baltimore

## Spot-breasted Oriole

juvenile

immature

white on tertials and at base of primaries

adult

orange "shoulder"

## ORIOLES II

Adult male orioles (pp. 366-369) can be identified by their striking color schemes. To ID other plumages—including males well into their second year—pay attention to overall heft and bill structure.

### Altamira Oriole

*Icterus gularis* | ALOR | L 10" (25 cm)

An oversize oriole with intense orange hues, the Altamira is near the top of the must-see list for many birders visiting the Lower Rio Grande Valley.

■ **APPEARANCE** Large oriole with powerful bill. Adult has orange wing panel, white on smaller Hooded Oriole (p. 366). Immature has narrow black bib like immatures of several other orioles; wing patch yellow-orange.

■ **VOCALIZATIONS** Song comprises slow, rich, low whistles; runs on for 5–10 seconds. Calls include monosyllabic whistles and squawks.

■ **POPULATIONS** Relatively recent addition to our avifauna, first recorded in Tex. in 1939; expansion has stalled or reversed in 21st century.

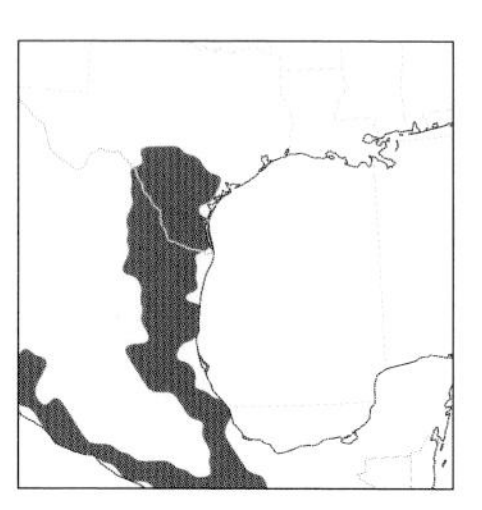

### Audubon's Oriole

*Icterus graduacauda* | AUOR | L 9½" (24 cm)

This less heralded of the two Tex. specialty orioles is more inclined to drier habitats—thickets and dry forest—than the showier Altamira Oriole.

■ **APPEARANCE** Lanky, fairly large oriole with average bill. Adult yellow and black like longer-billed Scott's, but the two species' ranges barely overlap. Audubon's has plain, dusky-yellow back in all plumages; back black on adult Scott's, streaked dusky on immature Scott's.

■ **VOCALIZATIONS** Song like Altamira's, but whistled notes even lower and richer, delivered more slowly. Varied calls include a rising, nasal squawk.

■ **POPULATIONS** Owing to secretive nature, hard to study, but likely in long-term decline in Tex. Less of a Lower Rio Grande Valley specialty than Altamira; regular north to San Antonio.

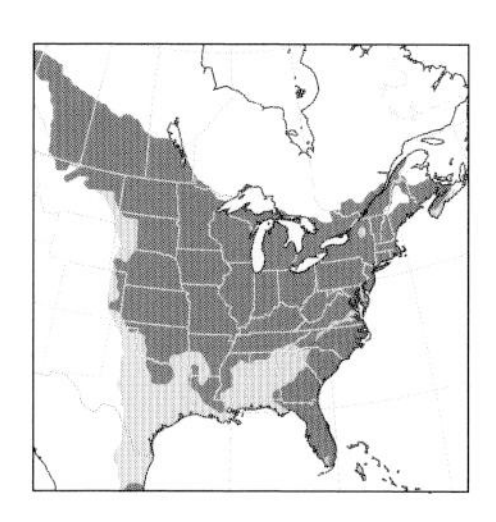

### Baltimore Oriole

*Icterus galbula* | BAOR | L 8¼" (21 cm)

Across much of the East, this is *the* oriole. It is common in diverse woodland types and readily visits feeding stations provisioned with orange slices and grape jelly.

■ **APPEARANCE** Of average build for an oriole, straight-billed and not particularly long-tailed. In most of East, Orchard (p. 366) is the only other oriole; in addition to different color schemes, Orchard is smaller, slighter-billed, longer-tailed, and more active. Compare with Bullock's Oriole (p. 366); all plumages of Bullock's more orange-faced than corresponding plumages of Baltimore. Adult male Baltimore has black head and limited white in wing; adult female Baltimore has orange tones overall and brown-black face; young Baltimore dusky-faced, body with orangish wash.

■ **VOCALIZATIONS** Song richer and flutier than Bullock's, usually lacking chattery elements of Bullock's. Chatter call and rising flight call like Bullock's.

■ **POPULATIONS** Expanded west in 20th century into Great Plains, bringing it into contact with Bullock's; hybrids not uncommon in zone of overlap.

### Scott's Oriole

*Icterus parisorum* | SCOR | L 9" (23 cm)

Beautiful black-and-lemon oriole of deserts and arid western woodlands; reaches East in Edwards Plateau of Tex. Compare with black-and-yellow Audubon's Oriole, with limited range overlap; Scott's longer-billed. Gurgling song like Western Meadowlark's (p. 364), but steadier, not descending.

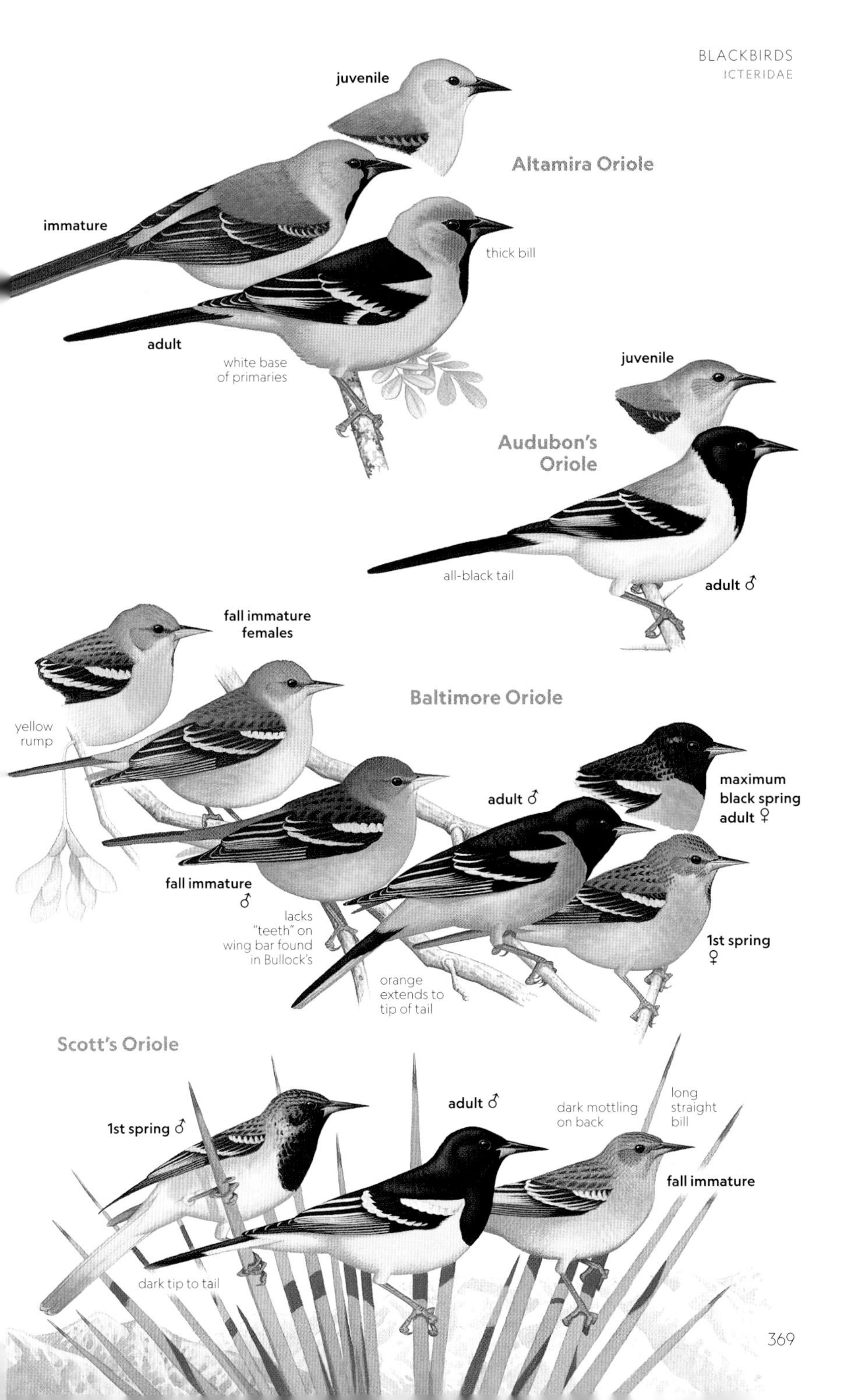
juvenile
Altamira Oriole
immature
thick bill
adult
white base
of primaries
juvenile
Audubon's
Oriole
all-black tail
adult ♂
fall immature
females
Baltimore Oriole
yellow
rump
maximum
black spring
adult ♀
adult ♂
fall immature
♂
lacks
"teeth" on
wing bar found
in Bullock's
1st spring
♀
orange
extends to
tip of tail
Scott's Oriole
1st spring ♂
adult ♂
dark mottling
on back
long
straight
bill
fall immature
dark tip to tail

## RED-WINGED BLACKBIRD, COWBIRDS

They are aggressive and maligned, even pestiferous, but the icterids on this spread are among the most behaviorally fascinating of all eastern birds.

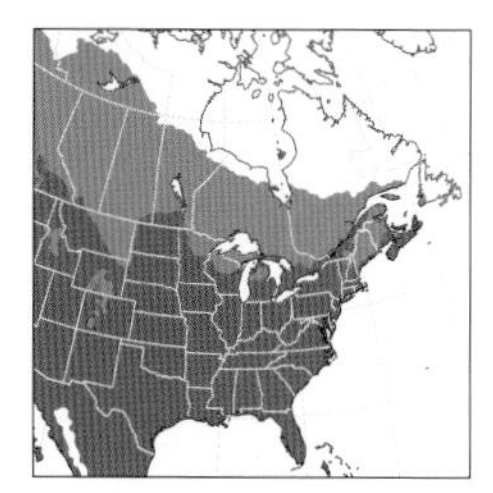

### Red-winged Blackbird

*Agelaius phoeniceus* | RWBL | L 8¾" (22 cm)

One of the surest signs of spring is a marsh full of singing Red-winged Blackbirds. Polygynous males are conspicuously territorial during the breeding season—and just as flocky and sociable when not breeding.

■ **APPEARANCE** Stout in build, but not as short-tailed as cowbirds; straight-billed. Adult male has red-and-yellow "epaulets," flared while singing, partially concealed otherwise. Sparrowlike female, smaller and streaked, frequently misidentified. (See sidebar, p. 372.)

■ **VOCALIZATIONS** Song, loud and grating, a few clicks, then a harsh trill, then a trailing hiss: *ch'ch'LEEEEEEEsss.* Calls include musical *cluck* and powerful, down-slurred whistle.

■ **POPULATIONS** Common to locally abundant; winter roosts persecuted for their supposed threat to agriculture.

### Shiny Cowbird

*Molothrus bonariensis* | SHCO | L 7½" (19 cm)

Cowbirds are obligate brood parasites: Females lay eggs only in other species' nests; young typically evict host parents' biological offspring from the nest. Arrived in Fla. in 1985 to initial alarm about impact on indigenous bird populations, but subsequent spread has been limited; not yet confirmed as breeder in U.S. Adult male glossy blue-purple all over, female gray-brown with weak eyebrow. Now regularly sighted across Fla. and nearby stretches of Gulf and Atlantic coasts, with vagrants much farther north and inland.

### Bronzed Cowbird

*Molothrus aeneus* | BROC | L 8¾" (22 cm)

Our largest cowbird, this is a Mid. Amer. counterpart of the widespread Brown-headed. The two overlap broadly across the southern tier of the U.S.

■ **APPEARANCE** Longer-billed and bulkier than Brown-headed. Adult male has red eye, puffed-out neck, and dark bronze cast. Gray-brown adult female also red-eyed, with neck "ruff" somewhat reduced.

■ **VOCALIZATIONS** Song high and wheezy; not as sharply slurred as Brown-headed's. Call a dull *chup.*

■ **POPULATIONS** Gradually expanding into U.S. Well established to central Gulf Coast; increasing in Fla.

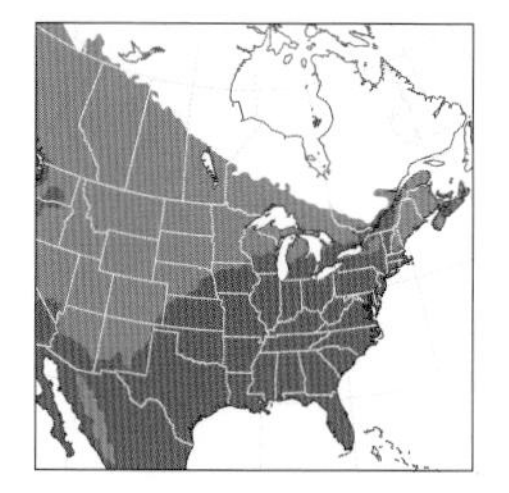

### Brown-headed Cowbird

*Molothrus ater* | BHCO | L 7½" (19 cm)

Few birds are more reviled than the Brown-headed Cowbird, by far the most widespread obligate brood parasite in the East.

■ **APPEARANCE** Smaller than Red-winged Blackbird, with short, conical, sparrowlike bill. Adult male has dull brown hood and metallic blue-green body. Adult female, plain grayish brown, is one of the most misidentified of all birds. Scaly juvenile often seen being fed by host parent.

■ **VOCALIZATIONS** Song a liquid gurgle rising sharply at end. Calls include a rattle and, in flight, an up-slurred whistle.

■ **POPULATIONS** Expanded in East with felling of forests in 19th century; parasitizes many hosts, including threatened species. Cowbird control controversial for ethical reasons, but also because its effectiveness is equivocal.

adult ♂
adult ♀
Red-winged Blackbird
heavily streaked underside
1st year ♂
pale supercilium
♀
sharp spike-like bill
Shiny Cowbird
♂
♂
Bronzed Cowbird
"ruff" often raised
♀
thick conical bill
juvenile
♂
parasitic female Brown-headed Cowbirds lay eggs in nests of other species
heavily streaked above and below
Yellow Warbler
juvenile
small conical bill
immature ♂ in molt
Brown-headed Cowbird
♀

## GENUS *EUPHAGUS*

These two are "grackles lite"—similar to grackles (p. 374) but smaller, with shorter tails, slighter bills, and less strident vocalizations.

### Rusty Blackbird

*Euphagus carolinus* | RUBL | L 9" (23 cm)

The entire annual cycle involves shady, swampy woods. Breeding, wintering, and even on migration, the Rusty Blackbird finds its way to well-wooded bogs, lakeshores, and streambanks.

■ **APPEARANCE** Slightly larger than Red-winged Blackbird (p. 370) with finer bill. Rusty plumage acquired in fall molt, wears down to glossy black (male) and dusky gray-brown (female) by spring. Eye yellow year-round; fall and winter birds sport broad buff supercilium and black face mask, fading by spring.

■ **VOCALIZATIONS** Song a rough gurgle followed by a rising whistle: *schwrlklwsh-WEEE*. Call a resonant *chenk*.

■ **POPULATIONS** Recently shown to be in decline, for reasons unknown; the finding caught conservation biologists off guard, for many eastern blackbirds are doing well.

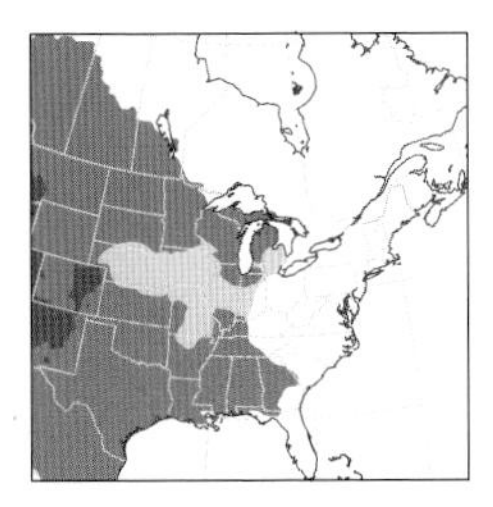

### Brewer's Blackbird

*Euphagus cyanocephalus* | BRBL | L 9" (23 cm)

During breeding season, found in diverse, mostly open habitats, typically with some water and vegetation. In winter in East, often makes its way to feedlots, where it enlists in mixed-species blackbird flocks; also has a well-known predilection for parking lots.

■ **APPEARANCE** Size and shape like Rusty Blackbird; bill thicker, straighter. Male dull glossy black in fall, wearing to bright glossy black by spring. Female gray-black with dark eye; female Rusty has pale eye.

■ **VOCALIZATIONS** Song a few clacks or clanks, then a tinny buzz: *tsick-tsicka wzzzzzzzz*. Call a dull, nasal *chent*.

■ **POPULATIONS** Breeding range spread east in 20th century, but expansion has stalled in recent decades. A few winter all the way to East Coast.

### Molt in Blackbirds

Male Rusty Blackbirds in fall and spring are notably different in appearance, and a natural inference is that they get their spring plumage by molt. But they don't. Instead, males in spring and summer acquire their breeding garb via feather wear: The rusty feather tips, acquired in the annual fall molt, abrade or break off during the winter, exposing the glossy black pigments beneath. It's an ingenious strategy, metabolically less demanding than growing a new coat of feathers. Several other birds in the East employ this same workaround, among them the European Starling (p. 310), House Sparrow (p. 320), and Snow Bunting (p. 334).

And so do blackbirds other than the male Rusty. For starters, take the female Rusty. Her freshly molted feathers are, on average, an even brighter rusty than the male's. The closely related Brewer's can also show a bit of rusty on the feather tips in fall, although it takes a close look to see this. Another "rusty blackbird" in fall is the familiar Red-winged Blackbird (p. 370), with first-winter males extensively scalloped, especially on the back and wing coverts; like the Rusty Blackbird, they lose the rusty scalloping during the winter. Fresh adult female Red-winged Blackbirds, as well as first-fall immatures (both males and females), likewise sport extensive rust, in addition to their complexly streaked and splotchy plumage. To get an appreciation for the variation among individuals of the same species, find a flock of blackbirds—ideally, accompanied by some hypervariable starlings—in early autumn, and take in the almost endless diversity in their colors and patterns.

Rusty Blackbird
distinctive head pattern with rusty crown and pale supercilium
gray rump
fall ♀
pale eye
breeding ♀
fall ♂
breeding ♂
Brewer's Blackbird
bill straighter than Rusty
plumage more glossy than Rusty at close range
wings uniformly dark
♂
female has dark eye
immature ♂
♀
grackles long-headed with long, keeled tails
Boat-tailed Grackle
Common Grackle
Brewer's/ Rusty Blackbird
Red-winged Blackbird
square tail, broad wings
small head; rather long, club-shaped tail
Great-tailed Grackle
Brown-headed Cowbird
Bronzed Cowbird
European Starling for comparison
short-tailed, with translucent, triangular wings
short-tailed and big-headed

## GRACKLES

These large icterids attract attention with their noisy displays. They strut on lawns, shriek from shrubs and trees, and fly over in disorganized flocks.

### Common Grackle

*Quiscalus quiscula* | COGR | L 12½" (32 cm)

Although it is our smallest grackle, the Common is large and loud and hard to miss. It might be the most frequently noted bird without a name in urban areas—seen daily by millions of commuters who have no idea what it is.

■ **APPEARANCE** Longer and larger than Red-winged Blackbird (p. 370), with longer tail; tail corners angled in flight like a sharply cut wedge. Adult male glossy, adult female less so. Juvenile (first summer into fall) dull brown-gray with dark eyes.

■ **VOCALIZATIONS** Song, given as male spreads wings and puffs feathers, a grating *wshrishAAAANK,* like a louder, more clanging Brewer's Blackbird (p. 372). Call a loud, toneless *chack.*

■ **POPULATIONS** An early-season, loosely colonial breeder, often nesting in conifers. Two groups, formerly treated as separate species: rather uniform "Purple Grackle" (nominate *quiscula*) of the southeastern coastal plain and widespread "Bronzed Grackle" (*versicolor*), mostly bronze-purple with glistening blue hood; the latter expanded well west in 20th century.

### Boat-tailed Grackle

*Quiscalus major* | BTGR | L 14½–16½" (37–42 cm)

An ecological doppelganger of the Laughing Gull (p. 142), the Boat-tailed Grackle nests mostly in and near coastal marshes, but gladly ranges to public beaches and strip malls.

■ **APPEARANCE** Larger overall than Common Grackle, with an even more ample tail; head dome-shaped. Adult male glossy blue-black. Adult female smaller-bodied and lighter-colored: dirty brown above, pale orangish brown below. Eye color geographically variable; from La. westward, eyes dark—a useful distinction from light-eyed Great-tailed.

■ **VOCALIZATIONS** Song utterly different from Common's, a series of harsh buzzes and short whistles in rapid succession: *zjurp zjurp ... jeep! jeep! ... jeer jeer jeer ....* Calls a solid *chuk* and sharp *kyark!*

■ **POPULATIONS** Breeds around marsh edges in loose colonies; in South, favors nest sites with alligators, who serve unwittingly as bodyguards. Males and females go their own way after breeding, occurring in single-sex flocks. Increased north along coast in 20th century, spreading inland in Fla.

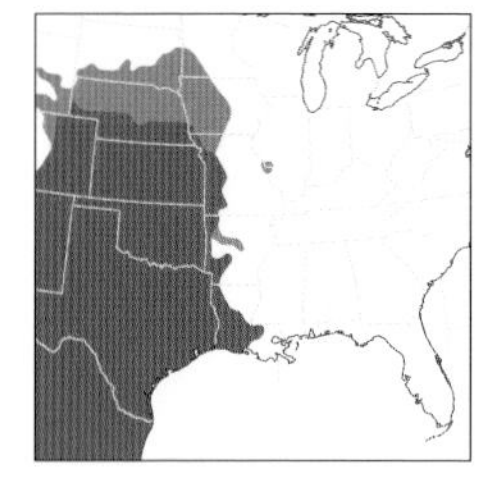

### Great-tailed Grackle

*Quiscalus mexicanus* | GTGR | L 15–18" (38–46 cm)

One of our largest and loudest songbirds, the Great-tailed Grackle is closely related to the Boat-tailed, replacing it on the central Tex. coast and southwestward.

■ **APPEARANCE** Flatter-headed and a bit larger than Boat-tailed. Adult male uniformly glossed deep purple; smaller, variable female usually shows pale eyebrow and dark face mask. Eyes pale from first fall onward; Boat-tailed Grackles in area of range overlap conveniently dark-eyed.

■ **VOCALIZATIONS** Full song highly complex, with well-spaced chirps, explosive rattles, wild wailing, and wheezy whistles. Calls like Boat-tailed's.

■ **POPULATIONS** Closely related to Boat-tailed and formerly treated as conspecific, but reproductively isolated. Rapidly invaded north in 20th century, still expanding.

"Purple Grackle"
quiscula
"Bronzed Grackle"
versicolor
bill shorter and thicker than Great-tailed
glossy purple
ll in good light
♂
♂
juvenile
Common Grackle

Boat-tailed Grackle
juvenile
1st fall ♂
eyes dusky to completely dark where range overlaps with Great-tailed
♀
western Gulf Coast adult ♂
long, keeled tail, slightly rounder than Great-tailed
juvenile
long thin bill
♀
no contrast between head and body as in Common Grackle
Great-tailed Grackle
adult ♂
very long tail, deeply keeled

## BROWN WARBLERS

The warblers, family Parulidae (pp. 376–399), active and insectivorous, are mostly colorful, many with at least some yellow. But not these four, clad in shades of brown.

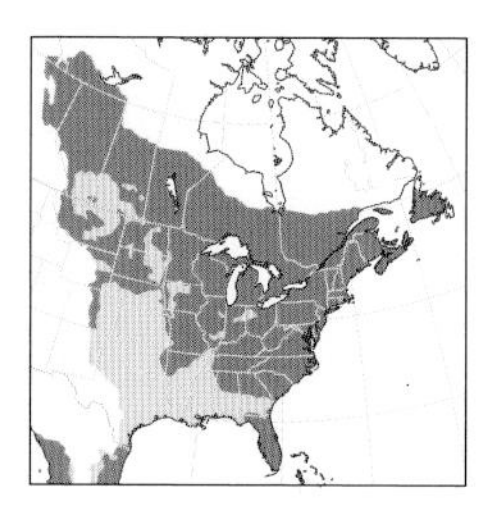

### Ovenbird

*Seiurus aurocapilla* | OVEN | L 6" (15 cm)

The most un-warblerlike of the warblers, the streak-breasted Ovenbird struts about the forest floor like a thrush.

■ **APPEARANCE** Large and plump for a warbler. Olive-brown above; white below with bold streaks. Orange crown stripe has black borders.

■ **VOCALIZATIONS** Rollicking song increases in amplitude: *er-teech er-teech er-TEECH er-TEECH!* Also gives soft, rambling flight song, typically at night. Call note a sharp *chip,* flight call a piercing *pseek!*

■ **POPULATIONS** Nests in mature hardwood and mixed hardwood-conifer forests; absent from small forest tracts.

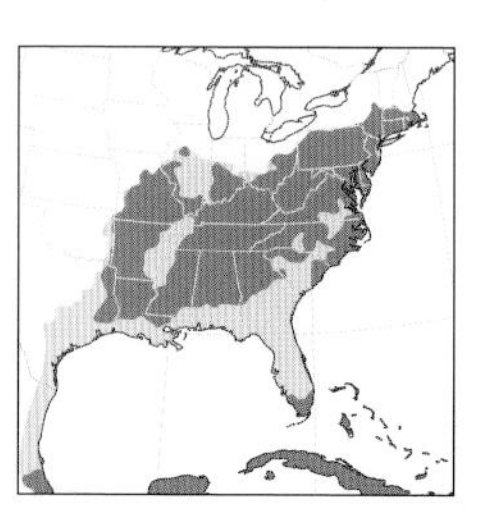

### Worm-eating Warbler

*Helmitheros vermivorum* | WEWA | L 5¼" (13 cm)

They don't eat worms, they don't warble, and their scientific name is misspelled (it should be *Helmintheros*). But Worm-eating Warblers are enchanting and much sought by birders.

■ **APPEARANCE** Long-billed, short-tailed. Head and breast warm buff; face has bold black stripes. Compare with Swainson's Warbler (p. 380).

■ **VOCALIZATIONS** Song a rapid trill; individual notes rise sharply like Pine Warbler's (p. 394), but trill faster. Compare also with Chipping Sparrow (p. 342). Call note like Ovenbird's, but flight call thinner, softer *zzzt.*

■ **POPULATIONS** Breeds in broadleaf forests, especially on drier, south-facing slopes; like many tropical songbirds, but few in the East, forages for invertebrates in dead-leaf clusters.

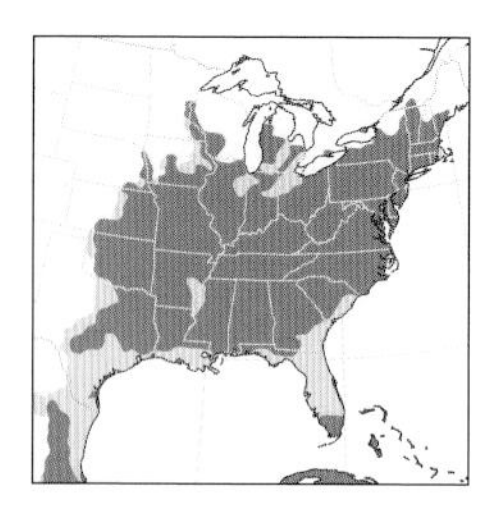

### Louisiana Waterthrush

*Parkesia motacilla* | LOWA | L 6" (15 cm)

The two tail-twitching waterthrushes love streams and pond edges. Though behaviorally distinct from other warblers, they are hard to tell apart.

■ **APPEARANCE** Both waterthrushes strong-legged, short-tailed, terrestrial; Louisiana larger-billed, longer-winged, shorter-tailed, and pumps tail in slow, circular movements. Eyebrow broad, long, and white; throat white and unmarked. Diffuse streaking on white below; flanks buff; feet pink.

■ **VOCALIZATIONS** Song, loud and ringing, a few descending whistles, then a short jumble. Call note a flat *spwit,* flight call a short trill.

■ **POPULATIONS** One of the earliest warblers in spring; departs early, on the move by midsummer. Favors streams and small rivers.

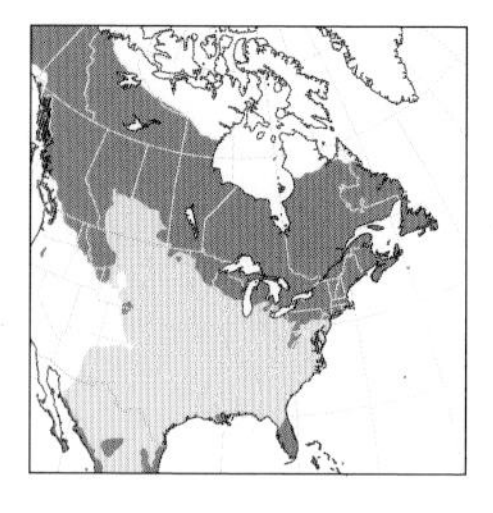

### Northern Waterthrush

*Parkesia noveboracensis* | NOWA | L 5¾ (15 cm)

Plumage and body shape differ weakly between the waterthrushes, but differences in timing, microhabitat, and vocalizations are significant.

■ **APPEARANCE** Slighter-billed than Louisiana; tail movements of Northern jerkier. Northern has dull yellow wash below, with fine streaking, extending to throat. Yellow-tinged eyebrow narrower than Louisiana's, feet less bright.

■ **VOCALIZATIONS** Explosive song drops in pitch: *CHEE CHEE chip chip chup chup.* Call note sharper than Louisiana's, flight call buzzier.

■ **POPULATIONS** Later spring and fall migrant than Louisiana; prefers bogs and pond edges to moving streams.

Ovenbird
olive-brown above
white eye ring
white below with streaky black spots
Worm-eating Warbler
golden-buff head with bold stripes
Louisiana Waterthrush
broad white eyebrow
large bill
unspotted white throat
buff rear flank
bright pink legs
Northern Waterthrush
narrow supercilium
smaller bill than Louisiana
denser, darker streaks than Louisiana
paler
dull pink legs
usually buff underparts and supercilium

Warblers

## THE "WINGED" WARBLERS

Despite their striking differences in appearance, these two are closely related. Both are constantly on the move as they probe dead leaves for arthropods.

### Golden-winged Warbler

*Vermivora chrysoptera* | GWWA | L 4¾" (12 cm)

Both of the "winged" warblers breed around the sunny, shrubby edges of second-growth woodlots. This species is a bit more northerly and generally less common where both occur.

■ **APPEARANCE** Like Blue-winged, a small and slight warbler. Adult male has striking black-and-white head; body mostly gray with yellow wing panel. Scheme of adult female same as adult male, but colors and contrast muted.

■ **VOCALIZATIONS** Song a short, high buzz, followed by two to three somewhat lower buzzes: *bee bzz bzz bzz*. Many variants, and some sing like Blue-winged. Call note a simple *zip*, flight call a rising *zee?*

■ **POPULATIONS** Increased during deforestation era of 19th century; has declined in past century with recovery of forests and, problematically, genetic swamping by Blue-winged Warbler.

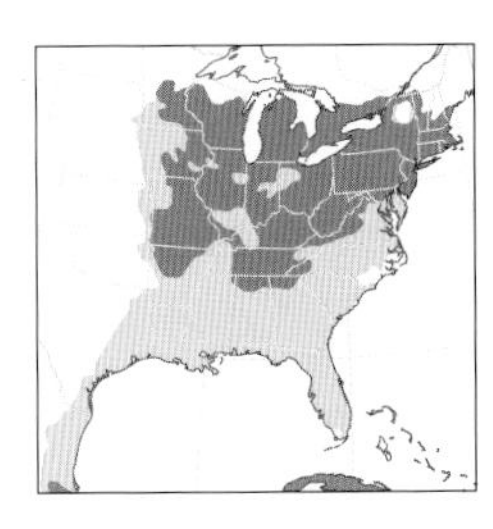

### Blue-winged Warbler

*Vermivora cyanoptera* | BWWA | L 4¾" (12 cm)

The more generalist of the two "winged" warblers, this species is also the more southerly. Somewhat more inclined than the Golden-winged to penetrate the forest interior, it is nevertheless a lover of roadsides and power-line cuts.

■ **APPEARANCE** Size and shape identical to Golden-winged. Adults are mostly yellow, with a black line through eye; blue-gray wings have white wing bars.

■ **VOCALIZATIONS** Lazy song, lower-pitched than Golden-winged's, an inhaled-sounding buzz, then a looser buzz: *hhheeee zzzzzz*. Like Golden-winged, sings weird variants. Calls like Golden-winged's.

■ **POPULATIONS** Along with Golden-winged, was a beneficiary of 19th-century clear-cutting. Continues to expand north and upslope, but species also declining in core range in Midwest and Ohio River Valley.

### "Brewster's" and "Lawrence's" Warblers

Until well into the 19th century, Blue-winged and Golden-winged Warblers were uncommon and largely allopatric, meaning their ranges did not overlap. They have since come into extensive contact, however, and considerable hybridization has ensued. Two striking hybrids, the relatively common "Brewster's Warbler" and the rarer "Lawrence's Warbler," were treated as full species when they were formally described to science in 1874.

Textbook "Brewster's" and "Lawrence's" Warblers, especially adult males, are distinctive, but backcrosses (for example, the offspring of a "Brewster's" and a Blue-winged) are legion. Quite often, careful study of an otherwise pure Golden-winged reveals a faint wash of yellow across the breast—an indication of Blue-winged ancestry. Further complicating matters, many apparently pure individuals have been shown to harbor extensive genetic material from the "wrong" species; so what a bird looks like in the field (its phenotype) often obscures its genetic material (or genotype).

Genetic swamping of Golden-winged Warblers by Blue-winged Warblers is widely conjectured to be an important factor in the former's decline, but proving population-level impacts is difficult. A phenotypic baseline is valuable, and birders with digital cameras are advancing our understanding of the status and distribution of seemingly pure, obviously hybridized, and subtly backcrossed Blue-winged and Golden-winged Warblers and their progeny.

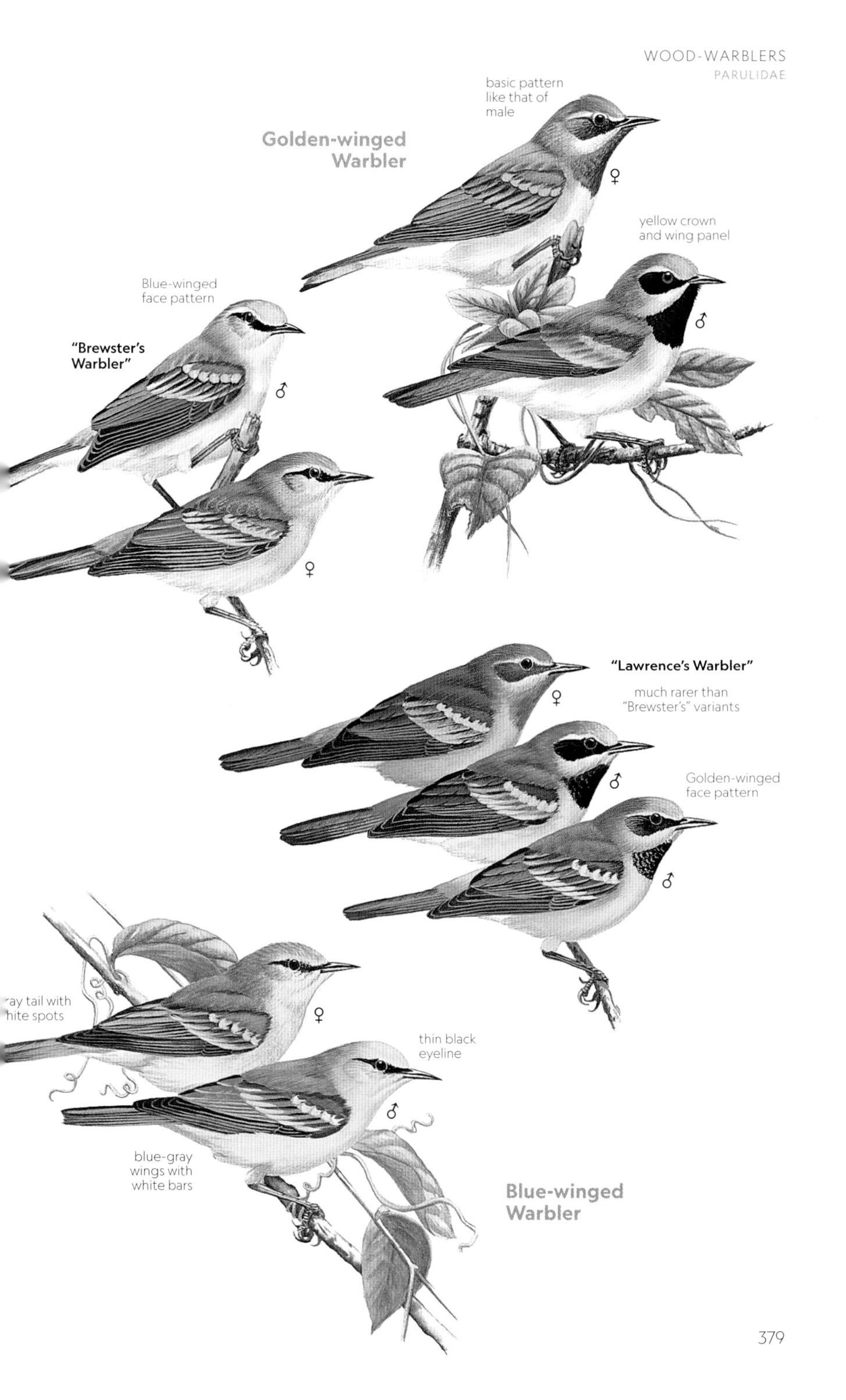
basic pattern like that of male
Golden-winged Warbler
♀
yellow crown and wing panel
♂
Blue-winged face pattern
"Brewster's Warbler"
♂
♀
"Lawrence's Warbler"
♀
much rarer than "Brewster's" variants
♂
Golden-winged face pattern
♂
ray tail with hite spots
♀
thin black eyeline
♂
blue-gray wings with white bars
Blue-winged Warbler

## THREE DISTINCTIVE WARBLERS

Each is in its own genus, but they form an evolutionarily cohesive unit. Multiple recent analyses place these three, and only these three, in a monophyletic group, or clade. All three are long-billed, with distinctive behaviors.

### Prothonotary Warbler

*Protonotaria citrea* | PROW | L 5½" (14 cm)

The bird gets its name from garb worn by Catholic scribes, although the origins for that etymology are murky at best. An alternate name, "Golden Swamp Warbler," still in wide use, is completely unambiguous.

■ **APPEARANCE** Underparts and entire head radiant yellow; long-billed and short-tailed. Plain blue-gray wings lack wing bars; tail flashes much white in flight. Compare with Blue-winged Warbler (p. 378); even the drabbest Blue-winged has white in wing and a bit of black through eye.

■ **VOCALIZATIONS** Song, ringing and clear, a pulsing *sweet sweet sweet sweet sweet* .... Call note a clear *tweet;* higher-pitched flight call also clear, a rising *chee?*

■ **POPULATIONS** Restricted to swamps and other forests with standing deadwood for nesting. Nests in holes, natural or artificial; especially attracted to bald cypress (*Taxodium distichum*), but also takes to human-proffered nest boxes. Arrives fairly early in spring and leaves breeding grounds early; even migrants on passage have a knack for finding a bit of water in the woods.

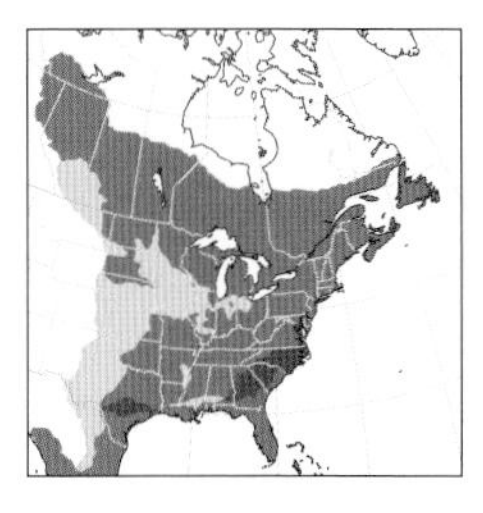

### Black-and-white Warbler

*Mniotilta varia* | BAWW | L 5¼" (13 cm)

More than any other warbler in the East, this species stays close to branches, even tree trunks. Foraging, it is more like a nuthatch than a warbler.

■ **APPEARANCE** Streaked black-and-white all over; bill long and thin. Adult male has black throat; female and young have mostly white face and throat. Compare with Blackpoll Warbler (p. 392); Blackpoll feeds in foliage, does not creep along trunks.

■ **VOCALIZATIONS** Song, high and thin, a chanting *wee-see wee-see wee-see wee-see* .... Call note a husky *chint,* flight call a high buzz.

■ **POPULATIONS** Among the first wave of warblers to return each spring, and among the commoner breeders in much of its range. Winters north regularly to Carolinas.

### Swainson's Warbler

*Limnothlypis swainsonii* | SWWA | L 5½" (14 cm)

No warbler is more strongly associated with the Deep South than the skulking, secretive Swainson's, heard far more often than it is seen. The species is notably terrestrial, tossing about leaf litter in the search for spiders and other arthropods.

■ **APPEARANCE** With its spiky bill and brown tones, suggests Worm-eating Warbler (p. 376). Swainson's gray-faced with bright brown crown and pale eyebrow; Worm-eating buffier overall, with black stripes on face.

■ **VOCALIZATIONS** Song, utterly different from Worm-eating's but similar to Louisiana Waterthrush's (p. 376), a few rich whistles followed by bright chirps: *s'wee seee seee* ... *syewt-sswewt-syit.* Call note a fairly low smacking sound. Flight call, high and hissing, *sssi.*

■ **POPULATIONS** Breeds in two well-delineated habitats: giant cane (genus *Arundinaria*) thickets in lowland swamps throughout the Southeast, and dense thickets, often with rhododendron (genus *Rhododendron*), in the southern Appalachians.

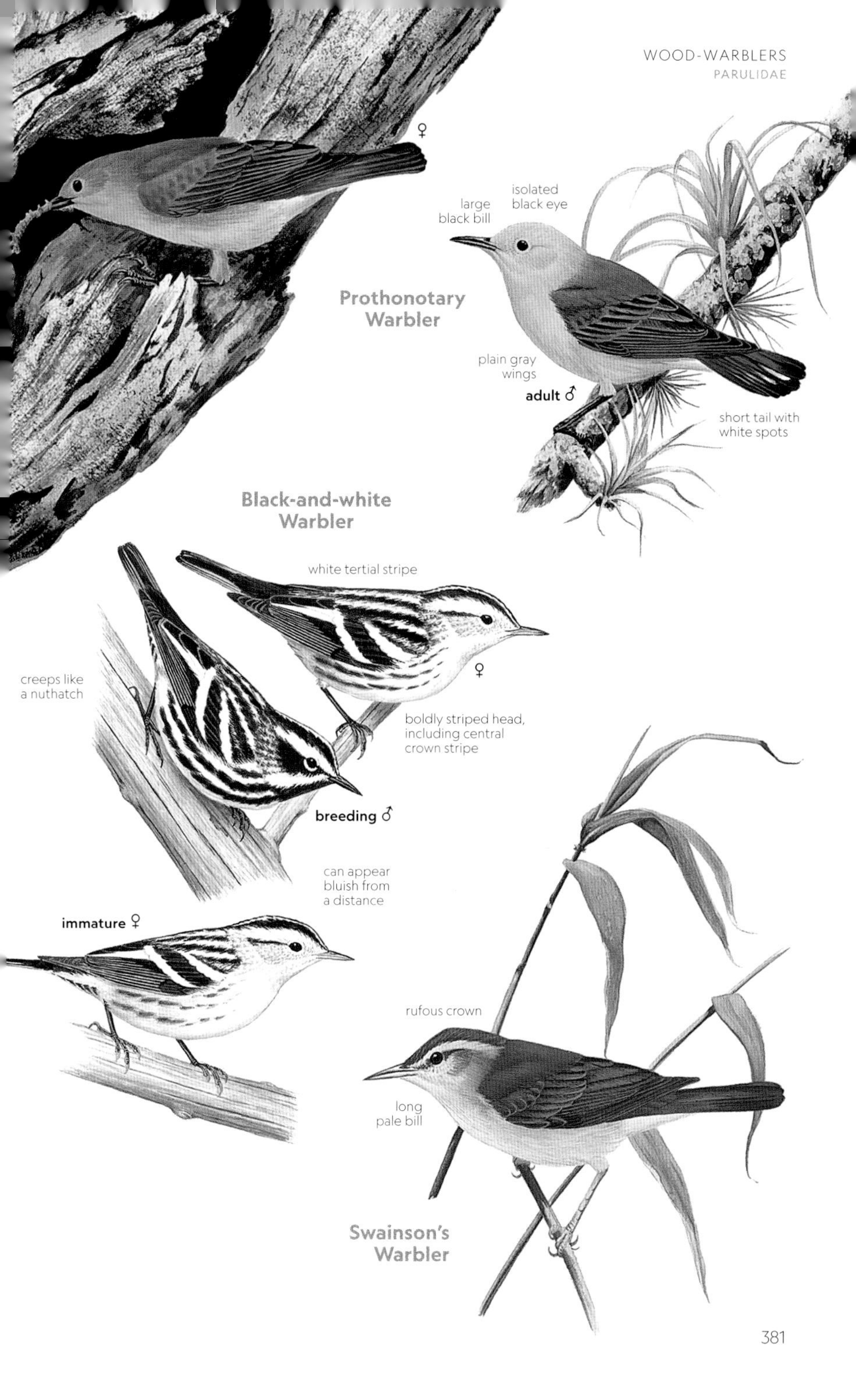
♀
large
black bill
isolated
black eye
Prothonotary
Warbler
plain gray
wings
adult ♂
short tail with
white spots
Black-and-white
Warbler
white tertial stripe
♀
creeps like
a nuthatch
boldly striped head,
including central
crown stripe
breeding ♂
can appear
bluish from
a distance
immature ♀
rufous crown
long
pale bill
Swainson's
Warbler

## GENUS *LEIOTHLYPIS*

This trio comprises small-bodied, slight-billed warblers that sing fast songs. They are fidgety foliage-gleaners.

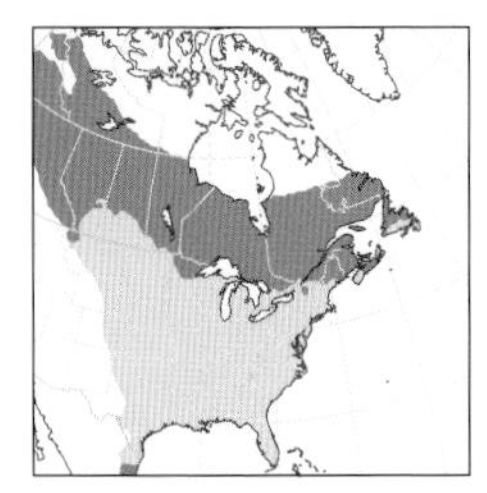

### Tennessee Warbler

*Leiothlypis peregrina* | TEWA | L 4¾" (12 cm)

A little gray-green sprite, the Tennessee is definitely in the heard-more-often-than-seen camp. Males sing loudly from the treetops, where they blend in with the foliage.

■ **APPEARANCE** Short-tailed, fairly long-winged. Breeding male mossy green above with blue-gray cap set off by white eyebrow and thin black line through eye; underparts, including undertail coverts, white. Breeding female olive green above, including crown, with white eyebrow and black eyeline; breast yellowish, undertail coverts white. Fall bird like Orange-crowned, but note stronger eyebrow and eyeline, richer color, and, especially, white undertail coverts of Tennessee. Compare with Philadelphia Vireo (p. 268).

■ **VOCALIZATIONS** Tripartite song builds in amplitude: *tik-tik-tik-twick-twick-twick-twick-TYEW-TYEW-TYEW-TYEW....* Call note, light and smacking, *tsinck;* flight call a clear, rising *tsee?*

■ **POPULATIONS** One of the spruce budworm specialists with interannual variation in abundance caused by prey availability.

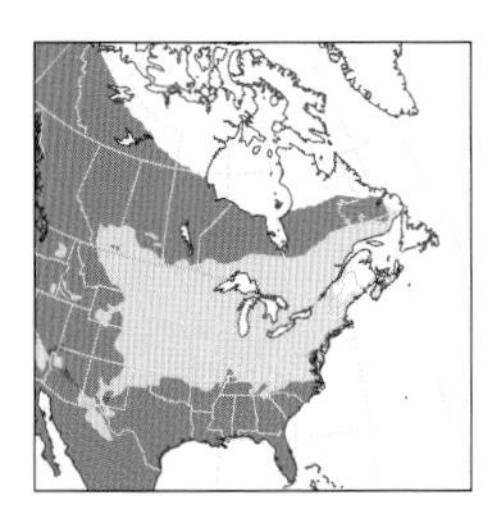

### Orange-crowned Warbler

*Leiothlypis celata* | OCWA | L 5" (13 cm)

Uncharacteristically drab for a warbler or a marvel of subtlety and intricacy? Both assessments apply to the Orange-crowned.

■ **APPEARANCE** Longer-tailed and shorter-winged than Tennessee, with slightly decurved bill; orange crown rarely visible. Dark olive-gray above, with thin eyebrow and trace of an eye ring; underparts mostly dusky yellow-gray, with undertail coverts brighter yellow. Tennessee is the classic point of confusion, but beware drab, grayish, late-season Yellow (p. 392).

■ **VOCALIZATIONS** Song a loose trill that drops off at end: *titititititit-tlu-tlu-tlu.* Call note a clanking *tswint,* flight call a Tennessee-like *tsee?*

■ **POPULATIONS** Uncommon in fall and rare in spring east of Appalachians, more common farther west. A late fall migrant, not showing up at mid-latitudes of U.S. until the end of Sept. Increasing in winter in recent years well north of mapped range.

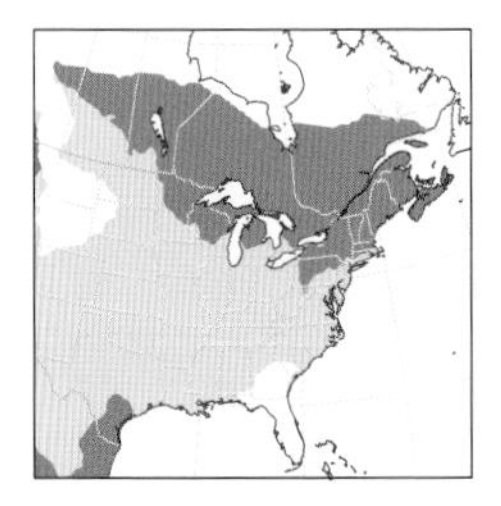

### Nashville Warbler

*Leiothlypis ruficapilla* | NAWA | L 4¾" (12 cm)

The Nashville Warbler is a lover of earlier-successional woods where sun-light gets down well into the understory—forested, but not too forested.

■ **APPEARANCE** Like Tennessee, short-tailed and straight-billed. Adult in spring has olive green upperparts, yellow underparts, and blue-gray head with white eye ring. Larger Connecticut Warbler (p. 384) has similar color scheme but differs in shape and behavior: Connecticut struts on sturdy legs; dainty Nashville flits in flowers and foliage. Immature Common Yellowthroat (p. 386), dark above and yellow below, with hint of eye ring, resembles immature Nashville; yellowthroat has much longer tail, shorter wings, and long legs.

■ **VOCALIZATIONS** Song bipartite: *s'bee-s'bee-s'bee titititititititi.* Call note a sharp *spink,* flight call like Tennessee's.

■ **POPULATIONS** Returns fairly early in spring; a circum-Gulf migrant, with correspondingly few sightings in southeastern coastal plain.

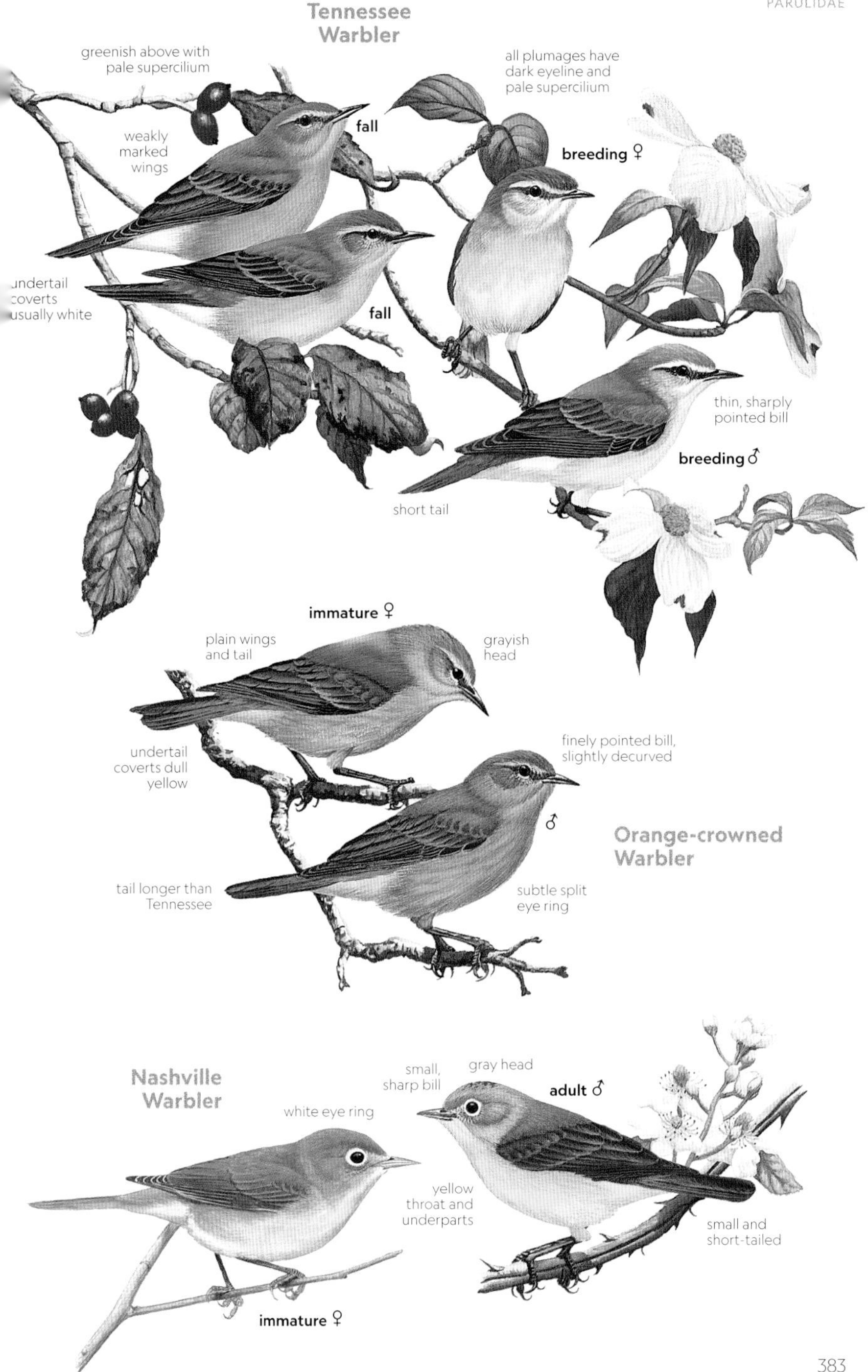
Tennessee Warbler
greenish above with pale supercilium
all plumages have dark eyeline and pale supercilium
weakly marked wings
fall
breeding ♀
undertail coverts usually white
fall
thin, sharply pointed bill
breeding ♂
short tail
immature ♀
plain wings and tail
grayish head
undertail coverts dull yellow
finely pointed bill, slightly decurved
♂
Orange-crowned Warbler
tail longer than Tennessee
subtle split eye ring
Nashville Warbler
small, sharp bill
gray head
adult ♂
white eye ring
yellow throat and underparts
small and short-tailed
immature ♀

## FOREST UNDERSTORY WARBLERS

These three are found in the lower levels of forests and forest edges, where they chant their rollicking songs. They are short-tailed with mostly dark heads and extensive yellow below.

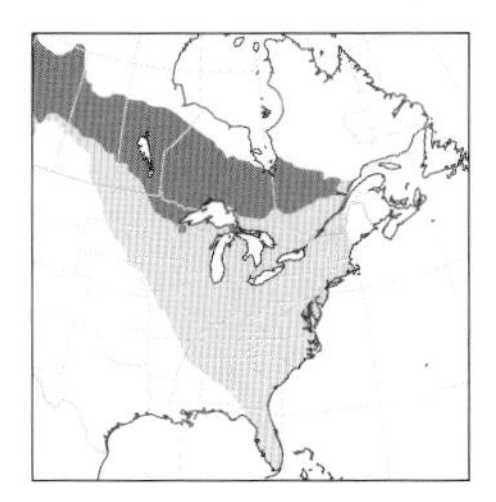

### Connecticut Warbler

*Oporornis agilis* | CONW | L 5¾" (15 cm)

Elusive and uncommon, the Connecticut is among the most sought-after of warblers. Timing, behavior, and microhabitat are important cues for field ID.

■ **APPEARANCE** Large, sturdy-legged; walks (does not hop) on fallen logs and forest floor. Bold eye ring stands out on dark hood; hood gray on breeding male, browner in other plumages. Smaller Mourning usually lacks complete eye ring; hood of male Mourning black at base. Compare with daintier Nashville Warbler (p. 382), which has yellow throat.

■ **VOCALIZATIONS** Song, chirpy and syncopated, builds in amplitude: *tee! ch'PEE-tee p'CHIP p'CHEE-CHIP!* Call note a dull *tyip,* flight call high and buzzy like Blackpoll Warbler's (p. 392).

■ **POPULATIONS** Spring migration, through Midwest, very late, into June; fall migration, farther east, also late. Spring migrants skulk in fully leafed-out understory vegetation but sing loudly; fall migrants frequent brushy, boggy places, often with jewelweed (genus *Impatiens*).

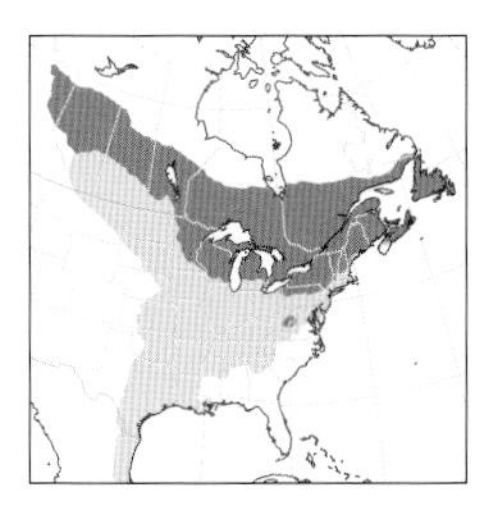

### Mourning Warbler

*Geothlypis philadelphia* | MOWA | L 5¼" (13 cm)

The bird's name is an allusion to funereal garb from a bygone era. Like the rarer and similarly attired Connecticut Warbler, the Mourning is a prized find, especially on migration.

■ **APPEARANCE** A bit smaller than Connecticut. Hood gray, with black around eyes and base of hood. Female, especially young, in fall only weakly hooded, sometimes showing thin eye ring; Connecticut always shows strong eye ring. Compare also with young Common Yellowthroat (p. 386), which has much longer tail, shorter wings, and completely yellow throat.

■ **VOCALIZATIONS** Bipartite song, breathy and burry, falls in pitch: *chree chree chree chroo chroo.* Call note a dull, smacking *tchook.* Flight call a short, rising *si?*

■ **POPULATIONS** Circum-Gulf migrant; rare at best in southeastern coastal plain. Late spring migrant. Nesters take well to clear-cuts and other disturbances in forested landscapes.

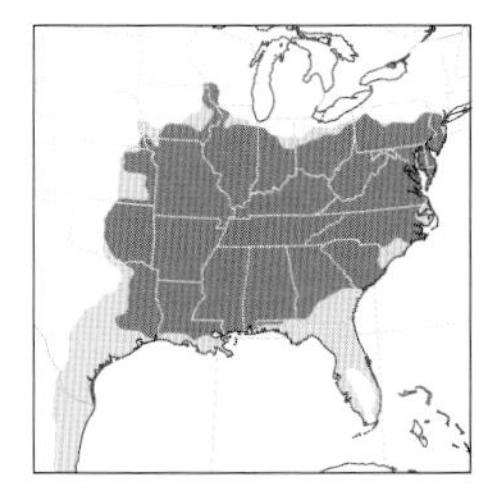

### Kentucky Warbler

*Geothlypis formosa* | KEWA | L 5¼" (13 cm)

A Carolinian counterpart of the more northerly Mourning Warbler, this species abounds in broadleaf woodlands, typically in valleys and floodplains.

■ **APPEARANCE** Built like Mourning. Dark olive above; bright yellow below, including throat. Face mostly dark, but with yellow above and behind eye, creating "spectacled" appearance.

■ **VOCALIZATIONS** Song, clear and ringing, a monotone *ch'ree ch'ree ch'ree ch'ree ....* At a distance, song can be confused with Ovenbird's (p. 376) or Carolina Wren's (p. 304). Call note a powerful *tsak,* flight call a rising, wavering *tzzee.*

■ **POPULATIONS** An area-sensitive species, absent from smaller tracts, especially those damaged by deer; like Mourning, a lover of dense understory. Range nudging slowly northward.

Connecticut Warbler
extensive dull brown hood
complete white eye ring
immature
large pink-and-gray bill
adult ♀
pale gray hood
very long undertail coverts
adult ♂
chalky yellow belly

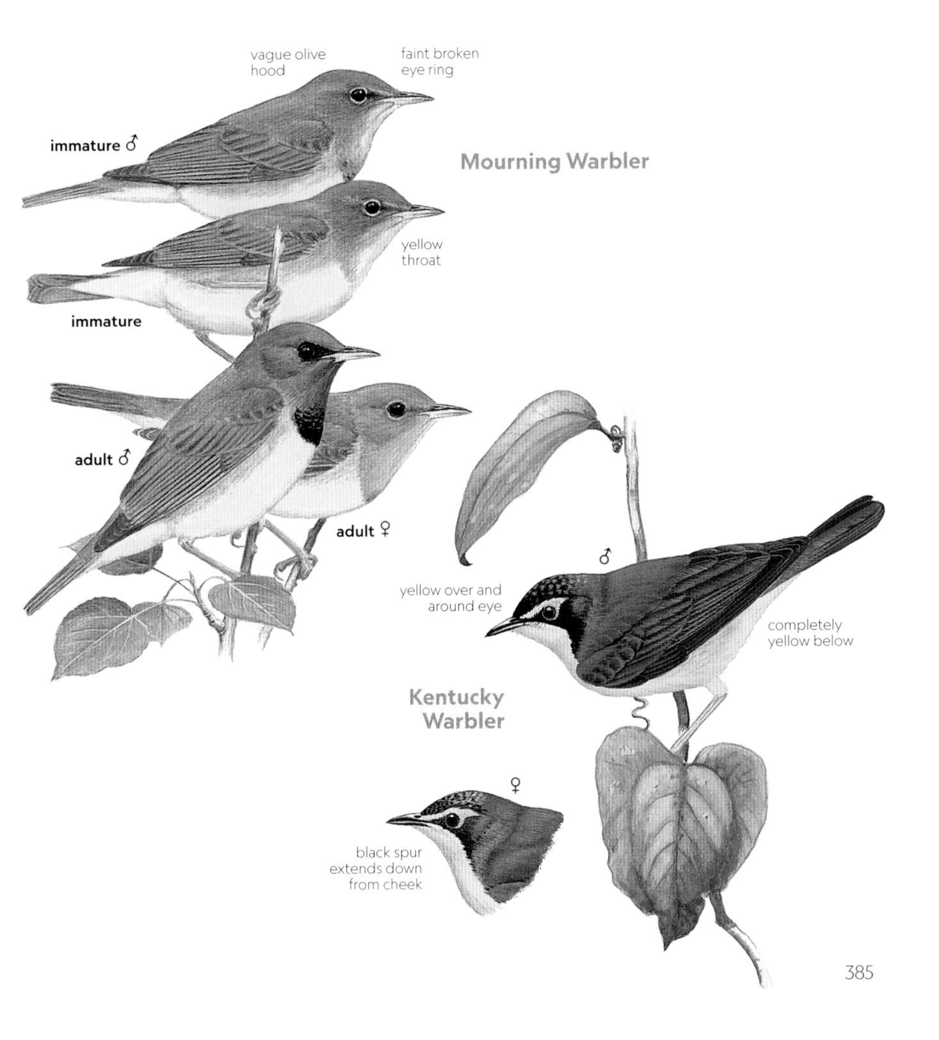
vague olive hood
faint broken eye ring
immature ♂
Mourning Warbler
yellow throat
immature
adult ♂
adult ♀
♂
yellow over and around eye
completely yellow below
Kentucky Warbler
♀
black spur extends down from cheek

## YELLOWTHROAT AND GENUS *SETOPHAGA* I

These three are among the most ingratiating of warblers: active, brightly colored, and easily attracted by pishing. They are common breeders around major human population centers in the East.

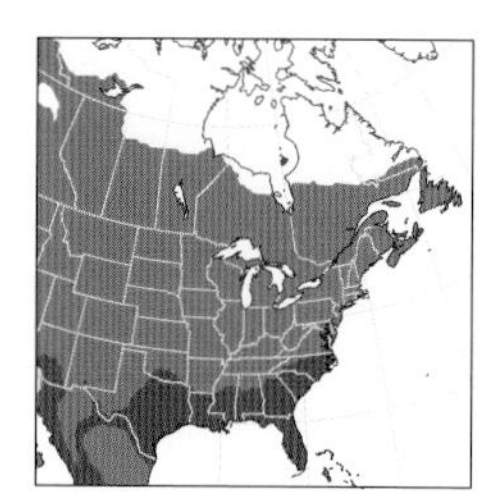

### Common Yellowthroat

*Geothlypis trichas* | COYE | L 5" (13 cm)

The yellowthroat, related to the Mourning and Kentucky Warblers (both on p. 384), suggests a wren. It has a long tail, frequently cocked, and it scolds from dense vegetation, especially cattails.

■ **APPEARANCE** Short-winged, long-legged, and long-tailed. Adult male has rectangular black face mask, edged bluish above. Other plumages have reduced or no black on face, but all have plain yellow throat, brownish flanks, and yellow undertail coverts.

■ **VOCALIZATIONS** Song, vigorous and repetitious, with chirpy quality like Connecticut Warbler's (p. 384): *witchity witchity witchity.* More complex flight song, with *witchity* elements and other notes, given day or night. Varied call notes include a rough *chap* and rapid chatter; flight call a low, rough *zzzrt.*

■ **POPULATIONS** Hardy; migrates late in fall, with some lingering into early winter well north of mapped range.

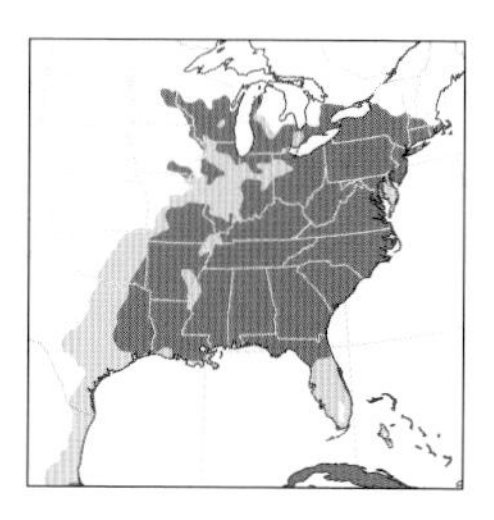

### Hooded Warbler

*Setophaga citrina* | HOWA | L 5¼" (13 cm)

Like the Kentucky Warbler (p. 384), the Hooded thrives in Carolinian forests, especially in bottomlands. Less of a heliophobe than the Kentucky, it often ventures into gaps and other forest clearings.

■ **APPEARANCE** Active; like redstart, often fans tail. All are olive green above, bright yellow below. Yellow face of male encircled by black cowl; cowl reduced in adult female, further reduced in young female. Compare with immature female Wilson's Warbler (p. 398), which has no white in tail.

■ **VOCALIZATIONS** Song, bright and ringing, has cadence of Magnolia Warbler's (p. 390), but richer, sweeter timbre: *s'wee s'wee s'WEE-chee-you.* Call note a high, metallic *tsaap.* Flight call like Common Yellowthroat's, but not as abrupt and buzzy.

■ **POPULATIONS** Requires large forest tracts for breeding; range expanding sporadically northward.

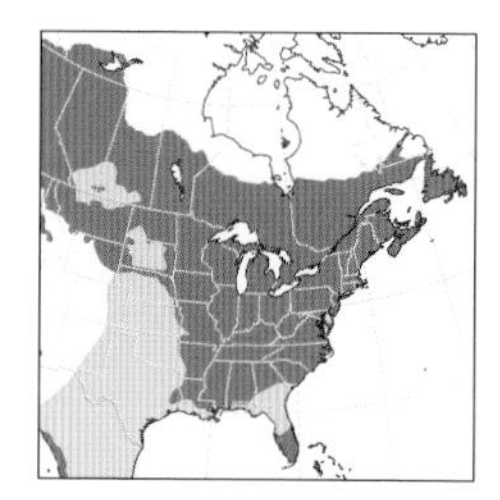

### American Redstart

*Setophaga ruticilla* | AMRE | L 5¼" (13 cm)

The foraging behavior of this redstart is distinctive: The bird constantly spreads its wings and tail, flashing orange or yellow to startle hidden insect prey into flushing.

■ **APPEARANCE** Small and active. Adult male plumage, not acquired until second fall, mostly black with white belly and bright orange highlights. Young and female gray overall with reduced yellow on wing and flanks; but tail edges, frequently fanned, are lemon yellow. Second-summer male (around one year of age) begins to show black of adult.

■ **VOCALIZATIONS** Song quite varied, but the basic pattern is a few bright chirps, then a lower buzz: *twee twee twee tzzyeeer.* Call note a rich, smacking *zick.* Flight call a clear *ts'weet,* subtly disyllabic.

■ **POPULATIONS** Common in much of range, although declines have been documented in several regions. Occurs in diverse forest types, mostly broadleaf, especially around edges and clearings.

black mask
with whitish
border
adult ♂
variable,
often very dull
pale yellowish
throat contrasts
with darker
cheek
♀
indistinct
dark mask
immature ♂
long tail and
short wings
yellow throat
and undertail
coverts
Common
Yellowthroat
amount of
black variable
adult ♀
constantly
flicks tail open
and closed
adult ♂
blank yellow
face with
dark lores
Hooded
Warbler
extensive
white in tail
immature ♀
distinctive
patches of
yellow
gray head
American
Redstart
♀
frequently fans tail
and spreads wings
adult ♂
variable, with
splotchy black
1st spring ♂

## GENUS *SETOPHAGA* II

Warblers in the genus *Setophaga* (pp. 386–399) occupy a special place in the pantheon of eastern birds. No other genus has nearly as many regularly occurring species in our region, and the spring migration of *Setophaga* warblers in the East is one of the greatest avian spectacles anywhere.

### Kirtland's Warbler

*Setophaga kirtlandii* | KIWA | L 5¾" (15 cm)

Our rarest warbler is, unsurprisingly, one of the most sought-after. Tourism devoted to this bird is big business near the breeding grounds, and the possibility of seeing a migrant Kirtland's draws thousands of birders to the Ohio-based Biggest Week in American Birding each May.

■ **APPEARANCE** Largest *Setophaga;* twitches tail. Adult slate blue above, with white eye crescents; yellow below, flanks streaked black. Immature brownish above. Female Magnolia (p. 390) has yellow rump, does not wag tail; female Prairie (p. 396), also a tail twitcher, has paler face with more yellow.

■ **VOCALIZATIONS** Song low and chirpy like Northern Waterthrush's (p. 376): *chut chut whip whip whip whee whee.* Call note a dry smack, flight call a short buzz.

■ **POPULATIONS** Near-threatened, but recovering thanks to intensive management of early successional pines for nesting.

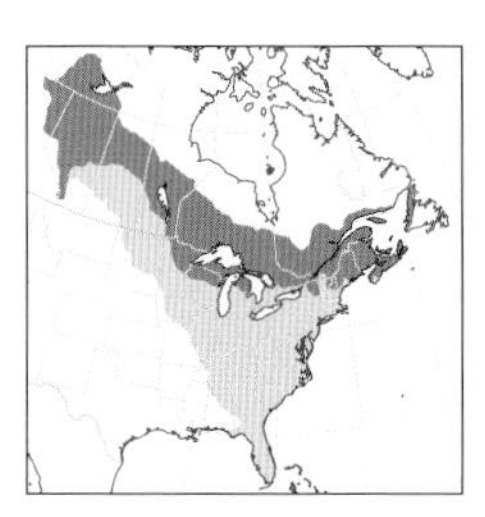

### Cape May Warbler

*Setophaga tigrina* | CMWA | L 5" (13 cm)

Because it feeds in the treetops and sings a high-pitched song, the Cape May is easy to miss on migration. Breeders, found in dense spruce-fir forests, aren't any easier to espy.

■ **APPEARANCE** Short-tailed with thin, decurved bill. Breeding male has orange patch on yellow face, large white wing panel, and black streaking on yellow breast. Other plumages told by yellowish rump, green-edged remiges, and bill shape.

■ **VOCALIZATIONS** Song a series of very high notes, sometimes on one pitch, sometimes increasing in amplitude: *si see see! SEEE!* Call note, also quite high, a clipped *teep;* flight call a descending *seeew.*

■ **POPULATIONS** One of the spruce budworm specialists, with much annual variation in numbers. Migrants funnel through Fla., with relatively few seen in southern Great Plains.

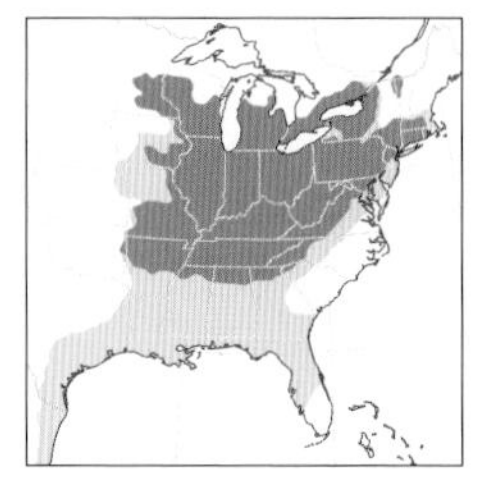

### Cerulean Warbler

*Setophaga cerulea* | CERW | L 4¾" (12 cm)

One of the avian icons of the Ohio River Valley, the Cerulean Warbler sings from the canopy of woods hosting oaks, hickories, and sycamores especially in floodplains and river valleys.

■ **APPEARANCE** Short-tailed and long-winged, with prominent wing bars at all ages. Adult male sky blue above with black "necklace" separating white throat from white belly. Adult female duskier below, with aqua upperparts. Immature female can suggest other wing-barred warblers, but note unstreaked moss-green back and bold eyebrow of Cerulean; compare with Bay-breasted (p. 390), Blackburnian, and Blackpoll (p. 392) Warblers.

■ **VOCALIZATIONS** Primary song a buzzy trill, then a higher trill: *z'z'z'z'z'z'z zeeee.* Similar to secondary song of Northern Parula (p. 390). Call note a flat *tsip,* flight call a short, hard buzz.

■ **POPULATIONS** Declining, in part due to degradation of wintering grounds; flourishes around shade-grown coffee operations.

Kirtland's Warbler
gray above, with thin wing bars
white eye arcs
spotty streaks on flanks
dult ♀
adult ♂
immature ♀
dark brownish above
twitches tail
Cape May Warbler
plain, but note green-edged remiges and blurry streaks below
ull yellow-reen rump
slender, slightly decurved bill
immature ♀
immature ♂
breeding ♀
entirely streaked breast
breeding ♂
white wing panel
Cerulean Warbler
bold white wing bars
fall immature ♂
bold eyebrow
unstreaked green back
immature ♀
short tail and long wings
adult ♂
adult ♀

**GENUS *SETOPHAGA* III**

The vocalizations of *Setophaga* warblers (pp. 386–399) trend fast and high, with buzzes and trills. Call notes are often hard and smacking, with flight calls mostly high and ringing.

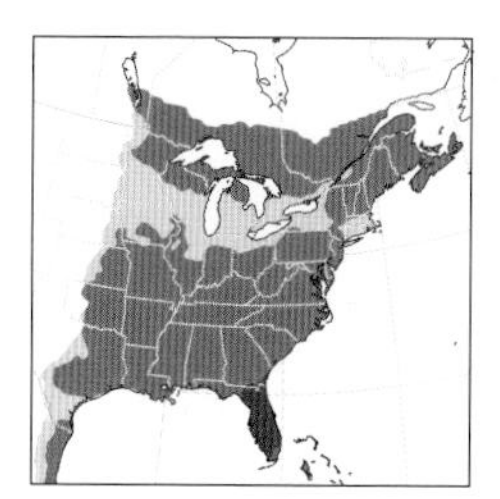

## Northern Parula

*Setophaga americana* | NOPA | L 4½" (11 cm)

This parula is a habitat generalist on migration, but it is fussy about where it nests: in woods with Spanish moss (a bryophytic air plant) in the South and in woods with beard moss (a lichen) farther north.

■ **APPEARANCE** Small and short-tailed, with bill yellow below. All are bluish above with greenish back, white eye arcs, white wing bars, and yellow throat and breast. Adult sports blue-and-rust breastband, fainter on female; immature, especially female, may lack breastband.

■ **VOCALIZATIONS** Primary song a rising trill, then an abrupt *chip*. Secondary song, comprising several short, rising trills, recalls Cerulean Warbler's (p. 388). Call note a smacking *tsick;* flight call a clear, descending *zeep*.

■ **POPULATIONS** Air pollution kills lichens and bryophytes used for nest construction, resulting in regional declines.

## Tropical Parula

*Setophaga pitiayumi* | TRPA | L 4½" (11 cm)

Widespread tropical counterpart of the Northern Parula, breeding north to Lower Rio Grande Valley. Dark-faced Tropical lacks eye arcs of Northern. Adult has orangish suffusion (not breastband) on breast. Vocalizations like Northern's. Some parulas nesting in South Texas appear to be hybrids.

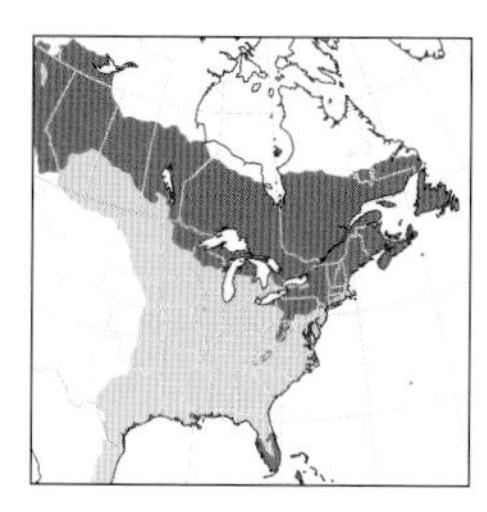

## Magnolia Warbler

*Setophaga magnolia* | MAWA | L 5" (13 cm)

"Hemlock Warbler" would have been a better name for this species—a common breeder in dark, moist coniferous forests.

■ **APPEARANCE** Tail from below shows broad white band at base, broad black band at tip. Breeding male yellow below with thick dark streaks; grayish wings have frosty panel. Immature, especially female, has mostly unmarked yellow underparts and blue-gray face with thin white eye ring; compare with Nashville (p. 382) and Kirtland's (p.388).

■ **VOCALIZATIONS** Short song, not unlike Hooded's (p. 386), clear and ringing, *swee swee swee swee-yee-oh*. Distinctive call note a muffled *shwrint;* flight call a buzzy *frrzz*.

■ **POPULATIONS** Highest densities in forests with lush midstory and understory, favored for foraging and especially nesting.

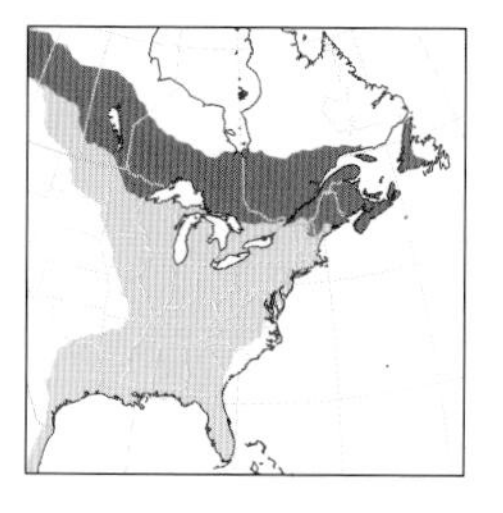

## Bay-breasted Warbler

*Setophaga castanea* | BBWA | L 5½" (14 cm)

A spruce budworm specialist, the Bay-breasted splits its year between Canada and S. Amer., sojourning en route in the eastern U.S.

■ **APPEARANCE** Adult has pale patch on neck and extensive chestnut (bay) on flanks. Immature has dark legs, thick white wing bars, and pinkish wash on flanks; undertail coverts buffy, breast plain, and overall hue yellow-green. Compare with immature Blackpoll (p. 392).

■ **VOCALIZATIONS** Song fast and very high, with notes often paired: *w'si w'si w'si*. Like Black-and-white's (p. 380), but song less leisurely. Call note a simple *tsip;* flight call a ringing *dzee*.

■ **POPULATIONS** Scarce many years, but numbers spike following spruce budworm outbreaks; most migrate earlier in fall than Blackpoll.

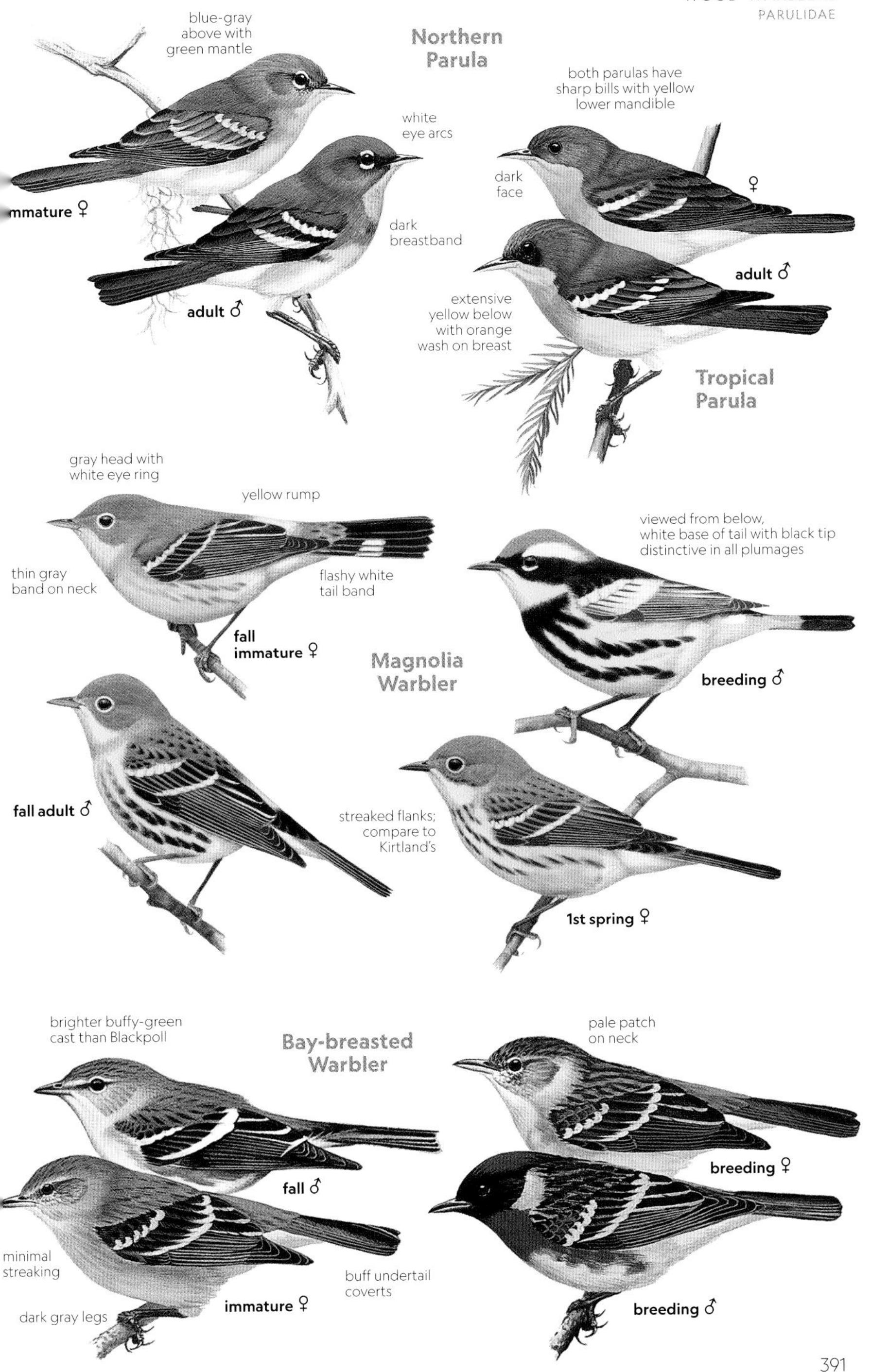
Northern Parula
blue-gray above with green mantle
white eye arcs
immature ♀
dark breastband
adult ♂
both parulas have sharp bills with yellow lower mandible
dark face
♀
adult ♂
extensive yellow below with orange wash on breast
Tropical Parula
gray head with white eye ring
yellow rump
thin gray band on neck
flashy white tail band
fall immature ♀
Magnolia Warbler
viewed from below, white base of tail with black tip distinctive in all plumages
breeding ♂
fall adult ♂
streaked flanks; compare to Kirtland's
1st spring ♀
brighter buffy-green cast than Blackpoll
Bay-breasted Warbler
fall ♂
minimal streaking
buff undertail coverts
dark gray legs
immature ♀
pale patch on neck
breeding ♀
breeding ♂

## GENUS *SETOPHAGA* IV

Most *Setophaga* warblers (pp. 386–399) exhibit seasonal, sex-related, and age-related variation in plumage. But even drab first-year females in fall typically show the overall pattern of the brightest breeding males.

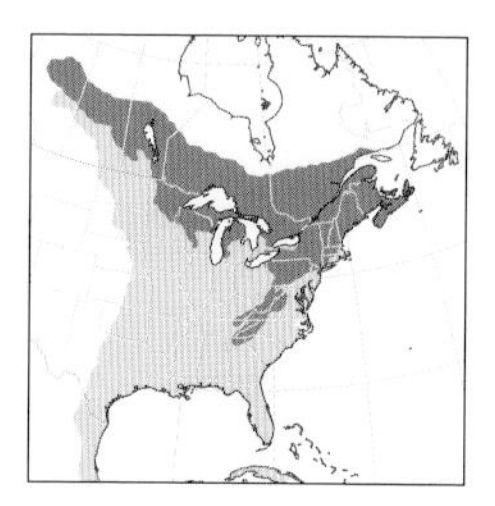

### Blackburnian Warbler

*Setophaga fusca* | BLBW | L 5" (13 cm)

The fabled "Firethroat" forages in the canopy. It breeds in and around coniferous forests and stops over in diverse woodlands on migration.

■ **APPEARANCE** Short-tailed for genus *Setophaga*. Breeding male unmistakable. Other plumages told by broad yellow eyebrow, yellow throat, and pale "braces" on dark back.

■ **VOCALIZATIONS** High-pitched song stutters up the scale: *ch'ch'ch'ch'-chee-chee-cheeeee*. Call note a bright *chip;* flight call a thin, high *dzi*.

■ **POPULATIONS** Trans-Gulf migrant, moving broadly across eastern U.S., but uncommon along southern Atlantic coastal plain.

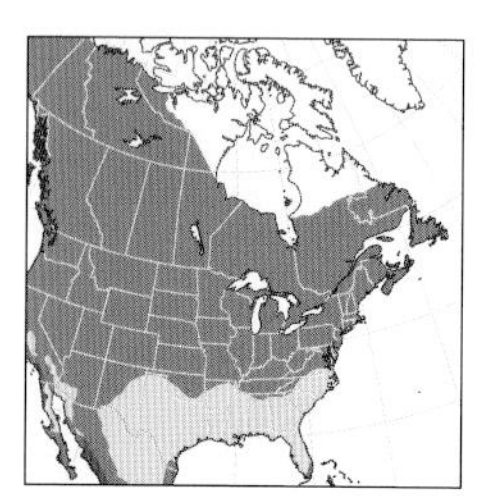

### Yellow Warbler

*Setophaga petechia* | YEWA | L 5" (13 cm)

One of the most common warblers, the Yellow occurs in almost any habitat on migration. It breeds in wetter habitats, often with dense shrubbery.

■ **APPEARANCE** Short-tailed; the yellowest *Setophaga*, with even an all-yellow tail. All have beady black eye on plain face. Adult male has rusty "pinstripes." Other plumages unmarked, with considerable range in brightness. Orange-crowned (p. 382) has thin eyebrow, short wings, long tail.

■ **VOCALIZATIONS** Variable song of chirpy down-slurred notes, ending with loud up-slurred note: *cheep cheep cheep ch'yeepy CHEET*. Call note a loud, bright *chip;* flight call a high buzz.

■ **POPULATIONS** "Golden" subspecies of southern Fla. has reddish cap; "Mangrove" subspecies, local in South Tex., strikingly rusty-hooded.

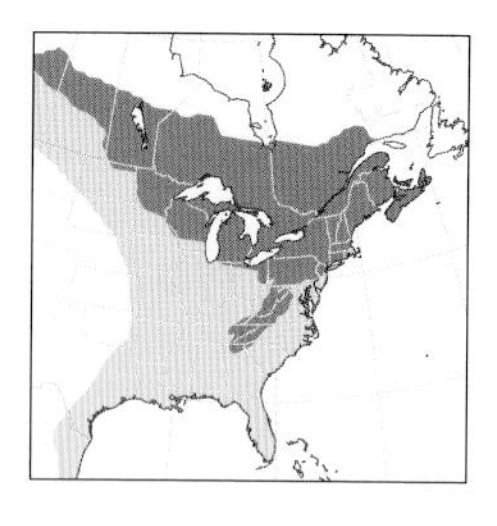

### Chestnut-sided Warbler

*Setophaga pensylvanica* | CSWA | L 5" (13 cm)

Like the Yellow Warbler, this species loves habitats with dense brush. But the latter goes for drier sites in and around early successional woods.

■ **APPEARANCE** Average *Setophaga* structure, but has distinctive habit of cocking tail. Adult combines yellow crown, white eye ring, black "whisker," yellow wing bars, and chestnut flanks. Trademark chestnut reduced on fall birds, but note yellow wing bars, white eye ring, and yellow-green crown.

■ **VOCALIZATIONS** Song the inverse of Yellow's: up-slurred notes with terminal down-slurred note. Call note rougher than Yellow's; flight call a fine buzz.

■ **POPULATIONS** Mostly circum-Gulf migrant in spring, but many migrate across Gulf in fall. Early fall migrant.

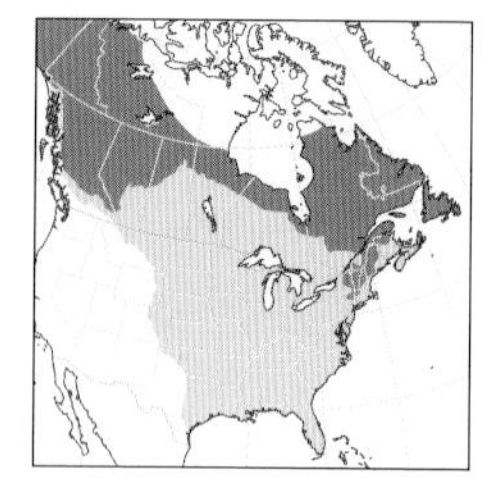

### Blackpoll Warbler

*Setophaga striata* | BLPW | L 5½" (14 cm)

From the Ohio River Valley eastward, hearing the first Blackpoll is bittersweet: This late migrant heralds the end of the spring warbler flight.

■ **APPEARANCE** Large and long-winged. Breeding male has black cap, white face. Young in fall similar to Bay-breasted (p. 390); Blackpoll streakier with unique yellow feet. Compare also with drab young Yellow-rumped (p. 396).

■ **VOCALIZATIONS** Primary song a series of very high *zi* notes. Call note a loud smack; flight call like Yellow's.

■ **POPULATIONS** Migrates later in fall than Bay-breasted; most traverse Atlantic Ocean to S. Amer.

## Blackburnian Warbler

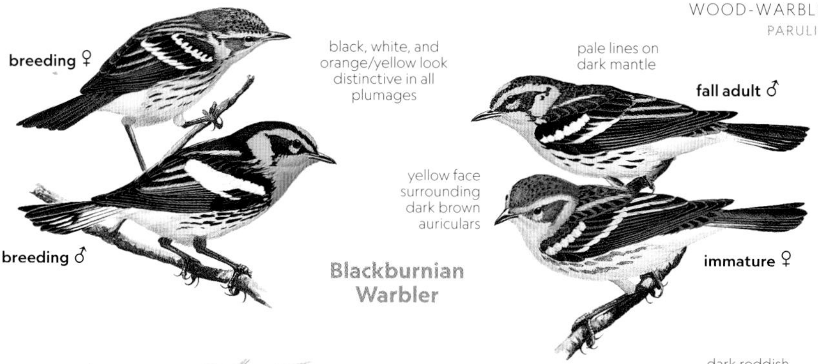

breeding ♀

black, white, and orange/yellow look distinctive in all plumages

pale lines on dark mantle

fall adult ♂

yellow face surrounding dark brown auriculars

breeding ♂

immature ♀

## Yellow Warbler

*aestiva*

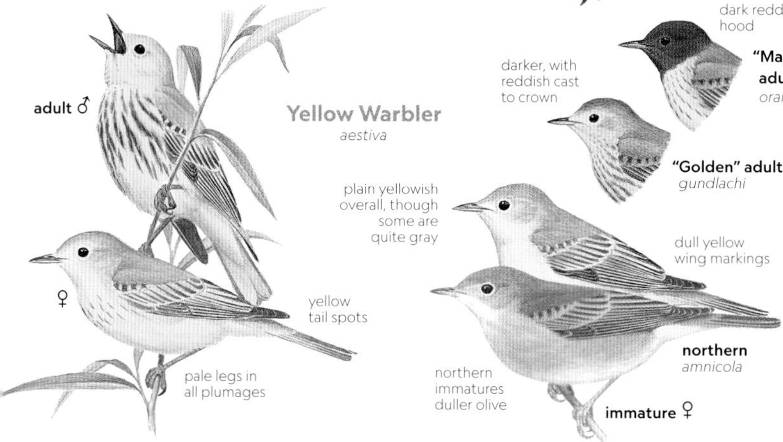

dark reddish hood

"Mangrove" adult ♂
*oraria*

darker, with reddish cast to crown

adult ♂

"Golden" adult ♂
*gundlachi*

plain yellowish overall, though some are quite gray

dull yellow wing markings

♀

yellow tail spots

northern
*amnicola*

pale legs in all plumages

northern immatures duller olive

immature ♀

## Chestnut-sided Warbler

white eye ring

yellow-green upperside

silvery-white face and underparts

immature

usually cocks tail

eding ♀

breeding ♂

## Blackpoll Warbler

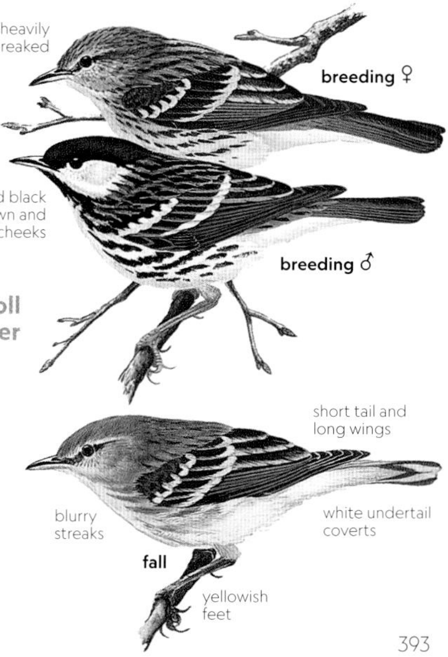

heavily streaked

breeding ♀

solid black crown and white cheeks

breeding ♂

short tail and long wings

blurry streaks

white undertail coverts

fall

yellowish feet

## GENUS *SETOPHAGA* V

Foraging behavior in *Setophaga* warblers (pp. 386–399), as well as in other warblers, is important in field ID. Note, in particular, how they catch food and the microhabitats in which they do so.

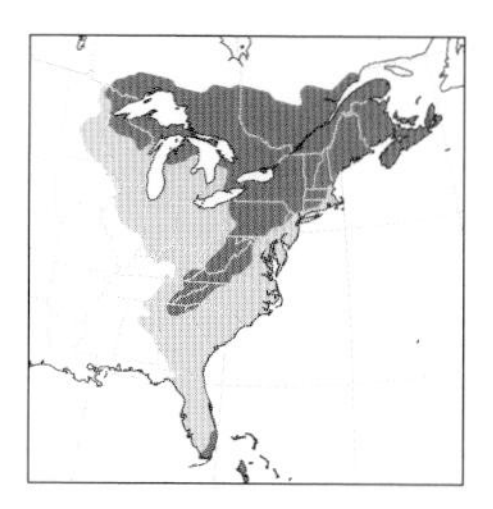

### Black-throated Blue Warbler

*Setophaga caerulescens* | BTBW | L 5¼" (13 cm)

Sexual dimorphism is stronger in the Black-throated Blue than in any other eastern warbler. Male and female were originally treated as different species, and the sexes were later shown to forage differently: males in the midstory, females in the understory.

■ **APPEARANCE** Active warbler of average build for its genus; gleans undersurface of leaves for caterpillars. Beige-and-olive female and black-and-blue male very different, but they do share one field mark: white at the base of the primaries, forming a "pocket handkerchief." Note also thin white eyebrow on female.

■ **VOCALIZATIONS** Primary song a few harsh whistles followed by a buzz: *beer beer beer bzzeeee*. Other songs, also buzzy but faster, suggest Cerulean's (p. 388). Call note a smacking *tsik* like Dark-eyed Junco's (p. 348); flight call also smacking, a rising *tsit*.

■ **POPULATIONS** Males of Appalachian subspecies, *cairnsi*, darker-backed than males of nominate subspecies from Pa. northward.

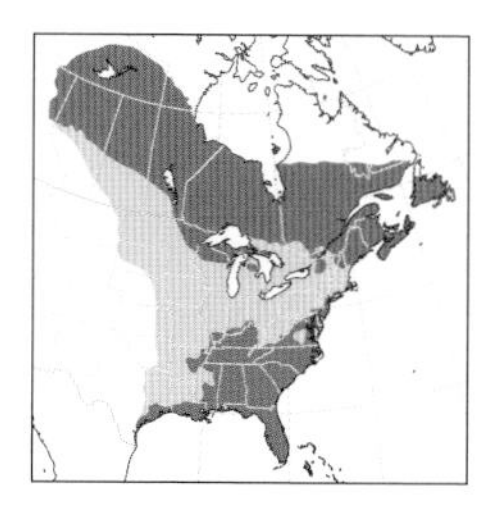

### Palm Warbler

*Setophaga palmarum* | PAWA | L 5½" (14 cm)

Despite its name, the Palm Warbler feeds on the ground in boreal bogs in summer, in shrubs and overgrown fields on migration and in winter.

■ **APPEARANCE** Ground-loving; pumps tail constantly. All have obvious eyebrow, streaking below, and yellow undertail coverts. Most breeders have obvious reddish cap, reduced or absent fall and winter.

■ **VOCALIZATIONS** Song a loose, slow trill, not unlike Swamp Sparrow's (p. 354). Call note sharp and smacking; flight call a relatively low *zeep*.

■ **POPULATIONS** Eastern breeders, *hypochrysea* ("Yellow Palm Warbler"), much yellower than western breeders, nominate "Western Palm Warbler"; dividing line roughly Appalachians (migrants) and Ont.-Que. border (breeders), but much overlap.

### Pine Warbler

*Setophaga pinus* | PIWA | L 5½" (14 cm)

Unlike the Palm Warbler, this species is well named. Whether in the Deep South or boreal woods, Pine Warblers—breeders in particular, but also migrants and winterers—find their way to pines where they glean the needles and probe the bark.

■ **APPEARANCE** Body structure unusual for genus *Setophaga:* large-billed, short-winged, and long-tailed. Wing bars prominent. Adult males stand out for their bright yellow underparts with heavy side streaking; compare with Yellow-throated Vireo (p. 266). Other plumages of Pine invite comparison with fall Bay-breasted (p. 390) and Blackpoll (p. 392) Warblers; even the drabbest Pine shows dark feet, white undertail coverts, unstreaked back, and good contrast between dark face and pale throat.

■ **VOCALIZATIONS** Song a loose trill, subtly changing pitch midway through; has speed and timbre of Dark-eyed Junco's (p. 348). Call note a flat *tip;* flight call a high, clipped *tsi*.

■ **POPULATIONS** One of the earliest warblers in spring. Winters across Southeast; relatively omnivorous in colder months, tends feeders.

whitish eyebrow and eye arc
white "handkerchief"
Black-throated Blue Warbler
caerulescens
♀
smooth olive-buff overall
black streaks on back
♂
♂
Appalachian
cairnsi
breeding
mostly brown, with yellow throat and undertail coverts
Palm Warbler
breeding
yellow underparts with rufous streaks
white tail corners
western
palmarum
eastern
hypochrysea
dark eyeline and pale eyebrow
ll variations
streaky warm brown
fall
more yellowish overall, especially on underparts and eyebrow
Pine Warbler
unstreaked back
adult ♀
immature ♀
pale throat wraps subtly around dark cheek
ng-tailed
immature ♂
yellow breast with olive streaks on sides
adult ♂
white undertail coverts
dark legs

## GENUS *SETOPHAGA* VI

Many *Setophaga* warblers (pp. 386–399) exhibit discernible geographic variation. Differences in plumage catch our attention, but vocalizations and habitat use may vary between subspecies too.

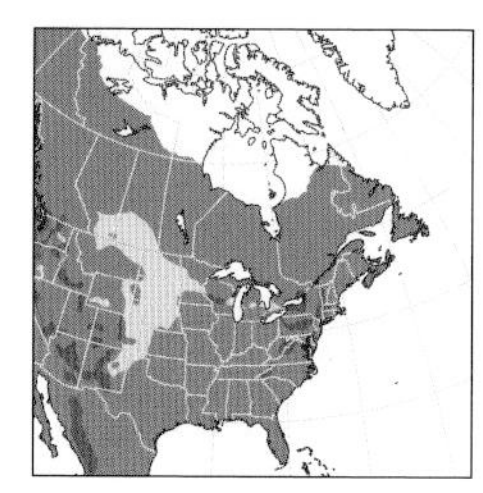

### Yellow-rumped Warbler

*Setophaga coronata* | YRWA | L 5½" (14 cm)

The most widely encountered of all eastern warblers, the Yellow-rumped is notably generalist. It gleans leaves for caterpillars, habitually hawks for flying insects, and is heavily frugivorous in winter.

■ **APPEARANCE** The familiar "Butterbutt," with all plumages yellow-rumped. "Myrtle" subspecies, widespread in East, has white throat that wraps under cheek; wing bars thin. Western "Audubon's," breeding to western Nebr. and S. Dak. and regular on migration to Great Plains, has yellow throat that does not wrap under cheek; wing has large white panel. Breeding adult, especially male, well-marked, but young in fall and winter, especially female, drab; note yellow rump, call note, and active demeanor.

■ **VOCALIZATIONS** Song a loose warble, shifting pitch toward end. Call note of "Myrtle" a hard, ringing *tchep;* call note of "Audubon's" more muffled, a rising *tchit*. Flight call of both a breathy *wheew.*

■ **POPULATIONS** Returns early in spring, leaves late in fall; winters farther north and inland than any other warbler. "Audubon's" annual to East Coast, fall and winter.

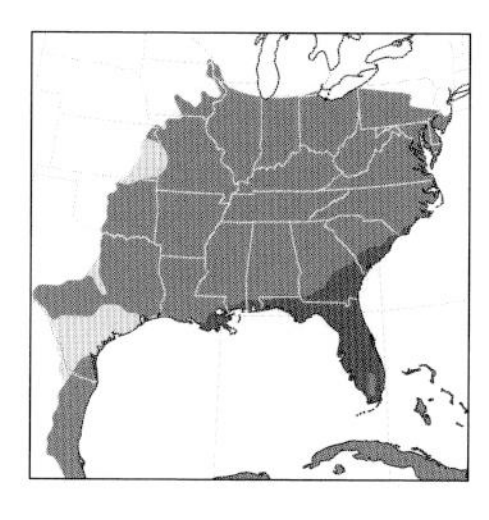

### Yellow-throated Warbler

*Setophaga dominica* | YTWA | L 5½" (14 cm)

Long-billed and inclined to forage on trunks, the Yellow-throated Warbler has behavioral affinities with the tree-creeping Black-and-white (p. 380).

■ **APPEARANCE** Boldly marked; all share general pattern of gray upperparts, bright yellow underparts, and high-contrast black-and-white on face. Inland and western subspecies *albilora* has shorter bill, more white in tail, and all-white eyebrow. Nominate *dominica,* more coastal and easterly, has longer bill, less white in tail, and eyebrow with yellow in front of eye.

■ **VOCALIZATIONS** Simple song of *chweet* notes, each slightly lower-pitched than the preceding. Call note a solid *chup;* flight call a descending *syeew.*

■ **POPULATIONS** Early spring migrant. Inland and western birds often in sycamores, coastal and easterly birds in pines and palms. Regional plumage and behavioral differences gradual (clinal), with boundary between *albilora* and *dominica* fuzzy.

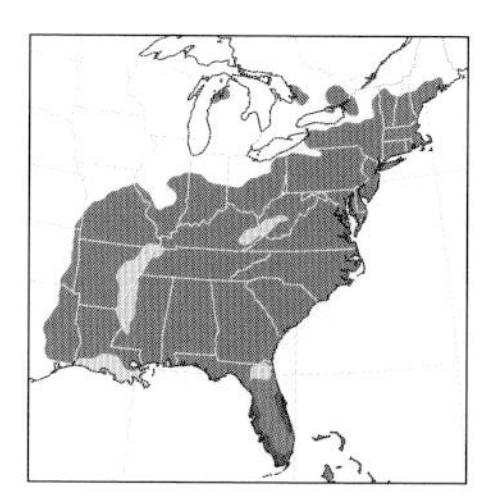

### Prairie Warbler

*Setophaga discolor* | PRAW | L 4¾" (12 cm)

Although unlikely to be found in unbroken prairie, this tail-twitching warbler is indeed a lover of open habitats: overgrown meadows, power-line cuts, and reclaimed strip mines.

■ **APPEARANCE** Small and slight; one of our yellowest warblers. Breeding adult, especially male, told by black and olive on yellow face, bright yellow underparts with black streaks on sides of breast, and red streaks on back. Fainter birds in fall and winter show hint of adult face pattern.

■ **VOCALIZATIONS** Song of short, high, fine buzzes goes up the scale. Smacking call note has slight lisp; flight call like Yellow-throated's, but finer and more clipped.

■ **POPULATIONS** Most migrate to West Indies, but subspecies *paludicola,* duller and larger, resident in Fla. mangrove swamps.

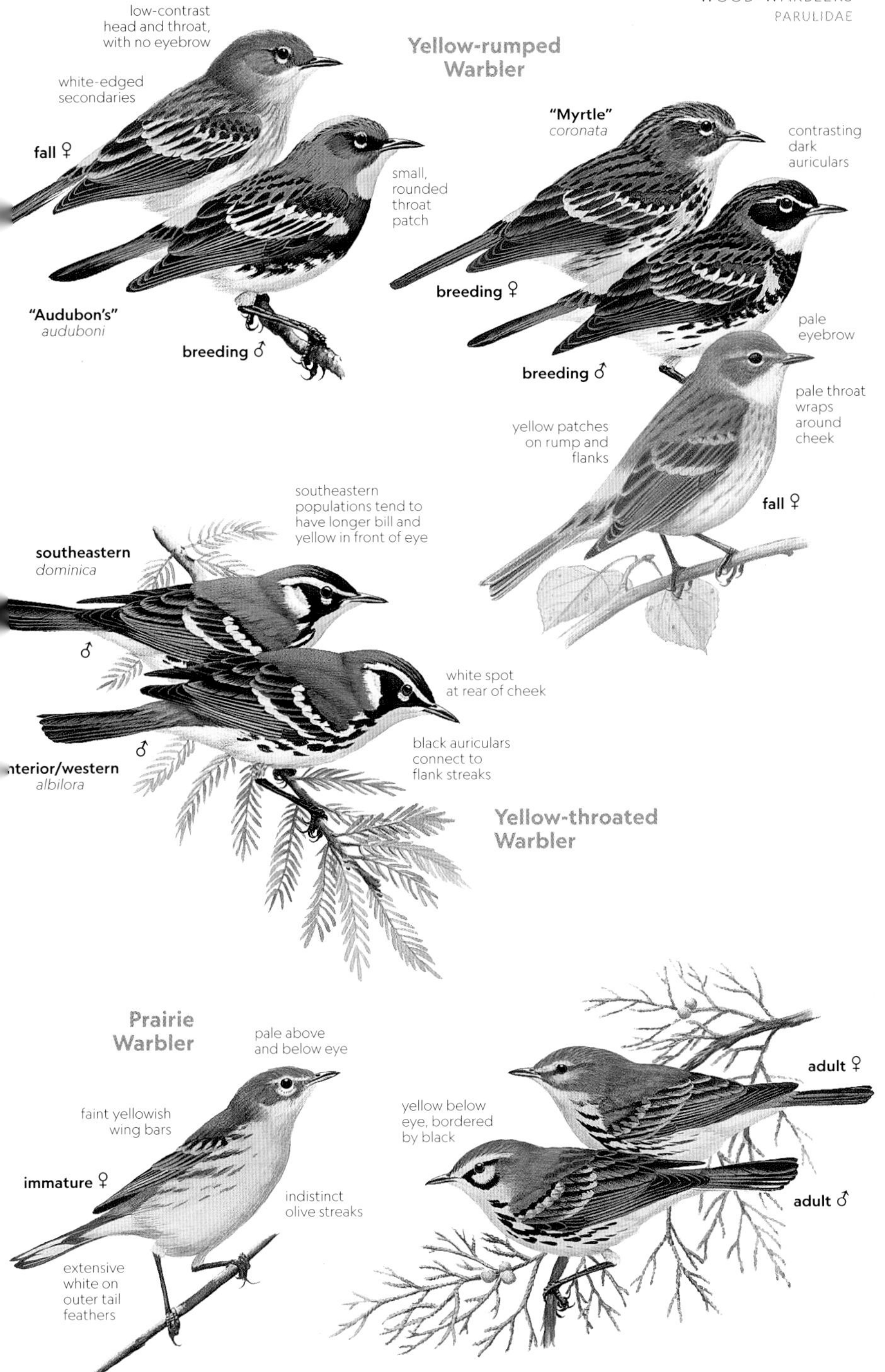
Yellow-rumped Warbler
low-contrast head and throat, with no eyebrow
white-edged secondaries
fall ♀
small, rounded throat patch
"Audubon's" *auduboni*
breeding ♂
"Myrtle" *coronata*
contrasting dark auriculars
breeding ♀
breeding ♂
pale eyebrow
pale throat wraps around cheek
yellow patches on rump and flanks
fall ♀
southeastern populations tend to have longer bill and yellow in front of eye
southeastern *dominica*
♂
white spot at rear of cheek
♂
black auriculars connect to flank streaks
nterior/western *albilora*
Yellow-throated Warbler
Prairie Warbler
pale above and below eye
faint yellowish wing bars
immature ♀
indistinct olive streaks
extensive white on outer tail feathers
yellow below eye, bordered by black
adult ♀
adult ♂

## GENUS *SETOPHAGA* VII AND GENUS *CARDELLINA*

Within the huge genus *Setophaga*, Golden-cheeked and Black-throated Green Warblers are closely related. Canada and Wilson's Warblers are in the genus *Cardellina*, close to *Setophaga*.

### Golden-cheeked Warbler

*Setophaga chrysoparia* | GCWA | L 5½" (14 cm)

Its breeding range restricted to central Tex., the endangered Golden-cheeked Warbler is holding on in juniper woodlands of the Edwards Plateau.

■ **APPEARANCE** Darker overall than Black-throated Green; Golden-cheeked has black eyeline. Adult male has little or no olive; female and young show some olive above like Black-throated Green, but none below.

■ **VOCALIZATIONS** Song of short buzzes ascends scale, ends in short whistle. Call note clinking like Black-throated Green's; flight call a short *zee*.

■ **POPULATIONS** Returns early, by mid-Mar.; most depart by Aug.

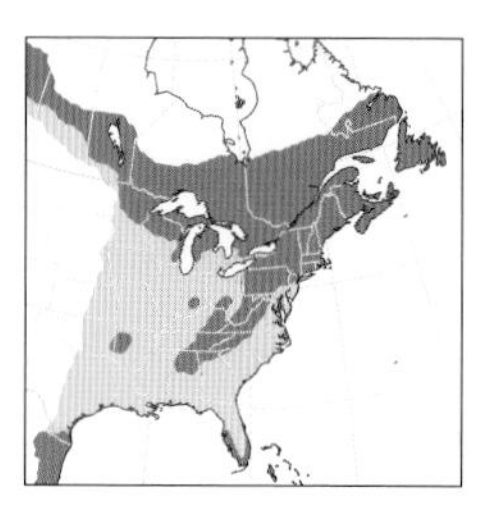

### Black-throated Green Warbler

*Setophaga virens* | BTNW | L 5" (13 cm)

A widespread generalist on migration, the Black-throated Green Warbler finds its way to coniferous forests of every sort during the breeding season.

■ **APPEARANCE** Adult male is indeed black-throated and green (more of an olive), but the bright yellow face is the most striking feature. All have some yellow below, especially on vent; Golden-cheeked lacks yellow below.

■ **VOCALIZATIONS** Two main songs, both comprising short buzzes and sharp whistles: fast *zi zi zi zoo zi* and slower *zooo zeee zoo zoo zeeee*. Call note a clinking, lisping *stik*; flight call a sharp *tzee*.

■ **POPULATIONS** Breeds widely in diverse coniferous forests, with outposts in Ark. and even cedar swamps of coastal Southeast.

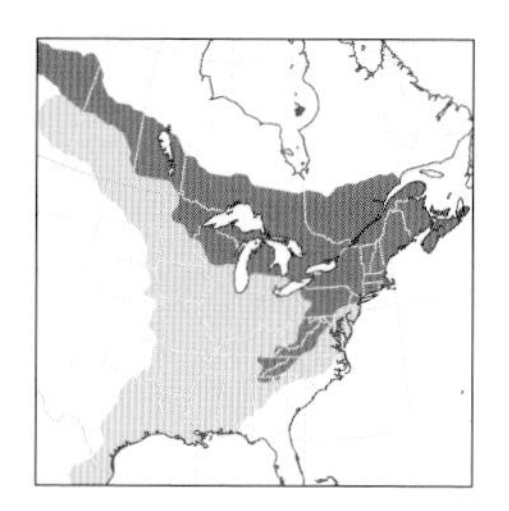

### Canada Warbler

*Cardellina canadensis* | CAWA | L 5¼" (13 cm)

This uncommon denizen of the dark forest understory breeds where there are ferns, uprooted trees, pitcher plants, and Winter Wrens (p. 302).

■ **APPEARANCE** Small and long-tailed; active but furtive. Plain blue-gray above, yellow with "necklace" of dark streaks below. All have yellow-and-white "spectacles" on dark face, fainter on female and young.

■ **VOCALIZATIONS** Song bright and chirpy; starts with *chip*, then goes into disorganized ramble. Call note a smacking *tip;* flight call low and hard, *tyit*.

■ **POPULATIONS** Late spring migrant; fairly early fall migrant. Declining overall but establishes well where forests are recovering.

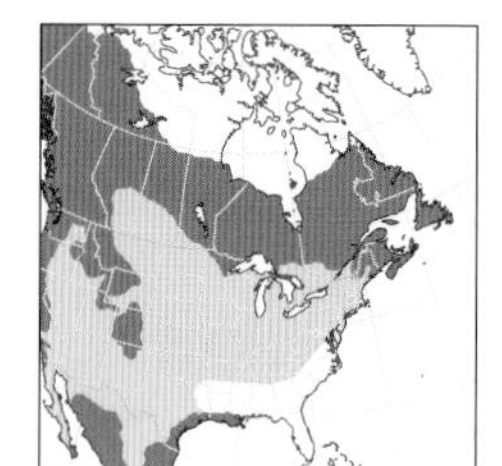

### Wilson's Warbler

*Cardellina pusilla* | WIWA | L 4¾" (12 cm)

Like the Canada Warbler, this very active bird forages close to the ground in dense cover. But it is less sun-averse, found in alder and willow thickets with no overstory.

■ **APPEARANCE** Very small; long tail often cocked. Adults, including older females, have black cap. Young wear a weak olive cap. Beady black eye stares from blank face; compare with Yellow Warbler (p. 392). Tail dark; compare with Hooded Warbler (p. 386).

■ **VOCALIZATIONS** Song a series of falling *chi* and *chet* notes, like a soft Northern Waterthrush's (p. 376). Call note a distinctive, nasal *wenk?* Flight call muffled and low, *schkwrit*.

■ **POPULATIONS** Spring migration fairly late, mostly west of Appalachians; more widespread on fall migration.

unmarked
yellow cheek in
all plumages
Golden-cheeked
Warbler
adult ♀
black back,
crown, eyeline,
and throat
adult ♂
Black-throated
Green Warbler
adult ♀
yellow face
with olive
auriculars
immature females
dark
eyeline
olive
auriculars
olive green from
crown to rump
adult ♂
Golden-cheeked
Black-throated Green
extensive
white in tail,
visible in flight
and from below
Canada
Warbler
plain gray above
immature ♀
younger birds
have little if any
black on crown
adult ♀
plain face
with isolated
dark eye
adult ♂
"necklace"
of streaks,
very faint gray
in immatures
plain wings
and tail
unmarked
lemon yellow
below
Wilson's Warbler
♂

**BIRDS CALLED TANAGERS**

In the East, the birds called tanagers—especially the searing Scarlet Tanager—are well known to nature lovers. But multiple lines of evidence point to a closer relationship with the familiar cardinal (p. 402), and all the "tanagers" in the East are now placed in the family Cardinalidae (pp. 400–409).

## Summer Tanager

*Piranga rubra* | SUTA | L 7¾" (20 cm)

The cheery songs and explosive calls of this tanager enliven sunny, shrubby, sandy pine-oak woods in the South. The species is admired for consuming copious quantities of stinging insects.

■ **APPEARANCE** Slightly larger than Scarlet Tanager, with larger, paler bill and peaked crown. Adult male rosy-red all over. Adult female mostly olive-yellow, but, unlike female Scarlet, often has splashes of dull red in plumage. One-year-old male variable; most have bright rosy head and blotchy red and olive-yellow otherwise.

■ **VOCALIZATIONS** Song a bright warble; has same cadence as hoarser Scarlet's, but rich quality of a robin (p. 318). Call a loud, fast chortle, descending in pitch: *t't'tk'tk'TUCK.* Flight call a soft whistle.

■ **POPULATIONS** Core breeding range in East in lower Mississippi Valley and southeastern coastal plain; a few wander well north every spring, reaching southern Canada.

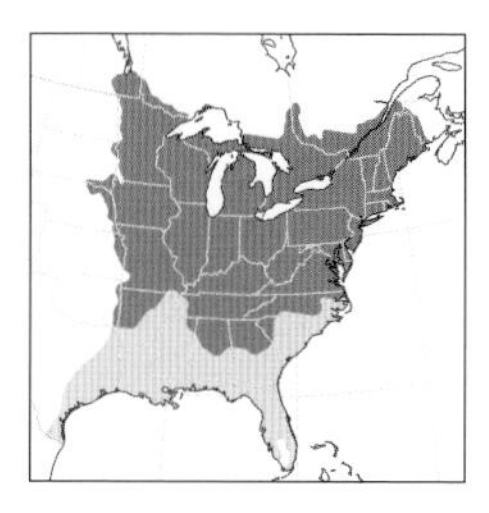

## Scarlet Tanager

*Piranga olivacea* | SCTA | L 7" (18 cm)

Birders joke that you should never stare directly at an adult male Scarlet Tanager, lest you incur permanent retinal damage. The bird can be oddly hard to spot, though, as it forages slowly and deliberately in the leafy treetops.

■ **APPEARANCE** Slighter-billed and a bit smaller overall than Summer Tanager. Breeding adult male is simply brilliant red and jet-black, but other plumages are more interesting. First-spring male (in its second calendar year) resembles breeding adult male, but wings tinged greenish. Male in fall mostly yellow-green with wings black (adult) or blackish tinged green (first-year). Female year-round like first-fall male; females of all *Piranga* tanagers vary in brightness and hue, and some female Scarlets and Summers can be the same color.

■ **VOCALIZATIONS** Song like a robin (p. 318) with a sore throat: a series of raspy and slurred notes, often in groups of four. Call note a far-carrying *CHIP! purrr;* first note higher than second. Flight call similar to Summer's.

■ **POPULATIONS** Area-sensitive breeder, absent from small forest tracts. Orange-variant male seen from time to time, yellow variant much rarer; variation in color in both sexes, unrelated to geography, caused by diet and genetics.

## Western Tanager

*Piranga ludoviciana* | WETA | L 7¼" (18 cm)

Spectacular *Piranga* tanager of western pine forests, with breeding outposts east to Sask., S. Dak., and Nebr. Uncommon migrant western Great Plains; rare but annual vagrant in winter to East Coast. Adult male in breeding plumage unmistakable, but most wanderers eastward are in plainer winter plumage. All show prominent wing bars, the upper yellowish; pale uppertail coverts and rump contrast with darker back. Song is like Scarlet Tanager's, but call note more similar to Summer's.

**Summer Tanager**

**1st spring ♂**

variable, with splotchy red and olive-yellow

**red-morph ♀**

some females with variable reddish feathers

ochre-yellow, including wings

olive wings and tail do not contrast with body

♀

large bone-colored bill

**adult ♂**

rosy-red overall, including wings

**Scarlet Tanager**

**1st spring ♂**

♀

darker tail and wings than female Summer

small grayish bill

**1st fall ♂**

usually brighter green than Summer

**fall adult ♂**

**breeding adult ♂**

**BIG-BILLED CARDINALIDS**

None of the cardinalids (pp. 400–409) have small bills, and some have huge appendages. The four on this spread are in the latter category. The cardinal and Pyrrhuloxia are in a grouping with *Piranga* tanagers (p. 400), the *Pheucticus* grosbeaks closer to the genus *Passerina* (pp. 404–407).

## Northern Cardinal

*Cardinalis cardinalis* | NOCA | L 8¾" (22 cm)

Practically all humans in our area are within walking distance of a splendid cardinal right now—from urban South Beach and Hyde Park and the Bronx to every farm in Appalachia and the Midwest.

■ **APPEARANCE** Slim, long-tailed, and crested. Adult male red except for black face mask. Adult female warm brown with ample reddish highlights; note large red-orange bill. Juvenile rufous-brown with dark bill.

■ **VOCALIZATIONS** Song, given by both sexes, comprises rich, repetitious whistles: *cheer! cheer! cheer!* or *birdy birdy birdy*.... Call a thin, smacking *tsit*.

■ **POPULATIONS** Nonmigratory, but range has been expanding steadily north for 100+ years.

## Pyrrhuloxia

*Cardinalis sinuatus* | PYRR | L 8¾" (22 cm)

The range of the "Desert Cardinal" in the East is completely subsumed by that of the more widespread Northern Cardinal. Where the two species co-occur, the Pyrrhuloxia is more likely to be out in arid mesquite scrublands.

■ **APPEARANCE** Differs structurally from Northern Cardinal by wispier crest and bill shape: upper mandible strongly curved, with hooked tip. Male gray with splotchy red highlights. Female grayer overall than cardinal; note also yellow bill, strongly curved above.

■ **VOCALIZATIONS** Song like Northern Cardinal's, but not as rich and vigorous. Call note, the opposite, is richer and more solid than Northern Cardinal's.

■ **POPULATIONS** Like Northern Cardinal, nonmigratory yet expanding north; range expansion has followed long-term spread of mesquite.

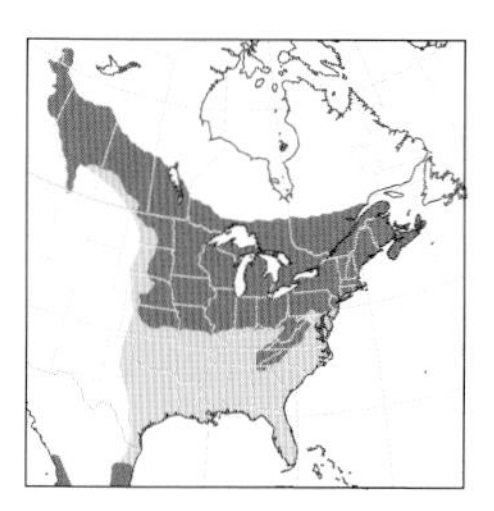

## Rose-breasted Grosbeak

*Pheucticus ludovicianus* | RBGR | L 8" (20 cm)

Somewhat tanager-like in its haunts and habitats, the Rose-breasted Grosbeak moves slowly and sings loudly in the canopy of broadleaf forests.

■ **APPEARANCE** Chunky, with huge pale bill. Adult male in spring and summer is black and white with variable inverted rosy triangle on breast. Brownish female, variably streaked below with boldly patterned face, suggests Purple Finch (p. 326), but grosbeak chunkier and bigger-billed. All have white in wing, prominent in flight.

■ **VOCALIZATIONS** Caroled song robinlike, but faster, thinner, and sweeter. Call note an abrupt squeak (like wet sneakers on a gym floor).

■ **POPULATIONS** Hybridizes with Black-headed Grosbeak where ranges overlap in Great Plains. Regular but uncommon migrant west to Rockies.

## Black-headed Grosbeak

*Pheucticus melanocephalus* | BHGR | L 8¼" (21 cm)

Western counterpart of Rose-breasted Grosbeak; breeds east to central Great Plains in broadleaf woods, especially along rivers. Breeding adult male, with extensive orange-umber, distinctive, but other plumages resemble Rose-breasted. Black-headed has bicolored bill; breast streaking of female finer and sparser than on Rose-breasted. Song much like Rose breasted's, but not as joyful; call note of Black-headed less squeaky.

black mask surrounds red bill in adult plumages
♂
♀
dark bill turns red gradually
Northern Cardinal
juvenile
tan with red wash on tail, wings, and crest
Pyrrhuloxia
long, thin crest
grayish overall with red highlights
♀
♂
yellowish bill, very rounded
winter ♂
white rump
breeding ♂
similar to Black-headed; note bill color, pink "wingpits"
Rose-breasted Grosbeak
1st fall ♂
breeding ♂
1st spring ♂
large pale bill
♀
breast whitish, completely streaked
both male *Pheucticus* grosbeaks have striking white wing patches, smaller in females and immatures
two-toned bill
buffy breast; streaks confined to flanks
♀
Black-headed Grosbeak
breeding ♂
1st fall ♂

**GENUS *PASSERINA***

On a bird-for-bird basis, the genus *Passerina* might be the most colorful in the East. (And check out the species in Mexico!) They occur in brushy habitats around forest edges, and, on top of their crazy colors, they sing loudly and exuberantly.

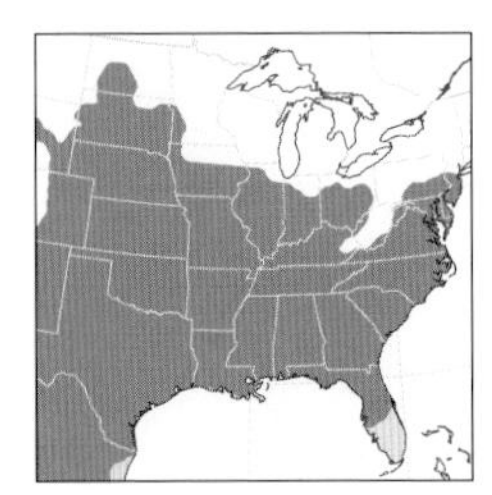

## Blue Grosbeak

*Passerina caerulea* | BLGR | L 6¾" (17 cm)

The name grosbeak, like the name bunting (see sidebar, p. 334), is an imprecise catchall, nonindicative of true taxonomic relationships. Ecologically and behaviorally, the Blue Grosbeak is perfectly typical of the genus *Passerina*.

■ **APPEARANCE** Largest *Passerina* bunting, with the biggest bill. Adult male deep indigo (even more so than Indigo Bunting, p. 406) with chestnut wing bars; first-spring male (in its second calendar year) splotchy indigo and rust. Female and immature cinnamon-brown overall; wing bars chestnut in summer, often whitening in fall.

■ **VOCALIZATIONS** Song a rushed, burry warble with little change in pitch; strangely similar in timbre to *swainsoni* ("Western") Warbling Vireo (p. 268). Call note flat and clinking, *plenk;* flight call a low buzz, *szzzzt.*

■ **POPULATIONS** A mostly southern species east of Mississippi R., but expanding north in Midwest. Movements in summer complex; late-season nesters, with dependent young into Sept., may have bred elsewhere earlier in the summer.

## Varied Bunting

*Passerina versicolor* | VABU | L 5½" (14 cm)

Occurs mostly in Mexico, ranging north to U.S. border states. Regular but sporadic in Tex., mostly in warmer months: some on Edwards Plateau; others along Rio Grande, but a bit upstream from Gulf Coast. Adult male in good light wears the colors of a fruit bowl, with plum, blueberry, and strawberry hues. Brownish female like Indigo Bunting, but breast lacks streaking of Indigo and wings are unmarked; upper mandible of Varied more rounded. Song bright and steady, more like Painted's than Indigo's.

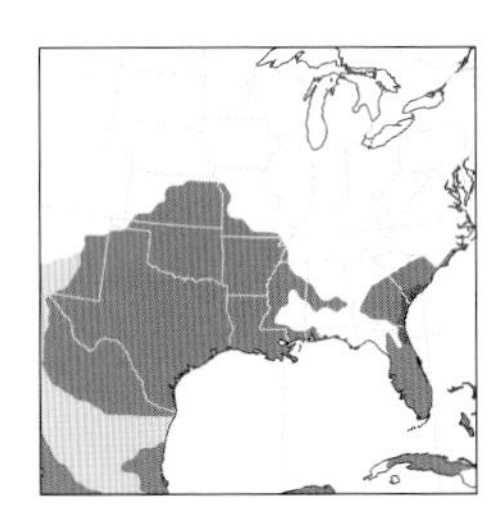

## Painted Bunting

*Passerina ciris* | PABU | L 5½" (14 cm)

The Wood Duck (p. 30) and Purple Gallinule (p. 92) give it a run for its money, but the Painted Bunting is absolutely a finalist for the title of most colorful bird in the East. A wonderful colloquial name is "Nonpareil"—without equal.

■ **APPEARANCE** Like other *Passerina* buntings (except Blue Grosbeak), about the size and shape of a sparrow. Adult male a riot of color. Other plumages, more uniform, sport a thin yet complete eye ring. Adult female bright lime-gray; younger birds grayer, with just a hint of green.

■ **VOCALIZATIONS** Song a bright warble, sweeter than song of Indigo Bunting (p. 406) and lacking paired elements. Call note an up-slurred *chep,* like "Audubon's" subspecies of Yellow-rumped Warbler (p. 392). Flight call, buzzy and scratchy, is lower than Indigo's.

■ **POPULATIONS** Breeding range has large gap along Gulf Coast; breeders west of the gap average smaller and paler than range-restricted eastern breeders in southeastern coastal plain; breeders east of the gap molt on or near breeding grounds, western breeders elsewhere. Declining range-wide, especially eastern breeders.

very large bill
♀
Blue Grosbeak
unstreaked brown
breeding ♂
rusty wing bars
immature
deep blue
variable splotches of blue and tan
tail black
1st spring ♂
dark purplish, often looks black
culmen more rounded than other buntings
♀
Varied Bunting
adult ♂
unmarked dull tan
blue rump
adult ♂
Painted Bunting
yellowish eye ring
pale green overall, brightest on back
adult ♀
very plain
always shows some green tinge around rump
juvenile

## LAZULI AND INDIGO BUNTINGS

See what looks like a bright blue sparrow singing lustily from a high perch? Depending on where you are, it's likely one of these two West-East counterparts. They flourish in woods that aren't too wooded and in clearings that aren't too clear.

### Lazuli Bunting

*Passerina amoena* | LAZB | L 5½" (14 cm)

This widespread and common westerner reaches our area in the northern Great Plains, especially in woods around waterways; also very rare but annual much farther east, all the way to coast, primarily late fall through winter and typically at bird feeders. Breeding male's orange, white, and blue scheme roughly mirrors that of Eastern Bluebird (p. 312), but body shape and vocalizations different; note also white wing bar of Lazuli. Song like Indigo Bunting's, but buzzier and with fewer paired phrases. Compare female with female Indigo.

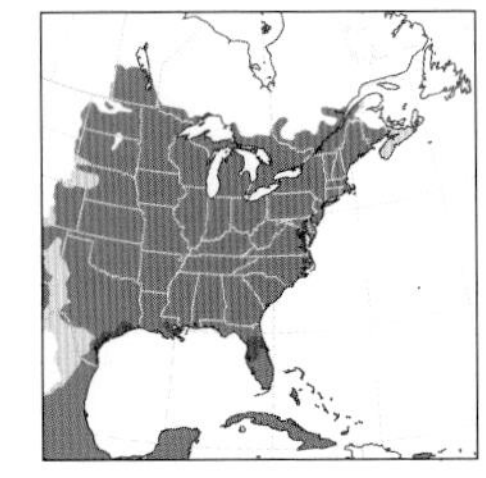

### Indigo Bunting

*Passerina cyanea* | INBU | L 5½" (14 cm)

In every forest clearing and along every power-line cut in the East, this impossibly blue songster is likely to be found. Just hop out of your car in semiwooded country in June, listen for not even a minute, and you'll probably hear one.

■ **APPEARANCE** Body structure identical to that of Lazuli Bunting; breeding ecologies also similar. Breeding male is the only completely blue bird ordinarily found in the East. Adult female gray-brown all over with paler throat, fine streaking on breast, and thin wing bars; female Lazuli similar, but shows unstreaked breast washed pale gray-buff, whiter belly and bolder wing bars than Indigo, and throat less contrasting than Indigo. Compare also with female Brown-headed Cowbird (p. 370). In both Lazuli and Indigo, first-spring male (in second calendar year) and adult male in winter like breeding adult of corresponding species, but blotchy brown all over.

■ **VOCALIZATIONS** Song comprises slurred elements, often paired, trailing off: *spit spit chew chew spit-it-out-chew.* Call note clinking like Northern Cardinal's (p. 402), but more robust; flight call a short, lazy buzz.

■ **POPULATIONS** Trans-Gulf migrant with sometimes massive fallouts late April to early May along central Gulf Coast. Very rare but annual in winter to Great Lakes and New England.

### "Lazigo" Buntings

Lazuli and Indigo Buntings will sometimes hybridize where their ranges overlap—in the Dakotas, for example. Hybrids are variable, but first-generation males exhibit a somewhat stereotyped phenotype: They have a Lazuli-like color scheme overall, but with blue (like Indigo), not orange (like Lazuli) on the breast. Hybrid females are probably not safely identifiable in the field.

Another dimension to the identification problem involves males, especially Indigos, in their first spring, when they are close to one year old. The splotchy blue and white of a first-spring male Indigo is variably expressed, and sometimes it matches rather closely that of a hybrid. Indigo Buntings, although rare in winter in most of the East, are far more likely at that time of year than Lazuli Buntings. A weird bird coming to a feeder then is probably an Indigo, not a hybrid.

A final consideration is song. Hybrids may sing the song of either species, or a blend of the two, but pure adults in the zone of range overlap may do likewise. A pure Lazuli can pick up the song of an Indigo, and vice versa. Take the time to track down an unseen songster; the bird will be beautiful, no matter what, and it might not be what you expected.

Lazuli Bunting
winter ♀
unstreaked warm buff
juvenile
Aug. to Sept.
faint streaks,
ilar to Indigo
sky blue
white wing bars, upper quite broad
breeding ♂
buff breast contrasts subtly with white belly
breeding ♀

Indigo Bunting
pale throat
breeding ♀
cool brown with blurry streaks
completely rich blue
breeding ♂
variable, patchy
1st spring ♂
winter ♂
females and immatures more richly brown in fall
fall ♀
Lazuli x Indigo hybrid
usually similar pattern to Lazuli, with blue breast and deep blue tones closer to Indigo

## DICKCISSEL, BANANAQUIT, SEEDEATER

These three have hopped around the checklist quite a bit of late. The Dickcissel has come to rest among the Cardinalidae (pp. 400–409), and the Bananaquit and Morelet's Seedeater turn out to be "true" tanagers, family Thraupidae.

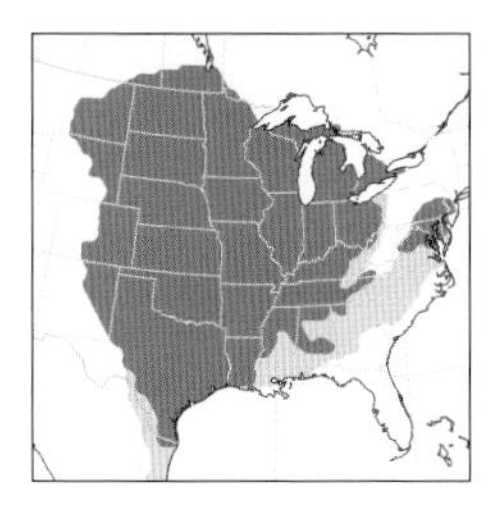

### Dickcissel

*Spiza americana* | DICK | L 6¼" (16 cm)

With yellow underparts marked with a broad black breast patch, the breeding male Dickcissel suggests a small meadowlark (p. 364). Like meadowlarks, it is a bird of pastures and prairie.

■ **APPEARANCE** Stout-billed like other cardinalids. Breeding male, yellow and black below, distinctive. Other plumages duller, resembling female House Sparrow (p. 320), but note yellowish wash across breast, rufous on wing, and gray face set off by buff-yellow eyebrow and white throat.

■ **VOCALIZATIONS** Song a few chips followed by short buzzes: *dick dick dick SISS SISS SISS*. Flight call an abrupt, flatulent *ffrrrt*.

■ **POPULATIONS** Most winter in northern S. Amer., but annual in East in winter, mostly near coast and typically at feeders with House Sparrows. Limits of breeding range vary annually.

### Bananaquit

*Coereba flaveola* | BANA | L 4½" (11 cm)

Familiar tanager in gardens and parks in the West Indies. Wanders less than annually to South Florida, but has never established—something of a mystery. A little blob of a bird, with a short tail and decurved bill; fidgets about flowering shrubs and fruiting trees. All are dark above and pale below, with a yellow rump and a yellow wash below; broad pale eyebrow contrasts with dark head. Songs and calls high, hissing, and squeaky.

### Morelet's Seedeater

*Sporophila morelleti* | MOSE | L 4½" (11 cm)

Mid. Amer. tanager; barely reaches South Texas, where it inhabits cane thickets along Rio Grande, especially from Falcon Dam northward. Small and chubby with bulbous bill. Adult male, mostly dirty buff, has black hood and dark wings with bright white flashes, suggesting Lesser Goldfinch (p. 330). Female plain gray-buff with thin wing bars; compare with female Lazuli Bunting (p. 406). Song, bright and chirpy, has timbre of Painted Bunting's (p. 404), but not as herky-jerky. Calls include flat *chint* like Blue Grosbeak's (p. 404).

### Tanager, Not a Tanager

Eight species on the American Birding Association (ABA) checklist of the birds of the U.S. and Canada are in the tanager family Thraupidae. They have the following names: Bananaquit, seedeater, finch, honeycreeper, grassquit (two species), and cardinal (two species in Hawaii). Meanwhile, all the ABA birds called tanager (three in the East, two others in the West) are in the cardinal family Cardinalidae; likewise, the Western Spindalis, formerly called the Stripe-headed Tanager, is not a tanager.

The very word "tanager" is an etymological muddle. It is a Tupi word that came to English through Portuguese and faux-Latin intermediaries. It surely meant something to the speakers of classical Tupi, a dynamic language of Indigenous peoples in present-day Brazil. But in 21st-century English, "tanager" is just a word, a useful placeholder, something to say (or not say) when one lays eyes on a luminous Scarlet Tanager or bewitching Bananaquit. Saying that word is the beginning, not the end point, of understanding and ultimately appreciating the bird in life.

Dickcissel
breeding ♀
yellowish breast
winter ♂
rufous "shoulders"
breeding ♂
breeding plumage acquired by feather wear, not by molt
pale eyebrow (compare female House Sparrow)
immature ♂
dark moustache line
immature ♀
Bananaquit
white eyebrow
juvenile
curved, tapered bill
adult
high-contrast pattern
drab olive-buff
♀
stubby, curved bill
1st winter ♂
adult ♂
distinctly patterned wings
Morelet's Seedeater

This appendix enumerates 241 species that occur less than annually in the East based on records compiled from previous editions of this guide, eBird data, and regional lists, and has been reviewed by experts in the field. The following lists are current as of summer 2023, but new information is always becoming available—sometimes related to quite old records—and of course, birds will continue to show up in unexpected places long after this book goes to press.

## SPECIES OF CASUAL AND PERIPHERAL OCCURRENCE IN THE EAST

**A casual species is one that appears less than annually in a region, yet not so infrequently as to be entirely unpredicted—in late fall, for example, or during irruption years. Peripherals are species just beyond their main range. Many birds of the Rocky Mountains extend out onto the High Plains a bit and are best viewed as being of peripheral occurrence in the East.**

Garganey

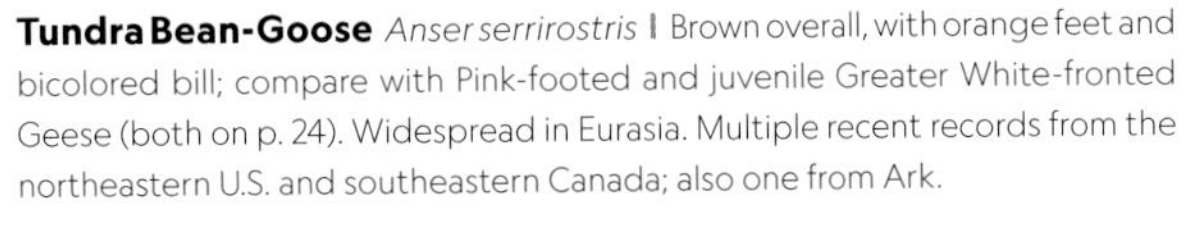

**Tundra Bean-Goose** *Anser serrirostris* | Brown overall, with orange feet and bicolored bill; compare with Pink-footed and juvenile Greater White-fronted Geese (both on p. 24). Widespread in Eurasia. Multiple recent records from the northeastern U.S. and southeastern Canada; also one from Ark.

**Garganey** *Spatula querquedula* | Breeding male striking, with huge white flare on head and face; female and nonbreeding male gray overall with diffuse head stripes. Widespread in Eurasia. Nearly annual in East, especially Atlantic Canada. Preponderance of records winter and spring, when adult males are easiest to identify.

White-cheeked Pintail

**White-cheeked Pintail** *Anas bahamensis* | Shaped like Northern Pintail (p. 38), with long neck, long bill, and very long tail. Drake has huge white cheek patch, bill with red base, and pale tail; female has lower-contrast plumage. Widespread in West Indies and S. Amer.; recorded annually in Fla., but status there complicated by escapes from captivity.

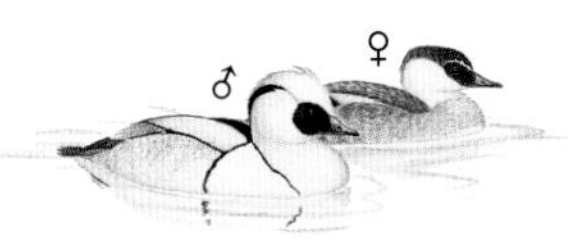


Smew

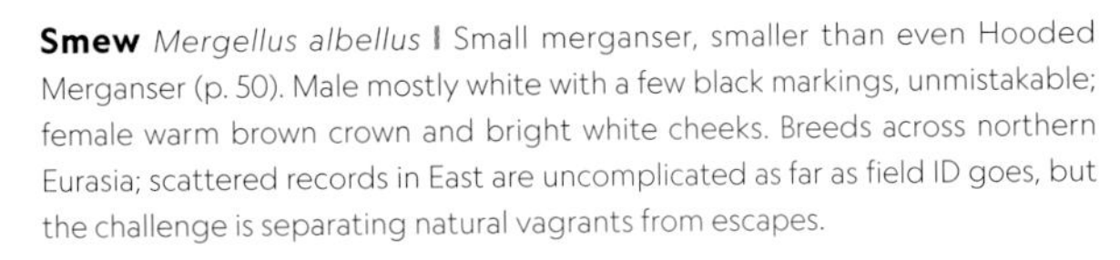

**Smew** *Mergellus albellus* | Small merganser, smaller than even Hooded Merganser (p. 50). Male mostly white with a few black markings, unmistakable; female warm brown crown and bright white cheeks. Breeds across northern Eurasia; scattered records in East are uncomplicated as far as field ID goes, but the challenge is separating natural vagrants from escapes.

**Band-tailed Pigeon** *Patagioenas fasciata* | Large pigeon of western forests, south well into S. Amer. Similar in build and color to familiar Rock Pigeon (p. 70); white band on hindneck more prominent than dull tail band. Most U.S. and Canada populations migratory, with numerous records to East; frequency of sightings has declined in 21st century.

Ruddy Ground Dove

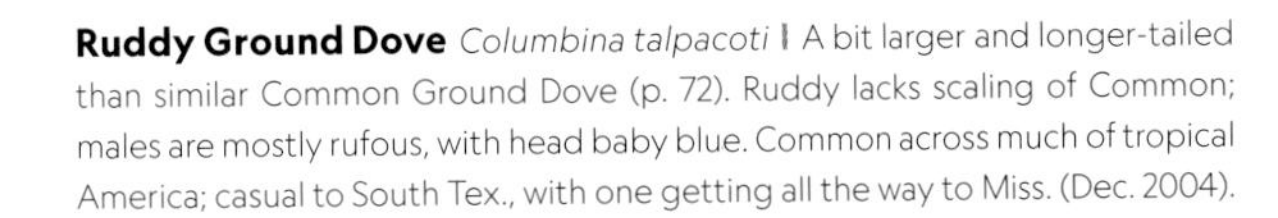

**Ruddy Ground Dove** *Columbina talpacoti* | A bit larger and longer-tailed than similar Common Ground Dove (p. 72). Ruddy lacks scaling of Common; males are mostly rufous, with head baby blue. Common across much of tropical America; casual to South Tex., with one getting all the way to Miss. (Dec. 2004).

Key West Quail-Dove

**Key West Quail-Dove** *Geotrygon chrysia* | Ground-dwelling dove (not a quail), exquisite in good light; forages on shady forest floor. Upperparts glossed red, purple, and green; face has broad white stripe below eye. West Indies specialty, formerly breeding to Fla. Keys (hence its name); today only a vagrant, to South Fla.

**White-collared Swift** *Streptoprocne zonaris* | Large and dark with deeply notched tail; white collar visible from afar. Widespread in New World tropics, mostly around forests in foothills and mountains, from Mexico and Cuba south to northern Chile and northern Argentina; records in East from Tex. and Fla., and even as far north as the Great Lakes (Mich., Ont.).

White-collared Swift

**Vaux's Swift** *Chaetura vauxi* | Similar to Chimney Swift (p. 84), but proportionately shorter-winged and smaller overall; rump and throat paler than Chimney's. Chatters in flight like Chimney; calls of Vaux's shriller, more insectlike. Breeds western N. Amer., winters mostly in tropics, but a few recorded, less than annually, Gulf Coast, mainly La. to northern Fla.

**Green-breasted Mango** *Anthracothorax prevostii* | Breast wholly dark in adult male, but other plumages recognized by black stripe down center of largely white breast; all have broad purplish to rufous tails, cornered with white in female and young. Tropical species ranging north regularly to Tamaulipas; almost annual in recent years to southern Tex.

Green-breasted Mango

**Calliope Hummingbird** *Selasphorus calliope* | Smallest hummingbird regular to the U.S. and Canada, and indeed one of the smallest birds on Earth. Other *Selasphorus* hummingbirds larger, longer-billed, and larger-tailed. Breeds northern and central Rockies, winters mostly western Mexico, but increasing and annual to East, especially Gulf Coast region.

**Broad-billed Hummingbird** *Cynanthus latirostris* | More conspicuous than the bill's broadness is its redness, especially on adult male. Adult male also has blue bib and green belly; female, gray below, has white eyebrow. A mostly Mexican species that reaches U.S. chiefly in Southeast Ariz.; regular in very small numbers in winter to La., casual elsewhere in East.

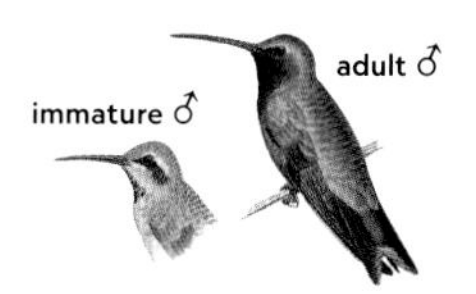

Broad-billed Hummingbird

**Corn Crake** *Crex crex* | Similar in structure to Sora (p. 90), but larger. Shows dark barring and spotting against buff-gray ground color; wings show extensive rufous. Widespread in Old World; particularly well-known in Europe. Formerly a rare vagrant coastally from mid-Atlantic northward; records dropped off in late 20th century but seem to have picked up a bit again recently.

Corn Crake

**Common Crane** *Grus grus* | Well-marked adult has mostly black head and neck with thick white stripe up neck and red on crown. Most records in N. Amer.—and almost all in East—are from Nebr. Sightings usually in association with large flocks of Sandhill Cranes (p. 96); both species breed in Russia, and individuals sometimes migrate with the "wrong" species.

Common Crane

**Northern Lapwing** *Vanellus vanellus* | Large, well-marked plover with a long crest. High-contrast black-and-white plumage is especially prominent in flight; flight call a squeaky *slwee-eet*. Almost annual vagrant from Eurasia, fall to spring, to Atlantic coast, especially Atlantic Canada. Habitats much like Killdeer's (p. 100), often found in fields in coastal lowlands.

**European Golden-Plover** *Pluvialis apricaria* | Similar to American Golden-Plover (p. 100), but does not usually overlap with that species: European nearly annual to Atlantic Canada in spring, when American is migrating north across Great Plains. European, larger than American, has white underwings; white side-stripe in breeding plumage runs length of body. Flight call a wavering whistle.

Northern Lapwing

Northern Jacana

**Pacific Golden-Plover** *Pluvialis fulva* | Alas. and Russian breeder; away from the breeding grounds, most records in N. Amer. from West Coast, but almost annual in recent years to Northeast. Pacific has slightly shorter wings, slightly longer legs, and slightly larger bill than American (p. 100). Flight call a powerful *weee* or *w'weee*.

**Northern Jacana** *Jacana spinosa* | The only shorebird of the family Jacanidae found in the East. Clambers about floating vegetation on preposterously long toes. Adult black and chestnut with yellow remiges; immature white below with boldly patterned face. Widespread and easily spotted in marshes in Mid. Amer. Erratic visitor to South Tex.; not annual.

Sharp-tailed Sandpiper

**Eurasian Curlew** *Numenius arquata* | Large curlew, widespread in Eurasia, with scattered records in Northeast. Nearly the heft of Long-billed Curlew (p. 106), not expected in Northeast. Compare with smaller Whimbrel (p. 106), especially white-rumped European subspecies, probably annual to East, which is shorter-billed, smaller, and with a more boldly patterned face than Eurasian.

**Sharp-tailed Sandpiper** *Calidris acuminata* | Breeds in Arctic Russia; a few stray each year to western N. Amer. and even to East, where annual in recent years. Plumage and body structure similar to Pectoral Sandpiper (p. 118); most vagrants to East are juveniles, with warmer tones, plainer breast, and bolder eyebrow than Pectoral's.

Red-necked Stint

**Red-necked Stint** *Calidris ruficollis* | Old World "peep"; breeds mostly Russian Far East, but also sparingly to Alas. Casual vagrant to East, especially mid-Atlantic and New England coasts. Long-winged, short-legged. Breeding adult distinctive; drab juvenile, gray-brown above with dusky breast, resembles Semipalmated Sandpiper (p. 118).

Little Stint

**Little Stint** *Calidris minuta* | Like Red-necked Stint, Old World vagrant to East with concentration of records mid-Atlantic and New England coasts. Breeding adult brightly colored like Red-necked. Bill of Little thick-based and slight at tip; sides of breast streaked. Juvenile has prominent white V on mantle. Field ID of vagrant stints/peeps is very challenging, so get photos!

**Spotted Redshank** *Tringa erythropus* | Old World relative of yellowlegs (p. 124); casual vagrant to East, with preponderance of records mid-Atlantic coast and eastern Great Lakes. Striking breeding garb mostly black. Grayer nonbreeding and juvenile plumages told by strong eyebrow, mostly orange legs, and long, thin bill with droopy tip and variable orange at base.

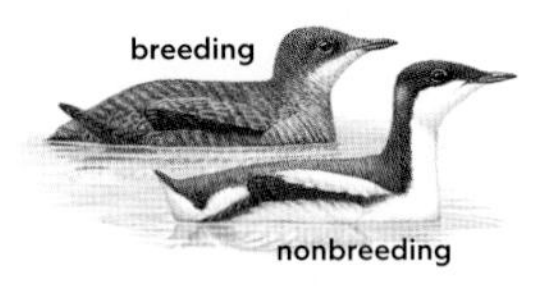

Long-billed Murrelet

**Long-billed Murrelet** *Brachyramphus perdix* | Breeds coastal northeastern Asia, yet vagrates deep into N. Amer., reaching East Coast. Most in East are from fall when in black-and-white nonbreeding plumage; an alcid well inland in East might well be this species. Note superficial similarity to Horned Grebe (p. 66) in winter plumage.

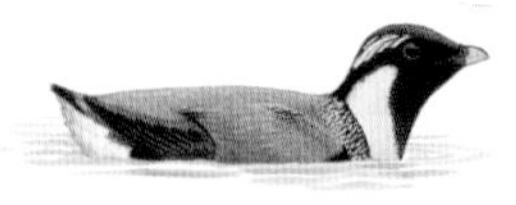
Ancient Murrelet

**Ancient Murrelet** *Synthliboramphus antiquus* | Like Long-billed Murrelet, a Pacific Rim breeder that vagrates well inland, typically fall; strongest concentration of records in East from Great Lakes. Plumper than Long-billed; bill pale and short. Mantle gray and head black; white on breast extends to sides of neck. Whitish streaking on head (thus "ancient") reduced in fall and winter plumage.

**Ross's Gull** *Rhodostethia rosea* | Scarce, high Arctic gull conjectured to winter near edge of pack ice. Has bred in our region in Churchill, Man.; reported not quite annually in East, mostly north of 40th parallel. Breeding adult has unique black collar; other plumages told by pointed wings and wedge-shaped tail. One of the most sought-after and iconic of all birds.

Ross's Gull

**Black-tailed Gull** *Larus crassirostris* | Characteristic gull of coastlands of East Asia; reaches East not quite annually. Just a bit larger than Ring-billed Gull (p. 144). Long bill has red-and-black tip on adult. Winter adult and immature sport white eye crescents. Black tail band of adult striking, but note that immature gulls of most species can show much black in tail.

Heermann's Gull

**Heermann's Gull** *Larus heermanni* | Striking gull of Pacific coast of N. Amer., relatively small for genus *Larus;* has propensity for wandering all the way to East Coast, with sightings increasing in recent years. Adult gray with red bill; head nearly white in breeding plumage, streaked dusky in nonbreeding plumage. Young birds uniformly dark with black-tipped, dull orangey bill.

Yellow-legged Gull

**Yellow-legged Gull** *Larus michahellis* | European relative of Herring Gull (p. 146); preponderance of records in East from Nfld., especially St. John's, in winter. Adult a bit darker above than Herring Gull, with distinctive yellow legs. First-winter, with thick black bill and salt-and-pepper upperparts, suggests first-winter Great Black-backed Gull (p. 150).

**Slaty-backed Gull** *Larus schistisagus* | Like Black-tailed Gull, characteristic of East Asian coastlands. More frequent vagrant to East than Black-tailed; annual in recent years, especially Great Lakes and Atlantic coast. Large-bodied and relatively large-billed. Winter adult dark-backed with dark smudge on face across eye; legs deep pink.

Slaty-backed Gull

**Glaucous-winged Gull** *Larus glaucescens* | Large gull of coastlands of the Pacific Northwest to northeastern Asia, nearly annual in recent winters to East, especially Great Lakes. As large gulls go, rather distinctive: On both young and adults, wing tips same color as mantle. Beware hybrids, especially with Western (p. 420) and Herring (p. 146) Gulls.

**Kelp Gull** *Larus dominicanus* | S. Amer. gull increasingly reported in East; formerly bred in La., where it hybridized with Herring Gull (p. 146). Most records in East from coasts and Great Lakes. Adult strikingly black-backed with large bill and dull greenish legs; first-year birds like Lesser Black-backed (p. 148) in plumage but bulkier and bigger-billed.

White-winged Tern

**White-winged Tern** *Chlidonias leucopterus* | An Old World "marsh tern" related to Black Tern (p. 154); numerous records from East, mostly mid-Atlantic coast northward and Great Lakes, but not annual. Build more compact than Black Tern's. Upperwings and tail of White-winged paler than on corresponding plumages of Black; breeding White-winged has black underwing linings.

**Elegant Tern** *Thalasseus elegans* | Pacific coastal tern of the Americas. Increasingly reported in East, especially Fla. and mid-Atlantic to New England. Royal Tern (p. 160) is larger and chunkier; Elegant has notably thin, slightly decurved bill. In nonbreeding plumage, Elegant more crested and with more black on head than Royal.

Elegant Tern

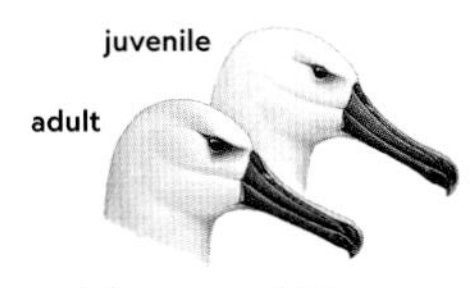


Yellow-nosed Albatross

**Yellow-nosed Albatross** EN *Thalassarche chlororhynchos* | Southern Ocean albatross that wanders rarely to East; most sightings offshore, mid-Atlantic northward. The yellow "nose" refers to a bright stripe of yellow-orange along upper ridge of upper mandible. Taxonomy fraught; birds to our area mostly or entirely pertain to nominate *chlororhynchos*, treated by some authorities as a distinct species.

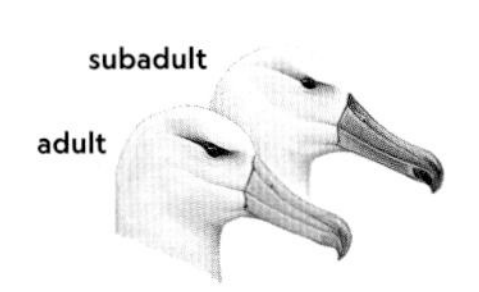


Black-browed Albatross

**Black-browed Albatross** *Thalassarche melanophris* | Like Yellow-nosed Albatross, a Southern Ocean species that wanders rarely to East. Sightings average farther north (off Nfld. and Lab.) than those of Yellow-nosed. Larger overall and stockier than Yellow-nosed.

**European Storm-Petrel** *Hydrobates pelagicus* | Occurs mostly eastern Atlantic Ocean, but a few range into our waters; records concentrated off N.C., where close to annual in recent years. Dark, with contrasting white rump and white underwing linings. Build compact: small overall with short wings, short tail, and short legs.

Barolo Shearwater

Gray Heron

**Barolo Shearwater** *Puffinus baroli* | Breeds on islands in eastern Atlantic Ocean off Europe and Africa. All recent records from East in small region of ocean near continental shelf well off New England. Similar to Audubon's Shearwater (p. 176) but even smaller; whiter below than Audubon's, with shorter tail and faster wingbeats.

**Jabiru** *Jabiru mycteria* | Colossal stork of tropical America, with multiple records to coastal Tex.; also to Okla., Miss., and La. Notably larger than Wood Stork (p. 178), with which it may be seen; bill of Jabiru massive and slightly upturned. Adult white with black head and red collar; dusky juvenile told by size and bill structure.

**Gray Heron** *Ardea cinerea* | Widespread Old World counterpart of closely related and ecologically similar Great Blue Heron (p. 186); numerous records coastally, especially Nfld., year-round. Increase in sightings reflects, at least in part, improving birder awareness. Gray is smaller than Great Blue, lacking rufous highlights of Great Blue.

Little Egret

**Little Egret** *Egretta garzetta* | Old World counterpart of Snowy Egret (p. 188). Annual to East Coast from mid-Atlantic states north to Nfld., with most records coastal. Probably increasing; may have bred in N. Amer., and hybrids with other *Egretta* herons possible. Most records from warmer months. Adult Little's two long head plumes a good point of distinction from Snowy.

**Western Screech-Owl** *Megascops kennicottii* | Nearly identical in appearance to Eastern Screech-Owl (p. 218); bill of Western gray (apple green on Eastern), streaking on breast more prominent on Western. Primary song of Western a "bouncing ball" series of mellow hoots, totally different from thin glissando of Eastern. Western reaches Morton Co., Kans.; Cimarron Co., Okla.; and the Edwards Plateau of Tex.

Thick-billed Vireo

**Thick-billed Vireo** *Vireo crassirostris* | Caribbean species, restricted mostly to the Bahamas; annual in recent years to southeastern Fla. mainland and Keys. Very similar to closely related White-eyed Vireo (p. 264) but larger-billed and a bit larger-bodied, duller and browner overall. The two sing similar songs.

**Brown Jay** *Psilorhinus morio* | Mid. Amer. corvid, formerly resident in woodlands of Lower Rio Grande Valley. After almost a decade of absence, returned to region in 2022. Large and long-tailed; dark brown above, paler gray on belly. Thick bill is black in adults, yellow in young. Call a short, nasal, descending scream.

Brown Jay

**Steller's Jay** *Cyanocitta stelleri* | Western counterpart of the Blue Jay (p. 274). Frequently wanders out onto plains a bit, but not by much; vagrant to Kans., Nebr. Crested like Blue Jay but darker: head, throat, and back blackish; rest of body deep blue. Calls include a raucous *shook shook shook.*

**Clark's Nutcracker** *Nucifraga columbiana* | Crowlike corvid of western mountains prone to periodic irruptions eastward, sometimes reaching Great Lakes; a few records to Southeast. Breeds in small numbers in Black Hills. Plumage mostly gray and black, but flashes white in wings and tail in flight. Far-carrying cry long and rasping.

Clark's Nutcracker

**Mountain Chickadee** *Poecile gambeli* | Common in western mountains. A few drift downslope every winter, regularly reaching western plains; records east to Nebr., Kans., and panhandles of Okla. and Tex. Grayer overall than Black-capped Chickadee (p. 282); black cap is broken by conspicuous white eyebrow. "Chickadee" call notably hoarser than that of Black-capped. Occasionally hybridizes with Black-capped where ranges overlap.

Mountain Chickadee

**Great Tit** *Parus major* | One of the most beloved birds in Europe. Started to show up around western Lake Michigan a few decades ago; now established in small numbers in Wis. Origins of this population unclear, but undoubtedly due to direct human agency. May recall Black-capped Chickadee (p. 282), but larger and much more colorful.

**Pygmy Nuthatch** *Sitta pygmaea* | Western counterpart of Brown-headed Nuthatch (p. 298); often common in yellow pine forests, especially ponderosa pine. Regular east to Black Hills of S. Dak. and Pine Ridge of Nebr.; occasionally wanders farther east, with records to Minn., Iowa. Sociable and noisy like Brown-headed; Pygmy has cold gray-brown crown (warm brown on Brown-headed). Call a sharp *peep,* often in frenzied outburst.

Pygmy Nuthatch

**Pacific Wren** *Troglodytes pacificus* | Western counterpart of Winter Wren (p. 302); found chiefly in Pacific coast region of U.S. and Canada, but some get well east. Numerous records for Black Hills, including summer. Vagrants, mostly fall to winter, elsewhere in S. Dak., as well as Nebr., Kans., Iowa. Plumage differences with Winter Wren slight, but songs and calls different.

Pacific Wren

**Crissal Thrasher** *Toxostoma crissale* | Desert-loving species similar in size and structure to Curve-billed Thrasher (p. 306); Crissal has longer bill and eponymous reddish crissum (under base of tail). At home in hot deserts of Mexico and Southwest, especially where there is mesquite and creosote bush; a few barely enter our area near Pecos R. in Tex.

**Fieldfare** *Turdus pilaris* | Old World relative of American Robin (p. 318), breeding as close as southern Greenland. A large thrush; frosty gray and pale brown above, with yellow bill and dense, dark chevrons below. Vagrants to East mostly in colder months, with most records from Atlantic Canada; a few to New England and Great Lakes.

Fieldfare

Redwing

**Redwing** *Turdus iliacus* | Similar to Fieldfare (p. 415), but smaller and browner, with distinct pale eyebrow. Name refers to reddish linings of the underwing, prominent in flight. Like Fieldfare, breeds as close to our area as Greenland, with vagrants mostly to northeastern U.S. and Atlantic Canada, especially Nfld.

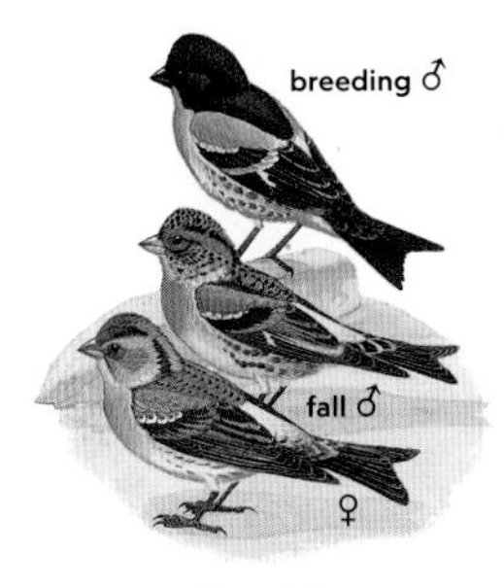

Brambling

**Brambling** *Fringilla montifringilla* | Well-marked finch of medium build; breeds in forests across much of northern Eurasia. Nearly annual to East, with bulk of records from northern U.S. and southern Canada. Breeding male has extensive black; other plumages grayer. All have orangey wash on breast and bright orangish wing patch.

**European Goldfinch** *Carduelis carduelis* | Popular songbird indigenous to Eurasia (mostly Europe) with long history of introductions worldwide. Recently established from Chicago, Ill., north to Milwaukee, Wis; also recently noted in southern New England and coastal N.Y. Colorful, with yellow on black wings; adult has red face.

**Golden-crowned Sparrow** *Zonotrichia atricapilla* | Close relative of White-crowned Sparrow (p. 346), found mostly in far western N. Amer. Annual, mostly fall and winter, to East, in very small numbers; a few reach Atlantic coast most years. Adults easy to identify, but beware look-alike first-winter White-crowned, with warmish buff crown suggesting young Golden-crowned.

Chihuahuan Meadowlark

**Chihuahuan Meadowlark** *Sturnella lilianae* | Found in arid grasslands of Mexico and Southwest. Recently split from Eastern Meadowlark (p. 364), but genetically distinct from both Eastern and Western Meadowlarks; regular in our area along far western edge of Tex. Panhandle. Has more white in tail than other meadowlarks; vocal array closer to Eastern's than Western's.

**Virginia's Warbler** *Leiothlypis virginiae* | Western interior counterpart of Nashville Warbler (p. 382), nearly annual in recent years to the East Coast; most are in late fall. All plumages mostly gray with prominent eye ring; even adult male drab, with a bit of rufous on crown and yellow below. Active; twitches tail constantly. Odd call, *pik*, sounds like a loose screw clanking about.

Gray-crowned Yellowthroat

**Gray-crowned Yellowthroat** *Geothlypis poliocephala* | Widespread in tropical N. Amer., this skulker is casual north to the Lower Rio Grande Valley; favors shrubby habitats, including those modified for agriculture. Long-tailed, thick-billed; has black lores, white eye arcs. Adult male sports trademark gray crown. Song rich and warbling, suggesting a *Passerina* bunting's.

MacGillivray's Warbler

**MacGillivray's Warbler** *Geothlypis tolmiei* | Sister species of Mourning Warbler (p. 384); a widespread Westerner that vagrates to eastern seaboard almost annually. Colorful adult male distinguished from Mourning by prominent white eye arcs and black lores, but less black on breast. Females and immatures more difficult to ID, and hybrids are known. Sightings concentrated in recent years in late autumn.

**Grace's Warbler** *Setophaga graciae* | A specialty of pinewoods in the Southwest, apparently expanding northward. Suggests a small Yellow-throated Warbler (p. 396) but less contrastingly black-and-white, with more yellow in eyebrow. Not commonly seen in migration even in core range, but there are several records well east of the Rockies, all the way to Ill., Ont., and N.Y.

**Black-throated Gray Warbler** *Setophaga nigrescens* | Among the western warblers that stray eastward, this species and Townsend's do it the most. The "BT Gray" is annual east to the Tex. coast, and there is a goodly cluster of records for the northeastern U.S. Male gray above with prominent black-and-white markings; female, especially immature, gray-and-white overall.

Black-throated Gray Warbler

**Townsend's Warbler** *Setophaga townsendi* | Like closely related Black-throated Gray Warbler, this species strays annually to the East Coast. Most records late fall to early winter, as with most other western vagrants in East. Black-and-yellow adult male distinctive, but other plumages, olive overall with muted yellow, can be passed off as Black-throated Green (p. 398).

Townsend's Warbler

**Hermit Warbler** *Setophaga occidentalis* | Even though it breeds in a thin slice of the Pacific Northwest, this lemon-headed warbler disperses well east on fall migration. Most don't get beyond the Great Basin, but a few reach the Rockies and a very few stray much farther east. Face of adult male brilliant yellow, but females, especially immatures, are drabber olive-gray and easily overlooked.

**Golden-crowned Warbler** *Basileuterus culicivorus* | Widespread Neotropical warbler with disjunct distribution; considered nonmigratory, but a few have wandered to Lower Rio Grande Valley, and one inexplicably strayed to eastern Colo. Eponymous golden crown bordered by black. Face boldly striped. Long-tailed and sturdy-legged; stays in cover low to the ground.

Golden-crowned Warbler

**Painted Redstart** *Myioborus pictus* | Striking warbler of montane woodlands in Mid. Amer., extending into Ariz. and N. Mex. Migratory in northern portion of range; vagrants almost annual well north and east of normal range. Adult strikingly red, black, and white. Young mostly dusky black with bold white wing panel. Unusual call is siskin-like.

**Hepatic Tanager** *Piranga flava* | Widespread "tanager" (technically a cardinalid) ranging from southwestern U.S. to northern Argentina. Breeds as far east as West Tex., wandering farther east in the state almost annually; accidental to Great Lakes, Midwest. Resembles Summer Tanager (p. 400), but duller and larger overall, with gray face and bill.

Crimson-collared Grosbeak

**Crimson-collared Grosbeak** *Periporphyrus celaeno* | Resident in northeastern Mexico, this stunning cardinalid has been showing up with increasing frequency in southern Tex. It is perhaps now best considered rare but regular in the region; most sightings in winter. Male's brilliant crimson contrasts with black face; on female, bright yellow-green contrasts with black face.

Blue Bunting

**Blue Bunting** *Cyanocompsa parellina* | Mid. Amer. species, annual in recent years to South Tex.; casual farther north along Gulf Coast to La. Easily confused with Indigo Bunting (p. 406), but both sexes darker, with curved culmen. Female Blue richer brown than female Indigo; male Blue mostly dark glossy blue with paler highlights. Lacks cinnamon wing bars of Blue Grosbeak (p. 404).

**Black-faced Grassquit** *Melanospiza bicolor* | One of the "true" tanagers, occurring widely in Caribbean region to northern S. Amer.; favors scrubby, often disturbed habitats. Annual vagrant to South Fla., especially Keys. Small and compact, with short tail and stubby bill. Adult male brown-green above, black otherwise. Female paler: olive-brown above, dirty gray below.

Black-faced Grassquit

## SPECIES OF ACCIDENTAL OCCURRENCE IN THE EAST

**The boundary between accidental and casual status is fuzzy. Accidentals to the East have occurred, on average, fewer than 5–10 times; casuals more than 10. But there are numerous exceptions. Species' ranges also change over time, and our knowledge improves. In addition to accidentals, this list includes a few recently established or currently establishing exotic species not yet on official state and provincial lists in the East.**

**West Indian Whistling-Duck** *Dendrocygna arborea* | Caribbean species. Most in East presumed escapes; a few from Southeast may be vagrants.

**Emperor Goose** *Anser canagicus* | Breeds mostly Alas., winters in very small numbers to Pacific coast of lower 48; one record in East: Nebr., Mar. 1997.

**Taiga Bean-Goose** *Anser fabalis* | Widespread in Eurasia, breeding farther south than Tundra Bean-Goose (p. 410). Records in East from Iowa, Nebr., Que.

**Whooper Swan** *Cygnus cygnus* | Common in Europe, especially Iceland. Most in East presumed escapes, but some (e.g., Nfld. and Lab.) probably natural vagrants.

**Ruddy Shelduck** *Tadorna ferruginea* | Afro-Eurasian species. Most in East presumed escapes, but six in Nvt. (2000) likely natural vagrants.

**Common Shelduck** *Tadorna tadorna* | Old World duck, common in captivity in East. Many are escapes, but some, especially Atlantic Canada, likely vagrants.

**Common Pochard** *Aythya ferina* | Old World diving duck resembling Canvasback and Redhead (both on p. 40). In East, several records from Que.

**Steller's Eider** *Polysticta stelleri* | High Arctic breeder restricted in New World mostly to Alas. A few records from western N. Atlantic region: Que., Nvt., Me., Mass.

**Common Wood Pigeon** *Columba palumbus* | Familiar Old World species, especially western Europe; one record for East: Que., May 2019.

**Scaly-naped Pigeon** *Patagioenas squamosa* | Widespread but local in West Indies; two old records from Fla., plus one from 2019.

**European Turtle-Dove** *Streptopelia turtur* | Old World species, breeding mostly Palearctic. Three records from East: Fla., Apr. 1990; Mass., July 2001; St. Pierre and Miq., May 2001.

**Ruddy Quail-Dove** *Geotrygon montana* | Widespread in Neotropics, regular north to Cuba, Tamaulipas. A few records for southern Fla. (as recently as Dec. 2020) and one from southern Tex. (Mar. 1996).

**Zenaida Dove** *Zenaida aurita* | Caribbean species, similar to Mourning Dove (p. 74). Vagrant to Fla., especially Keys, where it apparently bred in early 19th century.

**Common Cuckoo** *Cuculus canorus* | Famous "cuckoo clock" cuckoo of Old World. Annual to Alas.; recorded from East twice in Mass. and once in Que.

**Dark-billed Cuckoo** *Coccyzus melacoryphus* | Widespread S. Amer. species, with several records, mostly recent, to Gulf Coast region.

**Common Swift** *Apus apus* | Widespread Old World swift; scattered records, mostly coastal, N.C. to Nfld.

**Antillean Palm-Swift** *Tachornis phoenicobia* | Resident in Greater Antilles; has reached Fla. Keys twice (1972, 2019).

**Rivoli's Hummingbird** *Eugenes fulgens* | Mid. Amer. species, breeding north to Southwest; a few records from Southeast.

**Amethyst-throated Mountain-gem** *Lampornis amethystinus* | Resident in moist montane woodlands of Mid. Amer.; first record north of Mexico was, impressively, Que. (July 2016).

**Blue-throated Mountain-gem** *Lampornis clemenciae* | Widespread in oak woodlands in Mexico, breeding north into Ariz. Casual to coastal Tex., accidental northeast to Ga.

**Bahama Woodstar** *Nesophlox evelynae* | Small Bahamian hummingbird with a handful of records to Fla. and one remarkable stray to Pa. (Apr. 2013).

**Costa's Hummingbird** *Calypte costae* | Breeds mostly in deserts of West; scattered records to Tex., Midwest, and exceptionally farther east.

**White-eared Hummingbird** *Basilinna leucotis* | Mid. Amer. hummingbird that strays to Southwest; a few vagrants to Gulf Coast and one all the way to Mich. (Aug. 2005).

**Violet-crowned Hummingbird** *Ramosomyia violiceps* | Mostly Mexican hummingbird that reaches N. Mex. and Ariz.; several records to Tex. coast and one to Va. (June 2009).

**Berylline Hummingbird** *Saucerottia beryllina* | Mid. Amer. species that rarely reaches Ariz.; one made it to Mich. (Sept. 2014).

**Spotted Rail** *Pardirallus maculatus* | Widespread Neotropical rail with two records from Tex. (Aug. 1977; Dec. 2021) and one from Pa. (Nov. 1976).

**Paint-billed Crake** *Mustelirallus erythrops* | S. Amer. rail with two records from 1970s from East: Tex., Feb. 1972; Va., Dec. 1978.

**Eurasian Coot** *Fulica atra* | Widespread Old World relative of American Coot (p. 92) with several records from eastern Canada, none recent.

**Azure Gallinule** *Porphyrio flavirostris* | Occurs mostly in Amazonian lowlands of S. Amer.; one controversial record for East: N.Y., Dec. 1986.

**Hooded Crane** *Grus monacha* | East Asian species with records from Tenn., Ind., Nebr., and Idaho (Apr. 2010–Feb. 2012).

**Double-striped Thick-knee** *Burhinus bistriatus* | Large Neotropical shorebird widespread north of equator; one U.S. record: Tex., Dec. 1961.

**Eurasian Oystercatcher** *Haematopus ostralegus* | Palearctic species, breeds as close as Iceland; at least four records Nfld., all spring.

**Southern Lapwing** *Vanellus chilensis* | Neotropical species rapidly expanding northward; recorded in East in Fla., Md., and Mich.

**Eurasian Dotterel** *Charadrius morinellus* | Colorful Eurasian plover, formerly casual to Alas., where it even bred. One was photographed in southern Ont., 2015.

**Lesser Sand-Plover** EN *Charadrius mongolus* | East Asian shorebird, regularly reaching Alas.; scattered records to East, both inland and coastally.

**Collared Plover** *Charadrius collaris* | Widespread in Americas north well into Mexico; two records to Tex.: Uvalde Co., May 1992; Hidalgo Co., Oct. 2015.

**Slender-billed Curlew** CR *Numenius tenuirostris* | Old World species feared extinct, with one old record for the Americas: Ont., Oct. 1925.

**Black Turnstone** *Arenaria melanocephala* | West Coast "rockpiper"; has wandered east to coastal Tex. and even Wis.

**Great Knot** EN *Calidris tenuirostris* | Widespread East Asian and Australasian species, with two records for East: W.V., Aug. 2007; Me., July 2016.

**Surfbird** *Calidris virgata* | West Coast "rockpiper" ranging south to southern S. Amer.; scattered records east all the way to Pa. and Me.

**Broad-billed Sandpiper** *Calidris falcinellus* | Iconic Palearctic shorebird; one at urban Jamaica Bay Wildlife Refuge, N.Y. (Aug.–Sept. 1998), was a sensation.

**Jack Snipe** *Lymnocryptes minimus* | Widespread Old World sandpiper; two records from Nfld. and Lab. of this small snipe.

**Eurasian Woodcock** *Scolopax rusticola* | Large Palearctic counterpart of American Woodcock (p. 122). Scattered records from East, all from quite some time ago.

**Common Snipe** *Gallinago gallinago* | Widespread Old World counterpart of Wilson's Snipe (p. 122), regular to Alas. Several records in East from Nfld. and Lab.

**Terek Sandpiper** *Xenus cinereus* | Distinctively shaped Old World tail-bobber with several summer records along East Coast.

**Gray-tailed Tattler** *Tringa brevipes* | Widespread around western Pacific Rim. Records in East from Mass. (Oct. 2012), Me. (Aug. 2017), and Fla. (Nov. 2021).

**Wandering Tattler** *Tringa incana* | West Coast "rockpiper" that breeds in Alas. and western Canada; scattered records in East, both inland and coastally.

**Common Redshank** *Tringa totanus* | Rarer of the two redshanks in the New World, with just a handful of records from Nfld. and one from Mich.

**Wood Sandpiper** *Tringa glareola* | Old World species that suggests Solitary Sandpiper (p. 122). A few records from East, mostly mid-Atlantic to New England.

**Tufted Puffin** *Fratercula cirrhata* | Northern Pacific Rim species; several records for Me., N.B.

**Gray-hooded Gull** *Chroicocephalus cirrocephalus* | S. Hemisphere species with two records in East; one at Coney I., N.Y. (summer 2011), was one of the first viral internet superstars in the bird world.

**Belcher's Gull** *Larus belcheri* | Found on Pacific coast of S. Amer.; a few records from Fla. pertain to either this species or Olrog's Gull of eastern S. Amer.

**Western Gull** *Larus occidentalis* | Common West Coast species with a handful of records in East, including several to Tex.

**Large-billed Tern** *Phaetusa simplex* | S. Amer. species found mostly well inland. Several 20th-century records from East (Ill., Ohio, N.J.), plus two in Fla., summer 2023.

**Whiskered Tern** *Chlidonias hybrida* | Along with Black (p. 154) and White-winged (p. 413) Terns, a "marsh tern"; three records, all from Delaware Bay.

**Arctic Loon** *Gavia arctica* | Old World counterpart of Pacific Loon (p. 164) with scattered recent records from East, both inland and coastal.

**Black-bellied Storm-Petrel** *Fregetta tropica* | One of the Southern Ocean storm-petrels, with at least four records off N.C., late spring and summer.

**Swinhoe's Storm-Petrel** *Hydrobates monorhis* | Western Pacific and Indian Ocean species, with uncertain status in Atlantic; one record for East, from N.C., June 2017.

**Mottled Petrel** *Pterodroma inexpectata* | Pacific Ocean seabird with one old, bizarre record from East: well inland in N.Y., Apr. 1880.

**Zino's Petrel** EN *Pterodroma madeira* | Rare and endangered, nesting only on Madeira; recorded off N.C., Sept. 1995.

**Stejneger's Petrel** *Pterodroma longirostris* | Breeds off coast of Chile; one record for East: Tex., Sept. 1995.

**Bulwer's Petrel** *Bulweria bulwerii* | Found mostly tropical Pacific and eastern Atlantic oceans; two records off N.C., one for Va.

**Cape Verde Shearwater** *Calonectris edwardsii* | Closely related to Cory's Shearwater (p. 174). Breeds eastern equatorial Atlantic; has been recorded off Md., N.C., Mass.

**Wedge-tailed Shearwater** *Ardenna pacifica* | Indian and Pacific Oceans species with two recent records for East: off N.C., May 2021; Fla., Aug. 2021.

**Buller's Shearwater** *Ardenna bulleri* | Pacific Ocean species with only one record for the Atlantic Ocean: off N.J., Oct. 1984.

**Short-tailed Shearwater** *Ardenna tenuirostris* | Pacific Ocean species similar to Sooty Shearwater (p. 174). Records in East from Va., Fla., and northern Canada.

**Lesser Frigatebird** *Fregata ariel* | Occurs mostly in equatorial Indian and western Pacific Oceans, but two records from the East region: Me., July 1960; Mich., Sept. 2005.

**Great Frigatebird** *Fregata minor* | Like Lesser Frigatebird, occurs in equatorial Indian and Pacific Oceans; one record from the East region: Okla., Nov. 1975.

**Blue-footed Booby** *Sula nebouxii* | Occurs along Pacific coast of the Americas, mostly at low latitudes; three records to Tex., late 20th century.

**Bare-throated Tiger-Heron** *Tigrisoma mexicanum* | Widespread Mid. Amer. species, resident north to Tamaulipas; several recent Tex. records.

**Western Reef-Heron** *Egretta gularis* | Chiefly coastal heron from Africa, Middle East, India; scattered coastal records from mid-Atlantic to Nfld.

**Scarlet Ibis** *Eudocimus ruber* | Closely related to White Ibis (p. 194); occurs mostly northern S. Amer. Scattered Fla. peninsula records pertain mostly to escapes.

**Double-toothed Kite** *Harpagus bidentatus* | Hawk of low latitudes in Americas; two records from the East: Galveston Co., Tex., May 2011; Hernando Co., Fla., Oct. 2018.

**Western Marsh Harrier** *Circus aeruginosus* | Old World hawk, with three records, Aug.–Dec., to eastern seaboard (Me., N.J., Va.).

**Gundlach's Hawk** EN *Accipiter gundlachi* | Endemic to Cuba, but one was photographed in Lee Co., Fla., Jan. 2023.

**Eurasian Goshawk** *Accipiter gentilis* | Until very recently, treated as conspecific with American Goshawk (p. 204). One record for East (Lab., Nov. 1925).

**White-tailed Eagle** *Haliaeetus albicilla* | Eurasian relative of Bald Eagle (p. 206), breeding as close as Greenland; 20th-century records for Mass., N.Y.

**Steller's Sea-Eagle** *Haliaeetus pelagicus* | Magnificent eagle of coastal northern Asia; numerous recent records East, apparently all involving one exceedingly peripatetic individual.

**Crane Hawk** *Geranospiza caerulescens* | Widespread Neotropical raptor, occurring north well into Tamaulipas; one made it to South Tex. (winter 1987–88).

**Great Black Hawk** *Buteogallus urubitinga* | Widespread in Neotropics. Records for Tex. and Me. (2018–19) believed to be of the same individual! Several records for South Fla., thought to be of escapes.

**Roadside Hawk** *Rupornis magnirostris* | Widespread Neotropical raptor with several records to East, mostly recent, along and near Rio Grande of South Tex.

**Flammulated Owl** *Psiloscops flammeolus* | Small, secretive migratory owl of mountain regions in the West and in Mid. Amer.; some scattered records for Gulf Coast, Tex. to Fla.

**Mottled Owl** *Strix virgata* | Neotropical relative of Barred Owl (p. 222); one record for East: South Tex., Feb. 1983.

**Stygian Owl** *Asio stygius* | Little-known Neotropical counterpart of Long-eared Owl (p. 224); in East, two Dec. records for South Tex. (1994, 1996), one for South Fla. (June 2018).

**Elegant Trogon** *Trogon elegans* | Migratory Mid. Amer. species that reaches southeastern Ariz.; a few records in East from Tex., including one in metro San Antonio (Feb. 2018).

**Amazon Kingfisher** *Chloroceryle amazona* | Widespread Neotropical species expanding northward; several records, all recent, for Tex.

**Williamson's Sapsucker** *Sphyrapicus thyroideus* | Migratory woodpecker of western interior mountains; scattered records east to N.Y.

**Red-breasted Sapsucker** *Sphyrapicus ruber* | Found mostly in Pacific coast region, but wanders east, especially fall; one record as far as Iowa (Dec. 2006).

**Collared Forest-Falcon** *Micrastur semitorquatus* | Tropical raptor ranging north regularly to Tamaulipas; one record for South Tex. (Jan.–Feb. 1994).

**Eurasian Kestrel** *Falco tinnunculus* | Widespread Old World falcon with scattered records to East Coast from Atlantic Canada to Fla.

**Red-footed Falcon** *Falco vespertinus* | Highly migratory Old World species with one record to New World: Mass., Aug. 2004.

**Eurasian Hobby** *Falco subbuteo* | Old World falcon with well-supported records from Mass., Nfld.; sightings elsewhere in East possibly valid.

**Bat Falcon** *Falco rufigularis* | Widespread in Neotropics and spreading north; one wintered at Santa Ana N.W.R., Tex. (2021–22).

**Rose-ringed Parakeet** *Psittacula krameri* | Indigenous to Africa, South Asia. Widely established globally; most commonly sighted in East on the west coast of Fla.

**Barred Antshrike** *Thamnophilus doliatus* | Widespread Neotropical antbird (Thamnophilidae); one in South Tex. (Sept. 2006) was audio-recorded but not seen.

**Masked Tityra** *Tityra semifasciata* | Neotropical relative of Rose-throated Becard (p. 246) occurring north to Tamaulipas; one record for Tex. (1990).

**Greenish Elaenia** *Myiopagis viridicata* | Small flycatcher of mainland tropical America; one record for our area: Galveston Co., Tex., May 1984.

**Small-billed Elaenia** *Elaenia parvirostris* | Migratory S. Amer. flycatcher with records for Tex., Que., Ill. (two).

**White-crested Elaenia** *Elaenia albiceps* | Like Small-billed Elaenia, a migratory S. Amer. flycatcher; has strayed north to South Tex. and N. Dak.

**Caribbean Elaenia** *Elaenia martinica* | Dull flycatcher with disjunct range in Caribbean; an elaenia in Fla. (Apr. 1984) may have been this species.

**Dusky-capped Flycatcher** *Myiarchus tuberculifer* | Neotropical species breeding north regularly to Ariz., N. Mex.; a few vagrants to eastern Tex. and as far afield as Okla., La.

**Social Flycatcher** *Myiozetetes similis* | Widespread Neotropical species ranging well north into Mexico; several records for South Tex.

**Sulphur-bellied Flycatcher** *Myiodynastes luteiventris* | Mostly Neotropical species; breeds north to Ariz. Migratory; almost annual to East, with most records from Gulf Coast.

**Piratic Flycatcher** *Legatus leucophaius* | Widespread Neotropical species with records for Fla., Tex., and eastern N. Mex.

**Variegated Flycatcher** *Empidonomus varius* | Migratory Neotropical flycatcher similar to smaller Piratic; East records from Fla., Ont., Tex., Tenn., and Me.

**Crowned Slaty Flycatcher** *Empidonomus aurantioatrocristatus* | Migratory Neotropical flycatcher with a single record in our area: La., June 2008.

**Thick-billed Kingbird** *Tyrannus crassirostris* | Mostly Mexican species breeding north to Ariz., N. Mex. A few records in our area to Tex. and even N. Dak., Ont.

**Loggerhead Kingbird** *Tyrannus caudifasciatus* | Widespread in West Indies, with several recent records to Fla. Keys and Miami region.

**Tufted Flycatcher** *Mitrephanes phaeocercus* | Flycatcher of highlands in Neotropics, rarely reaching Southwest; one from coastal Tex., Apr. 2014.

**Greater Pewee** *Contopus pertinax* | Mostly Mid. Amer. flycatcher that reaches Southwest; multiple records for Tex. coastal plain region.

**Cuban Pewee** *Contopus caribaeus* | Occurs mostly Bahamas and Cuba, with scattered records to Fla. Keys and Miami region.

**Cuban Vireo** *Vireo gundlachii* | Cuban endemic with several recent records to Fla. Keys (all Apr.).

**Yucatan Vireo** *Vireo magister* | Range-restricted relative of Red-eyed Vireo (p. 270), with one record for U.S.: coastal Tex., spring 1984.

**Brown Shrike** *Lanius cristatus* | Old World species, found mostly South and East Asia; one strayed to Halifax, N.S. (Nov.–Dec. 1997).

**Eurasian Jackdaw** *Corvus monedula* | Old World crow that showed signs of establishing in Northeast in late 20th century; one was found recently in Lab., May 2023.

**Juniper Titmouse** *Baeolophus ridgwayi* | Drab but spirited resident of juniper woodlands in West; barely reaches East in Okla. Panhandle.

**Bahama Swallow** EN *Tachycineta cyaneoviridis* | West Indies relative of Tree Swallow (p. 288), mostly found in the Bahamas; scattered records to southern Fla.

**Mangrove Swallow** *Tachycineta albilinea* | Widespread from coasts to foothills of Mid. Amer.; one reached Brevard Co., Fla. (Nov. 2002).

**Blue-and-white Swallow** *Pygochelidon cyanoleuca* | Mostly S. Amer. species; one record in East: Tex. (July 2020).

**Brown-chested Martin** *Progne tapera* | S. Amer. swallow with scattered records, summer to fall, to Gulf and Atlantic coasts.

**Southern Martin** *Progne elegans* | Swallow of southern S. Amer.; one migrated all the way to Key West, Fla. (Aug. 1890).

**Gray-breasted Martin** *Progne chalybea* | Neotropical species breeding north to Tamaulipas; a couple of old records for South Tex., plus one in Brooklyn, N.Y. (Apr. 2021).

**Cuban Martin** *Progne cryptoleuca* | Breeds Cuba, disperses in winter throughout Caribbean; one old record for Fla. Keys (May 1895).

**Western House-Martin** *Delichon urbicum* | Widespread Old World swallow; one made it to St. Pierre and Miq. (May 1989).

**Yellow-browed Warbler** *Phylloscopus inornatus* | Old World species with penchant for vagrancy; recorded Wis. (Oct. 2006) and Ont. (Apr. 2021).

**Gray Silky-flycatcher** *Ptiliogonys cinereus* | Elegant waxwing relative from Mexico and Guatemala; one strayed to South Tex. (Oct.–Nov. 1985).

**Phainopepla** *Phainopepla nitens* | A silky-flycatcher of arid landscapes in Mexico and Southwest, with a few records to Tex. a bit east of Pecos R.

**Blue Mockingbird** *Melanotis caerulescens* | Mexican endemic with several records from U.S. border region, including South Tex.; some may be escapes from captivity.

**American Dipper** *Cinclus mexicanus* | Aquatic songbird of western rivers and streams. Resident in very small numbers in Black Hills, accidental elsewhere in East.

**Black Catbird** *Melanoptila glabrirostris* | Restricted to Yucatan Peninsula; records for Tex. and La. problematic for involving ship assistance.

**Orange-billed Nightingale-Thrush** *Catharus aurantiirostris* | Neotropical species ranging north to northern Mex; records in East from South Tex. and, remarkably, S. Dak.

**Black-headed Nightingale-Thrush** *Catharus mexicanus* | More restricted in range than Orange-billed; Mid. Amer. only. One record for South Tex. (May–Oct. 2004).

**Mistle Thrush** *Turdus viscivorus* | Large Palearctic relative of American Robin (p. 318); one strayed to N.B. (winter 2017–18).

**Eurasian Blackbird** *Turdus merula* | Despite the name, a thrush. Familiar Old World species, immortalized by the Beatles; vagrants to Ont., Nfld. and Lab.

**Song Thrush** *Turdus philomelos* | Old World species; suggests a small Mistle Thrush. One record for East: Que., Nov. 2006.

**White-throated Thrush** *Turdus assimilis* | Neotropical robin relative, found from northern S. Amer. to northern Mexico. Scattered records from South Tex., mostly 21st century.

**Rufous-backed Robin** *Turdus rufopalliatus* | Mexican species, annual to Southwest. Scattered records to southern and eastern Tex., including well inland.

**Red-legged Thrush** *Turdus plumbeus* | Widespread in West Indies; in East, multiple recent records for Fla.

**Aztec Thrush** *Ridgwayia pinicola* | High-elevation Mexican specialty; strays to U.S. borderlands, with several records in East from southern Tex.

**European Robin** *Erithacus rubecula* | Beloved European species, actually an Old World flycatcher. Four records, all recent, for New World: Pa. (Feb. 2015) and three in Fla. since 2018.

**Siberian Rubythroat** *Calliope calliope* | Old World flycatcher from Asia with one record for East: Ont., Dec. 1983.

**Asian Stonechat** *Saxicola maurus* | Asian species. Like European Robin and Siberian Rubythroat, an Old World flycatcher; one record for East: N.B., Oct. 1983.

**Northern Red Bishop** *Euplectes franciscanus* | Indigenous to equatorial Africa, popular in captivity; escapes widely noted in East. Breeding in Houston, Tex., area.

**Southern Red Bishop** *Euplectes orix* | Indigenous to southern Africa; less frequently reported in East than Northern Red Bishop.

**Bronze Mannikin** *Spermestes cucullata* | African species found in small numbers around Houston, Tex.

**Scaly-breasted Munia** *Lonchura punctulata* | Asian species introduced, established, and likely increasing in East, especially northern Gulf Coast.

**Tricolored Munia** *Lonchura malacca* | Old World species introduced to Cuba and now "naturally" invading South Fla., where increasing and likely breeding.

**Eastern Yellow Wagtail** *Motacilla tschutschensis* | Asian and Alas. breeder; one wandered to Brooklyn, N.Y. (Sept. 2008).

**Citrine Wagtail** *Motacilla citreola* | Mostly Asian species, with one remarkable record to East: Starkville, Miss., well inland, Jan.–Feb. 1992.

**White Wagtail** *Motacilla alba* | Old World species prone to vagrancy; wanderers to East mostly coastal and Great Lakes.

**Common Chaffinch** *Fringilla coelebs* | Old World species; records in East thought to be mostly escapes, but some occurrences in Atlantic Canada and New England likely pertain to wild birds.

**Eurasian Siskin** *Spinus spinus* | Old World counterpart of Pine Siskin (p. 330); scattered records East, but most or all believed to be escapes from captivity.

**Lawrence's Goldfinch** *Spinus lawrencei* | Small nomadic finch of foothills in West; has wandered east to far eastern Wyo. and southern Okla.

**Little Bunting** *Emberiza pusilla* | Eurasian species that wanders casually to Alas. and West Coast; one record for East (Fla., Mar. 2023).

**Yellow-breasted Bunting** CR *Emberiza aureola* | Asian species that wanders west to Europe; one record for East: Lab., Oct. 2017.

**Black-chinned Sparrow** *Spizella atrogularis* | Dashing sparrow of arid country in Mexico and Southwest; a few have strayed into our area (Tex., eastern Colo.)

**Yellow-eyed Junco** *Junco phaeonotus* | Sparrow of mountain forests in Mexico and Southwest; one strayed to west-central Kans., winter 2020–21.

**Sagebrush Sparrow** *Artemisiospiza nevadensis* | Found in sagebrush deserts of West; rare but regular just east of Rockies on migration, with strays to Mich., Ky.

**Abert's Towhee** *Melozone aberti* | Large, range-restricted sparrow of Desert Southwest; a well-attested record from Minn., summer 2022, defies logic.

**Black-vented Oriole** *Icterus wagleri* | Mexico and northern Mid. Amer. species; in our area, several records to eastern Tex., mostly recent; also to eastern N. Mex.

**Streak-backed Oriole** *Icterus pustulatus* | Mid. Amer. species; has strayed to eastern Tex. a couple times; also all the way to Wis. (Jan. 1998).

**Black-backed Oriole** *Icterus abeillei* | Endemic to Mexico; records from Pa. (Mar.–Apr. 2017) and Mass. (May 2017) likely were of the same individual.

**Tawny-shouldered Blackbird** *Agelaius humeralis* | Related to Red-winged Blackbird (p. 370), restricted mostly to Cuba; old record of two in Fla. (Feb. 1936).

**Lucy's Warbler** *Leiothlypis luciae* | Tiny, drab desert-dweller of western Mexico and Southwest; records for East to Tex., La., Va., Mass., Mich.

**Rufous-capped Warbler** *Basileuterus rufifrons* | Mostly Mid. Amer. species. Scattered records to Tex. mostly from on and around Edwards Plateau.

**Red-faced Warbler** *Cardellina rubrifrons* | Highland forests of Mid. Amer., regularly reaching Ariz. and N. Mex.; records for East from Tex., La., Ga.

**Slate-throated Redstart** *Myioborus miniatus* | Widespread in Neotropics, occasional to Ariz.; multiple records, all 20th century, from South Tex. and Tex. Panhandle.

**Flame-colored Tanager** *Piranga bidentata* | Mid. Amer. species, with multiple records, all 20th century, from southern Tex.; also a remarkable stray to Wis., spring 2023.

**Yellow Grosbeak** *Pheucticus chrysopeplus* | Mid. Amer. species with records from Uvalde Co., Tex. (early 2019) and Iowa (Dec. 1990, a possible escape).

**Red-legged Honeycreeper** *Cyanerpes cyaneus* | Striking tanager from Cuba and mainland Neotropics; migratory in Mexico. Multiple records from Gulf Coast in 21st century; major influx in Oct. 2022.

**Yellow-faced Grassquit** *Tiaris olivaceus* | Neotropical tanager with multiple records from South Tex. and South Fla.; some Fla. records pertain to escapes.

Up to nine species of birds either have gone extinct or are presumed to have gone extinct from the eastern United States and Canada in the past 400 or so years. We list all nine here, some of which are considered extinct EX by the International Union for Conservation of Nature (IUCN), others still labeled, hopefully, critically endangered CR. Humans are implicated in every one of these extinctions. Aware of the irrevocable harm we can cause to the species with which we share our planet, conservationists are racing to prevent further losses. But what about those species that can no longer be saved?

Extinction is forever, and there is virtually no chance that a birder will ever see any of these species in life. But field guides play an important role in preserving knowledge about what has been lost. In addition to their intrinsic worth, extinct species inform our present-day understanding of taxonomy and biodiversity. Maybe, with greater knowledge of what we have lost, we will work harder to preserve what remains.

## Labrador Duck EX

*Camptorhynchus labradorius* | LABD | L 22½" (57 cm)

With its white-and-black plumage, the drake Labrador Duck suggested a nonbreeding Long-tailed Duck (p. 48), a species with which it co-occurred—notably on the North Atlantic coast in the winter months. The Labrador Duck was closer in size to an eider, however, and its closest relative was likely the Steller's Eider (*Polysticta stelleri*), a declining species that occurs in North America chiefly in Alaska. The Labrador Duck's bill was long and colorful with odd lateral flaps, suggesting a specialized diet—and a possible reason for its extinction. Hunting and harvesting of eggs no doubt played a role, but heavy exploitation of shellfish on the wintering grounds may have been at least as important. John James Audubon reported that the Labrador Duck's wings whistled in flight, and that is the extent of our knowledge about the sounds this bird made. The last sightings were in the late 1870s.

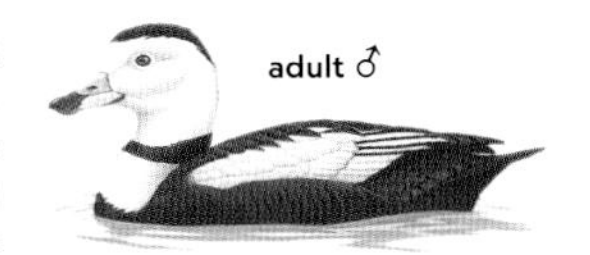


## Passenger Pigeon EX

*Ectopistes migratorius* | PAPI | L 15¾" (40 cm)

When it comes to superlatives in the world of birds, the Passenger Pigeon is at the top of the list. It was a large bird, two to three times the mass of the Mourning Dove (p. 74), which it resembled, but genetic data suggest that it was most closely related to pigeons in the genus *Patagioenas*, represented in the East by the White-crowned and Red-billed Pigeons (p. 70). The species was staggeringly abundant too, with multiple reports of single flocks exceeding one *billion* birds. Around their immense roosts, Passenger Pigeons were cacophonous, and the sound of huge flocks cutting through the air in distant flight, like the crashing surf or gale force winds, was soul-stirring. And then there were none. In the 19th century, North Americans slaughtered them by the millions. Passenger Pigeons were a species that couldn't survive if their numbers dropped below a certain threshold. They needed flocks of millions; hundreds, or even thousands, wouldn't do. The result was a precipitous collapse of populations during the last few decades of the 19th century, following many years of more gradual losses. The concomitant decline of the American chestnut also contributed to their demise. And there is recent evidence that disease may have played a role too. But hunting by humans was by far the main factor in the almost inconceivable extermination of what was once one of the most abundant bird species on Earth. The last known Passenger Pigeon, Martha, died in the Cincinnati Zoo in 1914.

## Eskimo Curlew CR

*Numenius borealis* | ESCU | L 14" (36 cm)

Whether the Eskimo Curlew has gone extinct is unresolved at the present time. The species is not flagged as extinct on the American Ornithological Society's *Checklist,* and the International Union for Conservation of Nature calls it "critically endangered (possibly extinct)." The last photographs are from the early 1960s. With no definitive records in 60+ years, should this curlew be declared extinct? Perhaps. But the scientific record is replete with "Lazarus species"—assumed to be extinct, only to be rediscovered. Patterned and proportioned like a miniature Whimbrel (p. 106), the Eskimo Curlew has (had?) an elliptical migration, moving north up the Great Plains, then south mostly over the Atlantic. At one time, it was common—and irresistible to hunters. It is darkly fitting that the last verified record is of a bird shot by a hunter on Barbados in 1963.

## Great Auk EX

*Pinguinus impennis* | GRAU | L 30" (76 cm)

This was the only fully flightless bird in the East, and indeed in all of continental North America, to survive into the modern era. The auk was huge, seven to eight times the mass of the Razorbill (p. 134), its closest extant relative and one of the largest alcids still with us. Great Auks in the East bred off Newfoundland and eastern Quebec; in winter, the species swam south regularly to New England, occasionally to the Carolinas. Its eggs were large and nutritious, and the bird's down was prized. The auk fell victim to human avarice, first in Europe and then in eastern North America. The last definitive record, from Iceland, was in 1844.

## Ivory-billed Woodpecker CR

*Campephilus principalis* | IBWO | L 19½" (50 cm)

The iconic "Lord God Bird" was one and a half times to almost twice the mass of the mighty Pileated Woodpecker (p. 236). The Ivorybill had more white on the upperwings than the Pileated, and a loud tooting call like a clarinet. The species was never common, but it lived securely in the sprawling virgin forests of the southeastern U.S. until the acceleration of deforestation in the 19th century. Hunting and collecting took their toll, but loss of virgin bottomland forests likely did the species in. The last definitive records in the East are from Louisiana in 1944, with sightings in Cuba until 1986. A purported rediscovery, in Arkansas, was announced with tremendous fanfare in 2005, but scientific support for the rediscovery quickly eroded. Given the Ivorybill's size and flamboyance, it is exceedingly unlikely that the species survives. Debate persists, however, at times acrimoniously.

## Carolina Parakeet EX

*Conuropsis carolinensis* | CAPA | L 13½" (34 cm)

Among the birds lost in the East in recent centuries, this parakeet was arguably the most distinctive. It was the only indigenous species in the East in the entire order Psittaciformes, occurring north regularly to at least the 40th parallel. The Carolina Parakeet was hunted for food and sport, but it was persecuted mostly because it was believed, with considerable exaggeration, to be harmful to agriculture. Disease may also have contributed to the species' demise. The last verified record in the wild is from 1904, and an individual in the Cincinnati Zoo lived on until 1918. Credible reports persisted for a couple more decades. But it is gone now, and so is a spectacle worth imagining: On snowy afternoons in the Ohio River Valley, amid the browns and grays of chickadees and juncos, squawking flocks of sociable parakeets in bright yellows and reds and glorious greens. Imagine recording one on your Christmas Bird Count!

## Dusky Seaside Sparrow

*Ammospiza [maritima] nigrescens* | DSSP | L 6" (15 cm)

Field guides and ornithology texts of the 1950s and 1960s described a species of bird called the Dusky Seaside Sparrow (*Ammospiza nigrescens*), endemic to the marshland along the central east coast of the Florida peninsula. Dredging and development shrank its habitat; pesticides used for mosquito control infiltrated its food supply, both reducing availability of and contaminating the insects that could be eaten. In the early 1970s, this distinctive bird was reclassified and treated as *Ammospiza maritima nigrescens*, a subspecies of the widespread Seaside Sparrow (p. 350). The U.S. Endangered Species Act extends protections to subspecies as well as to full species, but diminished public enthusiasm for the Dusky Seaside Sparrow, "demoted" to subspecies rank, may have hastened the demise of the bird. Efforts to support the sparrow began in 1979, but alas, too late. The last one, a partially blind male named Orange Band, died in a captive breeding facility in 1987.

## Bachman's Warbler CR

*Vermivora bachmanii* | BAWA | L 4¾" (12 cm)

This relative of the Blue-winged and Golden-winged Warblers (p. 378) bred in the low country of the southeastern United States and wintered chiefly in Cuba. The adult male sported a striking black throat and brilliant yellow underparts; females were less boldly marked. The species was well attested in the United States into the 1960s. The reason or reasons for the Bachman's Warbler's demise are poorly understood. The bird had a small wintering range, where habitat destruction was extensive; for its breeding grounds, it favored canebrakes, largely cleared for agriculture by the early 20th century. This gorgeous warbler may have begun its precipitous decline in the late 19th century, with little notice, and has now likely disappeared forever. The Bachman's Warbler was formally declared extinct by the U.S. Fish and Wildlife Service in 2023.

### Was There a Painted Vulture?

One of the challenges for extinction biology is uneven, incomplete knowledge. We know relatively little about the Labrador Duck, for example, yet a fair bit about the Passenger Pigeon. In the case of the Painted Vulture (*Sarcoramphus sacra*), we don't know what it was—or whether it even existed at all. There exist several 18th-century descriptions of it, impressively consistent with one another, one by renowned naturalist William Bartram. Of possible relevance are several other vulture species (from the West), well attested in the fossil record, that went extinct after the arrival of the first humans in the region. Nevertheless, there is no definitive evidence for the Painted Vulture's occurrence, no reliable images or museum specimens. The species does not appear on the American Ornithological Society's checklist of all birds documented to have occurred in North America since around 1600. It seems that the bird resembled the King Vulture (*Sarcoramphus papa*), a tropical species that is stunningly black and white with a parti-colored head. Possibly, it *was* the King Vulture. Possibly, although less likely, it was entirely fictive. And possibly, quite possibly, it was a U.S. endemic, extinguished before Western science ever caught up to it.

**ABA** American Birding Association, publisher of *Birding* magazine and the *ABA Checklist*, updated multiple times annually at *aba.org/aba-checklist*.

**ABA Area** North America north of Mexico and Hawaii, plus offshore waters.

**accidental** Generally denotes a species recorded five or fewer times overall in a region or fewer than three times in the past 30 years.

**advertising call** A loud, far-carrying vocalization, typically given in courtship.

**AOS** American Ornithological Society, publisher of the *Checklist of North and Middle American Birds*, updated at least annually at *checklist.americanornithology.org/taxa*.

**auriculars** Feather tract covering the ear (aural) openings, often bordered with contrasting stripes or lines; also called "ear coverts."

**axillaries** Feathers that make up the bird's "armpit," or "wingpit," visible in the field only in flight.

**backcross** A second-generation (or later) hybrid, typically between one hybrid parent and one "pure" parent.

**bare parts** Those areas of a bird's body completely without feathers, typically the eyes, bill, and feet.

**biodiversity** The variety of organisms in a given environment.

**body plan** A bird's overall build or structure.

**bosque** A grove of broadleaf trees, especially cottonwoods, along a river or pond.

**breeding plumage** The coat of feathers worn by many, but not all, birds when they are courting and raising young.

**brood parasitism** Mating system wherein a female lays its eggs in the nest of another species.

**call note** Short vocalization generally used to communicate alarm, contact, begging, or other messages between birds.

**carpal bar** A band on the inner wing formed by contrasting secondary coverts.

**casual** Generally denotes a species recorded ten or fewer times overall in a region, with at least three records in the past 30 years.

**cavity nester** A population or species that nests in holes, natural or artificial.

**character** A morphological trait.

**circum-Gulf migrant** A species that migrates along the Texas coast, not over the Gulf of Mexico.

**clade** A group of species or other taxa that share a common ancestor.

**class** In the Linnaean hierarchy, a clade of orders; the class Aves are the birds.

**cline** A gradual change in certain characteristics of individuals of the same population or species, evident in a geographic progression from one population to the next.

**colony** A densely packed, sometimes cooperative community of nesting birds.

**commensalism** A two-species relationship in which one benefits and the other is unaffected.

**common** Denotes a species easily found at a certain place and time of year.

**common name** A standardized English name set by the AOS.

**congener** A similar or closely related species.

**convergent evolution** The process whereby similar environments promote development of similar traits among distantly related species or groups of species.

**cooperative breeding** Mating system wherein individuals help to raise young that are not their direct offspring.

**coverts** Feathers that cover other feather tracts, especially on a bird's wing or tail.

**critically endangered** A species considered by the IUCN to be in likely danger of imminent extinction if corrective measures are not taken immediately.

**crown** The top of a bird's head, extending from the forehead to the nape.

**cryptic** Describing a bird that is difficult to detect in its environment due to morphological features and/or behavior.

**cryptic species** Two or more populations that look nearly identical but have considerable genetic differences and do not interbreed.

**culmen** The ridge of the upper mandible from the base to the tip.

**decurved** Of the bill, curving downward.

**digiscoping** Taking a photograph by holding a small camera or cellphone against the eyepiece of a spotting scope.

**display** A behavior presented in courtship or threat, often, but not always, delivered with song or other vocalizations.

**diurnal** Active during the day.

**eclipse plumage** The dull or plain plumage of most ducks, worn only in late summer.

**endangered** A species that is considered by the IUCN to be in likely danger of eventual extinction if corrective measures are not taken.

**endemic** Occurring only within a given geographic region.

**extinction** The permanent and complete disappearance of all the members of a population of an organism, especially a species.

**extirpation** The extinction of a population, especially a species, from a region.

**eye crescent** A partial eye ring visible either above or below—or both above and below—the eye.

**eye stripe** Synonym of *eyeline*, but typically thicker.

**eyeline** A thin patch of feathers extending horizontally in front of and behind the eye.

**eye ring** A circle of feathers around a bird's eye.

**family** In the Linnaean hierarchy, a clade of genera; ends in *-idae* for birds.

**feather tract** A group of feathers that typically share a function, for example, powering flight (primaries) or protecting another feather tract (coverts).

**feral** A formerly domesticated animal now living in the wild; also refers to a formerly domesticated population now established in the wild.

**field mark** A visible characteristic used to help identify a bird.

**flanks** The area of a bird's underparts on either side of its belly, usually visible just below the folded wing.

**fledge** To leave the nest for the first time.

**fledgling** A young bird that has left the nest but often still depends on its parents for food and other care.

**flight** A coordinated passage of birds on the wing, often on migration.

**flight call** Typically simple, often loud vocalization given by a bird on the wing, especially on nocturnal migration.

**flight feathers** The remiges (primaries, secondaries, tertials) and rectrices.

**flush** To flee from disturbance, usually by flying.

**forecrown** Feathering just above the eye.

**frontal shield** Unfeathered plate directly above the upper mandible.

**frugivorous** Feeding on fruit.

**gape** The juncture of the upper and lower mandible.

**generalist** Having broad engagement with the environment, for example, by using multiple habitats or eating diverse food types.

**genotype** An organism's genetic material.

**genus (pl.: genera)** In the Linnaean hierarchy, a clade of species.

**gestalt** An overall impression of the shape and size of a bird.

**gonys** A ridge formed where two segments of the lower mandible join.

**gorget** The brilliant patch of throat feathers on a hummingbird.

**granivorous** Feeding on seeds.

**greater coverts** The outermost and usually largest tract of feathers overlaying the secondaries.

**hybrid** The offspring of breeding between individuals of different species.

**immature** An individual that has not attained full adult plumage; can indicate juvenile or, especially, subsequent plumages.

**indigenous** Pertaining to a bird's geographic range prior to human introductions.

**inner primaries** The primaries farthest from the tip of the wing.

**insectivorous** Feeding on insects.

**intergrade** An individual showing a complex array of traits of two (or rarely more) species.

**irruption** An irregular seasonal movement often related to variation in food resources.

**IUCN** International Union for Conservation of Nature, an organization that provides guidance on understanding biological diversity and the threats it faces.

**juvenile** A bird in its first plumage, grown in around the time of fledging; in many species, the juvenile plumage is the first of two or more immature plumages.

**kettle** A group of soaring birds, especially hawks or vultures, often on migration.

**kingdom** In the Linnaean hierarchy, a clade of phyla.

**kleptoparasitism** Theft of food.

**lek** A communal courtship display.

**lesser coverts** The innermost tract of feathers overlaying the secondaries.

**leucistic** Having paler than normal plumage due to a partial lack of melanin pigments.

**lores** The area between the upper base of a bird's bill and its eyes.

**lower mandible** The lower half of the bill, also called just the "mandible."

**lumping** Reclassifying two or more bird species into a single species.

**malar** Feather tract extending from the base of a bird's bill downward along either side of its throat.

**mantle** Describes the back, scapulars, and upperwing coverts as a whole, often applied to gulls.

**median coverts** The feather tract between the greater and lesser coverts.

**median crown stripe** A contrasting line of feathers down the center of the top of the head.

**microhabitat** A pocket of differentiated habitat within a larger habitat type; for example, a wetter, leafier area of a forest.

**migrant** A bird in the process of moving from one place to another, often between breeding and nonbreeding territories.

**mimicry** The incorporation of other species' sounds into an individual bird's songs or calls.

**mobbing** Attacking or harassing a predator, sometimes as a group.

**molt** The process of growing new feathers to replace old ones.

**monophyletic** Referring to a group of organisms, especially a group of species, that share a common ancestor not shared with any other such group.

**morph** A population of birds with a consistent color difference from other populations in the same species.

**morphology** The physical, often structural, attributes of a bird; may exclude color and pattern.

**nail** The hard, hooked, often darkened tip of the upper mandible.

**nape** The back of a bird's head, below the crown and above the hindneck.

**naturalized** Referring to a population or species introduced, typically by human means, to a region but currently established and self-sustaining in the wild.

**nectivorous** Feeding on nectar.

**nominate subspecies** A subspecies whose name is the same as the name of its species; for example, *Zonotrochia leucophrys leucophrys.*

**nonbreeding plumage** A bird's coat of feathers when not courting and raising young.

**nonindigenous** Pertaining to the parts of a bird's geographic range where it was introduced, deliberately or unintentionally, by humans.

**obligate** Describing an individual, population, or species restricted to a particular niche or life strategy.

**occasional** Denotes a species that is found less than annually but nevertheless with a repeatable and predictable pattern of occurrence within a region.

**orbital ring** Circle of bare skin immediately surrounding the eye; compare *eye ring*.

**order** In the Linnaean hierarchy, a clade of families; ends in *-iformes* for birds.

**outer primaries** The primaries closest to the tip of the wing.

**outer rectrices** The rectrices farthest from the center of the tail.

**panel** A contrasting swatch of color, often paler than its surroundings, visible on the wing.

**parasitism** The practice of exploiting another individual or organism for food or other resources.

**passage transient** A bird in the process of migrating from one place to another.

**passerine** A species in the order Passeriformes, often called songbirds or perching birds.

**patagium** The leading edge of the inner wing; if this area is conspicuously dark, it forms a patagial bar.

**pelagic** Of the open ocean.

**phenology** Regular change in an individual or environment over the course of a season or year; molt in birds is an example.

**phenotype** The observable characters of a bird's genotype; examples include sex, color, and many behaviors.

**phylum** In the Linnaean hierarchy, a clade of classes.

**pishing** The human practice of mimicking the sound of birds' alarm calls *(psh psh psh)* to attract their attention.

**plumage** A bird's coat of feathers.

**population** A group of organisms; does not correspond neatly to species or subspecies.

**postocular stripe** A stripe extending back from the eye.

**primaries** The outermost, longest, and most powerful flight feathers of a bird.

**primary coverts** Feathers arranged in rows that overlay the bases of the primaries.

**rare** Refers to a species that occurs annually, but in very low numbers, sometimes only individually; scarcer than uncommon but more numerous than casual.

**record** Verified report of a bird's occurrence at a given time and place.

**rectrices (sing.: rectrix)** The strong flight feathers of the tail.

**regular** Denotes consistent, predictable occurrences over a particular time frame, like a year or season.

**remiges (sing.: remex)** The flight feathers—primaries, secondaries, and tertials—of the wing.

**resident** A population or species present in a region for a season or year-round.

**roost** The place a bird goes to rest during periods of inactivity.

**rufous** Reddish.

**rump** The part of a bird's body directly above the uppertail coverts.

**scapulars** The tract of feathers that overlays the area where the wing attaches to the body; sometimes called "shoulders."

**scavenger** An individual or species that forages on dead organic matter or human refuse.

**scientific name** The name for an organism, often derived from Latin or Greek and consisting of two words: the genus name, then the species name.

**secondaries** The inner, fairly long, fairly powerful flight feathers of a bird.

**secondary coverts** Feather tracts that overlay the secondaries.

**sexual dimorphism** Referring to male and female morphologies that are appreciably different to the human eye.

**solicitation whistle** A shrill sound intended to arouse the interest of another bird, typically of the same species and often a biological close relative.

**song** Patterned vocalizations given by males and many females, often to defend territory or attract mates.

**specialties** Species or subspecies known to occur in a certain area and sought after by birders.

**species** A population of organisms capable of producing fertile offspring with one another, reproductively isolated from other such populations.

**spectacles** Pale eye rings connected above the bill, giving the appearance of glasses.

**spectrogram** A visual representation of a bird sound, with frequency plotted on the vertical axis and time on the horizontal axis.

**splitting** Reclassifying birds within a single species as multiple species.

**subadult** A bird that has not attained full adult plumage; synonym of *immature*.

**subspecies** Within a species, groups of closely related yet morphologically distinct individuals from different geographic regions.

**supercilium** A tract of feathers above a bird's eye, extending from the base of the bill to behind the eye; synonym of "eyebrow."

**superspecies** A group of closely related species, often so close that frequent interbreeding can occur within the group.

**syrinx** A bird's vocal organ.

**taxon (pl.: taxa)** A taxonomic grouping.

**taxonomy** The scientific classification of organisms based on evolutionary relationships.

**tertials** The innermost flight feathers of a bird's wing, technically the innermost secondaries; sometimes distinct in color and shape from the rest of the secondaries.

**torpor** A condition of lowered metabolism, often brought on very rapidly, typically in response to environmental stresses like heat or cold.

**trans-Gulf migrant** A bird that migrates past the Gulf of Mexico by flying over it.

**uncommon** Denotes a species or population found in small numbers, but reliably so, at a certain place and time of year.

**undertail coverts** Feather tract covering the bases of the rectrices from below.

**underwing coverts** Tracts of feathers on the underside of the wing that cover the bases of the primaries and secondaries; "wing linings" is a synonym.

**undifferentiated (of plumage)** Referring to a bland, unpatterned, or otherwise unremarkable appearance.

**upper mandible** The upper half of the bill, also called the "maxilla."

**uppertail coverts** Feather tract covering the bases of the rectrices from above.

**upperwing coverts** Tracts of feathers on the upperside of the wing that cover the bases of the primaries and secondaries.

**vagrant** A bird in a location typically not inhabited or visited by its species.

**vent** Part of a bird where the belly meets the undertail coverts.

**very rare** Denotes a species found in extremely low numbers, but usually reliably so, at a certain place and time of year.

**visitant** A species that appears at a particular time of the year, especially winter.

**wing bar** A stripe of contrasting color across the middle of a bird's wing.

## ILLUSTRATIONS CREDITS

All maps created in partnership with the Cornell Lab of Ornithology, using eBird data. Visit *science.ebird.org* to explore interactive maps and data.

The following artists contributed the illustrations in this second edition: Jonathan Alderfer, David Beadle, Peter Burke, Andrew Guttenberg, Marc R. Hanson, Cynthia J. House, H. Jon Janosik, Donald L. Malick, Killian Mullarney, Michael O'Brien, John P. O'Neill, Kent Pendleton, Diane Pierce, John C. Pitcher, H. Douglas Pratt, David Quinn, Chuck Ripper, N. John Schmitt, Thomas R. Schultz, and Daniel S. Smith.

Front cover—Pratt; back cover (UP)—Pratt and (LO)—Pierce; 2—Pierce; 4—Cheryl Shank/Nature Friend Magazine; 6—George Ostertag/Alamy Stock Photo; 7—Nattapong/Adobe Stock; 10-11—Alderfer; 12—Phil Degginger/Alamy Stock Photo; 15 (UP)—Stan/Adobe Stock; 15 (LO LE)—George Grall/National Geographic Image Collection; 15 (LO RT)—Bernard Friel/DanitaDelimont/Adobe Stock; 16—Riverwalker/Adobe Stock; 17-8 (ALL)—Luke Franke; 19—Martha Allen; 21—House; 23—House; except Graylag Goose by Alderfer; 25—Schultz; except Pink-footed Goose by Mullarney; Brant by Alderfer; Barnacle Goose by House; 27—Schmitt; except Canada Goose (*canadensis*) by House 29—House; except Trumpeter and Tundra Swans (heads) by Mullarney; 31—House; except Egyptian Goose and Muscovy Duck (flight) by Schmitt; 33—House; except Cinnamon Teal (female) by Schmitt; 35—House; except wigeon hybrid by Schultz; 37—House; except American Black Duck (head) by Schmitt; Mottled Duck (*fulvigula*) by Schultz; 39—House; except Green-winged Teal (female) by Schmitt; 41-3—House; 45—House; except Common Eider (heads) and Harlequin Duck (adult males) by Alderfer; 47—House; 49—House; except goldeneye hybrid by Alderfer; 51—House; 53—House; except Plain Chachalaca by Schmitt; 55—Pendleton; except Northern Bobwhite (*floridanus* and *taylori*) by Schmitt; 57-9—Pendleton; 61—Pendleton; except Greater Sage-Grouse (displaying male) and Sharp-tailed Grouse (displaying male) by Alderfer; 63—Pendleton; except Indian Peafowl and Red Junglefowl by Guttenberg; 65—Alderfer; except American Flamingo (standing and swimming) by Pierce; 67—Alderfer; except Red-necked Grebe by Janosik; 69—Janosik; except Western and Clark's Grebe (heads) by Alderfer; 71—Pratt; 73—Alderfer; except Eurasian Collared-Dove by Schmitt and Alderfer; Inca Dove by Pratt; 75—Alderfer and Schmitt; except White-tipped Dove by Pratt; 77—Pratt; except Greater Roadrunner by Schmitt; 79—Pratt; 81—Schmitt; except Common Nighthawk by Schultz; 83—Schmitt; except all tails by Ripper; 85—Schmitt; except Mexican Violetear by Schmitt and Alderfer; 87—Schmitt and Alderfer; 89—Alderfer; except Buff-bellied Hummingbird by Schmitt and Alderfer; 91—Schultz; except Sora (standing figures and head) by Hanson; 93—Hanson; except Common Gallinule (breeding and juvenile) by Schultz; Purple Swamphen by Quinn; 95—Schultz; except Yellow Rail (standing figures) by Hanson; Sora (chick) by Guttenberg; 97—Pierce; except Limpkin (standing) by Hanson; Limpkin (flight) and Sandhill Crane (flight) by Schultz; 99—Janosik; 101—Alderfer; except Killdeer (standing figures by Mullarney; Killdeer (flight) by Smith; 103—Pitcher; except all flight figures by Smith; 105—Smith; except Wilson's Plover (standing figures) and Snowy Plover (standing figures) by Pitcher; Mountain Plover (standing figures and ventral flight) by Mullarney; 107—Smith; except Upland Sandpiper (standing figures) by Mullarney; 109—Alderfer; except Bar-Tailed Godwit (flight) and Marbled Godwit (flight) by Smith; 111—Pitcher; except Red Knot (standing figures) by Schultz; Red Knot (flight) by Alderfer; Ruff (standing figures) by Mullarney; Ruff (flight) by Smith; 113—Alderfer; except all flight figures by Smith; Curlew Sandpiper (standing figures) by Schultz; line drawings by Guttenberg; 115—Schultz; except Purple Sandpiper (standing figures) by Pitcher; all flight figures by Smith; 117—Pitcher; except all flight figures by Smith; 119—Pitcher; except Buff-breasted Sandpiper (standing figures) and Pectoral Sandpiper (standing figures) by Mullarney; Semipalmated Sandpiper (breeding female) by Schultz; all flight figures by Smith; 121—Alderfer; 123—Pitcher; except American Woodcock and Wilson's Snipe by Alderfer; 125—Pitcher; except Willet by O'Brien; both flying yellowlegs by Smith; 127—Mullarney; 129-31—Schultz; 133—Schultz; except Dovekie (breeding adults and swimming winter) by Ripper; Dovekie (flying winter) by Alderfer; 135—Ripper; except Common Murre (swimming) by Schmitt; Common Mure and Thick-billed Murre (flight) by Alderfer; 137—Ripper; except Black Guillemot (all *mandtii*) and Atlantic Puffin (flight) by Alderfer; 139-61—Schultz; 163—Janosik; 165-7—Quinn; 169—Alderfer; 171—Alderfer; except Northern Fulmar (swimming and ventral flight) by Guttenberg; Black-capped Petrel by O'Brien; 173—O'Brien; 175—Alderfer; except Sooty Shearwater by Hanson; 177—Alderfer; 179—Pierce; except Magnificent Frigatebird by Janosik; 181—Alderfer; 183—Alderfer; except Anhinga by Janosik; 185—Janosik; 187—Burke; except Great Blue Heron by Pierce; 189—Pierce; except Cattle Egret (flight) and Snowy Egret (flight) by Schultz; 191—Pierce; except Little Blue Heron (flight), Tricolored Heron (flight), and Reddish Egret by Schultz; 193—Burke; except Green Heron (standing) by Pierce; Green Heron (flight) by Schultz; 195—Pierce; 197—Burke; except flight figure and Glossy x White-faced hybrid (face detail) by Guttenberg; 199—Malick; except Osprey (*ridgwayi*) by Schmitt; 201—Malick; except Hook-billed Kite (perched adults) by Pendleton; Hook-billed Kite (large flight and perched juvenile) by Schmitt; 203—Malick; except Northern Harrier by Schmitt; 205—Malick; except three flying juveniles by Schmitt; 207—Malick; except Mississippi Kite (immature) and Snail Kite (perched) by Pendleton; Mississippi Kite (flying adults), Snail Kite (flight), and Bald Eagle (3rd year) by Schmitt;

209–Schmitt; except White-tailed Hawk (perched) by Malick; 211–Schmitt; except Zone-tailed Hawk (perched) by Malick; 213–Schmitt; except Swainson's Hawk (perched) by Malick; 215–Schmitt; except Short-tailed Hawk (perched) by Malick; 217–Schmitt; except Rough-legged Hawk (perched and flying juvenile) and Ferruginous Hawk (perched adult) by Malick; 219–Malick; 221–Malick; 223–Malick; except Ferruginous Pygmy-Owl by Schultz; 225–Malick; except Long-eared Owl (flight) and Short-eared Owl (*domingensis*) by Schmitt; 227–Malick; except Green Kingfisher (flight) by Schmitt; 229-31–Malick; 233–Malick; except American Three-toed Woodpecker (*bacatus* and *dorsalis*) by Schultz; 235–Malick; 237–Malick; except Northern Flicker (intergrade) by Schultz; 239–Malick; except Crested Caracara (flight) by Pendleton; American Kestrel (dorsal flight), Merlin (flight), and Aplomado Falcon (flight) by Schmitt; 241–Malick; except all flight figures by Schmitt; 243-5–Schmitt; 247–Schultz; except Rose-throated Becard by Alderfer; Northern Beardless-Tyrannulet by Beadle; Great Kiskadee (flight) by Pratt; 249–Burke; 251–Alderfer; 253–Schultz; except Fork-tailed Flycatcher (adult) and Scissor-tailed Flycatcher by Pratt; Fork-tailed Flycatcher (immature) by Schmitt; 255-61–Beadle; 263–Pratt; 265–Pratt; except White-eyed Vireo (immature) and Gray Vireo by Schultz; Hutton's Vireo by Schmitt; 267–Schultz; except Yellow-throated Vireo by Pratt; 269–Beadle; except Warbling Vireo (nest) by Guttenberg; 271–Schultz; except Red-eyed Vireo by Pratt; 273–Pratt; except flight figures by Schmitt; 275–Pratt; except Blue Jay (flight) by Schmitt; 277–Schmitt; except Woodhouse's Scrub-Jay (*texana*) by Schultz; Black-billed Magpie by Pratt; 279–Schmitt; 281–Schmitt; except Chihuahuan Raven (flight) and Common Raven (flight and calling) by Alderfer; 283– O'Brien; 285–O'Brien; except Tufted x Black-crested hybrid by Guttenberg; Verdin by O'Neill; 287–Beadle; 289–Pratt; except Bank Swallow and Violet-green Swallow (dorsal flight and perched adult male) by Schmitt; 291–Schmitt; 293–Pratt; except for Cliff Swallow (*pyrrhonota* dorsal flight) by Guttenberg; Cave Swallow (*pallida* flight) by Beadle; 295–Pratt; except Red-vented Bulbul by Guttenberg; Bohemian Waxwing (perched adult and flight) and Cedar Waxwing (flight) by Quinn; 297–Schmitt; except Bushtit by O'Neill; 299–Pratt; except White-breasted Nuthatch and Brown Creeper by Schultz; 301–Pratt; except Blue-gray Gnatcatcher (*obscura*) by Schultz; Rock Wren by Schmitt; 303–Schmitt; except House Wren by Pratt; 305–Pratt; except Cactus Wren (nest) by Schmitt; 307–Schultz; except Gray Catbird by Pratt; Curve-billed Thrasher by Schmitt; 309–Pratt; except Sage Thrasher and Northern Mockingbird (flight) by Schmitt; 311–Pratt; except European Starling (flight) by Guttenberg; Common Myna by Alderfer; 313–Pratt; 315-17–Schultz; 319–Pratt; 321–Schmitt; except Northern Wheatear by Quinn; 323–Quinn; except Horned Lark and Vesper Sparrow by Beadle; flying figures by Guttenberg; 325–Pierce; except Gray-crowned Rosy-Finch (*littoralis* and male *tephrocotis*) by Beadle; Gray-crowned Rosy-Finch (female *tephrocotis*) by Schmitt; 327–Schultz; 329–Pierce; 331–Pierce; except Pine Siskin (green morph) by Quinn; 333–Pierce; except wing tip figures by Schultz; 335–Pierce; except Thick-billed Longspur (wing tip figure) by Schultz; Snow Bunting (1st winter female) by Alderfer; 337–Schultz; 339–Schultz; except Olive Sparrow by Burke; 341–Pierce; 343–Schultz; 345–Schultz; except American Tree Sparrow by Pierce; 347–Pierce; except White-crowned Sparrow by Schultz; 349–Pierce; except Dark-eyed Junco (*cismontanus*) by Guttenberg; Dark-eyed Junco (flight) by Schmitt; Dark-eyed Junco (*shufeldti* and *mearnsi*) and Vesper Sparrow by Beadle; 351–Schmitt; except Seaside Sparrow by Pierce; 353–Pierce; except Savannah Sparrow by Schultz; 355–Pierce; except Song Sparrow (juvenile) and Lincoln's Sparrow by Schultz; 357–Burke; except Rufous-crowned Sparrow by Schultz; 359–Burke; 361–Burke; except Western Spindalis and Yellow-breasted Chat (female *auricollis*) by Schultz; 363–Pratt; except Bobolink by Schultz; all flight figures by Guttenberg; 365–Schultz; 367-9–Burke; 371–Schultz; except Red-winged Blackbird (males) and Bronzed Cowbird (female) by Pratt; Bronzed Cowbird (male and juvenile) by Beadle; Shiny Cowbird by Burke; Brown-headed Cowbird by Schmitt; 373–Pratt; except flight figures by Guttenberg; 375–Pratt; 377–Pratt; except Louisiana and Northern Waterthrushes by Schultz; 379–Pratt; 381–Pratt; except Black-and-white Warbler by Schultz; 383–Pratt; except Nashville Warbler (immature female) by Schultz; 385–Schultz; except Kentucky Warbler by Pratt; 387–Pratt; except Common Yellowthroat by Schultz; 389–Pratt; except Cerulean Warbler (immature male) by Schultz; 391–Pratt; except Magnolia Warbler by Schultz; Bay-breasted Warbler (fall male) by Beadle; 393–Pratt; except Yellow Warbler by Schultz; 395–Pratt; except Palm Warbler by Schultz; 397–Pratt; except Yellow-rumped Warbler (fall *coronata*) by Schultz; 399–Pratt; except Wilson's Warbler by Schultz; 401–Burke; 403–Pierce; 405–Schultz; except Blue Grosbeak by Pierce; 407–Schultz; except Lazuli x Indigo hybrid by Guttenberg; 409–Pierce; except Bananaquit by Pratt; Morelet's Seedeater (adult male) by Burke; 410–House; except Garganey by Mullarney; Ruddy Ground Dove by Alderfer; Key West Quail-Dove by Pratt; 411–Schultz; except White-collared Swift by Schmitt; Green-breasted Mango and Broad-billed Hummingbird by Schmitt and Alderfer; Northern Lapwing by Mullarney; 412–Pitcher; except Northern Jacana by Janosik; Sharp-tailed Sandpiper by Mullarney; Long-billed Murrelet by Alderfer; Ancient Murrelet by Schmitt and Alderfer; 413–Schultz; 414–Alderfer; except Little Egret by Quinn; Thick-billed Vireo by Schultz; 415–Pratt; except Mountain Chickadee by O'Brien; Pacific Wren by Beadle; Fieldfare by Quinn; 416–Schultz; except Redwing by Quinn; Brambling by Pierce; Gray-crowned Yellowthroat by Burke; 417–Pratt; except Black-throated Gray Warbler by Schultz; Golden-crowned Warbler and Crimson-collared Grosbeak by Burke; Blue Bunting by Pierce; 425–Alderfer; 426–Alderfer; except Eskimo Curlew by Smith; Ivory-billed Woodpecker by Malick; 427–Pierce; except Bachman's Warbler by Pratt; 435–Jeff Gordon.

## ABOUT THE AUTHOR

**TED FLOYD** is the longtime editor of *Birding* magazine, the award-winning flagship publication of the American Birding Association (ABA). He has written five bird books, including the *Smithsonian Field Guide to the Birds of North America* (HarperCollins, 2008) and *How to Know the Birds* (National Geographic, 2019). Floyd is also the author of more than 200 popular articles, technical papers, and book chapters on birds and nature, and he is a frequent speaker at bird festivals and ornithological society conferences worldwide. He has served on several nonprofit boards and is a recipient of the ABA Claudia Wilds Award for Distinguished Service.

## ACKNOWLEDGMENTS

More than 40 years ago, the National Geographic Society's *Field Guide to the Birds of North America* ushered in a new era of knowledge and enjoyment for bird lovers in the United States and Canada. That first edition (1983) was instantly hailed for being accurate, authoritative, and useful. Six subsequent editions (1987, 1999, 2002, 2007, 2011, 2017) built on that initial success, adding new species and expanded text, while keeping abreast of rapid changes in ornithological nomenclature and taxonomic relationships. Regional texts, pocket guides, and an encyclopedic *Complete Birds* followed.

Literally hundreds of birders, ornithologists, and editors have contributed to the excellence of the galaxy of National Geographic bird books, but two are deserving of special recognition. Jon L. Dunn and Jonathan Alderfer, who guided so many National Geographic birding books from conception to completion, were uncompromising in their precision, thoroughness, and attention to detail. Through their eyes, birders everywhere have learned to see the world of birds more clearly, more accurately, and, ultimately, more lovingly. All of us have been made better birders, directly or indirectly, through their legacy.

Many others have contributed their time and knowledge to the field guides preceding this one and its companions. All of us involved in the production of the present volume have benefited greatly from their many and varied contributions, which we acknowledge with great appreciation. Individuals who assisted in the preparation of this book's predecessors are listed in the acknowledgments of each edition.

The author thanks the following individuals, who provided facts, data, and other information with direct bearing on new content in this book: Jody Allair, Elisabeth Ammon, Eliana Ardila, George Armistead, Yousif Attia, Margaret Bain, Jen Ballard, Danielle Belleny, Lauryn Benedict, Yishai Blum, Kathi Borgmann, Mollee Brown, Klee Bruce, Joanna Burger, Peter Burke, Mike Burrell, Jared Clarke, Glenn Coady, Dominic Couzens, Leah Crenshaw, Amy Davis, Donna Dittmann, Diana Doyle, Chris Elphick, Laura Erickson, Bill Evans, Jodhan Fine, Marcel Gahbauer, Dan Gibson, Doug Gochfeld, Isaac Goes, Joseph Grzybowski, the late Mary Gustafson, Matt Hale, Paul Hess, Marshall Iliff, Jean Iron, Frank Izaguirre, Pete Janzen, the late Tom Johnson, Laura Kammermeier, Kenn Kaufman, Marc Kramer, Tony Leukering, Derek Lovitch, John Lowry, Ron Martin, Patrick Maurice, Holly Merker, Nick Minor, Bob Mulvihill, Joe Neal, Greg Neise, Adrianna Nelson, Ronan Nicholson, Michael O'Brien, Megan Jones Patterson, Wayne Petersen, Nathan Pieplow, the late Ron Pittaway, Bill Pranty, José Ramírez-Garofalo, Ryan Rodríguez, Rebecca Safran, Ioana Seritan, Kelly Smith, Tom Stephenson, Nate Swick, David Tønnessen, Claire Wayner, Scott Weidensaul, David Wilcove, Sheri Williamson, and Rick Wright.

The species accounts here are drawn primarily from the author's 30+ years of extensive field experience in the region, but several online references were indispensable and consulted practically daily during the writing of this guide. Foremost among these is the Macaulay Library *(ebird.org/catalog)*, with 56 million photos, 2.1 million audio recordings, and 271,000 videos of birds. Two other online resources of tremendous value were Xeno-Canto *(xeno-canto.org)* and *Flight Calls of Migratory Birds (oldbird.org)*. And for general guidance about behavior, ecology, and systematics, the author made use of the miraculously thorough Birds of the World *(birdsoftheworld.org)*.

The art here, created and refined over the years for National Geographic's birding field guides, represents the work of numerous artists, identified in the illustrations credits (pp. 433–434) and informed by numerous museum collections, cited in previous field guides. We continue to thank these artists and their research sources for this important collection of illustrations. The Denver Museum of Nature & Science and its Science Liaison, Jeff Stephenson, deserve special thanks for guiding the author's own research during the past two decades—including for this field guide.

The birding gods smiled on us when Andrew Guttenberg, one of the greatest ornithological illustrators of his generation, agreed to join the project. New illustrations for this edition have been created by Andrew, whose knowledge and aesthetics have added so much to this book, not only in the form of new art but also in layout arrangement, annotation composition, and deep grasp of ornithology. Thanks also to Marky Mutchler, who helped compose the art annotations, paying particular attention to the technical aspects of avian anatomy. Both Andrew and Marky also read the text and offered many corrections and clarifications.

Behind the scenes, a dedicated team of editors, designers, interns, assistants, and project managers tended to matters great and small, guiding this project from its earliest conception to the final finished product. Editorial intern Sienna Sullivan painstakingly researched morphometric data; editorial assistant Margo Rosenbaum managed a mountain of Google docs, Excel spreadsheets, and more; text editor Jennifer Seidel applied her trademark conscientiousness and grammatical acumen to a thorough review of the entire manuscript, going over trouble spots multiple times; graphic designer Carol Norton created page layouts, clean and clear, that present a considerable density of text and art in a manner that is aesthetically appealing and, more important, easily understood; cartographer Debbie Gibbons created state-of-the-art maps that combine the latest science with a novel and powerful way of visualizing birds' dynamic ranges in both space and time; and senior editor Susan Tyler Hitchcock hatched the wonderful plan for this and companion field guides, ambitious and sometimes crazy and ultimately deeply satisfying for all involved.

Last but certainly not least, project manager Adrienne Izaguirre brought order to chaos; it is not hyperbole to say that this field guide would have imploded for sure without Adrienne's preternatural discipline, almost freakish attention to detail, and unsurpassed excellence overall. Although it wasn't in her job description, Adrienne provided invaluable expertise in everything from copyediting to cartography; along the way, she made countless corrections and more than a handful of original contributions to the species accounts and other technical content in this book.

Special thanks to those essential to our partnership with the Cornell Lab of Ornithology, especially Miyoko Chu and Tom Auer. We also thank Steve Mlodinow, one of the foremost experts on bird ranges in the East and beyond, for his exacting review of the maps and text, bringing to bear his exhaustive knowledge of when and where birds occur in terms of both historical patterns and very recent range shifts.

Thanks go to Lisa Thomas, editorial director, who identified the need to reinvigorate National Geographic's line of bird field guides. Thanks also to Russell Galen, my literary agent—a wise mentor and tireless advocate.

Peter Pyle, Van Remsen, and Michael Retter read the entire manuscript, catching thousands of typos, omissions, misstatements, and other infelicities. Even more important, they read for clarity and coherence, the twin touchstones for the success of any field guide or other natural history reference. As we prepared the manuscript, we appreciated the help of Michael O'Connor, copy editor Jen Hess, and proofreader Mary Stephanos.

A field guide is a technical work, to be sure, but it is also intended for a contemporary readership. To put things in perspective, well over half the human population of the United States and Canada wasn't even born when the first edition of *National Geographic Field Guide to the Birds of North America* was published, in 1983. In this regard, the author frequently solicited the counsel of Zoomers Hannah Floyd and Andrew Floyd in matters of metaphor and idiom for a modern audience. Maya Izaguirre, representing Gen Alpha, born about the time the work on this book began, was too young to weigh in on style and sensibility, but she enlivened many a Zoom meeting with her constant cheer and companionableness.

Learning about birds is, more than ever, a crowdsourced and communitarian undertaking. It is not an exaggeration to say that literally thousands of bird lovers have contributed in one way or another to the content in this new edition of the *National Geographic Field Guide to the Birds of the United States and Canada—East*. To one and all: Thank you!

*—Ted Floyd*

# INDEX

The page number for the main entry for each species is listed in **boldface** type and refers to text page opposite the illustration.

---

Since 1888, the National Geographic Society has funded more than 14,000 research, conservation, education, and storytelling projects around the world. National Geographic Partners distributes a portion of the funds it receives from your purchase to National Geographic Society to support programs including the conservation of animals and their habitats.

National Geographic Partners, LLC
1145 17th Street NW
Washington, DC 20036-4688 USA

Get closer to National Geographic Explorers and photographers, and connect with our global community. Join us today at nationalgeographic.org/joinus



The Library of Congress cataloged the first edition as follows:
National Geographic field guide to the birds of eastern North America / edited by Jon L. Dunn and Jonathan Alderfer with Paul Lehman.
p. cm.
Includes index.
ISBN 978-1-4262-0330-5
1. Birds--East (U.S.)--Identification.
2. Birds--Canada, Eastern--Identification.
I. Dunn, Jon, 1954- II. Alderfer, Jonathan K. III. Lehman, Paul.
QL683.E27.N38 2008
598.0974--dc22
2007050769

ISBN: 978-1-4262-2277-1

Printed in China
24/RRDH/1